PLATELET-ACTIVATING FACTOR AND RELATED LIPID MEDIATORS 2

Roles in Health and Disease

ADVANCES IN EXPERIMENTAL MEDICINE AND BIOLOGY

Recent Volumes in this Series

Volume 407
EICOSANOIDS AND OTHER BIOACTIVE LIPIDS IN CANCER, INFLAMMATION, AND RADIATION INJURY 3
Edited by Kenneth V. Honn, Lawrence J. Marnett, Santosh Nigam, Robert L. Jones, and Patrick Y-K Wong

Volume 408
TOWARD ANTI-ADHESION THERAPY FOR MICROBIAL DISEASES
Edited by Itzhak Kahane and Itzhak Ofek

Volume 409
NEW HORIZONS IN ALLERGY IMMUNOTHERAPY
Edited by Alec Sehon, Kent T. HayGlass, and Dietrich Kraft

Volume 410
FRONTIERS IN ARTERIAL CHEMORECEPTION
Edited by Patricio Zapata, Carlos Eyzaguirre, and Robert W. Torrance

Volume 411
OXYGEN TRANSPORT TO TISSUE XVIII
Edited by Edwin M. Nemoto and Joseph C. LaManna

Volume 412
MECHANISMS IN THE PATHOGENESIS OF ENTERIC DISEASES
Edited by Prem S. Paul, David H. Francis, and David A. Benfield

Volume 413
OPTICAL IMAGING OF BRAIN FUNCTION AND METABOLISM II: Physiological Basis and Comparison to Other Functional Neuroimaging Methods
Edited by Arno Villringer and Ulrich Dirnagl

Volume 414
ENZYMOLOGY AND MOLECULAR BIOLOGY OF CARBONYL METABOLISM 6
Edited by Henry Weiner, Ronald Lindahl, David W. Crabb, and T. Geoffrey Flynn

Volume 415
FOOD PROTEINS AND LIPIDS
Edited by Srinivasan Damodaran

Volume 416
PLATELET-ACTIVATING FACTOR AND RELATED LIPID MEDIATORS 2: Roles in Health and Disease
Edited by Santosh Nigam, Gert Kunkel, and Stephen M. Prescott

---

PLATELET-ACTIVATING FACTOR AND RELATED LIPID MEDIATORS 2

Roles in Health and Disease

Edited by

Santosh Nigam
Gert Kunkel
Free University Berlin
Berlin, Germany

and

Stephen M. Prescott
University of Utah
Salt Lake City, Utah

PLENUM PRESS • NEW YORK AND LONDON

Library of Congress Cataloging-in-Publication Data

Platelet-activating factor and related lipid mediators 2: roles in health and disease / edited by Santosh Nigam, Gert Kunkel, and Stephen M. Prescott.
p. cm. -- (Advances in experimental medicine and biology ; v. 416)
"Proceedings of the Fifth International Congress on Platelet-Activating Factor and Related Lipid Mediators, held September 12-16, 1995, in Berlin, Germany"--T.p. verso.
Includes bibliographical references and index.
ISBN 0-306-45506-4
1. Platelet activating factor--Congresses. I. Nigam, S. K. (Santosh K.) II. Kunkel, Gert. III. Prescott, Stephen M. IV. International Congress on Platelet-Activating Factor and Related Lipid Mediators (1995 : Berlin, Germany) V. Series.
[DNLM: 1. Platelet Activating Factor--physiology--congresses. 2. Platelet Activating Factor--chemistry--congresses. W1 AD559 v.416 1996 / QU 93 P7161 1996]
QP752.P62P574 1996
612'.01577--DC21
DNLM/DLC
for Library of Congress 97-3871
CIP

Proceedings of the Fifth International Congress on Platelet-Activating Factor and Related Lipid Mediators, held September 12 – 16, 1995, in Berlin, Germany

ISBN 0-306-45506-4

A Division of Plenum Publishing Corporation
233 Spring Street, New York, N. Y. 10013

10 9 8 7 6 5 4 3 2 1

Printed in the United States of America

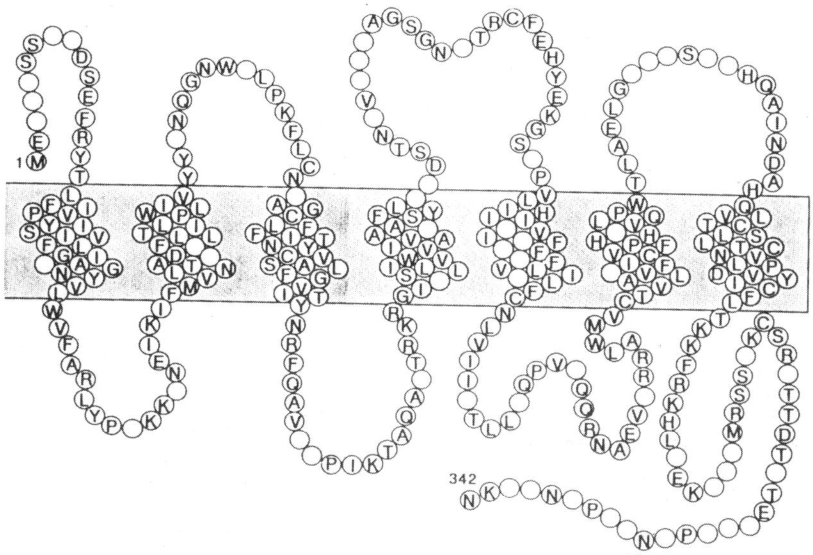

PAF-Receptor

PREFACE

In the last decade, research on platelet-activating factor (PAF) has expanded exponentially. Previous conferences on PAF in Paris, 1983, and the subsequent conferences in Gatlinburg, Tennessee, Tokyo, Snowbird, Utah, and Berlin, at three-yearly intervals, have chronicled the developments in the field of PAF. This volume records the proceedings of the Fifth International Congress on PAF and Related Lipid Mediators, held at the Free University Medical Hospital Benjamin Franklin in Berlin, from September 12–16, 1995. We are very much indebted to Free University Berlin for providing tremendous facilities and financial support. It was a great pleasure to have positive and generous input from the German Science Council (DFG), Bonn, Germany, and British Biotech, Oxford, United Kingdom. Their support was crucial in making the congress a scientific success. Twenty other organizations provided additional financial support, for which we extend our deepest appreciation. The editors would like to thank all of those who participated in this congress and the authors for their contributions.

The organization and planning of the Berlin Congress were carried out by an organizing committee. We gratefully acknowledge the support and assistance of the organizing committee members, especially Renate Nigam and Renate Roux for their untiring efforts to make the congress successful. Many colleagues also supported the congress with dedication, hard work, and expert input. We are grateful to them. We also wish to acknowledge the support of G. Sravan Kumar and Louis Kock for their efforts in producing this volume.

This volume is an up-to-date distillation of the current knowledge on the chemistry, biology, physiology, and pathology of PAF and reflects the sequence of the main sessions of the congress. The chapters discuss a wide range of topics encompassing *inter alia* PAF synthesis and transport, molecular structure and heterogeneity of PAF-receptor, regulation of gene expression by PAF, signal-transducing role of PAF and other phospholipids, role of PAF and PAF-receptor in infectious diseases, circulation, and reproduction. Particular attention has been devoted to addressing the roles and mechanisms of action of acetylhydrolase in human diseases and application of ether lipids in prevention of tumor growth. The last chapter of this volume presents the highlight of the congress, describing clinical trials on the application of PAF antagonists for the treatment of asthma and pancreatitis.

This volume will be pertinent to a great number of researchers in the field of bioactive lipids, including chemists, biochemists, biologists, pharmacologists, neuroscientists, physiologists, oncologists, gynecologists, and clinicians. We anticipate that all who read it will find much of interest and a substantial amount of new data.

Santosh Nigam
Gert Kunkel
Stephen M. Prescott

CONTENTS

1

THE CONTINUING BIOCHEMICAL CHALLENGE OF PAF AND CLOSELY RELATED LIPID MEDIATORS

Donald J. Hanahan

University of Texas Health Science Center
7703 Floyd Curl Drive, San Antonio, Texas 78284–7760

The discovery that platelet-activating factor (PAF) was a biologically active phosphoglyceride expanded dramatically our concept of the role of the phospholipids in cellular events. In the past fifteen years of study on PAF and related lipid mediators, it is evident that on the one hand they shared many features in common with the more conventional phosphoglycerides, but on the other hand, the bio-active phosphoglycerides which appear to be restricted to those with choline or hydrogen in the polar head (base) group, had certain characteristics unique unto themselves. It is appropriate then to review the similarities in these two classes of compounds together with a discussion of their differences.

In the first category the following points can be made:

1. In either class of compounds containing choline, there is a high degree of stereospecificty. All are of the sn-3 configuration. No sn-1 or sn-2 conformer has been detected.
2. A high degree of positional specificty is evident in each group. For example, over 90 percent of the long chain hydrocarbon chains at the sn-1 position are saturated (16:0 to 18:0). In the so called conventional phosphoglycerides, well over 90 percent of the long chain hydrocarbon chains located at the sn-2 position are of the unsaturated variety. As shall be discussed below the biologically active choline containing phosphoglycerides have only short chain saturated types at the sn-2 position.

In the second category for discussion, there are some decided differences in the characteristics of these two classes of compounds:

a. While the conventional phospholipids can have some ten different bases groups in the polar head group, the biologically active types contain only choline or hydrogen.
b. Even though the choline containing phosphoglycerides with biological activity exhibit a high degree of stereospecificity (only the sn-3 form is active), the same

is not true of the biologically active lipid phosphoric acids which appear not to require an asymmetric center for agonist activity.

It is hoped that this brief discussion will serve to illustrate certain unique characteristics of these phospholipids. In the following section attention will be centered specifically on the bioactive phosphoglycerides.

It is interesting to reflect on the fact that these International Congresses have been the outgrowth of a very simple finding—a phosphoglyceride containing an alkyl ether bond and a short chain acyl (i.e., an acetyl) moiety is a highly potent agonist. Though the alkyl ether bond had been found at the sn-1 position of phosphoglycerides many years before the observation that an acetyl group was located at the sn-2 position was indeed a very, very unique finding. Not only was the presence of such a substituent unique, the fact that the molecule possessed potent biological activity provided an exciting new venue for investigators in many different disciplines. Though some scientists have questioned the future of this field, a claim can be made that this area of investigation has many stimulating avenues to pursue. Some of those particularly related to more basic biochemical/physiological studies will be addressed here. The exciting aspects of the clinical approach must be addressed by those with such expertise. A number of interesting avenues of exciting research on PAF and related lipid chemical mediators and analogs are on the horizon:

First, attention will be focused on the interaction of PAF with its receptor. While important information on the nature of the PAF receptor has come from the laboratory of Shimizu and his associates in Japan, there is still a question as to whether there are any metabolic alterations to the PAF molecule on interaction with its specific receptor. Certainly the fact that a wide variety of biochemical and physiological events ensue, a question must be raised as to whether PAF may be converted rapidly to a derivative with potent (and diverse) biological activity. Hence a proposal is made that PAF upon interaction with its receptor is subjected to enzymatic attack by a combination of phospholipase A2 and phospholipase D to yield a lyso alkyl glycerophosphoric acid, which then is transported rapidly within seconds to an intracellular receptor(s) where it exerts its biological effects. The presence of intracellular receptors for PAF has been reported by Nigam and associates in Berlin. Certainly the more recent investigations of Moolenaar and colleagues in Holland, Tigyi and his associates in Tennessee and Sugiura and collaborators in San Antonio have attested to the significant biological activity that the acyl and alkyl(lyso) glycerophosphoric acids possess.

Second, the emergence of a clustering phenomena in PAF and analogous compounds requires serious attention. Two examples will be cited, the first of which centers on the PAF molecule formed in vivo. The important facets of this example centers on the consistent finding of an alkyl ether bond at the sn-1 position with the hydrocarbon chains ranging from 16:0, 17:0, 18:0 to 18:1 (a rather restricted pattern). Then there is the seemingly inviolate rule substitution (only in the alkyl derivative) of only an acetyl group at the sn-2 position for potent biological activity. A question to be asked is whether these individual species have different biological activities. The other "clustering" event occurs in the PAF analogues, i.e., wherein there is a long chain acyl group at the sn-1 position in place of the alkyl residue. In this class of compounds, significant variations in the chain length of the subsituent at the sn-2 position occurs as shown by Tokumura and associates in Japan. The sn-2 groups varied in chain length from 2.0 to 7.0 and were found in easily detectable amounts in unstimulated tissues. Inasmuch as these derivatives have little or no agonist activity towards platelets as the test cell, their role in the biological system is not clear at present. However this does present the interesting and challenging possibility that these derivatives may have a metabolic cycle among themselves and quite distinct form PAF as such as noted by Chilton and his collaborators. It is certainly possible that these acyl analogues could be converted to phosphatidic acids (PA) and lyso PA, by the combined action of Plase D and Plase A_2. Thus, it would be of interest to learn whether there is compartmentation of these analogues in the cell and what their biological role might be. Certainly a subject worth serious experimental exploration as these acyl analogues of PAF have been reported in several different cells.

A third and final area of research for discussion here concerns the long chain fatty acids liberated by attack of agonist stimulated phospholipase A2 on membrane phosphoglycerides. Certainly PAF can be an important stimulant and also can be a substrate. One end result is that these liberated fatty acids could initiate a significant intracellular Ca^{2+} movement. Several different laboratories have shown that free fatty acids exhibit many different regulatory effects. These can include moderation of channel activation, enzyme behavior and synaptic transmission and possibly as transcriptional regulators.

In summary, it is hoped that these brief comments will, in the words of George Bernard Shaw's medical doctor in the "Doctors Dilemma," stimulate the phagocytes of many investigators.

2

BIOSYNTHESIS OF PLATELET-ACTIVATING FACTOR AND ENZYME INHIBITORS

Fred Snyder, Veronica Fitzgerald, and Merle L. Blank

Medical Sciences Division
Oak Ridge Associated Universities
P.O. Box 117, Oak Ridge, Tennessee 37831

ABSTRACT

Platelet-activating factor (PAF) is known to be synthesized by either a remodeling or de novo pathway. The enzymes responsible have been extensively studied by a number of laboratories. All evidence indicates the remodeling route is activated during inflammation and other hypersensitivity responses, whereas the de novo pathway is thought to be the source of PAF required for physiological functions. This article provides an update of what is currently known adout the enzymatic systems that generate PAF as well as some preliminary findings we have obtained using potential inhibitors of the specific enzymes involved. Recent progress from our laboratory toward understanding the role of the CoA-independent and Co-A dependent transacylases in the formation of lyso-PAF and PAF is summarized.

1. INTRODUCTION

The existence of at least two completely different enzymatic pathways for the biosynthesis of PAF makes it essential to identify the specific enzymes responsible for cellular responses triggered by PAF in order to fully understand the biochemical mechanisms involved during cell activation. The ether bond originates in both biosynthetic routes through a reaction catalyzed by alkyldihdroxyacetone-P (alkyl-DHAP) synthase, whereby an intact long-chain fatty alcohol, including the oxygen, is substituted for the acyl moiety of acyl-DHAP. The remodelling and de novo enzymes have recently been reviewed in considerable depth[1,2]; references to the original investigations of these enzymes can be found in these articles. This report describes the various enzymes associated with PAF biosynthesis and discusses some preliminary results we have obtained in a project designed to identify an inhibitor(s) that might selectively prevent the biosynthesis of PAF via the remodeling pathway since this route has been shown to be the source of PAF and bioactive eicosanoid metabolites in a variety of inflammatory and other hypersensitivity responses.

2. METHODOLOGY

The reader is referred to the original work from our laboratory that describes the enzymatic assays for lyso-PAF acetyltransferase[3], the Co-A independent transacylase[4,5], the CoA-dependent transacylase[6,] and the de novo enzymes[7,8 9]. The experimental protocol for the testing of enzyme inhibitors is described in section 3.3 of this report.

3. RESULTS AND DISCUSSION

3.1 The De Novo Enzymes (See Fig. 1)

3.1.1 Acetyl-CoA:Alkyglycero-P Acetyltransferase. This enzyme initiates the sequence of reactions that forms PAF de novo. The acetate from acetyl-CoA is transferred to alkylglycero-P in much the same manner as when lyso-PAF is the substrate for the acetyltransferase in the remodeling pathway. However, the acetyltransferases associated with these two pathways have completely different properties. Since the de novo acetyltransferase catalyzes the first step in the sequence and also exhibits the lowest activity in all tissues/cells thus far studied, it is thought this reaction is rate-limiting in the de novo synthesis of PAF.

3.1.2 Alkylacetylglycero-P Phosphohydrolase. This phosphohydrolase appears to be highly specific for catalyzing the dephosphorylation of alkylacetyl glycero-P since its properties differ from phosphatidate phosphohydrolase and other known phosphatases. However, the enzyme is similar in that the end product is a diradylglycerol, albeit unique, which can be utilized by a cholinephosphotransferase to produce PAF.

3.1.3 CDP-Choline:Alkylacetylglycerol Cholinephosphotransferase {Dithiothreitol (DTT)-Insensitive}. The final step in the de novo synthesis of PAF is catalyzed by a novel cholinephosphotransferase that has distinctly different properties (e.g., substrate specificity, temperature sensitivity, etc.) from the cholinephosphotransferase responsible for the biosynthesis of phoshatidylcholine or its ether analogs. Perhaps the most significant difference is its insensitivity toward DTT since the choline phosphotransferase involved in the formation of phoshatidylcholine is strongly inhibited by DTT.

A related enzyme, cytidylyltransferase, is rate-limiting in forming CDP-choline, the cosubstrate for the DTT-insensitive cholinephosphotransferase. Thus, the regulatory role of cytidylyltransferase in PAF biosynthesis is the same as in phosphatidylcholine synthesis.

3.2 The Remodeling Enzymes (See Fig. 2)

In this pathway the acyl moiety (usually 20:4) of alkylacylglycerophosphocholine, a membrane-associated phospholipid, is replaced by the acetate group of acetyl-CoA to form PAF. The lyso-PAF intermediate can be formed via the direct action of a phospholipase A_2 or by either a CoA-independent transacylase or a CoA-dependent transacylase. The final step, acetylation of the lyso-PAF, is catalyzed by acetyl-CoA:lyso-PAF acetyl-

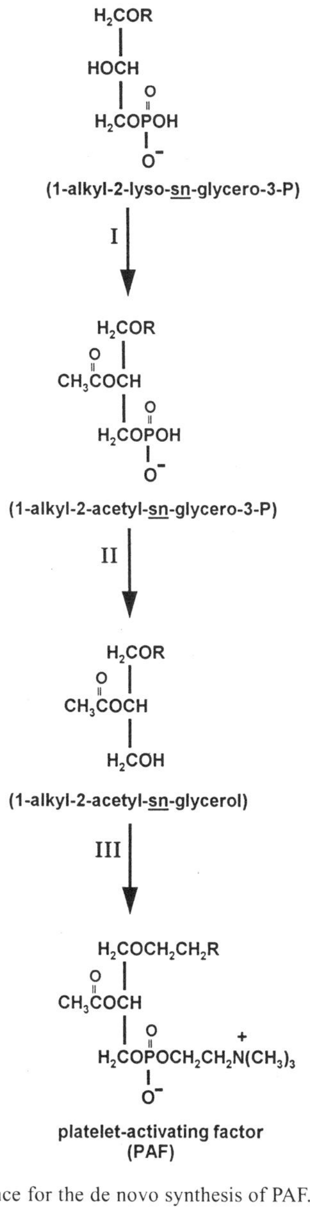

Figure 1. The enzymatic reaction sequence for the de novo synthesis of PAF. The Roman numerals designate the following enzymes: I - acetyl:CoA:alkylglycero-P acetyltransferase, II - alkylacetylglycero-P phosphohydrolase, and III - CDP-choline:alkylglycerol cholinephosphotransferase (DTT-insensitive).

transferase. A synopsis of the reactions catalyzed by these enzymes is provided in the subsequent subsections. During the past few years considerable progress has been made in understanding the role of the CoA-independent transacylases in the production of lyso-PAF, the immediate precursor of PAF[4–6,10–14]. Recent results from our laboratory indicate the CoA-dependent transacylase also can contribute to the formation of lyso-PAF[6]. Two review articles[15,16] are excellent sources of the available literature related to the overall function of the different transacylases involved in phospholipid metabolism

3.2.1 Phospholipase A_2. The simplest and most obvious way to form lyso-PAF from alkylacylglycero-phosphocholine is through the direct hydrolysis of the acyl moiety by a phospholipase A_2. However, no definitive proof for the involvement of this enzymatic step has yet been demonstrated during the activation of the remodeling pathway in PAF biosynthesis. Nevertheless , the existence of such a highly specific phospholipase A_2 for the hydrolysis of alkylacylglycerphosphocholine cannot be completely ruled out at the present time.

3.2.2 Transacylases.

3.2.2.1 CoA-Independent Transacylase. This type of transacylase transfers an *sn*-2 acyl group from a diacylglycerophosphatide to a lyso phospholipid in the absence of CoA, ATP, or any other known cofactors and most commonly occurs among the P-choline- and P-ethanolamine-containing glycerolipids. The initial lyso-phospholipid acceptor molecule is generated by a phospholipase A_2. Fig. 2 depicts the reaction sequence that produces lyso-PAF and PAF via the CoA-independent pathway.

3.2.2.2 CoA-Dependent Transacylase. Transacylation between intact phospholipid molecules induced by CoA is thought to represent the reversal of the well established acyl-CoA acyl transferase reaction. With this transacylase,CoA rather than a lyso-phospholipid serves as the acyl acceptor. When alkylacylglycerophosphocholine is the doner molecule and CoA the acceptor, the products of the reaction are lyso-PAF and acyl-CoA.

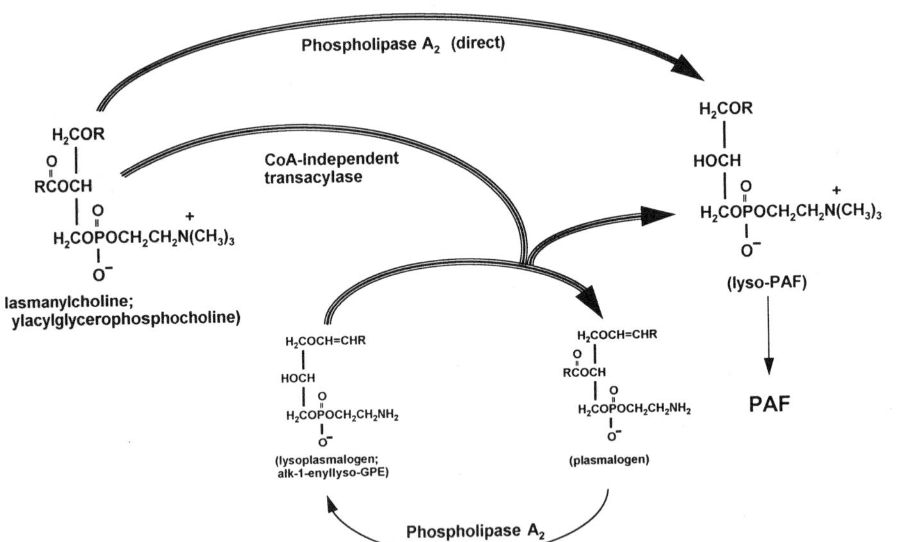

Figure 2. Biosynthesis of lyso-PAF and PAF via the remodeling pathway. The initial hydrolysis of the acyl moiety of alkylacylglycerophosphocholine to form lyso-PAF can be catalyzed by the action of a) a direct phospholipase A_2 or b) a Co-A- independent transacylase as shown in the reactions highlighted by the triple-lined arrows. Other lyso glycerophospholipids containing either ethanolamine or choline can substitute for lysopasmalogen as the acyl acceptor in the Co-A independent transacylase type of reaction. Alternately lyso-PAF can also be formed by a CoA-dependent transacylase in a reaction (not shown) that is similar to the one depicted in this figure. However, with the CoA-dependent transacylase, CoA serves as the acyl acceptor rather than the lysoplasmalogen. The lyso-PAF is converted to PAF by acetyl-CoA:lyso-PAF acetyltransferase.

We have recently described the potential importance of this CoA-dependent pathway as a source of PAF[6]. CoA-dependent transacylation was originally described by Irving and Dawson[17] and has been extensively studied by Suguira et al.[18] with other phospholipids unrelated to PAF synthesis.

3.2.3 Acetyl-CoA:Lyso-PAF Acetyltransferase. The final step in the formation of PAF involving the remodeling of alkylacylglycero-phosphocholine is the transfer of the acetate from acetyl-CoA to lyso-PAF by an acetyltransferase. This enzyme was originally described by our laboratory[3] and has been the focus of numerous other investiations[1,2]. As already mentioned, the acetyltransferases in the de novo and remodeling pathways have completely different properties and appear to represent separate catalytic proteins although neither have been purified for complete characterization.

3.3 Enzyme Inhibitors

A number of compounds were tested as potential inhibitors of the enzymes involved in PAF biosynthesis. Although the actual data obtained will be presented as a separate report, our preliminary conclusions are summarized in this article. The following compounds were tested: sanguinarine, N-methylcarbamyl-PAF, baicalein (5,6,7-trihydroxyflavone), 4,5,6,7-tetrahydroxyflavone, and luteolin (3′,4′,5,7-tetrahydroxyflavone). The purpose of this study was to determine if any of these compounds could selectivity block either the acetyltransferase or the transacylase activities associated with the remodeling pathway of PAF biosynthesis. The experimental conditions consisted of incubating the potential inhibitors (5 - 50 micromoles) for a period of 10 minutes with rat spleen microsomes (100 micrograms of protein) and then initiating the enzyme assays by adding appropriate concentrations of substrates (see references in Methods). Incubations were carried out over the linear ranges of time established in our earlier publications that describe the details for each enzymatic assay system.

Our results demonstrated that sanguinarine chloride was the only compound studied that exhibited a high degree of selectivity toward the inhibition of both the lyso-PAF acetyltransferase and the CoA-independent and CoA-dependent transacylase activities that form PAF via the remodeling route. Sanguinarine had little or no effect on the activities of the de novo enzymes under our experimental conditions. Therefore, this inhibitor would appear to be a very useful tool for distinguishing the role that the remodeling versus de novo pathways play in the biosynthesis of PAF during various cellular responses.

ACKNOWLEDGMENTS

Support for these studies was from the Department of Energy(DE-AC05-760R00033) and the NIHLB (HL 27109-14).

REFERENCES

1. Snyder, F. (1995) Biochim. Biophys. Acta 1254, 231–249.
2. Snyder, F. (1995) Biochem. J. 305, 689–705.
3. Wykle, R.L., Malone, B., and Snyder, F. (1980) J. Biol. Chem. 255, 10256–10260.
4. Uemura, Y., Lee, T-c., and Snyder, F. (1991) J. Biol. Chem. 266, 8268–8272.
5. Blank, M.L., Smith, Z.L., Fitzgerald, V., and Snyder, F. (1995) Biochim. Biophys. Acta 1254, 295–301.

6. Blank, M.L., Fitzgerald, V., Smith, Z.L., and Snyder, F. (1995) Biochem. Biophys. Res. Commun. 210, 1052–1058.
7. Lee, T-c., Malone, B., and Snyder, F. (1986) J. Biol. Chem. 261, 5373–5377.
8. Lee, T-c., Malone, B., and Snyder, F. (1988) J. Biol. Chem. 263, 1755–1760, 1988.
9. Woodard, D.S., Lee, T-c., and Snyder, F. (1987) J. Biol. Chem. 262, 2520–2527.
10. Nieto, M.L., Venable, M.E., Bauldry, S.A., Greene, D.G., Kennedy, M., Bass, D.A., and Wykle, R.L. (1991) J. Biol. Chem. 266, 18699–18706.
11. Sugiura, T., Fukuda, T. Masuzawa, Y., and Waku, K. (1990) Biochim. Biophys. Acta 1047, 223–232.
12. Venable, M.E., Nieto, M.L., Schmitt,J.D., and Wykle, R.L. (1991) J. Biol. Chem. 266, 18691–18698.
13. Winkler, J.D., Sung, C.-M., Hubbard, W.C., and Chilton, F.H. (1992) Biochem. Pharmacol. 44, 2055–2066.
14. Colard, O., Bidault, J., Berton, M., and Ninio, E. (1993) Eur. J. Biochem. 216, 835–840.
15. MacDonald, J.I.S. and Sprecher, H. (1991) Biochim. Biophys. Acta 1084, 105–121.
16. Snyder, F., Lee, T-c., and Blank, M.L. (1992) Prog. Lipid Res. 31, 65–86.
17. Irvine, R.F. and Dawson, R.M.C. (1979) Biochem. Biophys. Res. Commun. 91, 1399–1405.
18. Sugiura,T., Kudo, N., Ojima, T., Mabuchi-Itoh, K., Yamashita, A., and Waku, K. (1995) Biochim. Biophys. Acta 1255, 167–176.

3

INHIBITORS OF ARACHIDONATE METABOLISM AND EFFECTS ON PAF PRODUCTION

James D. Winkler,[1] Chiu-Mei Sung,[1] Lisa A. Marshall,[1] and Floyd H. Chilton[2]

[1]Division of Pharmacology
SmithKline Beecham Pharmaceuticals
King of Prussia, Pennsylvania 19406
[2]Section on Pulmonary and Critical Care Medicine and
Department of Biochemistry
Bowman Gray School of Medicine
Winston-Salem, North Carolina 27103

1. INTRODUCTION

Our understanding of the enzyme Coenzyme A-independent transacylase (CoA-IT) has increased dramatically over the last 10 years. The enzyme catalyses the removal of the fatty acyl group from the *sn*-2 position of glycerophospholipids (GPL) and transfers it into 1-radyl-2-lyso GPL.[1] The enzyme shows striking selectivity for transfer of arachidonate and other long-chain, unsaturated fatty acyl groups. It also shows strong preference for phosphocholine- and phosphoethanolamine-containing GPL, along with a preference for using 1-ether GPL as acceptors for the transferred arachidonate.[2] The mechanism of action of CoA-IT has yet to be defined at the molecular level, but CoA-IT is hypothesized to be a member of the family of tranferases typified by lecithin-cholesterol acyl transferase.[3] Based on these characteristics, CoA-IT has been presumed to play a role in the movement of arachidonate between GPL that occurs in inflammatory cells.[4–7]

Recent information points to a connection between CoA-IT, arachidonate movement and the formation of lyso platelet-activating factor (lyso PAF). The hypothesis is that when CoA-IT moves arachidonate out of 1-alkyl-2-arachidonoyl-*sn*-glycero-3-phosphocholine it forms lyso PAF as a product. Thus CoA-IT activity can contribute to lyso PAF, and hence PAF, production. Evidence that supports this hypothesis are the convincing connection between arachidonate content and the ability to produce PAF[8], the production of lyso PAF by addition of CoA-IT acceptor GPL[9], and the biochemical similarity between attributes of CoA-IT activity and lyso PAF production in broken neutrophils.[10]

One method to advance our knowledge about CoA-IT and its role in PAF production is to use tool compounds that inhibit CoA-IT activity. Blocking CoA-IT activity will help

Platelet-Activating Factor and Related Lipid Mediators 2
edited by Nigam *et al.*, Plenum Press, New York, 1996

freeze the dynamics of glycerophospholipid metabolism and allow study of the contribution of CoA-IT to arachidonate and lipid mediator biochemistry. Recently, we have reported on the first inhibitors of CoA-IT, SK&F 98625 and SK&F 45905.[11] In this report, we use these and other tools to provide additional insight to PAF production.

2. MATERIALS AND METHODS

Materials: [^{3}H]Acetic acid, sodium salt (50–100 Ci/mmol) was purchased from New England Nuclear (Boston, MA). Common laboratory chemicals were purchased from Sigma Chemical Co. (St. Louis, MO). Silica gel G plates were from Analtech Inc. (Newark, DE). SK&F 98625 is diethyl 7-(3,4,5-triphenyl-2-oxo-2,3-dihydro-imidazol-1-yl)heptane-phosphonate. SK&F 45905 is (2-[2-[3-(4-chloro-3-trifluoro methylphenyl)ureido]-4-trifluoro methyl phenoxy]-4,5-dichlorobenzenesulfonic acid).

Preparation of cells: Human monocytes were obtained from leukocyte-rich packs (Biological Specialties, Lansdale, PA) and prepared as previously described.[12] Monocytes were suspended in PBS at 5 x 10^6 / ml and stimulated as described. Mast cells were obtained from culture of bone marrow cells from CBA/J mice (Jackson Laboratories, Bar Harbor, ME) as previously described.[13] Mast cells were passively sensitized overnight by incubation with 20 μg/ml of mouse IgE anti-DNP and then stimulated with antigen BSA-DNP (2 ug/ml).

PAF production by intact cells: The incorporation of [^{3}H]acetate into 1-radyl-2-lyso-GPC during cell activation was used to quantitate PAF biosynthesis.[14,15] Cell suspensions in PBS containing 1 mM Ca^{2+} and 1.1 mM Mg^{2+} were incubated at 37° and exposed to drugs or vehicle for 5 min. Then a solution containing [^{3}H]acetic acid (30 μCi) with stimulus in PBS with Ca^{2+}, Mg^{2+} and 1 mg / ml BSA was added to the cell suspensions. After 10 min at 37°, the reactions were terminated by addition of one volume of chloroform / methanol (1:2, v/v) and the lipids were extracted and PAF determined as previously described.[15]

3. RESULTS AND DISCUSSION

3.1 Tool Inhibitors

We have recently described two inhibitors of CoA-IT activity, SK&F 98625 and SK&F 45905.[11] Both compounds inhibit CoA-IT activity in a broken cell assay system with IC_{50}s in the 5–10 μM range. Additionally, both compounds have some degree of selectivity for CoA-IT, having no effects on acetyltransferase or 14 kDa PLA_2 activities at 50–100 μM. Whereas SK&F 98625 has no effect on 85 kDa PLA_2 at 50 μM, SK&F 45905 did inhibit this enzyme with an IC_{50} of 5 μM. While the effects of these compounds on other enzyme systems are currently being evaluated, it is important to note that these two compounds have vastly divergent structures and will therefore be predicted to have different side effects on other enzymes. Their common effect is inhibition of CoA-IT.

One of the key roles hypothesized for CoA-IT is the mediation of the movement of arachidonate from 1-acyl-containing GPL into 1-ether-containing GPL. Using SK&F 98625 and SK&F 45905, we have demonstrated that both compounds block the movement of arachidonate into 1-ether GPL in human neutrophils.[11] This not only supports the hypothetical role of CoA-IT in arachidonate movement, but shows that the compounds have cellular entry.

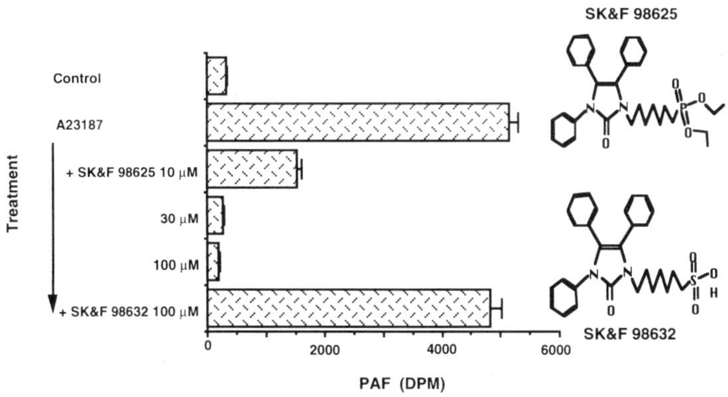

Figure 1. Effect of SK&F 98625 on PAF production in human monocytes. Monocytes were treated with vehicle, SK&F 98625 or SK&F 98632 for 5 min, followed by stimulation with 2 μM A23187. PAF production over 10 min was determined as described in Methods. The results are the means ± SE of triplicate determinations.

3.2 CoA-IT Inhibitors and PAF

We have used these tool compounds to explore the effects of inhibition of CoA-IT on the generation of PAF in stimulated inflammatory cells. Both SK&F 98625 and SK&F 45905 caused concentration-dependent reductions in PAF production in A23187-stimulated neutrophils.[12] In addition, SK&F 98625 reduced A23187-stimulated PAF production in human monocytes, whereas SK&F 98632, a structural analog that does not inhibit CoA-IT, failed to block PAF production (Fig 1). Pre-sensitized mouse bone marrow-derived mast cells respond to antigen with PAF production. Both SK&F 98625 and SK&F 45905 were able to block this production of PAF (Fig 2). Together, these data show that inhibitors of CoA-IT activity can block PAF production in a variety of inflammatory cells, under different stimuli. The blockade of PAF production was concentration-dependent and structurally specific for inhibition of CoA-IT.

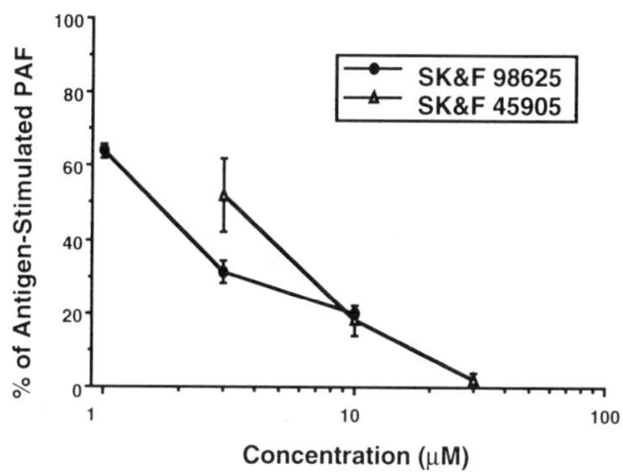

Figure 2. Effect of CoA-IT inhibitors on PAF production in mast cells. Mouse bone marrow-derived mast cells were treated for 5 min with vehicle or the indicated concentration of compound. PAF production over 10 min was determined. The results are the means ± SE from 3–5 determinations.

Table 1. Effects of PLA_2 inhibitors on PAF production in human neutrophils

Compound	Concentration (μM)	Effect on PAF (% of Stimulated Response)
Scalaradial	0.3	77
	1	64
	3	25
SB 203347	1	60
	3	11
	10	3
AA-CF3-ketone	1	102
	3	98
	10	92

Human neutrophils were treated for 5 min with the indicated concentrations of compounds, stimulated with 2 μM A23187 and PAF production determined after 10 min. The results are expressed as a % of the A23187 response and are the means of triplicate determinations.

In contrast to inflammatory cells, endothelial cells contain far less CoA-IT activity.[6] Also, whereas 50–75% of the GPL in inflammatory cells contain 1-ether linkages, the content of 1-ether-containing GPL is much lower in endothelial cells.[16] We have recently observed that this lower content of CoA-IT and 1-ether GPL results in a striking reduction in arachidonate remodeling over time (data not shown). As one would predict, in endothelial cells inhibition of CoA-IT had no effect on PAF production (data not shown). This result opens the possibility that inhibition of CoA-IT may have selective effects in inflammatory cells and warrants further study.

3.3 PLA_2 Inhibitors and PAF

Schemes of arachidonate movement into, through and out of GPL hypothesize that several enzymes are involved. One of the key enzymes is PLA_2, which is proposed to drive the remodeling pathway by releasing arachidonate.[7] There are two well-characterized and very different PLA_2 enzymes in human inflammatory cells, 14 kDa and 85 kDa PLA_2.[17] To determine the roles of these enzymes in PAF production, we have utilized inhibitors: scalaradial and SB 203347 are selective inhibitors of 14 kDa PLA_2[18,19]; arachidonate trifluoromethyl ketone is a selective inhibitor of 85 kDa PLA_2.[20]

The effects of these compounds on PAF production in human neutrophils is summarized in Table 1. The 85 kDa PLA_2 inhibitor had no effect on PAF production, whereas both 14 kDa PLA_2 inhibitors caused a concentration-dependent reduction in PAF production. Taken together, these data suggest that 14 kDa PLA_2 is coupled to CoA-IT activity and PAF production in human neutrophils. More study is needed to confirm this finding and expand it to other cells.

4. CONCLUSIONS

These studies show that inhibition of CoA-IT in inflammatory cells can block PAF production. Endothelial cells contain less CoA-IT activity and its inhibition is of less consequence. Data suggest that 14 kDa PLA_2 and CoA-IT act in a cooperative fashion in the production of PAF in inflammatory cells. A model for this notion of PAF production is shown is Fig 3.

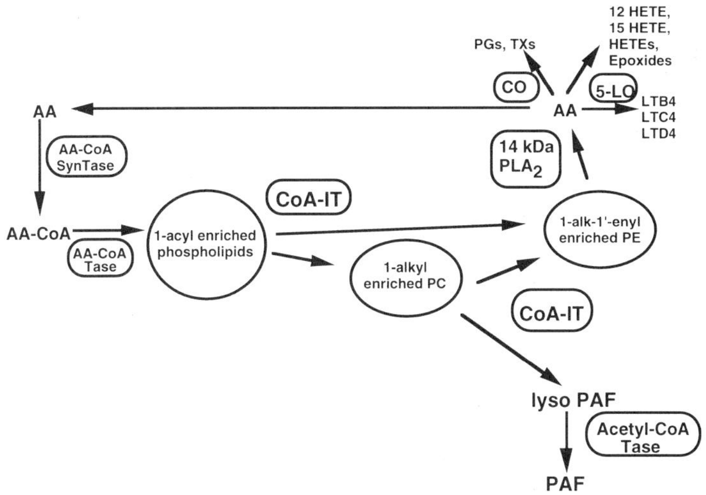

Figure 3. Model of the role of CoA-IT in PAF production.

REFERENCES

1. Kramer, R. M. and Deykin, D. (1983) *J. Biol. Chem.* **258**, 13806–13811
2. Sugiura, T., Masuzawa, Y., Nakagawa, Y., and Waku, K. (1987) *J. Biol. Chem.* **262**, 1199–1205
3. Winkler, J. D. and Chilton, F. H. (1993) *Drug News Perspec.* **6**, 133–138
4. MacDonald, J. I. S. and Sprecher, H. (1991) *Biochim. Biophys. Acta* **1084**, 105–121
5. Snyder, F., Lee, T. -C., and Blank, M. L. (1992) *Prog. Lipid Res.* **31**, 65–86
6. Winkler, J. D. and Chilton, F. H. (1995) in *Inflammation: Mediators and Pathways* (Ruffolo, R. and Hollinger, J., eds) pp. 147–171, CRC Press, Inc., Boca Raton, FL
7. Chilton, F. H., Fonteh, A. N., Surette, M. E., Triggiani, M., and Winkler, J. D. (1995) *Biochim. Biophys. Acta* (in press)
8. Suga, K., Kawasaki, T., Blank, M. L., and Snyder, F. (1990) *J. Biol. Chem.* **265**, 12363–12371
9. Sugiura, T., Fukuda, T., Masuzawa, Y., and Waku, K. (1990) *Biochim. Biophys. Acta Lipids Lipid Metab.* **1047**, 223–232
10. Winkler, J. D., Sung, C. -M., Hubbard, W. C., and Chilton, F. H. (1992) *Biochem. Pharmacol.* **44**, 2055–2066
11. Chilton, F. H., Fonteh, A. N., Sung, C. -M., Hickey, D. M. B., Torphy, T. J., Mayer, R. J., Marshall, L. A., Heravi, J. D., and Winkler, J. D. (1995) *Biochemistry* **34**, 5403–5410
12. Winkler, J. D., Fonteh, A. N., Sung, C. -M., Heravi, J. D., Nixon, A. B., Chabot-Fletcher, M., Griswold, D., Marshall, L. A., and Chilton, F. H. (1995) *J. Pharmacol. Exp. Ther.* (in press)
13. Triggiani, M., Fonteh, A. N., and Chilton, F. H. (1992) *Biochem. J.* **286**, 497–503
14. Mueller, H. W., O'Flaherty, J. T., and Wykle, R. L. (1983) *J. Biol. Chem.* **258**, 6213–6218
15. Winkler, J. D., Sung, C. -M., Hubbard, W. C., and Chilton, F. H. (1993) *Biochem. J.* **291**, 825–831
16. Takamura, H., Kasai, H., Arita, H., and Kito, M. (1990) *J. Lipid Res.* **31**, 709–717
17. Marshall, L. A. and Roshak, A. (1993) *Biochem. Cell Biol.* **71**, 331–339
18. Marshall, L. A., Winkler, J. D., Griswold, D. E., Bolognese, B., Roshak, A., Sung, C. -M., Webb, E. F., and Jacobs, R. (1994) *J. Pharmacol. Exp. Ther.* **268**, 709–717
19. Marshall, L. A., Hall, R. H., Winkler, J. D., Badger, A., Bolognese, B., Roshak, A., Flamberg, P. L., Sung, C. -M., Chabot-Fletcher, M., Adams, J. L., and Mayer, R. J. (1995) *J. Pharmacol. Exp. Ther.* (in press)
20. Street, I. P., Lin, H. -K., Laliberté, F., Ghomashchi, F., Wang, Z., Perrier, H., Tremblay, N. M., Huang, Z., Weech, P. K., and Gelb, M. H. (1993) *Biochemistry* **32**, 5935–5940

4

PRODUCTION OF PLATELET-ACTIVATING FACTOR BY BRAIN MICROVASCULAR ENDOTHELIAL CELLS

Kei Satoh, Masayuki Koyama, Hidemi Yoshida, and Shigeru Takamatsu

Department of Pathological Physiology
Institute of Neurological Diseases
Hirosaki University School of Medicine
Hirosaki, Japan

1. INTRODUCTION

Vascular endothelial cells produce platelet-activating factor (PAF) in response to various agonists, and endothelial PAF is one of the key molecules regulating the interactions between blood cells and the vascular wall.[1] To investigate the functional specificity of the vascular endothelium in brain, we have established pure cultures of porcine brain microvascular endothelial cells. Endothelial cells from porcine aorta were also cultured, and using these cultures we have conducted a comparative study on PAF production. Since endothelial PAF production is closely related to the adherence of white blood cells to the endothelium,[1] we also study the adhesion of human polymorphonuclear neutrophils (PMNs) to the cultured cells.

2. CULTURES OF ENDOTHELIAL CELLS

Culture of brain microvascular endothelial cells is a useful experimental model for studying functional specificity of the brain vascular wall, however, elimination of pericytes contaminating the cultures is a main obstacle for establishing a pure culture system. This problem has been dealt mainly by prolonged protease treatment of brain microvessels. We employed brief homogenization of pieces of fresh porcine cortical tissues followed by enzyme digestion, and this procedure enabled us to obtain pure cultures of porcine brain endothelial cells.[2] Aortic endothelial cells were separated by gentle scraping, with a scalpel, the inner surface of the aorta obtained from the same animal. Cells were cultured using modified MCDB131 medium supplemented with 2% (v/v) fetal calf serum, 10 ng/ml epidermal growth factor, 1 μg/ml hydrocortisone and 0.4% (w/v) bovine brain extract. Both types of cells assumed cobble stone appearance and brain microvascu-

Platelet-Activating Factor and Related Lipid Mediators 2
edited by Nigam *et al.*, Plenum Press, New York, 1996

Table 1. Production of PAF and PGI2 by endothelial cells from brain microvessels and aorta

	Brain microvascular endothelial cells		Aortic endothelial cells	
Agonists	PAF (dpm/10^6 cells)	6-ketoPGF1α (ng/10^6 cells)	PAF (dpm/10^6 cells)	6-ketoPGF1α (ng/10^6 cells)
None	222 ± 98	0.5 ± 0.5	534 ± 384	1.5 ± 1.4
A23187, 1 μM*	2892 ± 347	5.6 ± 5.3	8052 ± 2270†	16.0 ± 1.1†
Bradykinin, 1 μM	1760 ± 402	2.4 ± 1.5	3911 ± 2006	4.8 ± 1.8
Arachidonate, 33 μM	—	5.9 ± 2.5	—	27.4 ± 8.6†

PAF production was determined by the incorporation of [^{3}H]acetate into the phospholipid fraction that corresponded to PAF in thin-layer chromatography. PGI2 was quantified by RIA of 6-ketoPGF1α..

*Ca ionophore A23187.

†$p<0.01$: significantly different from the values of microvascular endothelial cells.

lar endothelial cells were thinner than aortic cells. They were positive in fluorescent immunostaining for von Willebrand factor and in binding fluorescence-labeled acetylated low density lipoprotein.

3. PRODUCTION OF PAF BY BRAIN MICROVASCULAR AND AORTIC ENDOTHELIAL CELLS

Production of PAF was assessed by the incorporation of [^{3}H]acetate into the phospholipid fraction that corresponded to PAF in thin-layer chromatography.[3)] Brain microvascular and aortic endothelial cells produce PAF and prostacyclin (PGI2) in response to the stimulation with Ca ionophore A23187 and bradykinin (Table 1). Brain microvascular endothelial cells produced a smaller amount of PAF or PGI2 as compared to aortic endothelial cells. Brain endothelial cells also produce less PGI2 in response to arachidonate, and these results suggest that microvascular endothelial cells have decreased activities of both phospholipase A2 and the step downstream to phospholipase A2. PAF and PGI2 antagonize each other in many aspects of activities, and the balance between these two autacoids may be of crucial importance in regulation of the vascular wall-blood cell interactions. In brain microvasculature this balance is maintained at a low level and this may predispose to impairment of the functional integrity of the vascular wall.

4. SYNTHESIZED PAF IS CELL-ASSOCIATED WITHOUT EXTRACELLULAR RELEASE

[^{3}H]PAF was always detected in lipid extract of the cells without any extracellular release . In experiments using the cells cultured on a culture insert, PAF was not released towards the basal side as well (Figure 1). Presence of human serum albumin or human serum did not enhance the extracellular release (PAF acetylhydrolase deficient serum[4)] was used to avoid possible breakdown of released PAF). PAF is vasoactive and endothelial PAF might be released to the basal side towards the smooth muscle layer, however, we did not detect any extracellular release of PAF even into the lower space of the culture insert. A factor in human serum is known to facilitate the extracellular release of PAF from

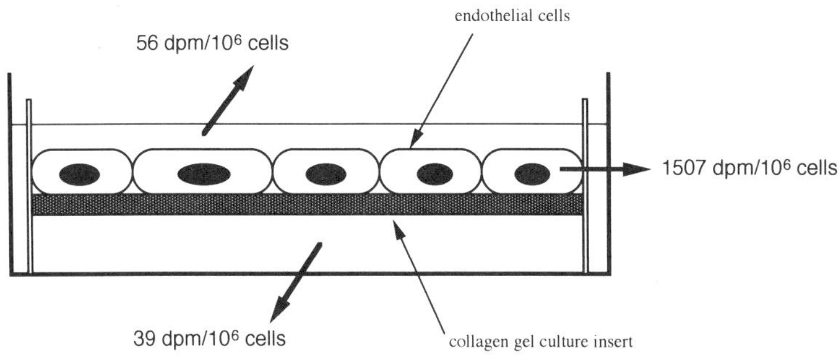

Figure 1. Intracellular retention of synthesized PAF.

PMNs,[5] however, such a factor is not effective on endothelial cells since addition of PAF acetylhydrolase-deficient human serum did not stimulate the extracellular release of PAF.

5. PMN ADHERENCE TO ENDOTHELIAL CELLS

Endothelial PAF is in close association with the cell surface expression of P-selectin, which, in turn, plays an important role in adherence and activation of PMNs to endothelial cells.[6] We next examined adherence of PMNs to cultured endothelial cells. As shown in Figure 2, in aortic endothelial cells both Ca ionophore A23187 and bradykinin stimulated PMN adherence. In brain microvascular endothelial cells, bradykinin failed to stimulate the adherence. These results do not correlate precisely with those of PAF production in these cells, and if this may be due to the lower PAF production in brain microvascular endothelial cells is not clear. P-selectin is a member of membrane proteins of the Weibel-Palade bodies.[7] Since microvascular endothelial cells are characterized by the absence of the Weibel-Palade bodies and cell surface expression of P-selectin in such cells

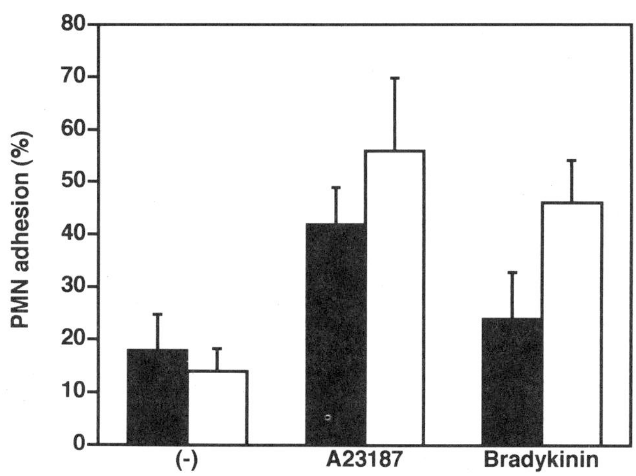

Figure 2. Adherence of PMNs to endothelial monolayers.

may be regulated in the manner different from that described for endothelial cells of larger vessels.[7] These points have to be addressed by further studies.

6. CONCLUSION

Cultured endothelial cells from brain microvessels and aorta produce PAF and PGI2 in response to the same agonists. However the amount of these two compounds produced in brain microvascular endothelial cells was smaller than that in aortic endothelial cells. This may partly related to the lower PMN adherence to brain endothelial cells. The anti-thrombogenic property of the endothelium is maintained by the balance among vaious factors, and the relation between PAF and PGI2 is one instance of such important functions. In brain micorvessels, the balance between these two autacoids is kept at a lower level as compared to larger vessels.

REFERENCES

1. Zimmerman GA, Prescott SM, McIntyre TM: Endothelial cell interactions with granulocytes: tethering and signaling molecules. *Immunol Today* **13**:93–100, 1992
2. Satoh K, Yoshida H, Imaizumi T, Koyama M, Takamatsu S: Production of platelet-activating actor by porcine brain microvascular endothelial cells in culture. *Thromb Haemostas* , in press
3. Prescott SM, Zimmerman GA, McIntyre TM: Human endothelial cells in culture produce platelet-activating factor (1-alkyl-2-acetyl-*sn*-glycero-3-phosphocholine) when stimulated with thrombin. *Proc Natl Acad Sci USA* **81**:3534–3538, 1984
4. Miwa M, Miyake T, Yamanaka T, Sugatani J, Suzuki Y, Sakata S, Araki Y, Matsumoto M: Characterization of serum platelet-activating factor (PAF) acetylhydrolase. Correlation of serum PAF acetylhydrolase and respiratory symptoms in asthmatic children. *J Clin Invest* **82**:1983–1991, 1988
5. Miwa M, Sugatani J, Ikemura T, Okamoto Y, Ino M, Saito K, Suzuki Y, Matsumoto M: Release of newly synthesized platelet-activating factor (PAF) from human polymorphonuclear leukocytes under in vivo conditions: contribution of PAF-releasing factor in serum. *J Immunol* **148:**872–880, 1992
6. Bonfanti R, Furie BC, Wagner DD: PADGEM (GMP-140) is a component of Weibel-Palade bodies of human endothelial cells. *Blood* **73**:1109–1112, 1989
7. Bowman PD, du Bois M, Dorovini-Zis K, Shivers RR: Microvascular endothelial cells from brain. *In*: Cell Culture Techniques in Heart and Vessel Research. Piper HM, ed., Springer-Verlag, Berlin, FRG, pp 140–157, 1990

5

PAF-SYNTHESIZING ENZYMES IN NEURAL CELLS DURING DIFFERENTIATION AND IN GERBIL BRAIN DURING ISCHEMIA

Ermelinda Francescangeli,[1] Louis Freysz,[2] and Gianfrancesco Goracci[1]

[1]Institute of Medical Biochemistry
University of Perugia
Via del Giochetto, 06100 Perugia, Italy
[2]Centre de Neurochimie
5 Rue Blaise Pascal, 67084 Strasbourg, Cedex, France

1. INTRODUCTION

Platelet-Activating Factor (PAF) is present in mammalian brain[1] and its cerebral origin has been demonstrated[2]. Furthermore, the capability of nervous tissue to produce PAF is supported by the observations that neural cells in culture synthesize this lipid mediator[3,4].

In the nervous tissue, PAF is involved in multiple and distinct phenomena as indicated by various studies. First of all, the production of PAF is stimulated by neurotransmitters in chick retina[5,6] and in human foetal brain cells in culture[3]. In addition, PAF modulates the release of neurotransmitters[7,8]. These findings, together with the presence of high-affinity PAF receptors in synaptosomes of rat brain cortex[9], strongly suggest its involvement in intercellular communications. An intracellular mediator role has been also suggested for PAF, since it is able to stimulate immediate early gene expression most likely by a mechanism triggered by its binding to intracellular receptors[10]. These and other observations have indicated PAF as a potential retrograde messenger in long term potentiation[11,12] and memory[13].

Several studies have demonstrated the involvement of PAF in pathological situations, too. In fact, tissue levels of PAF increase during brain or spinal cord ischemia[14,15] and during chemical convulsions[2]. Thus, it is well accepted the concept that this potent autacoid mediator participates to physiological and pathological phenomena in the nervous tissue and that its concentration, beside to the sites of its production, might be the discriminating factor between PAF-mediated normal and abnormal actions. In this contest, it is of great importance the definition of the factors controlling the rates of synthesis and degradation of this molecule in the nervous tissue.

As in other cells, two distinct pathways may operate for the synthesis of PAF in mammalian brain (for review, see 16) because it possesses the necessary enzymes and substrates. In fact, the presence of the following enzymes of the *de novo* route for PAF biosynthesis has been demonstrated: acetyl-CoA:1-alkyl-*sn*-glycero-3-phosphate acetyl-transferase[17,18], 1-alkyl-2-acetyl-*sn*-glycero-3-phosphate phosphohydrolase[19] and CDP-choline : 1-alkyl-2-acetyl-*sn*-glycerol phosphocholinetransferase (PAF-PCT)[20]. The other route (*remodelling pathway*) utilizes 1-alkyl-sn-glycero-3-phosphocholine (lyso-PAF) generated from 1-alkyl-2-acyl-sn-glycero-3-phosphocholine, a phospholipid present in neural membranes, by the action of a PLA_2 [21] or by a CoA-independent transacylase[22]. Lyso-PAF is then acetylated by acetylCoA:1-alkyl-*sn*-glycero-3-phosphocholine acetyl-transferase (lysoPAF-AcT) to produce PAF[23]. Since a cytosolic acetylhydrolase, which hydrolyses this mediator into the biologically inactive lysoPAF, is also present in brain tissue[24], the activity of this enzyme has to be also considered as a factor involved in the regulation of PAF concentration.

In the past, we have studied the activities and properties of CDP-choline : 1-alkyl-2-acetyl-sn-glycerol phosphocholinetransferase (PAF-PCT) and acetylCoA:1-alkyl-*sn*-glycero-3-phosphocholine acetyltransferase (lysoPAF-AcT), the two enzymes catalysing the last steps of PAF synthesis by the *de novo* and *remodelling* pathways, respectively. The aim of these studies was to get insight into the relative contribution of the two routes of PAF synthesis in function of the physiological status of the cells. Here, we provide an updated review of our results and suggest possible mechanisms influencing the rate of PAF synthesis during ischemia and reperfusion.

2. SYNTHESIS OF PAF DURING PROLIFERATION AND DIFFERENTIATION OF NEURAL CELLS

PAF interacts with hybrid neural cells (NG108–15) inducing, at nmolar concentrations, an arrest of growth of the cells followed by morphological differentiation with neurite extension[25]. This observation suggests that PAF acts as a mediator in the physiology of neuronal development. This role is supported by our data showing changes of the activities of PAF-synthesizing enzymes in chick embryo primary cultures during development[26,27].

During the development of neuronal cultures, the specific activity of PAF-PCT decreased, reached a minimum at the 3rd day and then, during the period of neuronal maturation, its activity increased up to the 6th day by about 8 fold[26] . The specific activity of lysoPAF-AcT was essentially constant during the first three days of culture and increased thereafter reaching values similar to those of the PAF-PCT.

Since both enzyme activities increased tremendously during the period of neuronal maturation, the formation of cellular contacts and synaptic like junctions, it seems likely that, in chick neurons, the expression of PAF-synthesizing enzymes or their transformation to an activated form may be relevant for controlling PAF levels in relation with cellular development. Thus, we might speculate that the relative contribution of the *de novo* and *remodelling pathways* to PAF synthesis would be mainly dependent on the relative concentrations of their substrates (i.e. alkylacetylglycerol and lysoPAF respectively) and on the concentration of intracellular Ca^{2+}, which inhibits PAF-PCT[20] whereas it is required for lysoPAF AcT[28,29]. The increase of both the enzymatic activities during the period of cell maturation is consistent with the hypothesis that PAF may be involved in some mechanism of neuronal differentiation and further support a role for PAF in the processes of synaptic transmission.

The complex mechanism of the regulation of PAF synthesis during differentiation of neuronal cells is further complicated by the observations that the treatment of human LA-N-1 neuroblastoma cells with retinoic acid, which induces their differentiation, caused a rapid decrease of PAF-PCT activity without any effect on the lysoPAF -AcT activity[30].

In chick embryo glial cell cultures (astrocytes), PAF-PCT activity remained almost constant during the development whereas lysoPAF-AcT increased up to the 12th day and then did not change significantly thereafter[26]. Thus, in these cells, the activity of the lysoPAF-AcT increased during the period of cellular proliferation without any further change when the cultures became confluent and the cells formed contacts. The activity of the lysoPAF-AcT was about 5 to 6 fold higher than that of PAF-PCT in mature glial cells and this indicates that in these cells PAF might be preferentially synthesized through the *remodelling pathway*.

This hypothesis becomes of particular interest considering that the cytosolic Ca^{2+}-dependent PLA_2 is mainly associated with astrocytes[31] . Thus, the activation of this enzyme could produce the substrate for PAF synthesis by this route. However, caution is recommended for the interpretation of these results because the relative activities of PAF-synthesizing enzymes may vary with the animal species.

The overall indications emerging from these observations indicate that: 1) PAF may be synthesized by both neuronal and glial cells and that the relative contribution of the *de novo* and *remodelling pathways* to PAF synthesis depends on cell type and on their physiological status; 2) the *remodelling pathway* may have a predominant role on the synthesis of PAF in astrocytes; 3) PAF synthesized by the *de novo pathway* may play a relevant role in the processes related to the synaptic transmission. The last hypothesis is supported by the early observations that the stimulation of chick retina with acetylcholine or dopamine induces PAF production by increasing the activity of PAF-PCT[5].

3. PAF-SYNTHESIZING ENZYMES DURING ISCHEMIA AND REPERFUSION

PAF levels, assayed in different brain areas by radioreceptor or aggregation assay was practically undetectable but much higher values were obtained when lipid extracts were acetylated before isolation of PAF by HPLC[32]. This observation suggests that PAF synthesized in various brain regions is rapidly degraded to lysoPAF by PAF-acetylhydrolase. Since after acetylation the highest levels of PAF were detected in the hippocampus, it seems likely that this area should also contain relative high activities of PAF-synthesizing enzymes.

Recent studies have shown that the concentration of PAF in brain tissue increases during ischemia. Particularly, Domingo et al.[33] have reported that, in gerbils subjected to the occlusion of both common carotid arteries, PAF levels were near 10 times more elevated with respect to sham-operated controls. This effect should be consequent to the activation of the biosynthetic routes or to the inhibition of the acetylhydrolase or to both events.

These observations have prompted us to investigate the activities of PAF-synthesizing enzymes in gerbil brain areas during ischemia and during the reperfusion[34]. The results can be summarized as it follows: 1) the legation of carotid arteries for 6 min caused a transient increase of PAF-PCT and lysoPAF-AcT activities in the cerebellum even if this area was not ischemic; 2) in the other brain areas PAF-PCT activity did not change during ischemia whereas that of lysoPAF-AcT increased in hippocampus; 3) during the first 5 minutes of reperfusion, a decrease of PAF-PCT activity, below the basal level, was observed in all brain areas examined but, in hippocampus only, the activity of the enzyme returned to normal levels within the following 60 minutes; 4) in opposite, lysoPAF-AcT activity remained elevated during reperfusion. Thus, in the hippocampus, both PAF-synthesizing enzymes were particularly active from 60 min to 3 days reperfusion.

4. CONCLUSIONS

The experiments using cell cultures have demonstrated the presence of the enzymes, catalyzing the last reactions of PAF synthesis by the *de novo* and the *remodelling pathways*, in neuronal and glial cells and that their activities undergo changes depending on the developmental status of cells. Although the problem concerning the relative contribution of the two routes to PAF synthesis during the different developmental stages of the cells is not yet solved, we have provided evidence that both PAF-PCT and lysoPAF-AcT should be subjected to regulation which may concern the expression of the enzymes and/or modifications of their activities by other mechanisms.

The latter possibility is supported by the observation that during ischemia and reperfusion rapid changes of enzyme activities occur. Since in these experiments, PAF-synthesizing enzyme activities were measured at their optimal conditions and with saturating concentrations of their substrates, it is possible that variations of their activities are the consequence of structural changes of the enzymes, for instance by phosphorylation-dephosphorylation mechanisms.

Preliminary studies, have indicated that rat brain microsomal lyso-PAF AcT might be activated by phosphorylation[35] as reported in other cell types[36]. Since protein kinase C is activated during ischemia[37,38] , the persistent activation of lysoPAF-AcT during reperfusion might be due to its phosphorylation by this enzyme.

Although the specific activity of PAF-PCT is much higher than that of lysoPAF-AcT, it seems unlike that the *de novo* pathway might provide a major contribute to PAF synthesis during ischemia *in vivo*. This assumption is based on the observations that intracellular Ca^{2+} concentration increases and that of ATP rapidly decreases at onset of brain ischemia[39,40] . It has been also demonstrated that Ca^{2+} is an inhibitor of PAF-PCT[20] and that ATP is required to maintain CDPcholine concentration sufficiently elevated for the synthesis of cholinephosphoglycerides, including PAF. Furthermore, it has been also postulated that the accumulation of AMP and CMP, as the consequence of energy deficiency in brain tissue, might favour the degradation of cholinephosphoglycerides through the reversal of phosphocholinetransferase reaction[41]. CMP reduces PAF-PCT[20] activity and therefore it is possible that this enzyme catalyses a reversible reaction as well. Thus its direction would depend on the relative concentration of CDPcholine and CMP. This enzyme may contribute again to PAF synthesis during reperfusion, when intracellular Ca^{2+} goes back to normal values and ATP levels are restored.

One of the early biochemical events taking place at the onset of ischemia is the accumulation of free fatty acids and diglycerides[42] which reflects the activation of phospholipases[43] and the CMP-dependent degradation of phosphoglycerides[44]. Thus, it seems likely that their production might have a correlation with PAF production.

PLA_2 becomes activated during ischemia and contributes to brain injury[45]. This enzyme catalyses the first step for the synthesis of PAF by the *remodelling pathway* producing lyso-PAF from the endogenous alkylacyl-GPC. Brain alkylacyl-GPC is a substrate for a cytosolic PLA_2 which is activated and translocated to the membrane upon an increase in intracellular calcium concentration[46]. Thus, ischemia might activate Ca^{2+}-dependent PLA_2, produce lysoPAF and consequently provide the substrate for lysoPAF -AcT. Furthermore, the activation of PKC observed during ischemia might lead to the phosphorylation of the enzyme thus leading to the full activation of PAF synthesis by the *remodeling pathway*.

Recently it has been shown that brain possesses a CoA-independent transacylase which catalyses the deacylation of alkylacyl-GPC in the presence of 1-alk-1-enyl-2-lyso-*sn*-glycero-3-phosphoethanolamine (1-Alkenyl-2-lyso-GPE; Ethanolamine lysoplasmalogen), forming lysoPAF and ethanolamine plasmalogens[22]. 1-Alkenyl-2-lyso-GPE acts as an acceptor molecule for the transfer of sn-2-acyl groups from alkylacyl-GPC and must be produced by the action of a PLA_2. A Ca^{2+} -independent PLA_2 specific for ethanolamine plasmalogen has been recently purified from bovine brain[47]. This enzyme seems to be involved in the release of free fatty acids in ischemia and to lead to the accumulation lysoplasmalogens and, consequently, triggering the biosynthesis of PAF through the CoA-independent transacylase mechanism. Thus, the activation of Ca^{2+}-dependent and/or Ca^{2+}-independent plasmalogen-selective PLA_2, during ischemic injury, may increase the lysoPAF concentration, the substrate of the acetyltransferase for the biosynthesis of PAF by the *remodeling pathway* .

This route for PAF synthesis should be particularly important for the generation of PAF in hippocampus because in this area, containing neurons highly sensitive to ischemia[48] , lysoPAF-AcT activity of gerbil brain increases rapidly onset ischemia.

ACKNOWLEDGMENTS

This work has been supported by a grant from Consiglio Nazionale delle Ricerche, Roma (n. 9502741CT04/11513770).

REFERENCES

1. Tokumura A., Kamiyasu K., Takanchi K., and Tsukatani H. (1987). Evidence for the existence of various homologues and analogues of platelet activating factor in a lipid extract from bovine brain. Biochem. Biophys. Res. Commun. 145, 415–425.
2. Kumar R., Harvey S.A.K., Kester M., Hanahan D.J. and Olson MS. (1988). Production and effects of Platelet-activating factor in the rat brain. Biochim. Biophys. Acta 963, 375–383.
3. Sogos V., Bussolino F., Pilia E., Torelli S. and Gremo F. (1990). Acetylcholine-induced production of Platelet-activating factor by human fetal brain cells in culture. J. Neurosc. Res. 27, 706–711.
4. Yue T.L., Lysko P.G. and Feuerstein G. (1990). Production of platelet activating factor from rat cerebellar granule cells in culture. J. Neurochem. 54, 1809–1811.
5. Bussolino F., Gremo F., Tetta C., Pescarmona G.P. and Camussi G. (1986) Production of platelet-activating factor by chick retina. J. Biol. Chem. 261, 16502–16508.
6. Bussolino F., Pescarmona G., Camussi G. and Gremo F. (1988a). Acetylcholine and dopamine promote the production of platelet activating factor in immature cells of chick embryonic retina. J. Neurochem. 51, 1755–1759.
7. Bussolino F., Tessari F., Turrini F., Braquet P., Camussi G., Prosdocimi M. and Bosia A. (1988b). Platelet-activating factor induces dopamine release in PC-12 cell line. Am. J. Physiol. 255, (Cell physiol) 24, 1755–1759.
8. Clark G.D., Happel L.T., Zorumski C.F. and Bazan (1992). Enhancement of hippocampal excitatory synaptic transmission by platelet activating factor. Neuron 9, 1211–1216

9. Marcheselli V.L., Rossowska M.J., Domingo M.T., Braquet P. and Bazan N.J. (1990). Distinct Platelet-activating factor binding sites in synaptic endings and in intracellular membranes of rat cerebral cortex. J. Biol. Chem. 265, 9140–9145.
10. Marcheselli V.L. and Bazan N.G. (1994). Platelet-activating factor is a messenger in the electroconvulsive shock.induced transcriptional activation of c-fos and zif-268 in hipocampus. J. Neurosci. Res. 37, 54–61.
11. 11.Kato K., Clark G.D., Bazan N.G. and Zorumski C.F. (1994). Platelet-activating factor as a potential retrograde messenger in CA1 hippocampal long-term potentiation. Nature 367, 173–179.
12. Wierasko A., Li G., Kornecki E., Hogan M.V. and Ehrlich Y.H. (1993) Long term potentiation in the hippocampus induced by platelet-activating factor. Neuron 10, 553–557.
13. Izquerdo I., Fin C., Schmitz P.K., Da Silva R.C., Jerusalinsky D., Quillfeldt J.A., Ferreira M.B.G., Medina J.H. and Bazan N.G. (1995). Memory enhancement by intrahippocampal, intraamygdala, or intraentorhinal infusion of platelet-activating factor measured in an inhibitory avoidance task. Proc. Natl. Acad. Sci. USA 92, 5047–5051.
14. Braquet P., Paubert-Braquet M., Koltei M., Bourgain R., Bussolino F. and Hosford D. (1989). Is there a case for PAF antagonist in the treatment of ischemic states? TIPS 10, 23–30.
15. Lindsberg P.J., Yue T-L., Frerichs K.U, Hallenbeck and Feuerstein G. (1990). Evidence for Platelet-activating factor as a novel mediator in experimental stroke in rabbits. Stroke 21, 1452–1457.
16. Goracci G. (1990). PAF in the nervous system: biochemistry and pathophysiology. In: Pharmacology of cerebral ischemia, (Krieglstein J. and Oberpichler H., Eds.), Wissenschaftliche Verlagsgesellshaft, Stuttgart, pp. 377–390.
17. Lee T-c., Malone B. and Snyder F. (1986). A new *de novo* pathway for the formation of 1-alkyl-2-acetyl-*sn*-glycerols, precursors of Platelet Activating Factor. J.Biol. Chem., 261, 5373–5377..
18. Baker,R.R. and Chang H-y. (1993). The potential for Platelet-activating factor synthesis in brain: properties of cholinetransferase and 1-alkyl-*sn*-glycero-3-phosphate acetyltransferase in microsomal fractions of immature rabbit cerebral cortex. Biochim. Biophys. Acta 1170, 157–164.
19. Lee T-c., Malone B. and Snyder F. (1988). Formation of 1-alkyl-2-acetyl-*sn*-glycerols via de novo biosynthetic pathway for Platelet Activating Factor. J.Biol.Chem., 263, 1755–1760.
20. Francescangeli E. and Goracci G. (1989) The de novo biosynthesis of platelet-activating factor in the rat brain. Biochem. Biophys. Res. Commun. 161, 107–112.
21. Woelk H., Goracci G. and Porcellati G. (1974). The action of brain phospholipase A_2 on purified specifically labeled 1,2-diacyl, 2-acyl 1-alk-1'anyl and 2-acyl-1-alkyl-*sn* glycero-3 phosphorylcholine. Hoppe-Seyler's Z. Physiol. Chem. 355, 75–81.
22. Blank M.L., Smith Z.L., Fitzgerald V. and Snyder F. (1995). The CoA-.independent transacylase in PAF biosynthesis: tissue distribution and molecular species selectivity. Biochim. Biophys. Acta 1254, 295–301.
23. Goracci G. and Francescangeli E. (1991) Properties of PAF-synthesizing phosphocholinetransferase and evidence for lysoPAF acetyltransferase activity in rat brain. Lipids 26, 986–991
24. Hattori M., Arai H. and Inoue K. (1993). Purification and characterization of bovine brain Platelet-activating factor acetylhydrolase. J. Biol. Chem. 268, 18748–18753.
25. Kornecki E. and Ehrlich Y.H. (1988) Neuroregulatory and neuropathological actions of the ether-phospholipid platelet-activating factor. Science 240, 1792–1794.
26. Francescangeli E., Freysz L., Dreyfus H., Boila A. and Goracci G. (1993). Biosynthesis of 1-alkyl-2-acety-*sn*-glycero-3-phosphocholine (Platelet activating factor) in cultured neuronal and glial cells. In: Phospholipis and Signal Transmission (Massarelli R., Horrocks L.A., Kanfer J. N., Loffelholz K., eds.), Springer-Verlag, Berlin Heidelberg, NATO Asi, vol.H 70, pp373–385.
27. Francescangeli E., Lang D., Dreyfus H., Boila A., Freysz L. and Goracci G (1996). Activities of enzymes involved in the metabolism of platelet-activating factor in neural cell cultures during proliferation and differentiation. Submitted
28. Ninio E., Mencia-Huerta J.M. and Benveniste J. (1983). Biosynthesis of platelet-activating factor (PAF-acether). V. Enhancement of acetyltransferase activity in murine peritoneal cells by calcium ionophore A23187. Biochim. Biophys. Acta 751, 298–304.
29. Gomez-Cambronero J., Inarrea P., Alonso F. and Sanchez-Crespo H. (1984). The role of calcium ions in the process of acetyltransferase activation during the formation of platelet-activating factor (PAF-acether). Biochemistry 219, 419–424
30. Francescangeli E., Goracci G., Dreyfus H., Boila A. and Freysz L. (1993). Synthesis of Platelet-activating factor (PAF) during differentiation of the human neuroblastoma cell LA-N-1. J.Neurochem. 61, Suppl., S239.
31. Stephenson D.T., Manetta J.V., White D.L., Chiou X.G., Cox L., Gitter B., May P.C., Sharp J.D., Kramer R.M. and Clemens J.A. (1994). Calcium-sensitive cytosolic phospholipase A_2 ($cPLA_2$) is expressed in human brain astrocytes. Brain Res. 637, 97–105.

32. Tiberghien C., Laurent L., Junier M.P. and Dray F. (1991). A competitive receptor binding assay for Platelet-activating factor (PAF). Quantification of PAF in rat brain. J. Lipid Med. 3, 249–266.
33. Domingo M.T., Spinnewyn P., Chabrier E. and Braquet P. (1994). Changes in [^{3}H]PAF binding and PAF concentrations in gerbil brain after bilateral common carotid artery occlusion: a quantitative autoradiographic study. Brain Res. 640, 268–276.
34. Francescangeli E., Domanska-Janik K. and Goracci G. (1996). Relative contribution of the *de novo* and the *remodelling* pathways to the synthesis of Platelet activating factor in brain areas and during ischemia. J.Lipid Med. in press.
35. Francescangeli E., Freysz L. and Goracci G. (1994). Regulation of Platelet-activating factor (PAF) metabolism in nervous tissue and in cultured neural cells. J. Neurochem 63, suppl. 1, 22.
36. Snyder F. (1995). Platelet-activating factor: the biosynthetic and catabolic enzymes. Biochem. J. 305, 689–705.
37. Onodera H., Araki T. and Kogure K. (1989). Protein kinase C activity in the rat hippocampus after forebrain ischemic autoradiographic analysis by [^{3}H]phorbol-12,13-dibutyrate. Brain Res.481, 1–7.
38. Domanka-Janik K and Zalewaka T. (1992). Effect of brain ischemia on protein kinase C. J. Neurochem. 58, 1432–1439.
39. Siesjö B.K. and Bengtsson F. (1989). Calcium fluxes, calcium antagonist and calcium-related pathology in brain ischemia, hypoglycemia and spreading depression. A unifying hypothesis. J. Cereb. Blood Flow Metab. 9, 127–140.
40. Ljunggren B., Schultz H. and Siesjo B.K. (1974) Changes in energy state and acid-base parameters of the rat brain during complete compression ischemia. Brain Res. 73, 277–289 .
41. Goracci G., Francescangeli E., Mozzi R., Porcellati S. and Porcellati G. (1985). Regulation of phospholipid metabolism by nucleotides in brain and transport of CDP-choline into brain. In: Novel Biochemical, Pharmacologiocal and clinical aspects of cytidinediphosphocholine. (Zappia V., Kennedy E.P., Nilsson B.J. and Galletti P. eds.), Elsevier, New York, pp105–116
42. Bazan N.G. (1970). Effect of ischemia and electroconvulsive shock on free fatty acid pool in the brain. Biochim.Biophys.Acta, 218, 1–10
43. Farooqui A.A., Hirashima Y., Forooqui T. and Horrocks L.A. (1992). Involvement of calcium, lipolytic enzymes and free fatty acids in ischemic brain trauma. In: Neurochemical correlates of cerebral ischemia, (Bazan NG, Braquet P and Ginsberg MD eds) Plenum Press , New York, pp 117–138.
44. Goracci G., Francescangeli E., Horrocks L.A. and Porcellati G. (1983). The effect of CMP on the release of free fatty acids of rat brain in vitro.. Neurochem. Res. 8, 971–981.
45. Rordorf G., Uemura Y. and Bonventre J.V. (1991). Characterization of Phospholipase A_2 (PLA_2) activity in gerbil brain: enhanced activities of cytosolic, mitochondrial, and microsomal forms after ischemia and reperfusion. J. Neurosc. 11, 1829–1836.
46. Bonventre J.V. and Koroshetz W.J. (1993). Phospholipase A_2 (PLA_2) activity in gerbil brain: characterization of cytosolic and membrane-associated forms and effects of ischemia and reperfusion on enzymatic activity. J. Lipid Med. 6, 457–471.
47. Yang H-C., Farooqui A.A. and Horrocks L.A..(1995). Purification and characterization of a calcium-independent phospholipase A_2 from bovine brain. (1995) J. Neurochem. 65 Suppl. S177B.
48. Kirino T., and Saito, K. (1984). Selective vulnerability in the gerbil hippocampus following transient ischemia. Acta Neuropathol. 62, 201–208.

6

PLATELET-ACTIVATING FACTOR PRODUCTION IN THE STIMULATED MACROPHAGES IS ENHANCED BY THE CYCLOOXYGENASE INHIBITORS

Masateru Yamada,[1] Masako Watanabe,[1] Suetsugu Mue,[2] and Kazuo Ohuchi[1]

[1]Department of Pathophysiological Biochemistry
Faculty of Pharmaceutical Sciences, Tohoku University
Sendai, Miyagi 980-77, Japan
[2]Department of Health and Welfare Science
Faculty of Physical Educations, Sendai University
Funaoka, Miyagi 989-16, Japan

1. INTRODUCTION

In inflammatory cells, PAF is considered to be synthesized via the remodeling pathway[1)]. Activation of phospholipase A_2 releases sn-2 fatty acids, predominantly arachidonic acid, which is metabolized to prostanoids by cyclooxygenase and lipoxygenase. The two different lipid mediators, PAF and prostanoids, are produced from a common precursor in membrane phospholipids when PAF is produced via the remodeling pathway. Because PAF and prostanoids are potent, biologically active mediators, we considered that their production is interdependent.

We previously reported that when the rat peritoneal macrophages are stimulated by the endomembrane Ca^{2+}-ATPase inhibitor thapsigargin and thapsigargicin, or the protein kinase C (PKC) activators such as 12-O-tetradecanoylphorbol 13-acetate (TPA), teleocidin and aplysiatoxin, the release of arachidonic acid from membrane phospholipids and prostaglandin E_2 (PGE_2) production increase[2,3)]. Furthermore, we found that the stimulation of the rat peritoneal macrophages by thapsigargin or TPA increases the content of PAF in the cells[4)]. Therefore, we examined the possible modulatory role of the simultaneously produced PGE_2 in PAF production under these conditions.

2. MATERIALS AND METHODS

The rat peritoneal macrophages collected 4 days after intraperitoneal injection of a 5% solution of soluble starch and bacto peptone[2)] were incubated for the indicated periods in Eagle's minimal essential medium (+ 10% calf serum) containing drugs. PGE_2 concentrations in the conditioned medium were radioimmunoassayed[3]. Total lipids in the cells and in the conditioned medium were extracted, respectively, and PAF in the lipid fraction was partially purified using the immunoaffinity mini-column for PAF[4], and the levels of PAF were radioimmunoassayed using a commercially available kit (New England Nuclear, U.S.A.).

3. RESULTS AND DISCUSSION

3.1 The Effects of Cyclooxygenase Inhibitors on Thapsigargin-Induced Production of PAF and PGE_2

When the macrophages were incubated in the presence of thapsigargin (46.1 nM, 30 ng/ml), the PAF contents in the cells reached a maximum 10 min after incubation; the levels then declined time-dependently. In the conditioned medium of the thapsigargin-treated or the non-treated macrophages, the amount of PAF was below the limits of detection (< 1.0 pmol/ml). In thapsigargin-treated macrophages, PGE_2 production increased time-dependently. In contrast, without stimulation with thapsigargin, the production of PAF and PGE_2 was low. When the macrophages were incubated for 10 min in the medium containing thapsigargin (46.1 nM) and various concentrations of indomethacin, the thapsigargin-induced PGE_2 production was inhibited by indomethacin in a concentration-dependent manner (Figure 1)[5]. In contrast, the thapsigargin-induced PAF production was further enhanced by indomethacin in a concentration-dependent manner (Figure 1)[5]. Naproxen (1 lM) and ibuprofen (3 lM) also inhibited PGE_2 production and enhanced the thapsigargin-stimulated PAF production at 10 min. In the absence of thapsigargin, the cyclooxygenase

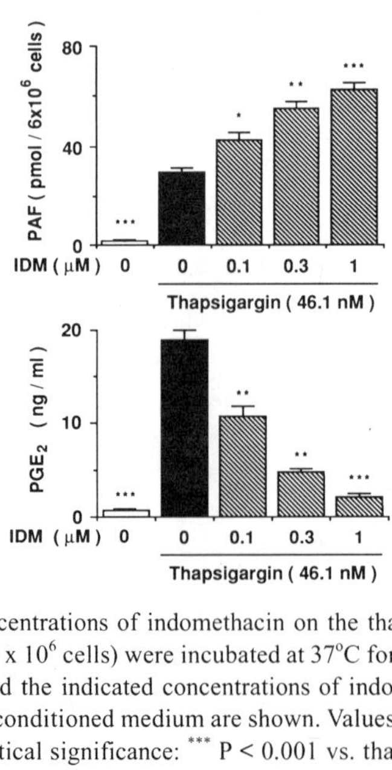

Figure 1. The effects of various concentrations of indomethacin on the thapsigargin-induced production of PAF and PGE_2. Peritoneal macrophages (6 x 10^6 cells) were incubated at 37°C for 10 min in 4 ml of medium containing thapsigargin (46.1 nM, 30 ng/ml) and the indicated concentrations of indomethacin (IDM). PAF contents in the cells and PGE_2 concentrations in the conditioned medium are shown. Values are the means from four samples with S.E.M. shown by vertical bars. Statistical significance: *** $P < 0.001$ vs. thapsigargin control. N.D. means not detectable.

inhibitors did not enhance PAF production. These findings indicate that the simultaneously produced PGE_2 down-regulates the thapsigargin-induced PAF production.

3.2 The Effects of Indomethacin on the Thapsigargicin-, A23187-, TPA-, Aplysiatoxin-, or Teleocidin-Induced Production of PAF and PGE_2

After incubation of the macrophages for 10 min in the medium containing the Ca^{2+}-ATPase inhibitors thapsigargin (46.1 nM) or thapsigargicin (48.2 nM, 30 ng/ml), the Ca^{2+} ionophore A23187 (1 lM), or the PKC activators TPA (48.6 nM, 30 ng/ml), aplysiatoxin (44.7 nM, 30 ng/ml) or teleocidin (67.5 nM, 30 ng/ml), both the contents of PAF in the cells and PGE_2 in the conditioned medium increased. When the macrophages were incubated for 10 min in the presence of indomethacin (1 lM), the PAF production induced by each of these stimulants was further enhanced in accordance with the inhibition of PGE_2 production.

3.3 The Effects of Exogenous PGE_2 on the Cyclooxygenase Inhibitor-Induced Enhancement of PAF Production

As shown in Figure 2[5], the addition of PGE_2 at 100 nM abolished the cyclooxygenase inhibitor-induced enhancement of PAF production at 10 min in the thapsigargin-stimulated macrophages, and the PAF contents in the cells decreased to the levels lower than that in the cells incubated with thapsigargin alone. These cyclooxygenase inhibitors inhibited the PGE_2 production at 10 min. The PAF production in the macrophages stimulated by thapsigargin alone at 10 min was also significantly inhibited by the addition of exogenous PGE_2 (100 nM). Because the enhancement of PAF production by the cyclooxygenase inhibitors was counteracted by exogenous PGE_2, it was suggested that the cyclooxygenase inhibitors enhance PAF production by decreasing PGE_2 levels.

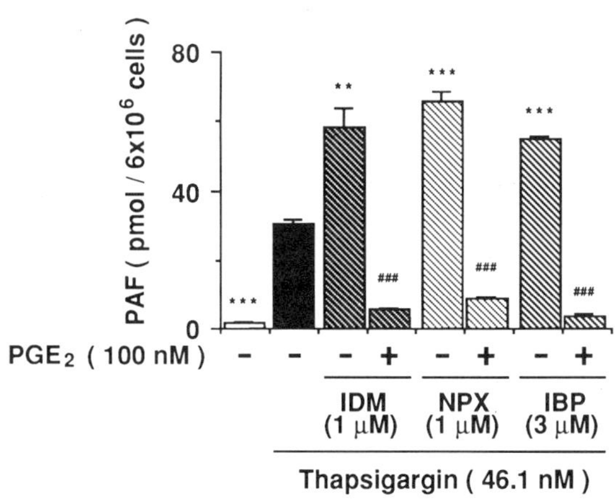

Figure 2. The effects of exogenous PGE_2 on the cyclooxygenase inhibitor-induced enhancement of PAF production. Peritoneal macrophages (6 x 10^6 cells) were incubated at 37°C for 10 min in 4 ml of medium containing thapsigargin (46.1 nM, 30 ng/ml) and the indicated concentration of indomethacin (IDM), naproxen (NPX), or ibuprofen (IBP) with or without PGE_2 (100 nM). PAF contents in the cells are shown. Values are the means from four samples with S.E.M. shown by vertical bars. Statistical significance: ** $P < 0.01$, *** $P < 0.001$ vs. TPA control; +++ $P < 0.001$ vs. corresponding control. N.D. means not detectable.

3.4 The Effects of Arachidonic Acid on the Thapsigargin-Induced Production of PAF and PGE_2

PAF production stimulated by thapsigargin (46.1 nM) at 10 min was inhibited by the addition of exogenous arachidonic acid (1 to 10 lM)[5], whereas the thapsigargin-induced production of PGE_2 increased. These results also indicated that the concurrently produced PGE_2 plays a role in the negative feedback regulation of the thapsigargin-induced PAF production.

3.5 The Effects of Dibutyryl cAMP on PAF Production Induced by Thapsigargin Alone or by Thapsigargin Plus Indomethacin

To obtain further insight into the mechanism of the inhibition of PAF production by the simultaneously produced PGE_2, the effects of dibutyryl cyclic AMP were examined because PGE_2 has an activity to increase the level of cyclic AMP. Treatment with dibutyryl cyclic AMP (0.1 to 1 lM) inhibited PAF production at 10 min induced by thapsigargin (46.1 nM) alone or by thapsigargin (46.1 nM) plus indomethacin (1 lM)[5]. The enhanced PAF production in the presence of indomethacin was much more effectively inhibited by dibutyryl cAMP than that stimulated by thapsigargin alone.

The findings obtained by the pharmacological modulations of PGE_2 production suggested that the PGE_2 produced by several stimulators plays a role in down-regulation of PAF production, probably due to increasing the adenylate cyclase activity and cAMP levels. Thus cyclooxygenase inhibitors induce PAF production by inhibiting PGE_2 production.

4. REFERENCES

1. Albert, D. H. and Snyder, F.: J. Biol. Chem., 258: 97–102. 2. Ohuchi, K., Sugawara, T., Watanabe, M., Hirasawa, N., Tsurufuji, S., Fujiki, H., Christensen, S. B. and Sugimura, T.: Br. J. Pharmacol. 94: 917–923, 1988.
3. Ohuchi, K., Sugawara, T., Watanabe, M., Hirasawa, N., Tsurufuji, S., Fujiki, H., Sugimura, T. and Christensen, S. B.: J. Cancer Res. Clin. Oncol. 113: 319–324, 1987.
4. Watanabe, M., Watanabe, T., Hirasawa, N., Mue, S.,Muramatsu, S., Matsushita, Y., Takahagi, H., Braquet, P., Broquet, P., Levine, L. and Ohuchi, K.: J. Chromatogr. 597: 309–314, 1992.
5. Watanabe, M., Yamada, M., Mue, S. and Ohuchi, K.: Br. J. Pharmacol. 116: in press, 1995.

7

THE ROLE OF PEROXISOMES IN ETHER LIPID SYNTHESIS

Back to the Roots of PAF

H. van den Bosch, E. C. J. M. de Vet, and A. W. M. Zomer

Department Biochemistry of Lipids
Centre for Biomembranes and Lipid Enzymology
Institute of Biomembranes, Utrecht University
Padualaan 8, 3584 CH Utrecht, The Netherlands

1. PEROXISOMES

Peroxisomes belong to a family of subcellular organelles called microbodies that play indispensable roles in cellular metabolism in protozoa, fungi, plant and mammalian cells[1]. These organelles show a greater variety in enzyme content than other subcellular organelles. Well over 50 enzymes have been reported to be located in microbodies of some kind, but no microbody contains all of them. This is illustrated in table I for a few major microbody processes. Whereas fatty acid β-oxidation is present in all microbody types, various oxidases producing H_2O_2 and catalase are not present in glycosomes, the microbody originally discovered in *Trypanosomatidae*[2]. The glyoxylate cycle is only found in the microbodies of plant seeds and those of yeast and Tetrahymena. Photorespiration and glycolysis are restricted to only one type of microbody, *i.e.* plant leaf peroxisomes and glycosomes, respectively. Ether lipid synthesis is confined to mammalian peroxisomes[3] and, as we recently demonstrated, to trypanosomal glycosomes[4], but is not found in plant and yeast microbodies.

Mammalian peroxisomes are ubiquitous organelles occurring in essentially all cells with the exception of erythrocytes. They are surrounded by a single membrane and are characterized by a relatively high equilibrium density of 1.23–1.25 g/cm^3, which facilitates the purification of these organelles that usually constitute only 1 to 2 percent of the total cellular protein. The membrane phospholipid/protein ratio of 250 nmol P-lipid/mg protein is comparable to that of mitochondria[5] and thus cannot explain the observation that isolated peroxisomes are leaky to molecules up to molecular weights of thousand[6]. The organelle contains no DNA and no glycoproteins as a first indication that proteins are nuclear-coded and do not get incorporated via the ER/Golgi route.

Platelet-Activating Factor and Related Lipid Mediators 2
edited by Nigam *et al.*, Plenum Press, New York, 1996

Table I. Main biochemical processes in microbodies

	Organism					
		Plants		Yeast	Protozoa	
Nomenclature Biochemical process	Mammals Peroxisomes	Seeds Glyoxysomes	Leaves Peroxisomes	Peroxi- somes	Tetrahymena Peroxisomes	Trypanosoma Glycosomes
FA-β-oxidation	+	+	+	+	+	+
H_2O_2 metabolism	+	+	+	+	+	–
Glyoxylate cycle	–	+	–	+	+	–
Photorespiration	–	–	+	–	–	–
Glycolysis	–	–	–	–	–	+
Ether lipid synthesis	+	–	–	–		+

Symbols: +, present; –, absent; blank, not yet reported.

Hints for the involvement of mammalian peroxisomes in cellular processes have originally been obtained frequently from cell fractionation experiments. In many cases these initial findings have now been confirmed and extended by studies on a series of inborn errors of metabolism that are now called peroxisomal disorders. This will be illustrated by work from our laboratory on one of the major peroxisomal biosynthetic processes, *i.e.* ether lipid biosynthesis.

2. ETHER PHOSPHOLIPIDS

Ether lipids can be distinguished in two types. Alkyl/acyl-phospholipids have a saturated ether linkage at the *sn*-1-position of the glycerol backbone, whereas alkenyl/acyl-phospholipids have an α,β-unsaturated ether linkage at that position. The latter class is also denoted by their trivial name plasmalogens. Plasmalogens occur widespread in nature, especially in the mammalian kingdom and are usually much more abundant in the ethanolamine glycerophospholipids than in the choline glycerophospholipids. The overall physical-chemical properties of ether lipids[7] are similar to those of the diacyl phospholipids, although some important differences can be noted as well (Table II).

As can be seen in table II, there are differences in the phase transition temperatures for the ethanolamine phospholipids, especially in the transition from lamellar to hexagonal phase and much less so in the gel to liquid crystalline transition. Differences in the cross-sectional areas that the different species, *i.e.* diacyl-, alkyl/acyl- and alkenyl/acyl-, occupy in monolayers can also be noted. These differences in packing densities are not found for ethanolamine phospholipids where polar headgroup interactions are predominant and result in smaller cross-sectional areas than observed for choline phospholipids. In the latter class the plasmalogen species show the smallest cross-sectional area. This can be explained by their structures as determined by NMR[8]. The characteristic differences are a different orientation of the glycerol backbone which is perpendicular to the plane of the membrane in case of the diacyl species and bent in case of the plasmalogen species. Due to a bent in the proximal end of the *sn*-2-acyl chain in diacyl species, there is a large spatial difference in the position of the carbonyl carbons of the *sn*-1- and *sn*-2-ester bonds. In plasmenylcholine the aliphatic chains are in much closer spatial proximity with the carbon

Table II. Physico-chemical properties of choline- and ethanolamine glycerophospholipid subclasses

Phospholipid class	Subclass	Phase transition (°C) $L \rightarrow H_{II}$ [a]	Phase transition (°C) $G \rightarrow L$ [b]	Cross-sectional area (A^2 at 30 dynes/cm) [c]
Diradyl-GPC				
	diacyl-	—	11.0	67.7
	alkylacyl-	—	9.4	61.2
	alkenylacyl-	—	—	59.9
Diradyl-GPE				
	diacyl-	68	30[a]	55.0
	alkylacyl-	53	29.5[a]	55.0
	alkenylacyl	30	26[a]	55.0

Abbreviations: L, liquid crystalline lamellar phase; H_{II}, hexagonal phase; G, gel lamellar phase; -GPC, glycerophosphocholine; -GPE, glycerophosphoethanolamine.
[a]Data from ref. 10; mainly 1-16:0/18:0-2-18:1 species.
[b]Data from ref. 11; 1-16:0-2-18:1 species.
[c]Data from ref. 12; mainly 1-16:0/18:0-2-18:1 species.

numbers of the aliphatic chains linked up and explaining the reduced cross-sectional area. The above differences in size and physical behaviour have, however, provided little clues as to the function of ether lipids. Other approaches have suggested putative roles for ether lipids. Plasmalogens may act as membrane-associated antioxidants against certain oxidative stresses[9]. Ether lipid species with a saturated ether linkage have been reported for inositol anchors to associate proteins to membranes[13] and in a low molecular weight lipidic modulator of glucocorticosteroid receptors[14]. The best known ether lipid in terms of biological function is platelet-activating factor, a phospholipid with a variety of biological activities for which the ether linkage is highly important as evidenced by the findings that the close structural analog with an ester linkage is several hundred fold less active[15].

3. ETHER LIPID SYNTHESIS AND PEROXISOMES

The biosynthesis of ether lipids starts with the acylation of dihydroxyacetone phosphate (DHAP) by a specific acyltransferase. The glycero-ether linkage is then formed in a reaction in which the acyl group is replaced by a long-chain alcohol catalyzed by alkyl-DHAP synthase. In cell fractionation experiments these enzymes were localized in peroxisomes[3]. The keto-group in alkyl-DHAP is then reduced to a hydroxyl group by a reductase that is present in both peroxisomes and endoplasmic reticulum. Acylation of this hydroxyl group and subsequent introduction of the phosphocholine or phosphoethanolamine group then proceeds in the endoplasmic reticulum and is catalyzed by the same enzymes that carry out these reactions in the synthesis of diacylphospholipids. The final step then is a desaturase reaction to yield the α,β-unsaturated ether linkage in plasmalogens. This reaction takes place in alkyl-acyl ethanolamine phospholipids only. For the synthesis of choline plasmalogens an additional cycle involving removal of the phosphoethanolamine group and introduction of the phosphocholine group in the plasmalogenic diglyceride analog is required[16].

The presence of the two enzymes involved in the synthesis of glycero-ether bonds, *i.e.* DHAP acyltransferases and alkyl-DHAP synthase, in rat and guinea pig liver peroxisomes[3], in conjunction with the discovery of a human inborn error of metabolism called Zellweger syndrome, in which peroxisomes could not be detected morphologically[17], led to the question whether such patients were deficient in ether lipids.

Phospholipid analyses of tissues from controls and Zellweger patients confirmed the virtual absence of ether lipids in the patients[18]. This showed that also in humans peroxisomes are indispensable for ether lipid synthesis and that this process cannot be catalyzed anywhere else in the cell.

The ether lipid deficiency in Zellweger syndrome was then applied to develop diagnostic tests for this disease. This was done by studying *de novo* ether lipid synthesis from the radioactive precursor hexadecanol. In control fibroblasts, 45% and 5%, respectively, of the total radioactivity incorporated into phospholipids was found to be present in the alkenyl chain of ethanolamine and choline plasmalogens. This indicated that fibroblasts are capable of synthesizing their own plasmalogens. This process appeared to be severely impaired in Zellweger fibroblasts. Similar findings were made in amniotic fluid cells. Thus, such hexadecanol incorporation studies can differentiate between normal and disease state and could be used for postnatal and prenatal diagnosis[19–21]. Further studies then revealed that *de novo* plasmalogen biosynthesis is not only deficient in Zellweger syndrome, but also in rhizomelic chondrodysplasia punctata, neonatal adrenoleukodystrophy and infantile Refsum disease[19,21]. Based on these studies on ether lipid biosynthesis and many other biochemical assays it became clear that Zellweger syndrome was only the prototype of a newly discovered class of inborn errors of metabolism caused by deficiencies in peroxisomal enzymes. This category of so-called peroxisomal disorders now comprises already some fifteen members[1,20].

4. PLATELET-ACTIVATING FACTOR SYNTHESIS IN PEROXISOMAL DISORDERS

The finding of ether lipid deficiency in a number of peroxisomal disorders led us to investigate the capacity for platelet-activating factor (PAF) synthesis in these patients. Initial experiments with polymorphonuclear leukocytes from Zellweger patients indicated that calcium ionophore-induced PAF synthesis was more severely deficient in younger patients than in older patients. This appeared to correlate with the ethanolamine plasmalogen level in erythrocytes from these patients which was likewise more deficient in the younger patients. This suggested that PAF production was related to the residual ether phospholipid level in the patients. This was corroborated in a study with rhizomelic chondrodysplasia punctata (RCDP) patients, in which the capacity for ionophore-induced PAF synthesis in polymorphonuclear leukocytes was directly compared with the residual level of alkyl/acyl-glycerophosphocholine from which PAF is produced by the remodelling pathway (Table III).

The results indicated that in comparison to controls PAF synthesis was reduced 600-, 7- and 17-fold, respectively, in three different patients. Indeed, a clear example of biological variation. The advantage of this was that a good correlation with the residual levels of the PAF precursor alkyl/acyl-glycerophosphocholine became evident. These were reduced 360-, 7- and 24-fold, respectively. Thus, the residual capacity for PAF synthesis in peroxisomal disorders with a deficiency in ether lipid synthesis is likely to be determined by the residual cellular content of alkyl/acyl-glycerophosphocholine.

Table III. PAF synthesis in polymorphonuclear leukocytes from an ether lipid-deficient peroxisomal disorder

Source	PAF synthesis nmoles/15 min/10^8 cells	Alkyl/acyl-GPC nmoles/10^8 cells
Controls		
mean (n=4)	5.8	58
range	3.3–7.3	37–81
RCDP		
patient 1	0.01 (600 x)	0.16 (360 x)
patient 2	0.80 (7 x)	8.0 (7 x)
patient 3	0.35 (17 x)	2.4 (24 x)

Abbreviations: -GPC, glycerophosphocholine; RCDP, rhizomelic chrondrodysplasia punctata. Numbers in parenthesis indicate the fold reduction compared to control values.

5. ETHER LIPID SYNTHESIS AND GLYCOSOMES

Within the family of microbodies ether lipid synthesis is not confined to mammalian peroxisomes and was recently shown to occur also in *Trypanosomatidae* glycosomes[4]. Glycosomes of *Trypanosoma brucei* contained both DHAP-acyltransferase and alkyl-DHAP synthase (Fig. 1).

Using gradient purified fractions both enzymes were clearly identified by analysis of their reaction products by thin-layer chromatographic and alkaline hydrolysis procedures, thus fully identifying for the first time the pathway of ether lipid synthesis in Trypanosomes. Alkyl-DHAP synthase was completely resistant to trypsin treatment in intact glycosomes but not in detergent disrupted glycosomes[4]. We tentatively concluded from these data that the active site of the enzyme is located at the inner aspect of the glycosomal membrane in full accord with what was previously found for this enzyme in mammalian peroxisomes[22,23].

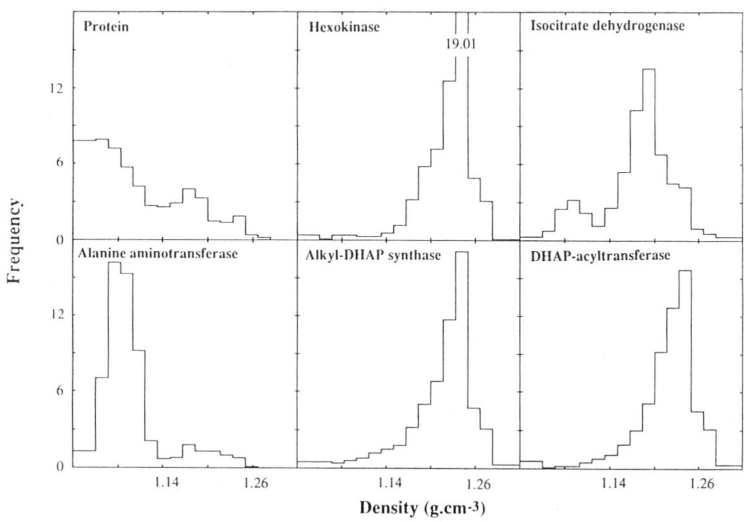

Figure 1. Distribution of protein and the indicated enzymes in a post-large granular homogenate of *T. brucei* after centrifugation on a linear sucrose gradient. Reproduced with permission from reference 4.

6. MEMBRANE ASSOCIATION AND PURIFICATION OF ALKYL-DHAP SYNTHASE

The carbonate pelleting procedure developed by Lazarow and coworkers[24] is a convenient method to distinguish between integral and peripheral membrane-associated proteins. It can only be applied, however, in case specific methods such as antibody detection are available to analyze the distribution of the protein under investigation over pellet and supernatant fractions. Since the procedure destroys most enzymatic activities it cannot usually be applied if one has to rely on enzymatic activity to determine the distribution. Therefore, we adopted the milder Triton X-114 phase separation procedure developed by Bordier[25] to obtain information on the type of membrane association of alkyl-DHAP synthase in mammalian peroxisomes. Briefly, in this procedure a protein sample in a solution containing Triton X-114 is layered on top of a sucrose solution and phase separation is induced by increasing the temperature to 30°C. Centrifugation then yields a buffer phase containing soluble and peripheral membrane proteins and a small detergent phase containing integral membrane proteins in highly concentrated form. When applied to purified peroxisomes matrix proteins such as catalase and thiolase were mainly recovered in the water phase whereas palmitoyl-CoA synthase and alkyl-DHAP synthase were almost exclusively recovered in the detergent phase, suggesting their integral membrane protein character[5].

The procedure required to solubilize the alkyl-DHAP synthase from guinea pig liver peroxisomes prior to purification of the enzyme provided an additional argument for the integral membrane character of this enzyme[26]. The enzyme was not solubilized by extracting a peroxisome-enriched membrane fraction with high salt solutions. Triton X-100 was required for solubilization and best results were obtained in mixtures containing 0.2% Triton X-100 and 0.2 M KCl. After several column chromatographic procedures, the enzyme was finally purified to homogeneity on a Concanavalin A column using an elution with a gradient of detergent Triton X-100 (Fig. 2).

As can be seen in Fig. 2, this procedure yielded fractions containing a singly 65 kDa protein band, the intensity of which correlated with enzymatic activity. Table IV summarized some of the properties of this enzyme that synthesizes glycero-ether bonds that constitute an essential structural element in the biological activity of PAF.

Analysis of the intact protein and of several fragments obtained after cyanogen bromide cleavage yielded the N-terminal and three internal sequences consisting of 21, 15, 21 and 25 amino acids, respectively. Using synthetic oligonucleotides based on these sequences and PCR technology allowed the cloning of two overlapping cDNA fragments encoding the complete sequence of the mature enzyme. The enzyme consists of 600 amino acids and the cyanogen bromide peptides sequenced at the protein level are confirmed and are preceded by

Table IV. Properties of alkyl-DHAP synthase from guinea pig liver

Molecular weight	65 kDa
N-terminal sequence	KARRA
Isoelectric point	5.9–6.4
pH optimum	7.5
K_M palmitoyl-DHAP	68 µM
K_M hexadecanol	72 µM
V_{max}	350 mU

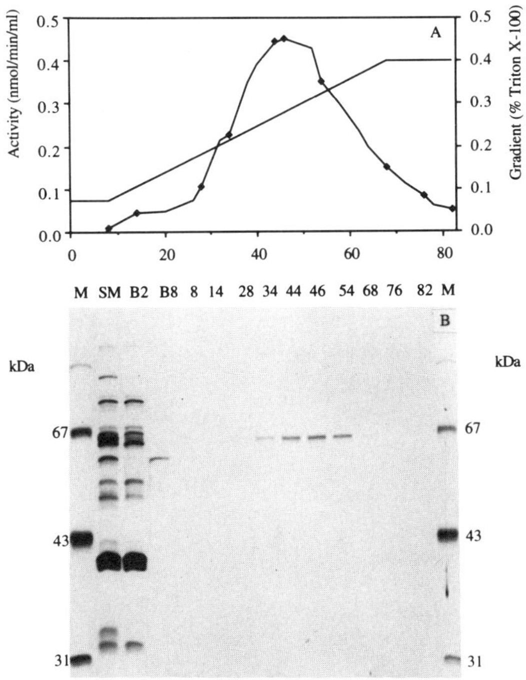

Figure 2. Purification of alkyl-DHAP synthase on a Concanavalin A column and analysis of fractions by SDS-PAGE. A: Column chromatography of partially purified fractions using a gradient of Triton X-100. B: SDS-PAGE of selected column fractions. Abbreviations: M, molecular weight markers; SM, starting material; B_2 and B_8, breakthrough fractions containing most of the contaminating proteins, but lacking the 65 kDa enzyme band that is eluted in fractions 28 to 68. Reproduced with permission from reference 26.

methionines as predicted. Remarkably, a hydropathy analysis (Fig. 3) gave no distinct indications for the presence of transmembrane segments, although these cannot be completely excluded either for residues 165–185. It is clear from these data that the nature of membrane association of alkyl-DHAP synthase requires further investigation.

Northern blot analysis (Fig. 4) indicates that the enzyme is synthesized from an approximately 4.2 kbase transcript. The knowledge of the complete amino acid sequence is currently being applied to obtain anti-peptide antibodies. These will be essential to obtain further information on one of the major biosynthetic processes in mammalian peroxisomes, *i.e.* ether lipid synthesis. Such antibodies will also be instrumental to unravel the topography of this

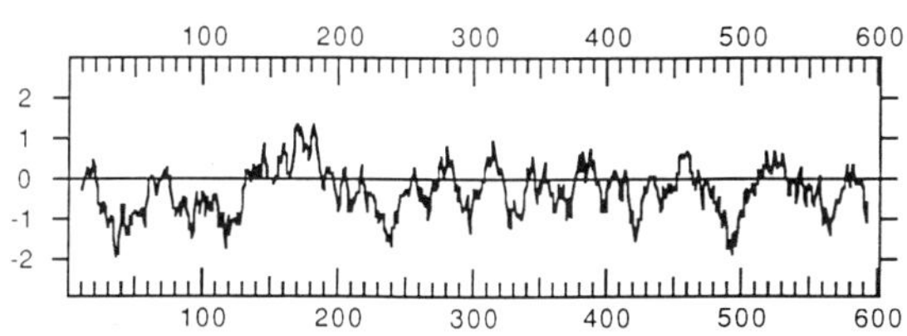

Figure 3. Hydropathy analysis of alkyl-DHAP synthase. The hydropathy profile was obtained according to ref. 27 using a span of 17 consecutive residues.

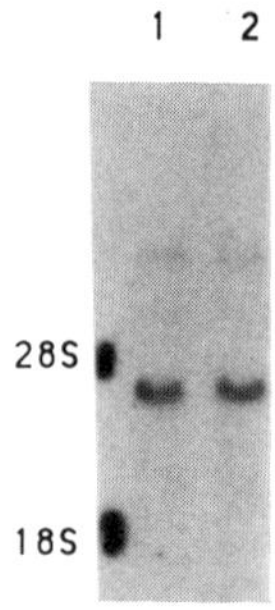

Figure 4. Northern blot analysis of total guinea pig liver RNA. The blot was probed with a ^{32}P-labelled 1100 basepair PCR fragment of alkyl-DHAP synthase cDNA.

process in peroxisomal membranes, the targeting of the enzyme to peroxisomes and the fate of this enzyme in peroxisomal disorders deficient in ether lipid synthesis.

REFERENCES

1. Van den Bosch, H., Schutgens, R.B.H., Wanders, R.J.A. and Tager, J.M., Ann. Rev. Biochem. 61 (1992) 157–197.
2. Opperdoes, F.R., Trends Biochem. Sci. 13 (1988) 255–260.
3. Hajra, A.K. and Bishop, J.E., Ann. NY Acad. Sci. 386 (1982) 170–182.
4. Zomer, A.W.M., Opperdoes, F.R., and van den Bosch, H., Biochim. Biophys. Acta 1257 (1995) 167–173.
5. Hardeman, D., Versantvoort, C., van den Brink, J.M. and van den Bosch, H., Biochim. Biophys. Acta 1027 (1990) 149–154.
6. Van Veldhoven, P.P., Just, W.W. and Mannaerts, G.P., J. Biol. Chem. 262 (1987) 4310–4318.
7. Paltauf, F., Chem. Phys. Lipids 74 (1994) 101–139.
8. Han, X and Gross, R.W., Biochemistry 29 (1990) 4992–4996.
9. Morand, O.H., Zoeller, R.A. and Raetz, C.R.H., J. Biol. Chem. 263, 11597–11606.
10. Lohner, K., Hermetter, A. and Paltauf, F., Chem. Phys. Lipids 34 (1984) 163–170.
11. Lee, T-C. and Fitzgerald, V., Biochim. Biophys. Acta 598 (1980) 189–192.
12. Smaby, J.M., Hermetter, A., Schmid, P.C., Paltauf, F. and Brockman, H.L., Biochemistry 22 (1983) 5808–5813.
13. Ferguson, M.A.J., Curr. Opinion Struct. Biol. 1 (1991) 522–529.
14. Schulman, G., Bodine, P.V. and Litwack, G., Biochemistry 31 (1992) 1734–1741.
15. Hanahan, D.J., Ann. Rev. Biochem. 55 (1986) 483–509.
16. Blank, M.L., Fitzgerald, V., Lee, T-C and Snyder, F., Biochim. Biophys. Acta 1166 (1993) 309–312.
17. Goldfischer, S., Moore, C.L., Johnson, A.B., Spiro, A.J., Valsamis, M.P., Wisniewski, H.K., Ritch, R.H., Norton, W.T., Rapin, I. and Gartner, L.M., Science 182 (1973) 62–64.
18. Heymans, H.S.A., Schutgens, R.B.H., Tan, R., van den Bosch, H. and Borst, P., Nature 306 (1983) 69–70.
19. Schrakamp, G., Schalkwijk, C.G., Schutgens, R.B.H., Wanders, R.J.A., Tager, J.M. and van den Bosch, H., J. Lipid Res. 29 (1988) 325–334.
20. Wanders, R.J.A., Schutgens, R.B.H., Barth, P.G., Tager, J.M. and van den Bosch, H., Biochimie 75 (1993) 269–279.
21. Van den Bosch, H., Schrakamp, G., Hardeman, D., Zomer, A.W.M., Wanders, R.J.A. and Schutgens, R.B.H., Biochimie 75 (1993) 183–189.
22. Hajra, A.K., Ghosh, M.K., Webber, K.O. and Datta, N.S., In: Enzymes of Lipid Metabolism II (Freysz, L., Dreyfuss, H., Massarelli, R. and Gatt, S., eds.) 1986, pp. 199–207, Plenum Publ. Corp., New York.
23. Hardeman, D. and van den Bosch, H., Biochim. Biophys. Acta 963 (1988) 1–9.
24. Fujiki, Y., Fowler, S., Shio, H., Hubbard, L. and Lazarow, P.B., J. Cell Biol. 93 (1982) 103–110.
25. Bordier, C., J. Biol. Chem. 256 (1981) 1604–1607.
26. Zomer, A.W.M., de Weerd, W.F.C., Langeveld, J. and van den Bosch, H., Biochim. Biophys. Acta 1170 (1993) 189–196.
27. Kyte, J. and Doolittle, R.F., J. Mol. Biol. 157 (1982) 105–132.

8

PLATELET-ACTIVATING FACTOR SYNTHESIS AND ITS ROLE IN SALIVARY GLANDS

T. Dohi, K. Itadani, H. Yamaki, Y. Akagawa, K. Morita, and S. Kitayama

Department of Pharmacology
Hiroshima University School of Dentistry
Kasumi 1-2-3, Minami-ku, Hiroshima 734, Japan

1. INTRODUCTION

Platelet-Activating Factor (PAF) was originally described as a substance released from activated basophils. It was then detected in a variety of inflammatory and immune-related cells, e.g. polymorphonuclear leucocytes, monocytes, macrophages and plateletes. PAF is considered to be a mediator in anaphylaxis and inflammatory responses. Moreover, PAF production has been demonstrated in a wider variety of cells and tissues, and it has been suggested that PAF plays various physiological and pathological roles. For example, PAF may play a role in stimulation-secretion coupling including neuron tissues in which PAF increases intracellular free Ca^{2+} concentration($[Ca^{2+}]i$)[1–6] and modulates neurotransmitter release[2,3,6–8], and is suggested its contribution to long-term potentiation fomation in the hippocampus[9,10] or enhanced neurotransmitter release in pathological conditions such as ischemia or convulsions. Söling et al.[11] has reported that PAF stimulates amylase release in the exocrine salivary gland and the pancreas. The capacity of cells from exocrine glands to produce PAF in response to physiological stimulations[12,13] and the presence of PAF in human saliva[14] have also been demonstrated. The present study examines the regulation of PAF biosynthesis and its role in secretory responses in salivary glands.

2. REGULATION OF PAF BIOSYNTHESIS IN SALIVARY GLANDS

Our previous study has shown that PAF is produced by the stimulation of muscarinic cholinergic receptors in the submandibular gland of dogs[13]. Dispersed cells from the parotid gland and the sublingual gland synthesized PAF in response to acetylcholine(ACh) and ionomycin. Norepinephrine and phenylephrine also stimulated PAF synthesis in submandibular gland cells and the effects were blocked by phentolamine but not by propranolol. Isoproterenol was without effect, suggesting the involvement of α-adrenergic receptors in PAF synthesis in response to norepinephrine. PAF synthesis induced by ACh and norepinephrine was diminished by removing Ca^{2+} from the medium and increased by increasing the concentration of Ca^{2+} in the medium, accompanied with the increase in liberation of ^{14}C-arachidonic acid. Lyso-PAF : acetyl-CoA acetyltransferase activity in cells stimulated by ACh sharply increased. ACh-induced activation of enzyme activity was markedly reduced by removing extracellular Ca^{2+}. Domenech et al.[15] reported that acetyltransferase activation by carbamylcholine in guinea pig parotid lobules is due to the activation via phosphorylation by Ca^{2+}/calmodulin-dependent protein kinase. The authors have also shown that catalytic subunits of cyclic AMP-dependent protein kinase activated acetyltransferase in the microsomal fraction from unstimulated guinea pig parotid glands but isoproterenol did not activate acetyltransferase in intact parotid lobules[15]. In the present study, norepinephrine, isoproterenol and 8Br-cAMP all activated the acetyltransferase activity of dog submandibular gland cells, and this activation was little affected by removal of extracellular Ca^{2+}. PAF can be synthesized by either remodelling or de novo pathways. The former pathway is Ca^{2+}-dependent and responsible for the sysnthesis of PAF during stimulation-mediated cell activation. Evidence that Ca^{2+} increased PAF production in digitonin-permeabilized submandibular gland cells and the inhibitors of phospholipase A_2 blocked the systhesis[13], and that the production of PAF, the liberation of arachidonic acid and activation of acetyltransferase by ACh and norepinephrine were dependent on the presence of extracellular Ca^{2+} suggests a critical role of remodelling pathways in PAF production in salivary glands. Our previous report shows that ACh failed to stimulate PAF synthesis in the absence of extracellular Ca^{2+}, even though ACh still increased the liberation of ^{14}C-arachidonic acid[13]. On the other hand, the addition of isoproterenol and 8Br-cAMP to submandibular cells activated acetyltransferase and this activation was not accompanied with the production of PAF. These results suggest that mere activation of phospholipase A_2 or acetyltransferase is not sufficient, but the concurrent increase of the supply of precursor, lyso-PAF and activation of acetyltransferase are required for PAF synthesis in response to the stimulation of submandibular gland cells. In cells stimulated by ACh, Ca^{2+} is involved in the activation of phospholipase A_2 and acetyltransferase. Norepinephrine, by stimulating α-adrenergic receptors, increases $[Ca^{2+}]i$ both through activation of Ca^{2+} entry via plasma membrane pathways and mobilization of Ca^{2+} from intracellular stores by stimulating phosphoinositide turnover. On the other hand, by stimulating β-adrenergic receptors, it stimulates the formation of cyclic AMP. Therefore, an increase in PAF sysnthesis by norepinephrine is mediated by both Ca^{2+}-dependent activation of phospholipase A_2 and activation of acetyltransferase by Ca^{2+} and/or cyclic AMP. CoA-independent and CoA-dependent transacylase have been shown to play a role in the formation of lyso-PAF. Protein kinase C also regulates phospholipase A_2, acetyltransferase and CoA-independent transacylase, and regulates PAF synthesis in such cells as leukocytes[16], monocytes[17] and mast cells[18]. However, none of the activators and inhibitors of protein kinase C tested e.g. TPA, mezerein and staurosporine affected PAF production in non-stimulated and ACh-stimulated submandibular gland cells. Purified kinase C did not activate acetyltransferase in microsome of guinea pig parotid glands[15]. As inability of the activation of protein kinase C to stimulate PAF synthesis by human neutrophils[19] and endotherial cells[20], the role of protein kinase C in PAF synthesis in salivary glands is not obvious.

3. THE ROLE OF PAF RECEPTORS IN SECRETION OF SALIVARY GLAND

PAF (0.5 - 10 nM) dose-dependently increased mucin release from slices of guinea pig submandibular glands although PAF is much less potent than ACh. The PAF-induced release was blocked by a PAF receptor antagonist, BN 50739, and reduced by removal of extracellular Ca^{2+}. Söling et al.[11] have shown that PAF stimulated amylase release from isolated pancreatic or parotid lobules of guinea pigs, and increased $^{45}Ca^{2+}$ uptake and stimulated phosphoinositide turnover in the lobules. These results agree with those which show that PAF increased the formation of inositol trisphospate(IP_3) and increased $[Ca^{2+}]i$ in guinea pig submandibular glands. It is generally accepted that mucin release is primarily mediated by the β-adrenergic recepror-cyclicAMP system and that α-adrenergic and cholinergic mechanisms are not a prerequisite for mucin release. However, we have shown that α-adrenergic- and muscarinic cholinergic-receptor-mediated Ca^{2+}-dependent mechanisms play an important role in mucosecretion in submandibular acini[21]. These results suggest that PAF interacts with PAF-receptors on plasma membranes in salivary and pancreas glands and may function in exocrine secretion.

PAF receptor expression in salivary glands was further explored by examining PAF receptor gene expression using the reverse transcriptase polymerase chain reaction (RT-PCR) technique. Primers were designed and synthesized based on the sequence of PAF receptor cDNA from guinea pig lungs[22]. The specific amplification of PAF receptor-specific PCR fragment from guinea pig parotid and submandibular glands, as well as lung mRNA, was detected. These results clearly demonstrated the expression of PAF receptor mRNA in guinea pig salivary glands, which is homologous to guinea pig's. However, any amplification was not detected in dog parotid and submandibular glands, although the presence of PAF receptors in dog adrenal glands has been demonstrated by RT-PCR with adrenal mRNA[23] ,PAF-receptor specific ligand [^{3}H]WEB 2086 binding in the plasma membrane of adrenal medulla and functional modulation of catecholamine release from chromaffin cells in dogs[6]. The difference in PAF receptor gene expression in salivary glands between dog and guinea pig, taken together with the fact that PAF synthesized in submandibular gland is not released out of cells in dogs but released in guinea pigs, might suggest the autocrine role of PAF in the exocrine secretion of guinea pig salivary glands.

4. INTRACELLULAR ACTIVITY OF PAF

Systhesized PAF is not transported out of cells or large part of PAF retained intracellulary in many types of cells. Novel intracellular effects of PAF have been supposed by a number of studies e.g., in human PMN function[19], human T-cell proliferation[24], eicosanoid generation in guinea pig resident peritoneal macrophages[25,26] and in bovine cultured aortic endothelial cells[26,27] ,and in changes of $[Ca^{2+}]i$ in human neutrophils[28], and in endothelial cell-associated PAF as a signal for an intercellular adhesion[29]. It has also been suggested that the non-receptor-mediated effects of PAF in human airway epithelial cells[30] and the occurrence of intracellular binding sites[31], favour the concept of PAF as an intracellular messenger[32]. It has been reported that PAF stimulates Na^+-Ca^{2+} exchange in rat ileal plasmalemma[33] and brain synaptosomes[34], and in addition, affects Ca^{2+} - lipid interactions in phospholipid vesicles[35]. On the other hand, PAF inhibits Na^+,K^+-ATPase of adrenal medullary plasma membranes[23]. Na^+,K^+-ATPase is located on the basolateral membrane of salivary glands and controls ionic transport during fluid secretion. Actually, PAF inhibited Na^+,K^+-ATPase in plasma membrane fractions of dog submandibular glands in concentrations around the tens micromolar range. Kinetic analysis showed that this inhibition was not due to competition at K^+- or Na^+ -binding sites on the enzyme, but has complex inhibitory feature. In addition, PAF, 1 -10 μM, increased Ca^{2+} release from digitonin-permeabilized submandibular gland cells. Since Na^+,K^+-ATPase and ion exchange pathways are important in secretory responses of acinar cells, it may be that PAF regulates the secretory function of acinar cells through modulating ionic homeostasis as an intracellular messenger, although such intracellular role associated PAF has yet to be established.

REFERENCES

1. Bito, H., Nakamura, M., Honda, Z., Izumi, T., Iwatsubo, T., Seyama, Y., Ogura, A., Kudo, Y. and Simizu, T. (1992) Neuron 9, 285–294.
2. Bussolino, F., Tessari, F., Turrini, F., Braquet, P., Camussi, G., Prosdocimi, M. and Bosia, A. (1988) Am. J. Physiol. 255, C559-C565.
3. Kornecki, E. and Ehrlich, Y. H. (1988) Science 240, 1792–1794.
4. Yue, T.-Y., Gleason, M.M., Gu, J.-L. Lysko, P.G., Hallenbeck, J. and Feuerstein, G. (1991) J. Pharmacol. Exp. Ther. 257, 374–381.
5. Yue, T.-Y., Gleason, M.M., Hallenbeck, J. and Feuerstein, G. (1991) Neuroscience 41, 177–185.
6. Morita, K., Suemitsu, T., Uchiyama, Y., Miyasako, T. and Dohi, T. (1995) J. Lipid Mediators Cell Signalling 11, 219–230.
7. Clark, G.D., Happel, L.T., Zorumski, C.F. and Bazan, N.G. (1992) Neuron 9, 1211–1216.
8. Jiang, Z.-G., Yue, T.-L. and Feuerstein, G. (1993) Neurosci. Lett. 155, 132–135.
9. Wieraszko, A., Li, G., Kornecki, E., Hogan, M.V. and Ehrlich, Y.H. (1993) Neuron 10, 553–557.
10. Kato, K., Clark, G.D., Bazan, N.G. and Zorumski, C.F. (1994) Nature 367, 175–179.
11. Söling, H.-D., Eibl, H. and Fest, W. (1984) Eur. J. Biochem. 144, 65–72.
12. Söling, H.-D. and Fest, W. (1986) J. Biol. Chem. 261, 13916–13922.
13. Dohi, T., Morita, K., Kitayama, S. and Tsujimoto, A. (1991) Biochem. J. 276, 175–182.
14. Cox, C.P., Wardlow, M.L., Jorgensen, R. and Farr, R.S. (1981) J. Immunol. 127, 46–50.
15. Domenech, C., Domenech, E.M.-D. and Söling, H.-D. (1987) J. Biol. Chem. 262, 5671–5676.
16. Hayashi, M., Imai, Y. and Oh-ishi, S. (1991) Lipids 26, 1054–1059.
17. Elstad, M.R., Prescott, S.M., McIntyre, T.M. and Zimmerman, G.A. (1988) J. Immunol. 140, 1618–1624.
18. Joly, F., Vilgrain, I., Bossant, M.J., Benveniste, J. and Ninio, E. (1990) Biochem. J. 271, 501–507.
19. Sisson, J.H., Prescott, S.M., McIntyre, T.M. and Zimmerman, G.A. (1987) J. Immunol. 138, 3918–3926.
20. Whatley, R.E., Nelson, P., Zimmerman, G.A., Stevens, D.L., Parker, C.J., McIntyre, T.M. and Prescott, S.M. (1989) J. Biol. Chem. 264, 6325–6333.
21. Dohi, T., Yamaki, H., Morita, K., Kitayama, S., Tsuru, H. and Tsujimoto, A. (1991) Archs Oral Biol. 36, 443–449.
22. Honda, Z., Nakamura, M., Miki, I., Minami, M., Watanabe, T., Seyama, Y., Okado, H., Toh, H., Ito, K., Miyamoto, T. and Shimizu, T. (1991) Nature 349, 342–346.
23. Dohi, T., Morita, K., Imai, Y. and Kitayama, S. (in press) in Platelet-Activating Factor and Related lipid Mediators in Health and Disease (Nigam, G., Kunkel, G., Prescott, S.M. and Vargaftig, B.B., eds), Plenum Press, New York.
24. Patrignani, P., Valitutti, S., Aiello, F. and Musiani, P. (1987) Biochem. Biophys. Res. Commun. 148, 802–810.
25. Stewart, A.G. and Phillips, W.A. (1989) Br.J. Pharmacol. 98, 141–148.
26. Stewart, A.G., Dubbin, P.N., Harris, T. and Dusting, G.J. (1990) Proc. Natl. Acad. Sci. USA 87, 3215–3219.
27. Stewart, A.G., Dubbin, P.N., Harris, T. and Dusting, G.J. (1989) Br. J. Pharmacol. 96, 503–505.
28. Tool, A.T.J., Verhoeven, A.J., Roos, D. and Koenderman, L. (1989) FEBS Letters 259, 209–212.
29. Zimmerman, G.A., McIntyre, T.M., Mehra, M. and Prescott, S.M. (1990) J. Cell Biol. 110, 529–540.
30. Stoll, L.L., Denning, G.M., Kasner, N.A. and Hunninghake, G.W. (1994) J. Biol. Chem. 269, 4254–4259.
31. Marcheselli, V.L., Rossowska, M.J., Domingo, M.-T., Braquet, P. and Bazan, N.G. (1990) J. Biol. Chem. 265, 9140–9145.
32. Henson, P.M. (1987) in Platelet-Activating Factor and Related Lipid Mediators (Snyder, F., ed.), pp.255–271, Plenum Press, New York.
33. Kester, M., Kumar, R. and Hanahan, D.J. (1986) Biochim. Biophys. Acta 888, 306–315.
34. Kumar, R., Harvey, S.A.K., Kester, M., Hanahan, D.J. and Olson, M.S. (1988) Biochim. Biophys. Acta 963, 375–383.
35. Bratton, D.L.,Harris, R.A., Clay, K.L. and Hensor, P.M. (1988) Biochim. Biophys. Acta 943, 211–219.

9

PHYSIOLOGICAL ACTION OF PAF IN YEAST *SACCHAROMYCES CEREVISIAE*

Reiko Nakayama,[1] Cheolwon Yun,[2] Hisanori Tamaki,[2] Kunihiko Saito,[3] and Hidehiko Kumagai[2]

[1]Department of Food Science, Kyoto Women's University
Kyoto, 605, Japan
[2]Department of Food Science and Technology, Kyoto University
Kyoto, 606, Japan
[3]Department of Medical Chemistry, Kansai Medical University
Osaka, 570, Japan

INTRODUCTION

In mammalian cells, PAF (platelet-activating factor) is involved in allergy and inflammation[1–3]. It is now apparent that PAF also plays a role in a number of normal physiological processes, including reproduction and the lowering of blood pressure[4,5]. We have been reported PAF is generated not only by inflammatory cells but also by many normal tissues, such as those of the uterus, heart and stomach[6].

Recently, PAF has been detected in the lower eucaryote, *Dictyostelium discoideum*[7], a protozoan *Tetrahymena pyriformis*[8] and a slug *Incilaria bilineata*[9].

We found that single cell eucaryote, yeasts belonging to the genus *Saccharomyces* had high PAF productivity[10]. These findings suggest that PAF exists in a wide variety of living organisms and plays a general and fundamental physiological role in their living cells. In this paper, we describe the PAF production and its physiological action in yeast *Saccharomyces cerevisiae* in relation to cell growth.

1. PAF PRODUCTION IN YEAST CELLS

1.1. Evidence for PAF Production in Yeast Cells

The yeast PAF was purified by HPLC and characterized by platelt aggregation activity and the inhibition by the PAF antagonist. Furthermore, the ether- and ester-linked molecular species and PAF precursor were identified by selected ion monitoring (SIM) on gas chromatography-mass spectrometry (GC-MS)[10]. As shown in Fig.1, the ratio of acylPAF to PAF in yeast was larger than that in mammalian sources, such as uterus and heart[6].

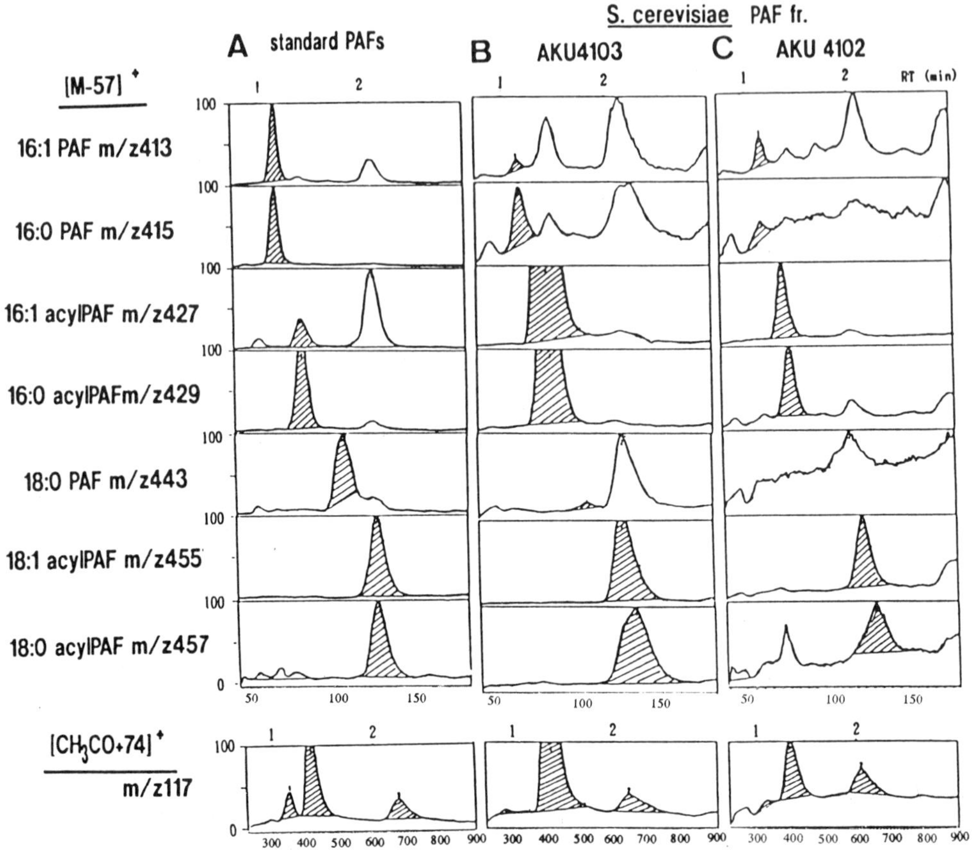

Figure 1. GC-EI-SIM traces for the various PAF molecular species of tBDMS derivatives in yeast[10].

1.2 PAF Production in Response to a Calcium Ionophore A23187

PAF production by a calcium ionophore A23187-stimulation was demonstrated in a large variety of types of cells, especially neutrophils[11].

We found a significant PAF production in *S. cerevisiae* cells in response to A23187[12]. As shown in Fig.2, the amount of PAF production increased A23187-concentration and maximun PAF synthesis was observed when the yeast cells treated with 2 μM of A23187 for 5 min. The amount decreased at the higher concentrations of A23187 and for the longer incubation time. However, PAF accumulated by the addtion of PMSF, and the result sugests that the presence of an enzyme responsible for PAF degradation in the yeast cells. Newly generated PAF was not detected in the extracellular fraction, suggesting that it was cell-bound, similarily to mammalian cells. The amount of PAF induced by A23187 in the yeast cells was more than 10-fold that of the basal PAF content presented in the non-stimulated condition.

1.3 Changes in PAF Production during Cell Growth

Fig.3 shows the time course changes in PAF production in response to A23187, and the total lipid contents of the whole cells. The cells reached to the stationary phase 18–24

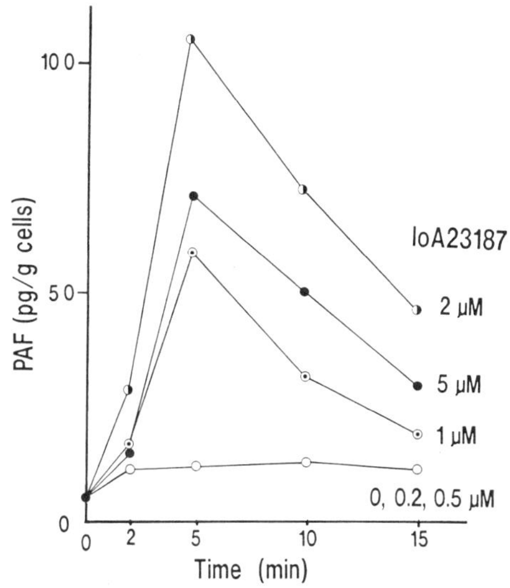

Figure 2. Concentation- and time-dependence of ionophore A23187-stimulated PAF synthesis in yeast cells[12].

h after the begining of cultivation. Total lipid content increased in parallel with the cell growth. In contrast, A23187-induced PAF productivity was not detected in the logarithmic phase, and increased in the early stationary phase and reached the maximum at 24 h[12].

1.4 Changes in PAF Production during Yeast Cell Cycle

Furthermore, we examined the changes in PAF, its precursor and the synthesizing enzyme acetyltransferase (AT) activity of cells in various cell division cycle.

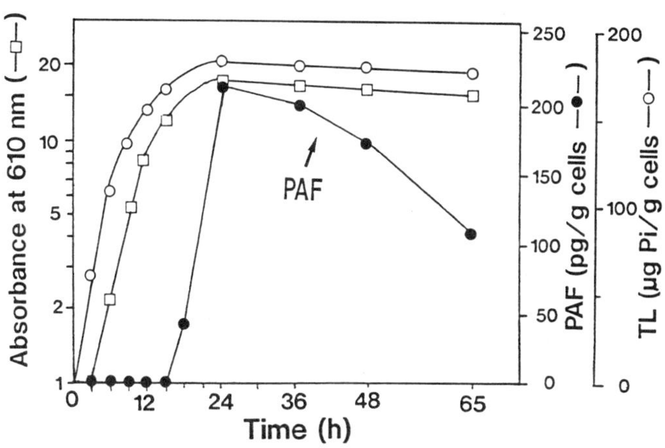

Figure 3. Changes in PAF productivity in yeast cells during cultivation. -□-, absorbance at 610 nm; -●-, PAF (pg per g wet weight cells); -O-, total lipids of whole cells (μg lipid-phosphorus per g wet weight cells).

The yeast cells were synchronized at various cell cycle phase and the cells were stimulated by A23187. Cells at G1 and M phases had the highest productivity of PAF in responce to A23187 and it was 20 times higher than that of cells at S phase. Without stimulation, the basal PAF contents of cells at G1 and M phases were also higher than that of S phase. The amount of PAF precursor of M and G1 phases were higher than that of S phase cells and the ratio of PAF to precursor was the highest at G1 phase.

We also examined the changes of AT activity during cell cycle, using microsomal fractions prepared from the yeast cells in various cell cycle phase. Both specific activity and total activity of the enzyme were the highest in G1 phase, and total activity of S phase cells were 1/10 of those of G1 phase. These results suggest that changes in PAF productivity in cell cycle were relative to precursor content and AT activity in those cells.

These findings suggest that yeast produces PAF in response to some physiological stimuli and that there is a relationship between PAF production and cell growth.

1.5 PAF Production during Mating Process

The mating hormones (a- and α-factor) play essential roles in the initial stage of the mating process, acting on target haploid cells (MATα and MATa), and causing the following biochemical and morphological changes[13]. The hormones rapidly alter the pattern of gene expression, induce cell surface agglutinins, and arrest mitotic cell division in the G1 phase.

We found that the yeast produces PAF in early stage of mating process in response to the mating hormone. MATa cells were treated with various concentrations of α-factor and the time course of PAF synthesis was followed. Maximal PAF synthesis occured on treatment with 3 μM of α-factor for 90 min incubation. Platelet aggregation activity was completely inhibited by PAF antagonist. We found that the mating hormone-induced PAF production is Ca^{2+} dependent.

Furthermore, changes in PAF production were followed during mating process, in which MATa cells were cultivated with MATα cells. In this time course study, we found the significant stimulation of PAF production in early stage of the mating process. PAF production reduced when Ca^{2+} was omitted from culture medium at this time period. Little is known about the role of Ca^{2+} in the mating hormone response pathway except that Ca^{2+} influx in MATa cells is stimulated by α-factor after a lag of 30–40 min in a dose-dependent manner[14]. Our time course study suggests that PAF productivity during early stage of mating process is stimulated by Ca^{2+} influx caused by mating factor. The relationship between PAF production and mating factor-induced Ca^{2+} influx and physiological meaning of PAF in signal transduction of mating process still remains to be investigated.

2. EFFECTS OF PAF ON YEAST CELL GROWTH

2.1 Inhibition of Cell Growth by PAF

To investigate the possible action of PAF on cell growth, PAF was added to culture medium and yeast cell growth was monitored by measuring the absorbance at 610 nm. We found that PAF inhibited yeast cell growth dose-dependent manner. Furthermore, we examined the effects of PAF and its analogs on yeast growth at final concentration of 5 uM. PAF showed strong inhibition and N-methylcarbamoyl PAF (C-PAF) had the same activity as PAF. LysoPAF showed a half level of activity of PAF. In contrast, lysophosphatidylcholine had no effect on cell growth. We also examined the effect of PAF antagonist on PAF-induced growth inhibition. By pretreatment with antagonist, PAF-induced growth inhibition was partly reversed. These results suggest that growth inhibition by PAF was specific to its molecule, not derived from physical character of phospholipids, such as a detergent-like action.

2.2 Reduction of Intracellular cAMP Level by PAF

It is reported that cAMP is a positive effector in G1 of *S. cerevisiae* cell division cycle[15]. We found that intracellular cAMP level of yeast decreased by PAF, concentration-dependently. The reduction of cAMP caused by PAF was completely or partly reversed by pretreatment with PAF antagonist.

2.3 Inhibition of Membrane Adenylate Cyclase by PAF

Furthermore, we investigated the effect of PAF on adenylate cyclase, a synthesizing enzyme of cAMP. PAF inhibited membrane adenylate cyclase activity dose-dependent manner, and PAF antagonist prevented partly the PAF-induced inhibition of the enzyme activity.

These results suggest that the inhibition of yeast cell growth by PAF is specific to PAF molecule, and itis related to the reduction of intracellular cAMP level by inhibiting membrane adenylate cyclase activity.

The present findings suggest that PAF is a regulator of yeast *S. cerevisiae* cell growth. The mechanism of this action of PAF still remains to be investigated.

ACKNOWLEDGMENTS

We are grateful to Mr.Hiroaki Udagawa, Ms.Fumiko Murase and Ms.Mika Ashihara for their techinical assistance. This work was supported in part by a Grant-in-Aid for Scientific Research (3858021, 6806012, 7806013) from the Ministry of Education, Science, and Culture of Japan, and by the Science Research Promotion Fund from Japan Private School Promotion Foundation (1994, 1995) and by the Fugaku Trust for Medicinal Research Foundation.

REFERENCES

1. J.Benveniste, E.Joulvin, E.Pirotzky, B.Arnoux, J.M.Mencia-Huerta, R.Roubin and B.B.Vargaftig, Int. Arch. Allergy Appl. Immunol. 66 (Suppl.1) 121–126 (1981)
2. R.N.Pinckard, L.M.McManus and D.J.Hanahan, in "Advances in Inflammation Research", vol.4, ed.by G.Weissman, Raven Press, New York, pp.147–180 (1982)
3. J.T.O'Flaherty, Lab.Invest., 47, 314–329 (1982)
4. M.J.K.Harper, Biol.Reproduc., 40, 907–913 (1989)
5. E.E.Muirhead, J.A.Stirman and F.Jones, J.Clin.Invest., 39, 266–281 (1960)
6. K.Saito, R.Nakayama, K.Yasuda, K.Satouchi and J.Sugatani, in "Biological Mass Spectrometry" eds.by A.L.Burlingame and J.A.McCloskey, Elsevier, Amsterdam, pp.527–547 (1990)
7. F.Bussolino, C.Sordano, E.Bebfenati and S.Bozzaro, Eur.J.Biochem., 196, 609–615 (1991)
8. M.Lella, A.D.Tselepis and D.Tsoukatos, FEBS Lett., 208, 52–55 (1986)
9. T.Sugiura, T.Ojima, T.Fukuda, K.Satouchi, K.Saito and K.Waku, J.Lipid Res., 32, 1795–1803 (1991)
10. R.Nakayama, H.Kumagai and K.Saito, Biochim.Biophys.Acta, 1199, 137–142 (1994)

11. T-C.Lee, in "Platelt-activating Factor and Related Lipid Mediators", ed.by F.Snyder, Plenum Press, New York, pp.115–133 (1987)
12. R.Nakayama, H.Udagawa, S.Mitsui and H.Kumagai, Biosci.Biotech.Biochem., 58, 1115–1119 (1994)
13. S.A.Moore, J.Biol.Chem., 258, 13849–13856 (1983)
14. Y.Ohsumi and Y.Anraku, J.Biol.Chem., 260, 10482–10486 (1985)
15. T.Ishikawa, I.Uno and K.Matsumoto, BioEssays, 4, 52–56 (1985)

10

FREE RADICAL-INDUCED OXIDATION OF GLYCEROPHOSPHOCHOLINE LIPIDS AND FORMATION OF BIOLOGICALLY ACTIVE PRODUCTS

Robert C. Murphy

Division of Basic Sciences
National Jewish Center for Immunology and Respiratory Medicine
1400 Jackson Street, Denver, Colorado 80206

1. INTRODUCTION

Lipid mediators are a diverse family of biologically active compounds derived from phospholipids and are thought to play important roles both in physiological processes as well as pathophysiologic events. Perhaps the best studied class of lipid mediators are those derived from arachidonic acid, termed eicosanoids. Several enzymatic systems are known to be involved in their formation including prostaglandin H synthase which catalyzes the formation of cyclic endoperoxides, the direct precursor of prostaglandins, thromboxane, and prostacyclin[1]. Leukotrienes are formed from arachidonic acid by the enzyme 5-lipoxygenase which initially introduces oxygen into the arachidonic acid backbone[2]. Other biologically active eicosanoids are formed by 15-lipoxygenase[3], 12-lipoxygenase[4], and cytochrome P-450[5]. These enzymes use nonesterified arachidonic acid as substrate. Another important class of lipid mediator is platelet-activating factor which is derived from ether phospholipids and has acetate at the *sn*-2 position. The formation of the precursor lyso glycerophosphocholine as well as the subsequent acetylation process are enzymatically mediated[6].

The biological activity of these lipid mediators are largely, if not exclusively, mediated through specific cellular receptors for these unique chemical structures. Recent advances in the cloning of the receptors for prostaglandins has revealed these to be G-protein linked with putative 7-membered transmembrane regions typical for such a class of receptor[7]. The receptor for platelet-activating factor has also been recently cloned and falls into the same general class of G-protein linked receptors[8]. Numerous advances and understanding the role that such lipid mediators play in disease processes has been a result of the development of specific antagonists for these receptors, particularly for leukotrienes and platelet-activating factor[9].

Platelet-Activating Factor and Related Lipid Mediators 2
edited by Nigam *et al.*, Plenum Press, New York, 1996

1.1 Oxygen Free Radical Mediated Reactions

Recent investigations have revealed alternative mechanisms for the formation of biologically active lipid mediators separate from the enzymatic processes mentioned above. Oxygen free radical species are known to be generated by a variety of enzymatic as well as nonenzymatic reactions and such reactive oxygen species possess a range of chemical reactivity related to the initial radical species. The most reactive free radical species produced *in vivo* is the hydroxyl radical, which can indiscriminately react with any organic molecule by removing a hydrogen atom upon a single collision[10]. One of the least reactive oxygen species is superoxide anion. Between these reactivity extremes, other oxygen centered radicals such as the alkoxide radical, peroxy radical, and hydroperoxy radical (which is a protonated superoxide anion) are known to exist. The initial event following the reaction of hydroxyl radical with polyunsaturated fatty acids esterified to glycerophospholipids is the formation of a carbon centered pentadienyl radical (Figure 1). The outer plasma membrane is the initial barrier which protects the cell from chemically reactive species such as hydroxyl radical, but this barrier is altered in the process and must be repaired. While lipid peroxidation has been largely assessed through the measurement of low molecular weight products such as pentane, malonyldialdehyde and 4-hydroxynonenal[11], it is now realized that more complex structures do result from the initial hydrogen atom abstraction process.

Initial hydrogen atom abstraction by hydroxyl radical could occur at any carbon atom of a phospholipid leading to a reactive carbon centered radical species. However, an important feature of lipid peroxidation is eventual formation of the rather stable 1,5-pentadienyl radical from those fatty acyl groups having more than two double bonds interrupted by a methylene group or bisallylic. For example, arachidonic acid has three bisallylic methylene groups at carbon atoms 7, 10, and 13. Calculations have shown that the bond strength of a typical alkyl carbon-hydrogen bond is approximately 101 Kcal/mol

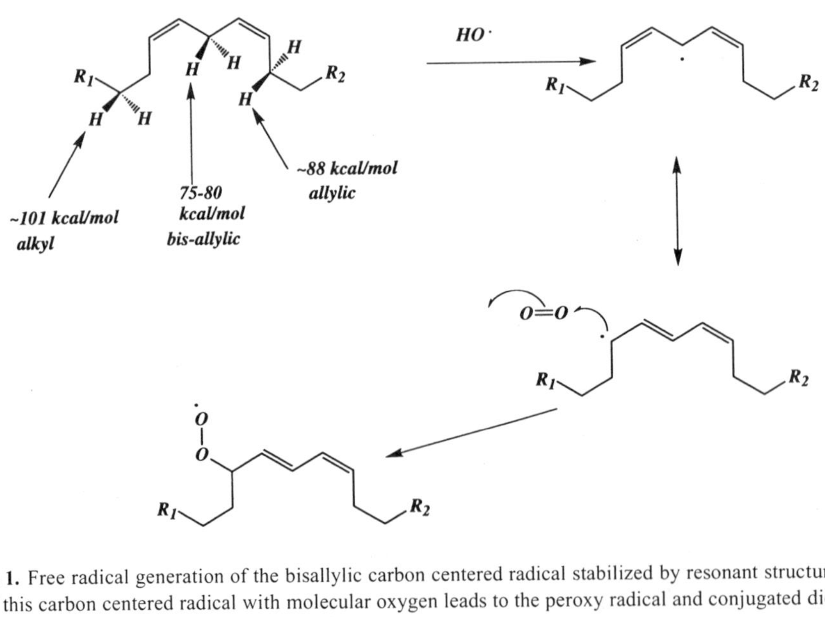

Figure 1. Free radical generation of the bisallylic carbon centered radical stabilized by resonant structures. Reaction of this carbon centered radical with molecular oxygen leads to the peroxy radical and conjugated diene structure.

while the bond strength of an allylic carbon hydrogen bond is approximately 88 Kcal/mol[12]. However, the bisallylic carbon hydrogen bond is significantly lower, being 75 - 80 Kcal/mol[13]. Furthermore, the rate of propagation of lipid peroxidation was found to be exponentially related to the total number of bisallylic protons in the polyunsaturated fatty acid[14]. The removal of a hydrogen atom and formation of the pentadienyl radical results in a carbon centered radical with a sufficiently long half-life (10^{-3} sec) that it can react with oxygen diffusing into the lipid bilayer to form the peroxy radical species seen in Figure 1. These peroxy radical species are themselves more stable with an estimated half-life of up to 10 sec. Rearrangement of these peroxy radicals and further reactions lead to the observed products of lipid peroxidation.

1.2 Isoprostanes

In 1988, Roberts and co-workers reported the appearance of a large number of isomeric prostaglandin $F_{2\alpha}$ compounds in both plasma and urine of human subjects which were thought initially to be metabolites of prostaglandin D_2[15]. These investigators subsequently found that upon storage of either plasma or urine at -20°C for one month, isomers of $PGF_{2\alpha}$ substantially increased in concentration suggesting that they were derived by a nonenzymatic processes, most likely autooxidation of precursors present in urine or plasma[16]. It then became clear that these isomeric $PGF_{2\alpha}$-like compounds were generated by free radical mediated processes that was not dependent upon cyclooxygenase. Furthermore, as many as 64 different isomers could be generated[17]. Most importantly, interest in these isomeric prostaglandins (F_2-isoprostanes) was generated when it was found that one specific epimer (8-epi-$PGF_{2\alpha}$), had significant biological activity[18,19]. It was found to extremely potent in constricting the renal artery of the rat and effecting a reduction in glomerual filtration rate of the rat kidney[18]. These biological effects on the kidney could be abolished or even reversed by a thromboxane A_2 receptor antagonist, suggesting that this F_2-isoprostanes may work through the endogenous receptor for thromboxane (TP-receptor) present in various cell types[19]. The mechanism of formation of these F_2-isoprostanes is outlined in Figure 2 summarizing the diverse family of nonenzymatic prostaglandin-like compounds that can be generated.

1.3 Isoleukotrienes

Studies of *in vitro* oxidation of 1-hexadecanoyl-2-arachidonoyl-GPC using conditions that would generate hydroxyl radicals revealed the formation of a family of compounds isomeric to leukotriene B_4. These metabolites were termed B_4-isoleukotrienes because they were 5,12-dihydroxy-6,8,10,14-eicosatetraenoic acids[20]. Importantly, several of the B_4-isoleukotrienes that were generated were found to possess significant biological activity in terms of elevating intracellular calcium when added to suspensions of human polymorphonuclear leukocytes (PMN). Since the human PMNs play a central role in inflammation and host defense reactions, such compounds could play an important role in activating this circulating cell. The biological activity of the B_4-isoleukotrienes could be blocked by a specific LTB_4 receptor antagonist as shown in Figure 3. These results suggested that the free radical generated B_4-isoleukotrienes exerted biological activity through the endogenous LTB_4 receptor (B-LT receptor) embedded in the human polymorphonuclear leukocyte plasma membrane.

As outlined in Figure 4, one mechanism for the formation of the B_4-isoleukotrienes would involve initial formation of 5-hydroperoxyeicosatetraenoic acid (5-HPETE) esteri-

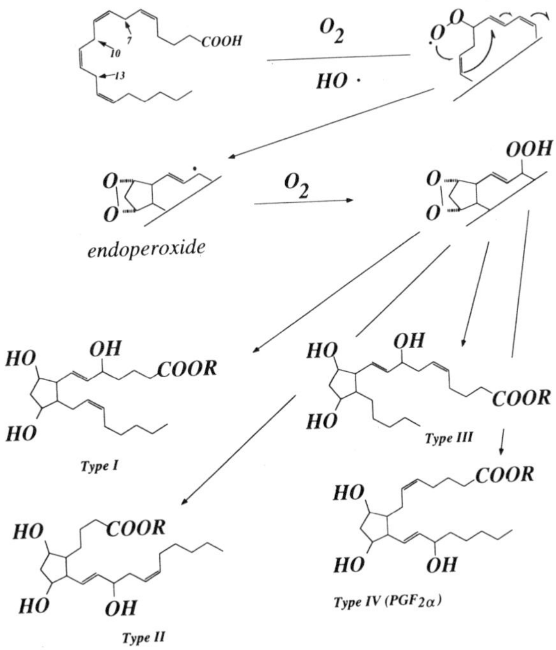

Figure 2. Proposed mechanism for formation of F_2-isoprostanes starting from free radical oxidation of arachidonic acid with the addition of two oxygen molecules. Rearrangement of the hydroperoxy endoperoxide results in four family regioisomers which differ as to the position of the two side chains and side chain hydroxl moiety attached to the cyclopentane ring.

fied to glycerophospholipids in the plasma membrane. Such products have been observed previously in the oxidation of polyunsaturated fatty acyl glycerophosphocholine lipids as abundant initial products[21]. The structure drawn in Figure 4 is for glycerophosphocholine, but other lipid classes are just as likely to be sources. The formation of this 5-HPETE ester would introduce a unique structural feature in that the methylene group at carbon-10 of arachidonate is now flanked on one side by a double bond and on the other side by a con-

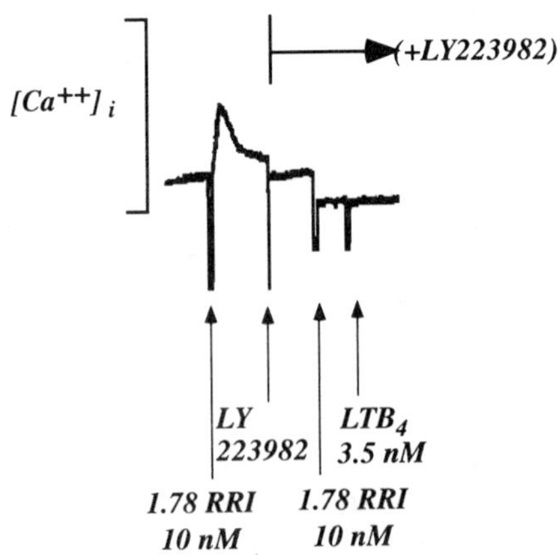

Figure 3. Increase in fluorescence in human neutrophils containing Indo-1 following addition of a B_4-isoleukotriene (final concentration 10 nM) with an HPLC relative retention index[20] of 1.78. The B-LT receptor antagonist LY223982[27] (10 nM) blocked both the response to the B_4-isoleukotriene as well as LTB_4 (3.5 nM).

jugated diene. One would expect that the strength of the carbon-hydrogen bonds in this methylene group would be substantially lower than even the bisallylic carbon-hydrogen bonds found in arachidonic acid. Thus, the 5-HPETE ester should form an even lower energy hepatrienyl radical illustrated in Figure 4 following a second free radical reaction. Such a radical species may possibly result directly from the initially formed peroxy or alkoxyl radical. Addition of oxygen to this radical species could then result in the 5,12-dihydroperoxyeicosatetraenoic acyl ester where a conjugated triene would exist between the two hydroperoxy groups. Since this free radical process would not be under stereochemi-

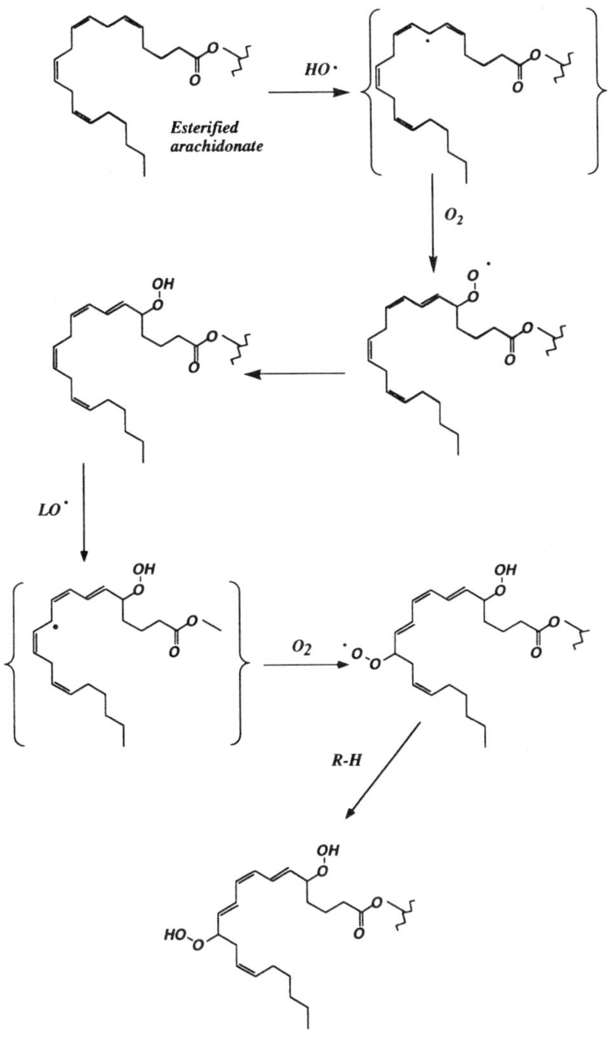

Figure 4. Proposed mechanism for the formation of B_4-isoleukotrienes from arachidonic acid esterified to glycerophosphocholine lipids. This process involves an initial radical removal of a hydrogen atom at a bisallylic methylene group, addition of oxygen followed by a second radical attack on the weakest carbon-hydrogen bond which is both allylic to an isolated double bond and a conjugated diene. The resultant radical species could attack molecular oxygen leading to the dihydroperoxy isoleukotriene.

cal control, a family of 2^5 or 32-members of the B_4-isoleukotriene family would be expected.

2. OXIDATIVELY FRAGMENTED PHOSPHOLIPIDS

Free radical oxidation of polyunsaturated fatty acids in glycerophospholipids and in particular, glycerophosphocholine lipids is now known to yield a large number of chain-shortened diacyl or alkyl, acyl lipids through the reactions of hydroperoxy radical decomposition[22]. Some of these structures which have been identified include short-chained monocarboxylic acid glycerophosphocholines, chain-shortened ω-hydroxy glycerophosphocholines, chain-shortened ω-oxo-glycerophosphocholines and chain-shortened dicarboxylic acid ester glycerophosphocholines (structures outlined in Figure 5). The positions of the hydroxyl groups, aldehyde moieties and carboxylic acid moieties are a result of radical reactions centered around the initial site of the double bond in the polyunsaturated fatty acid[22]. Interest in these oxidatively fragmented phospholipids has increased because of the discovery of the biological activities of such compounds. For example, using intravital fluorescence microscopy, administration of oxidized phospholipids to the hamster was found to stimulate interaction of polymorphonuclear leukocytes with the endothelium[23]. This interaction could be attenuated by pretreatment with a specific PAF-receptor antagonist (WEB2170)[23]. Other activities of these products of oxidatively fragmented phospholipids include causing hypotension in rats after intravenous injection[24], induction of neutrophil adhesion to gelatin matrix[25], and aggregation of washed rabbit platelets[26]. While the exact structures of the active metabolites are not known at the present time, many are known to exert their effects through the PAF-receptor. Furthermore, the biological activity of these products can be terminated by the enzyme PAF-acetylhydrolase[6]. This latter enzyme removes the *sn*-2 acetate from PAF, but does not hydrolyze long chain fatty acyl groups such as those found in diacylglycerophosphocholine lipids. A growing body of evidence therefore exists to suggest that biologically active glycerophosphocholine lipids derived from lipid peroxidation of polyunsaturated phospholipids have significant biological activity mediated through the PAF-receptor.

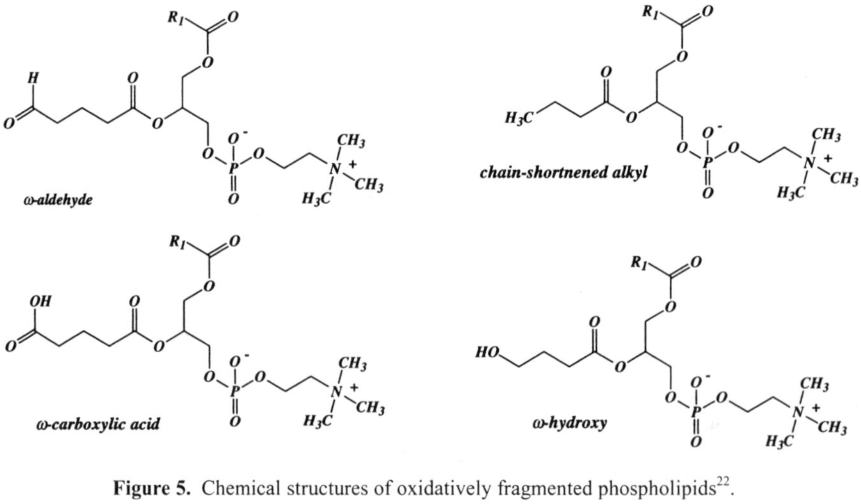

Figure 5. Chemical structures of oxidatively fragmented phospholipids[22].

3. EARLIEST LIPID MEDIATORS

It is perhaps not a coincidence that the products of lipid peroxidation, whether these products are isoprostanes, isoleukotrienes, or oxidatively fragmented glycerophospholipids, were found to be biologically active. Such products are likely to result from the chemical reactions of active oxygen species and propagation of the free radical chemistry taking place in the lipid environment of a cell membrane containing polyunsaturated fatty acyl groups. The formation of such products would constitute a signal indicating oxidative stress. Cells which developed mechanisms, namely protein receptors, for such signals could have a biological advantage if they mounted an appropriate response that increased survival. Thus, the various free radical generated lipid products could have been the earliest form of lipid mediator generated in evolving systems. Later, specific enzymes that formed prostaglandins, thromboxane, prostacyclin, leukotrienes, and platelet activating factor would impart a selective advantage to an organism since the randomness of the free radical process would be removed through the formation of specific products under stereochemical control. Sophistication of the mediator receptor protein could further develop a selective advantage since only specific signals would be recognized rather than generic signals of oxidative stress.

Free radical-mediated lipid peroxidation in cells leads to the formation of a complex mixture of products which include biologically active compounds isomeric to prostaglandins and leukotrienes. The chain-shortened glycerophosphocholine species produced by such free radical events leads to a mixture of compounds which exert biological activity mediated by the PAF-receptor. The importance of these molecules is only beginning to be recognized and the diverse biological activities so far observed suggest that such molecules may play an important role in processes such as ischemia reperfusion injury, inflammatory diseases, cancer, and aging. Interestingly, many of the biological activities thus far discovered are mediated through endogenous receptors suggesting an important role for pharmacologic antagonists of these receptors in such disease processes. The above molecules are also of interest as more specific markers of free radical-mediated lipid peroxidation events in cells.

ACKNOWLEDGMENTS

This work was supported, in part, by a grant from the National Institutes of Health (HL34303).

REFERENCES

1. Smith, W. L. (1992) *Am. J. Physiol.* **263**, F181-F191
2. Ford-Hutchinson, A. W., Gresser, M., and Young, R. N. (1994) *Annu. Rev. Biochem.* **63**, 383–417
3. Sigal, E. (1991) *Am. J. Physiol.* **260**, L13-L28
4. Yamamoto, S. (1992) *Biochim. Biophys. Acta* **1128**, 117–131
5. Fitzpatrick, F. A. and Murphy, R. C. (1989) *Pharmacol. Rev.* **40**, 229–241
6. Prescott, S. M., Zimmerman, G. A., and McIntyre, T. M. (1990) *J. Biol. Chem.* **265**, 17381–17384
7. Narumiya, S. (1995) *Adv. Prostaglandin, Thromboxane, Leukotriene Res.* **23**, 17
8. Nakamura, M., Honda, Z, Izmi, T., Sakanaka, C., Mutoh, H., Minami, M., Bito, Y., Seyama, Y., Matsumoto, T., Noma, M., and Shimizu, T. (1991) *J. Biol. Chem.* **266**, 20400
9. Summers, J. B. and Albert, D. H. (1995) *Adv. Pharmacol.* **32**, 67–168

10. Pryor, W. A. (1986) *Annu. Rev. Physiol.* **48**, 657–667
11. Halliwell, B. and Gutteridge, J. M. C. (1989) *Free Radicals in Biology and Medicine* Oxford University Press, Oxford, UK.
12. Gardner, H. W. (1989) *Free Radical. Biol. Med.* **7**, 65–75
13. Koppenol, W. H. (1990) *FEBS Lett.* **264**, 165–167
14. Wagner, B. A., Buettner, G. R., and Burns, C. P. (1994) *Biochem.* **33**, 4449–4453
15. Wendelborn, D. F., Seibert, K., and Roberts, L. J., II. (1988) *Proc. Natl. Acad. Sci. USA* **85**, 304–308
16. Morrow, J. D., Hill, K. E., Burk, R. F., Nammour, T. M., Badr, K. F., and Roberts, L. J., II (1990) *Proc. Natl. Acad. Sci. USA* **87**, 9383–9387
17. Morrow, J. D. and Roberts L.J., II (1991) *Free Radical. Biol. Med.* **10**, 195–200
18. Morrow, J. D., Minton, T. A., and Roberts, L. J., II (1992) *Prostaglandins* **44**, 155–163
19. Takahashi, K., Nammour, T. M., Fukunaga, M., Ebert, J., Morrow, J. D., Roberts, L. J., II, Hoover, R. L., and Badr, K. F. (1992) *J. Clin. Invest.* **29**, 136–141
20. Harrison, K.A. and Murphy, R.C. (1995) *J. Biol. Chem.* **270**, 17273–17278
21. Murphy, R.C. and Harrison, K.A. *Biological Mass Spectrometry* Humana Press, Totowa, NJ, in press
22. Tokumura, A. (1995) *Prog. Lipid Res.* **34**, 151–184.
23. Lehr, H. A., Seemüller, J., Hüber, C., Menger, M. D., and Messmer, K. (1993) *Arterioscler. Thromb.* **13**, 1013–1018
24. Yoshida, J., Tokumura, A., Fukuzawa, K., Terao, M., Takauchi, K., and Tsukatani, H. (1986) *J. Pharm. Pharmacol.* **38**, 878–882.
25. Smiley, P. L., Stremler, K. E., Prescott, S. M., Zimmerman, G. A., and McIntyre, T. M. (1991) *J. Biol. Chem.* **266**, 11104–11110.
26. Tanaka, T., Imori, M., Tsukatani, H., and Tokumura, A. (1994) *Biochim. Biophys. Acta* **1210**, 202–208.
27. Jackson, W. T., Boyd, R. J. Froelich, L. L., Mallett, B. E., and Gapinski, D. M. (1992) *J. Pharmacol. Exp. Ther.* **263**, 1009–1014

11

A DISCOVERY TRIP TO COMPOUNDS WITH PAF-LIKE ACTIVITY

Constantinos A. Demopoulos and Smaragdi Antonopoulou

Department of Chemistry
University of Athens
Athens GR 15 771, Greece

As I had mentioned in the Second International Conference on PAF in Tennessee on 1986, PAF seems to belong to a family of lipid molecules, which despite their different origins, share similar biological activities [1].

During these years, the potent biological role of PAF and its implication in various pathological conditions has been well established. The inability to explain the pathophysiological role of PAF in most cases, leads to the assumption that other molecules with PAF-like activity are also involved.

In this study, peculiar lipid compounds with PAF-like activity are illustrated. These compounds have been isolated from natural sources or semi-synthesized.

Some of the first lipid molecules, already reported by our team, about ten years ago, were derived either by substituting the phosphate group of the molecule of PAF with a phosphono group[2] or substituting the phosphocholine with the triphenyl-group[3] or by adding acetyl group to common lipids. For example an acetyl-group to sphingomyelin[4], an acetyl-group to cardiolipin [5] and acetyl-groups to gangliosides[4]. It was also suprising the occurence in Pinus pollen of a neutral glycerylether lipid without an acetyl group, but showing PAF-like activity[6].The above lipids not only cause platelet aggregation, but they also induce glycogenolysis in the protozoon Tetrahymena pyriformis and cause a decrease of systolic blood pressure in rats[4].

A number of compounds with PAF-like activity has been reported from several groups, for example substitution of the polar head group of the molecule of PAF by the team of Dr. Hanahan[7].

Recently, we have isolated several lipid compounds, with PAF-like activity from natural sources, such as human blood and Urtica dioica (nettle).

In human blood from patients with end-stage renal failure, we have detected about ten compounds, which induced washed rabbit platelet aggregation, belonging to neutral, glyco- and phospholipids. Illustrative examples are:

1. A phosphatidylinositol analog (fig.1), which induces dose-dependent platelet aggregation with a PAF-like tracing. This lipid acts through PAF receptors. This

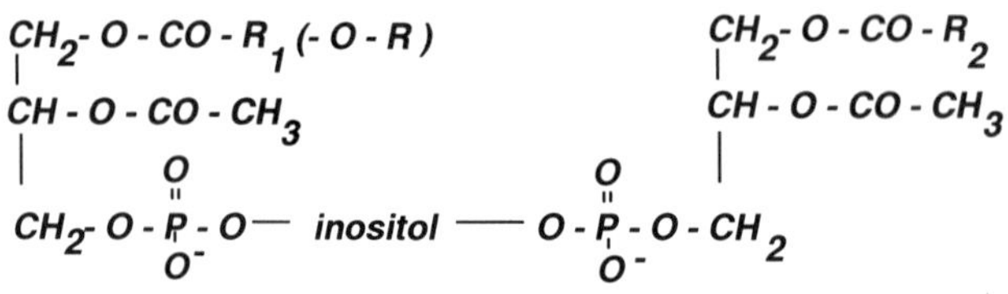

Figure 1. Natural phosphatidylinositol analog from human blood.

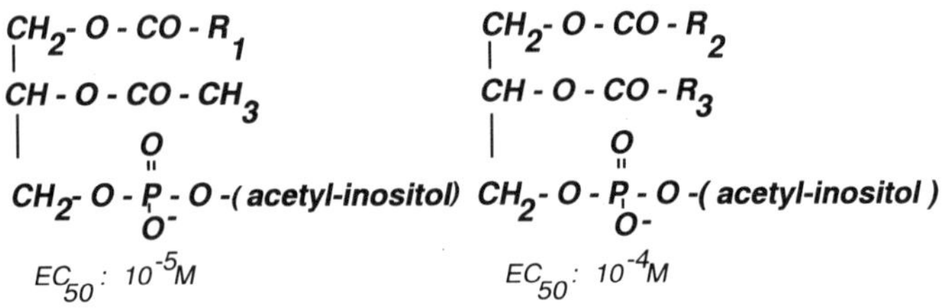

Figure 2. Semi-synthetic phosphatidylinositol analogs.

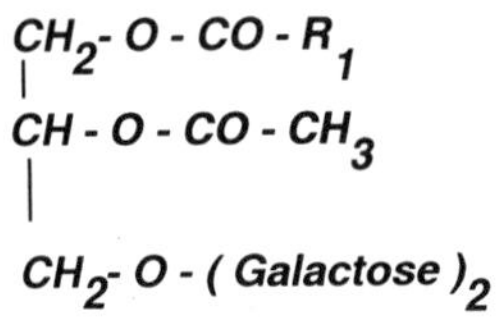

Figure 3. Natural digalactosyldiglyceride analog.

conclusion is based on : a) Desensitization and cross-desensitization experiments with itself, PAF and thrombin and b) Experiments with the specific inhibitors of the three aggregation pathways. Namely, indomethacin, the enzymic system CP/CPK and BN 52021. Furthermore, treatment with human plasma acetyl-hydrolase leads to complete inactivation of the phosphatidylinositol analog, with a rate similar to that of PAF.

In order to confirm the above structure, we prepared by semi-synthesis two phosphatidylinositol analogs (fig.2), one of them acetylated in sn-two position and in inositol group, while the other analog has a fatty acid in sn-two position .and is acetylated in inositol group.

These synthetic compounds induce washed rabbit platelet aggregation, but they are less active than PAF. Platelet aggregation characteristics are the same with the natural compounds. As it was expected, acetyl-hydrolase inactivates the sn-two acetyl-analog with the same rate as PAF. In addition, acetyl-hydrolase inactivates the sn-two fatty acid analog, but the rate is slower than that of PAF hydrolysis. Full inactivation of the sn-two fatty acid analog is accomplished in twice the time needed for PAF.

2. A digalactosyldiglyceride analog, which should have the proposed structure in figure 3. This analog causes platelet aggregation through PAF receptors, as indi-

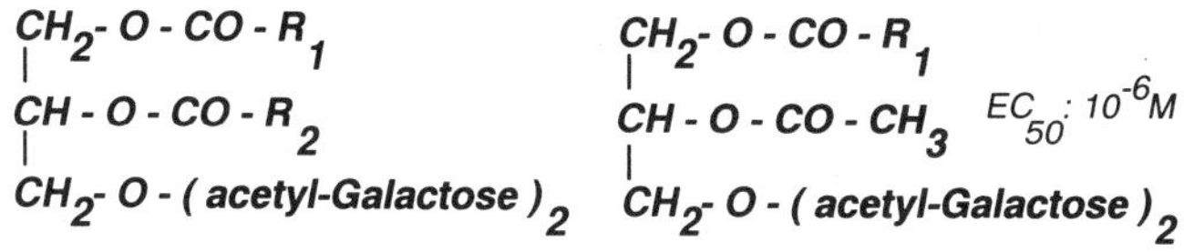

Figure 4. Semi-synthetic digalactosyldiglyceride analogs.

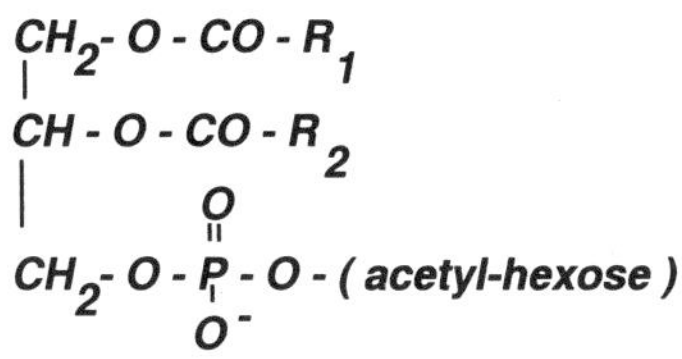

Figure 5. Natural phosphoglycoglycerolipid from Urtica dioica.

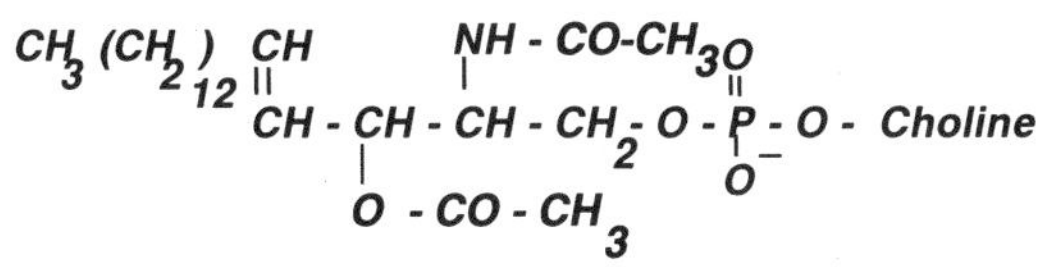

Figure 6. Natural sphingomyelin analog from Urtica dioica.

cated from the desensitization and cross-desensitization experiments and from the experiments with the specific inhibitors.
This lipid is also a good substrate for acetylhydrolase, since the enzyme inactivates the molecule with the same rate as PAF.

We prepared by semi-synthesis two digalactosyldiglyceride analogs (fig.4). One was acetylated only to the galactose group, while in the other analog the fatty acid in the sn-two position was also substituted by an acetyl group. Only the last one induces platelet aggregation.

Preliminary results show that this glycolipid may induce aggregation through PAF way, as indicated from the cross-desensitization experiments with PAF and the observed inhibition by BN 52021.

In Urtica dioica, apart from the existence of PAF, we have also detected more than ten molecules which induced washed rabbit platelet aggregation. These compounds belong to neutral, glyco- and phospholipids. Illustrative examples are:

1. A phosphoglycoglycerolipid (fig.5), which induces dose-dependent platelet aggregation with a PAF-like tracing. This lipid, desensitizes platelets against itself and PAF. Also, BN 52021 inhibits the aggregation induced by this phospho-glyco-glycerolipid. Treatment with acetyl-hydrolase causes a partial inactivation of the molecule.

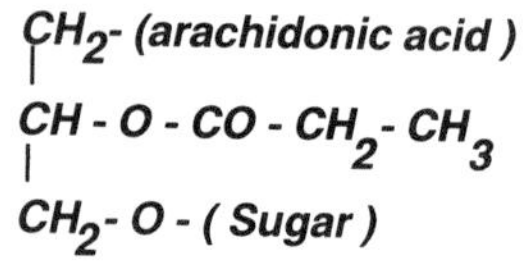

Figure 7. Natural glycoglycerolipid from Urtica dioica.

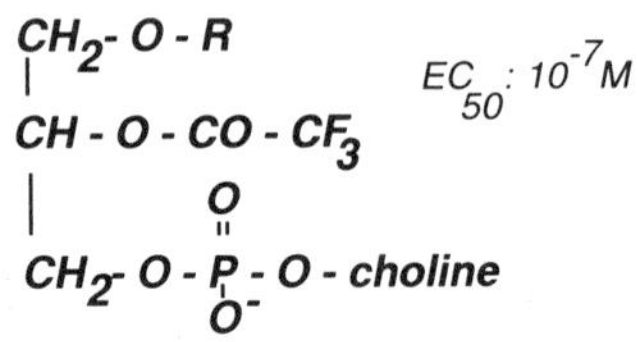

Figure 8. Trifluoroacetyl-analog of PAF.

2. A N-acetyl, O-acetyl sphingosylphosphocholine (fig.6), which induces platelet aggregation with a PAF-like tracing. This lipid acts through PAF receptors. This conclusion is based on : a) Desensitization and cross-desensitization experiments with itself, PAF and thrombin and b) Experiments with the specific inhibitors of the three aggregation pathways. Namely, indomethacin, CP/CPK and BN 52021.

   Furthermore, acetyl-hydrolase inactivates this di-acetyl-analog of sphingosylphosphocholine, but the observed rate of hydrolysis is lower than that of PAF.

   We have previously synthesized this lipid, which was found to exhibit PAF-like activity. More details on the biological activity of this lipid along with the biological activities concerning structural analogs have been published[8].
3. A glycoglycerolipid (fig.7), which induces dose-dependent platelet aggregation with a PAF-like tracing. This glycolipid seems to act through PAF receptors as indicated from the desensitization and cross-desensitization experiments and from the experiments with the already mentioned specific inhibitors.

   Treatment with acetyl -hydrolase leads to the inactivation of this molecule, but the rate of hydrolysis is slower than that of PAF. This may be due to the existence of a propionic group in the sn-two position instead of the acetyl group.

The above data, enforce the conception that PAF is a member of a large family, consisting of biologically active lipids, but it is important to notice that PAF seems to be the most potent lipid of this family. Consequently, the research should be focused not on PAF only, but on PAF-like activity, induced by different compounds, apparently having different physiological roles in animals, plants and monocellular organisms.

Finally, it is important to emphasize two more points.

One of them is that BN 52021 inhibits all the compounds acting through PAF receptors.

The other one is that human plasma acetylhydrolase modifies the structure of several biologically active lipids. Compounds which have acetyl group but not glycerol- backbone, compounds which has glycerol- backbone and the acetyl group in the sugar molecule, compounds with a low molecular weight fatty acid in the glycerol-backbone and a sugar molecule as a polar head group, compounds which are phosphatidylinositol analogs with the acetyl group in the glycerol-backbone, compounds which are phosphatidylinositol analogs with the acetyl group in the glycerol- backbone and in the inositol molecule, or in the inositol molecule only. Therefore, the specificity of this enzyme must be reconsidered. From these data, we can conclude that acetyl-hydrolase does not show high specificity either against the polar head group at the sn-three position of the molecule or against the glycerol-backbone, or finally, against the position of the acetyl group in the glycerol-backbone or in the polar head group . But, on the other hand, acetyl-hydrolase, which can hydrolyse low molecular weight fatty acids, seems to be highly specific for the acetyl group, since substitution of the hydrogen atoms by fluorine atoms on the acetyl group of PAF molecule (fig. 8) leads to a lipid with PAF-like activity[9] but which is not at all a substrate for the acetyl-hydrolase.

REFERENCES

1. Demopoulos, C.A. Second International Conference on Platelet-Activating Factor and Structurally Related Alkyl Ether Lipids, Tennessee, USA, Oct. 26–29, 1986.
2. Moschidis, M.C., Demopoulos, C.A. and Kritikou, L.G. Chem. Phys. of Lipids 33: 87, 1983.
3. Kritikou, L.G., Demopoulos, C.A., Moschidis, M.C., Siafaka, A. and Kapoulas, V.M. 5th Balkan Biochem. Biophys. Days, Thessaloniki, Greece, May 11–13, 1983.
4. Tournis, S., Demopoulos, C.A., Siafaka, A., Mavris, M., Tsangaris, G.Th., Tselepis, A. and Kapoulas, V.M. Second International Conference on Platelet-Activating Factor and Structurally Related Alkyl Ether Lipids, Tennessee, USA, Oct. 26–29, 1986.
5. Tsoukatos, D., Demopoulos, C.A., Tselepis, A., Moschidis, M.C., Donos, A., Evangelou, A. and Benveniste, J. Lipids 28: 1119, 1993.
6. Andrikopoulos, N.K., Siafaka-Kapadai, A., Yanovits-Argyriadis, N. and Demopoulos, C.A. Zeitschrift fur Naturforschung, Section B, 41: 396, 1986.
7. Satouchi, K., Pinckard, R.N., McManus, L.M. and Hanahan, D.J. J. Biol. Chem. 256:4425, 1981.
8. Zanglis, A., Lianos, E.A. and Demopoulos, C.A. Int. J. Biochem. Cell Biol. in press, 1995.
9. Kritikou, L.G., Moschidis, M.C., Siafaka, A. and Demopoulos, C.A. Platelet- Activating Factor, INSERM Symposium No 23, Eds. Benveniste J. and Arnoux B., Elsevier Science Publishers B.V., 65, 1983.

BIOLOGICALLY ACTIVE LIPIDS FROM *S. SCOMBRUS*

John Rementzis,[1] Smaragdi Antonopoulou,[2] Dimitris Argyropoulos,[2] and Constantinos A. Demopoulos[2]

[1]Biological Research Centre
Greek Army
P. Pedeli, Athens
[2]Department of Chemistry
University of Athens, Athens
GR 15771, Greece

1. INTRODUCTION

S. scombrus has been implicated in the disease known as Scombroid Food Poisoning (histamine intoxication), hypotension and hemocondensation being the main symptoms. Until now the above disease was attributed to the existence of high levels of histamine in this fish which were derived from histidine decarboxylation by the fish microbial greenfinch (Proteus, Klebsiella e.t.c.) [1]. However, some investigators claim that Scombroid Food Poisoning should not be caused by histamine only.

Platelet-Activating Factor, PAF, (1-O-alkyl-2-acetyl-*sn*-glycero-3-phosphocholine), biologically active ether phospholipid, is biosynthesized by a variety of animal cell types and exhibits a wide spectrum of biological actions being the most potent platelet aggregating factor known today [2].

In this study we isolated and purified neutral and phospholipids from *S. scombrus* muscles and studied their ability to induce platelet aggregation or inhibit PAF and thrombin induced aggregation.

2. MATERIALS AND METHODS

All reagents and chemicals were of analytical grade supplied by Merck. High-Performance Liquid Chromatography (HPLC) solvents were purchased from BDH, LABSCAN. Lipid standards of HPLC grade were obtained from Supelco. Semisynthetic PAF (80% C-16PAF and 20% C-18PAF) was synthesized in our laboratory as previously described [3].

Platelet-Activating Factor and Related Lipid Mediators 2
edited by Nigam *et al.*, Plenum Press, New York, 1996

2.1 Instrumentation

HPLC was performed on a Varian Star model, supplied with a 250ìl loop Rheodyne (7125) injector. A Varian model UV-VIS/9050 spectrophotometer was used as detector (205–210 nm). The spectrophotometer is connected to a Metter GA17 integrator-plotter. The following columns were used: a cation exchange column SS 10ìm Partisil 25cm x 4.6mm I.D., PXS 10/25 SCX and a reverse phase column Bondapack C18 25cm x 4.6mm I.D., 10ìm. The PAF-induced aggregation was measured in a Chrono-Log aggregometer coupled to an Omniscribe recorder. GC-MS was performed on a FISONS MD800 with a column DB1 50m, 0.22ìm I.D., 0.1ìm film thickness, ionisation source 70eV. NMR was performed on a AC200 Brucker connected with a computer Aspect 3000. The biologically active phospholipids were subjected to mass spectroscopy analysis. The Electrospray Mass Spectroscopy (ESMS) was recorded on a Fisons VG Quattro instrument with a VG Biotech Electrospray source, having a hexapole lens. Nitrogen 99.99% pure was used as the nebulizing and bath gas at flows of 20 and 150 dm^3 min^{-1} respectively. The sample was injected in the flow of solvent (10 ìL/min) of a Varian 9012 solvent delivery system, via a Fisons interface with a Reodyne 7125 injector. The capillary voltage was optimum at about 3 kV for negative ions and 3.30 kV for positive ions. The high voltage lens potential was kept at 0.56 kV. The focus and skimmer

lenses voltages were 40 and 45 V, respectively, in the majority of the measurements, as these values produced the highest peak intensities and minimum fragmentation. HPLC grade methanol/water (70:30, v/v) 0.01M in ammonium acetate was used as solvent.

2.2 Experimental Procedure

Lipids were extracted from homogenized fish muscles with Bligh-Dyer method [4] and separated into neutral and phospholipids with current counter distribution [5].

Neutral lipids were further fractionated on HPLC [6]. Briefly, neutral lipids were separated into classes and species with a stepped gradient elution with the following solvents: A, methanol/water 4/1 (v/v); B, acetonitrile/methanol 7/5 (v/v); C, acetonitrile/ tetrahydrofuran 98/2 (v/v); D, iso-propanol/acetonitrile 98/2 (v/v) and E, *c*-hexane. A linear gradient from solvent A to solvent B in 10 min, a hold for 5 min in B, then a linear gradient to solvent C in 10 min followed by a second hold in C for 5 min, then a linear gradient to solvent D in 10 min, a third hold in solvent D for 5 min and finally a linear gradient to solvent E in 10 min followed by a hold in E.

Phospholipids were further fractionated on HPLC [7]. Briefly, phospholipids were separated on a cation exchange HPLC column using a gradient elution system consisted of the following solvents: A. acetonitrile/methanol/water 75/20/5 (v/v/v); B. acetonitrile/methanol/water 50/40/10 (v/v/v). A hold in A for 10 min followed by a linear gradient to B in 30 min.

2.3. Biological Assay

PAF or sample-induced aggregation was examined towards washed rabbit platelets according to Demopoulos *et al* [3]. Experiments with specific inhibitors (creatine phosphate (CP) 0.7 mM/creatine phosphate kinase (CPK) 13 units per ml, indomethacin 10 ìM and BN 52021 0.1 mM), were also performed. These inhibitors are added to washed rabbit platelets 1 min prior to the addition of the examined sample into the aggregometer cuvette.

2.4 Desensitization Experiment

This experiment was carried out according to Lazanas *et al* [8].

2.5 Treatment with Acetylhydrolase

In a test tube prewarmed at 37 °C, phosphate buffer (0.08 M, pH 7.5) was added along with human serum acetylhydrolase, purified by Dr. D. Stafforini and the examined sample in 2.5 mg bovine serum albumin (BSA) per ml saline. The enzymic system is incubated at 37 °C and at different times aliquots were taken and tested for their ability to induce washed rabbit platelet aggregation.

3. RESULTS AND DISCUSSION

Neutral lipids were separated into classes and species as shown in Fig. 1.

The area of triglycerides of the above separation was rechromatographed under the same conditions (Fig. 2).

We detected a number of neutral lipids which inhibited PAF- induced aggregation as well as thrombin-induced aggregation but in higher concentrations. Structural studies revealed that one of these inhibitors (fraction 17 of Fig. 1) should be an 1-O-alkyl-2-methoxy- glycerol. This assumption was based on the chromatographic behaviour of this fraction since it was eluted in glycerylether area and on NMR and MS data (Fig. 3 and 4).

We also detected another neutral lipid which inhibited PAF and thrombin-induced aggregation (fraction 30 of Fig. 1). From structural studies with NMR and GC-MS (data not shown), it seems that this lipid belongs to the class of triglycerides.

Phospholipids were fractionated onto a SCX HPLC column (Fig. 5).

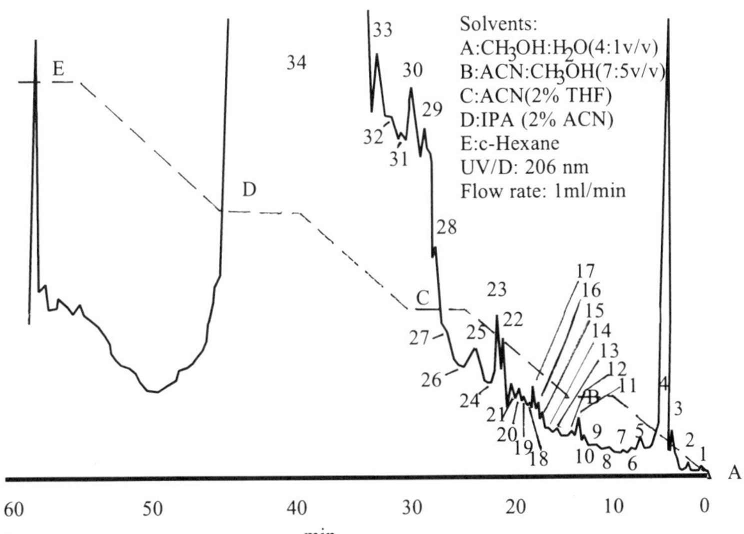

Figure 1. Neutral lipids fractionation onto a C18 HPLC column. Chromatographic conditions are described in the text.

Three phospholipids were detected inducing washed rabbit platelet aggregation. These phospholipids were eluted in the area between phosphatidylcholine and lysophosphatidylcholine. The first one (fraction 7 of Fig. 5) showed a dose-dependent aggregation with a pattern similar to PAF. Desensitization and cross-desensitization experiments with PAF and thrombin showed that this phospholipid should act through PAF receptors since it desensitized platelets only against itself and PAF. This conclusion is also supported by the results from the studies with phosphocreatine/creatine phosphokinase, indomethacin and BN 52021 where only BN 52021 inhibited the aggregation induced by fraction 7. Incubation of this fraction with human plasma acetylhydrolase resulted in its time dependent inactivation, suggesting the possible existence of an acetyl group at the sn-2 position of the glycero-backbone. The ESMS analysis revealed that this phospholipid is an analog of phosphatidylcholine.

The other active phospholipid (fraction 8 of Fig. 5) induced a dose-dependent washed rabbit platelet aggregation. This phospholipid desensitized platelets against itself and PAF. Also, the aggregation induced by fraction 8 was inhibited by BN 52021 and CP/CPK. It seems that fraction 8 acts through the receptors of PAF as well as of ADP as shown from the above results. Incubation of this fraction with human plasma acetylhydrolase resulted in its time dependent inactivation, suggesting the possible existence of an acetyl group at the sn-2 position of the glycero-backbone. Its structure is under investigation.

Fraction 9 of Fig. 5 also induced aggregation with a PAF-like platelet aggregation tracing. Studies with CP/CPK, indomethacin and BN 52021 as well as desensitization and cross-desensitization experiments showed that the above biologically active phospholipid should interfere with platelet membrane affecting mainly PAF receptor since the aggrega-

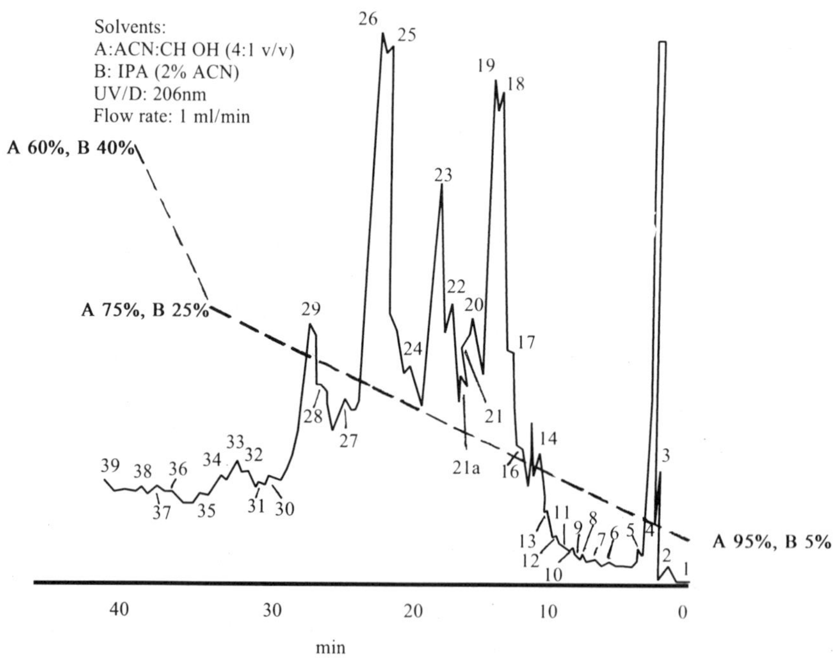

Figure 2. Further purification of triglyceride fraction onto a C18 HPLC column. Chromatographic conditions are described in the text.

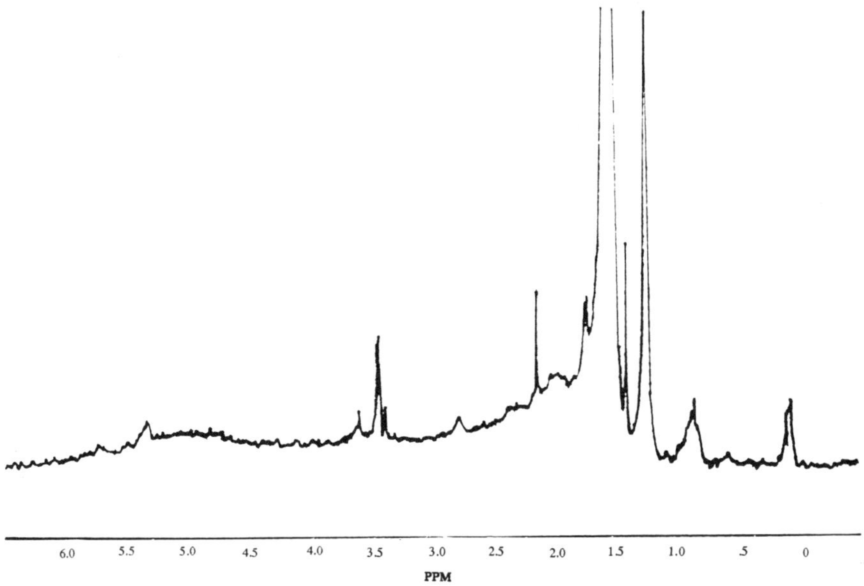

Figure 3. NMR spectrum of fraction 17.

tion induced by fraction 9 is inhibited only by BN 52021 and this phospholipid desensitized platelets only against itself and PAF. AH inactivated fraction 9, indicating the presence of an acetic acid in a glycero-backbone. The ESMS analysis (Fig. 6) revealed that this phospholipid is an analog of phosphatidylcholine.

We also detected phospholipid fractions which inhibited PAF-induced aggregation. One of them (fraction 5b of Fig. 5) showed a dose-dependent inhibition of PAF-induced

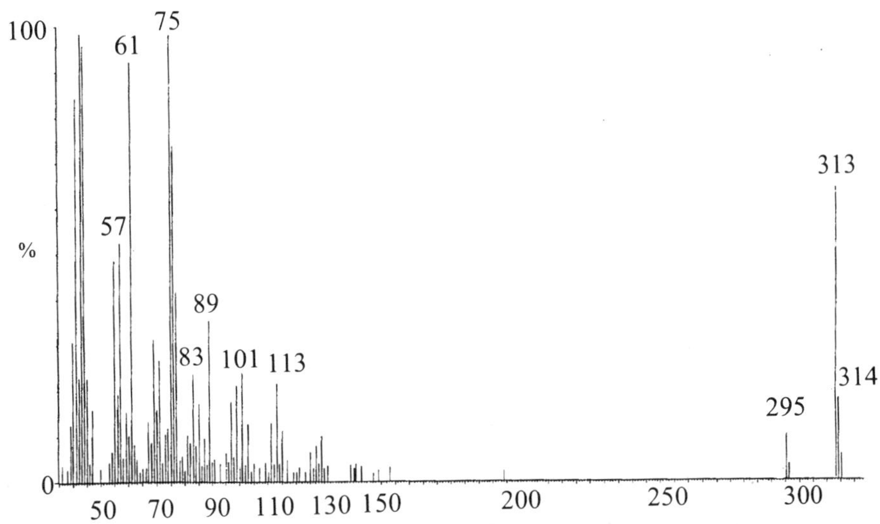

Figure 4. GC-MS spectrum of fraction 17.

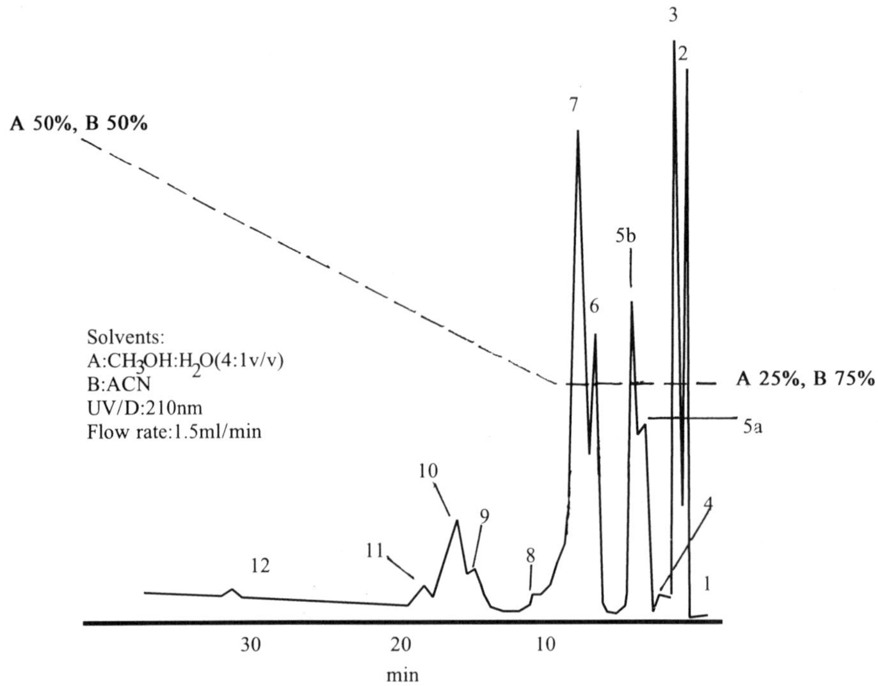

Figure 5. Phospholipids fractionation onto a SCX HPLC column. Chromatographic conditions are described in the text.

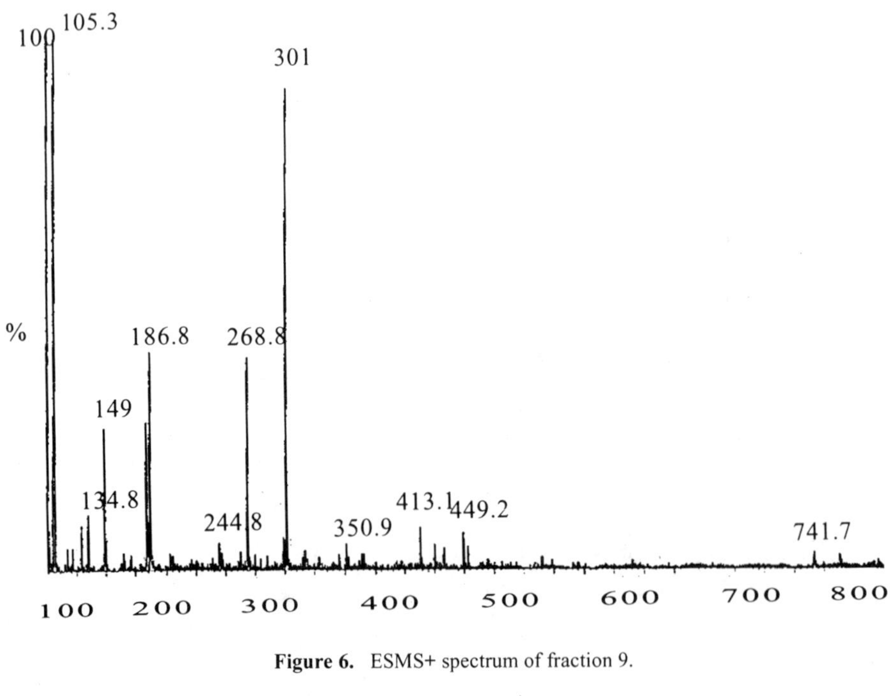

Figure 6. ESMS+ spectrum of fraction 9.

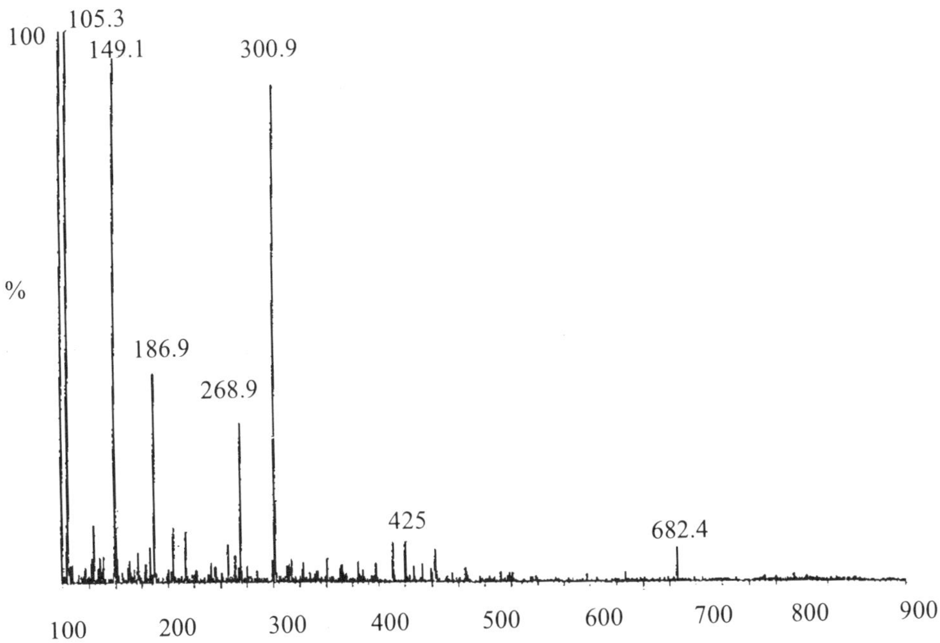

Figure 7. ESMS+ spectrum of fraction 5b.

aggregation as well as thrombin-induced aggregation but in higher concentration. This phospholipid should be an analog of phosphatidylethanolamine. This assumption is based on its chromatographic behaviour and on ESMS analysis data (Fig. 7 and 8).

Fraction 6 of Fig. 5 also inhibited PAF-induced aggregation. It seems to be a specific inhibitor of PAF since it did not cause any inhibition against thrombin-induced aggregation. The elucidation of its structure is under investigation.

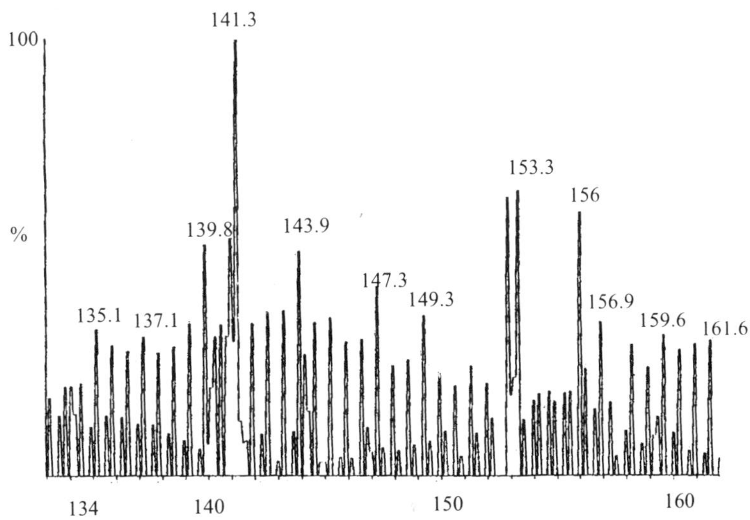

Figure 8. ESMS- spectrum of fraction 5b.

In conclusion, in this study we detected a number of biologically active lipids belonging to the classes of neutral lipids and phospholipids. Some of them inhibited washed rabbit platelet aggregation while others induced platelet aggregation. The last ones may be implicated in the Scombroid Food Poisoning caused by this fish.

REFERENCES

1. Ferencik M. (1970) J. Hyg. Epidemiol. Microbiol. Immunol., 14:52.
2. Lee T. C., Snyder F. (1985) In Phospholipids and Cellular Regulation; Editors Kuo J.F. and Boca R.; Vol. 2, pp.1–39; CRC Press Inc.; Boca Raton, Fla.
3. Demopoulos C. A., Pinckard R. N., Hanahan D. J. (1979) J. Biol. Chem., 254: 9355.
4. Bligh E. G., Dyer W. J. (1959) Can. J. Biochem. Physiol., 37:911.
5. Galanos D. S., Kapoulas V. M. (1962) J. Lipid Res., 3:134.
6. Antonopoulou S., Andrikopoulos N. K., Demopoulos C. A. (1994) J. Liquid Chrom., 17:633.
7. Andrikopoulos N. K., Demopoulos C. A., Siafaka-Kapadai A. (1986) J. Chromatogr., 363:412.
8. Lazanas M., Demopoulos C. A., Tournis S., Koussissis S., Labrakis-Lazanas K., Tsarouxas X. (1988) Arch. Dermatol. Res., 280:124.

13

PURIFICATION AND PARTIAL CHARACTERIZATION OF THE PAF ANTAGONIST FROM THE SALIVA OF THE LEECH *HIRUDO MEDICINALIS*

Miriam Orevi,[1] Amiram Eldor,[2] Maria-Elisabeth Gödeke,[1] and Meir Rigbi[1]

[1]Department of Biological Chemistry
The Hebrew University of Jerusalem
91904, Jerusalem, Israel
[2]Department of Hematology
Tel-Aviv Sourasky Medical Center
The Sackler Faculty of Medicine
Tel-Aviv University, Israel

We have found in the saliva of the leech *Hirudo medicinalis* a low-molecular weight fraction which powerfully inhibits platelet aggregation induced by PAF and thrombin, and by no other aggregating agent (1).

We obtain dilute leech saliva (DLS) by phagostimulating starved leeches to suck a solution of dilute arginine in saline through a membrane of modified collagen. The leech increases in weight about seven-fold. The ingested solution containing saliva is then forced out by squeezing the animal towards the mouth.

FRACTION II

Pooled lyophilized DLS was fractionated by gel permeation chromatography on a column of Bio-Gel P-2. The cut-off fraction (Fraction I, > 2kDa) contained apyrase, collagenase, hyaluronidase, hirudin, eglin (2), and the inhibitor of coagulation Factor Xa (3). The elution profile of the low-molecular weight fractions is shown in Figure 1. Fraction II (absorbance 260nm > absorbance 280 nm) was passed through a thrombin affinity column and traces of hirudin were removed. The eluate inhibited platelet aggregation induced by PAF and thrombin but not by ADP, collagen epinephrine or arachidonic acid. No other low molecular weight DLS fraction inhibited PAF-induced platelet aggregation. Fraction II did not inhibit thrombin coagulant activity.

Platelet-Activating Factor and Related Lipid Mediators 2
edited by Nigam *et al.*, Plenum Press, New York, 1996

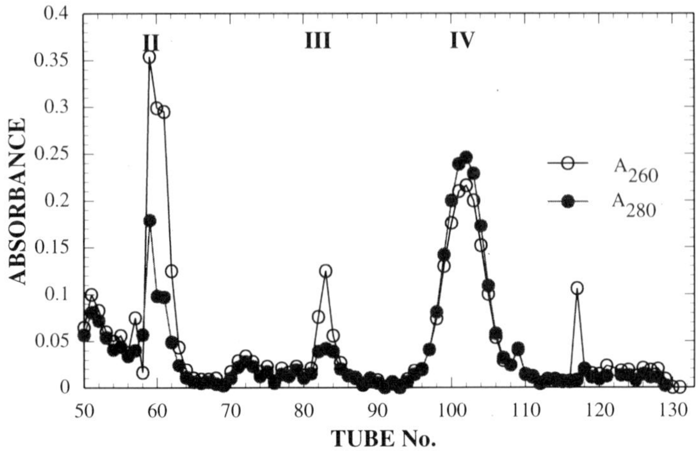

Figure 1. Elution of low-molecular weight fractions of dilute leech salive on Bio-Gel -2 at 4^{0}C. A 2 x 73 cm column was used with .2M ammonium bicarbonate pH 7.9 as eluent.

A decrease in platelet aggregation is generally accompanied by a decrease in thromboxane synthesis from released arachidonic acid. This is shown in Table 1 for the inhibition by Fraction II of PAF- and thrombin-induced platelet aggregation.

The inhibition of PAF-induced platelet aggregation is shown in Figure 2. As may be seen, the PAF antagonist (PAFA) is a competitive inhibitor of PAF-induced platelet aggregation.

PAF is an important mediator of the inflammatory response. Incubation of neutrophils with formyl Met-Leu-Phe or ionophore A23187 stimulates PAF formation with subsequent release of 0^{-}_{2} radicals (4). Fraction II strongly inhibits superoxide anion formation in activated neutrophils as shown in Table 2. The inhibition of superoxide generation by PAFA suggests that PAFA is able to penetrate neutrophils. BN-52021 does not inhibit superoxide generation in neutrophils (5).

Lyophilized Fraction II was dissolved in water and extracted with chloroform-methanol (6). Both its anti-PAF and anti-thrombin activities were distributed almost equally between the upper aqueous phase and the lower organic phase. The inhibitor is therefore amphipathic. We were also unable to distinguish between the two activities by chromatography on LH-Sephadex. PAF is more hydrophobic than PAFA, since on extraction under similar conditions all of it enters the organic phase.

Table 1. The effect of Fraction II on platelet aggregation and thromboxane synthesis induced by PAF and thrombin

	Control			Fraction II		
Aggregating agent	Percent aggregation	TxB_2, ng/ml		Percent aggregation	TXB_2, ng/ml	
PAF, 0.1 μM	46 ± 5	9.4 ± 5.1	(8)	21 ± 9	1.1 ± 0.2	(3)
Thrombin, 0.1 NIH U/ml	57 ± 10	29.6 ± 15.9	(5)	6 ± 2	1.7 ± 1.4	(2)

Human platelet rich plasma was used. Percent aggregation is calculated as the percent change in light transmission of the full range between platelet poor plasma and platelet rich plasma. The number of determinations (in parentheses) and the standard deviation are given. Figures for Fraction II are significantly different from those of the control.

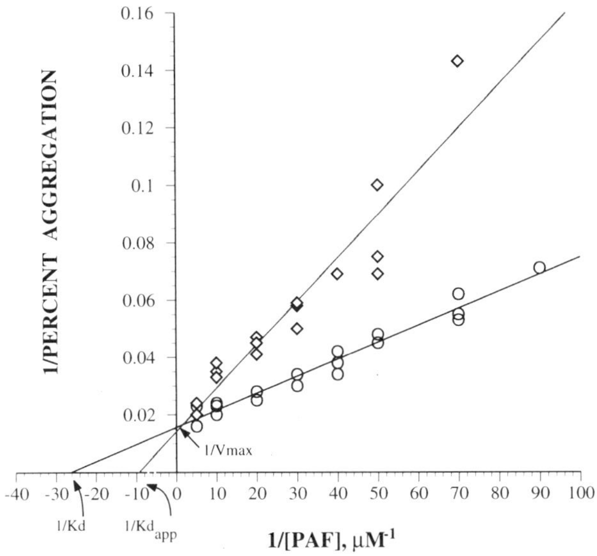

Figure 1. The inhibition of PAF-induced platelet aggregation by Fraction II. Circles indicate absence, and diamonds indicate presence of Fraction II.

THE PAF ANTAGONIST

As PAFA is amphipathic, its purification by liquid-liquid partitioning on Extrelut with an organic solvent as the mobile phase seemed desirable. This method was preferred to chromatography on Bio-Rad P-2 or solvent extraction in bulk for its rapidity and simplicity. The organic solvent was dichloromethane:isopropanol, 85:15 (v/v). Partitioning on Extrelut was followed by HPLC and then by RPLC. Purifed PAFA emerged as a single symmetrical peak with the absorbance at 260nm greater than at 280 nm, as in Fraction II. It inhibited PAF-induced aggregation only. In some batches, however, thrombin antagonist activity was also observed.

We speculated that PAFA might be an analogue of PAF, a phosphoglyceride. We therefore incubated PAFA with phospholipases (see Figure 3) at pH 8.0 for 30 min. at 37^{0}C. The reaction was stopped with chloroform. In order to separate between hydrophilic and hydrophobic products, methanol:water was added and the reaction mixture was parti-

Table 2. The inhibition by PAFA of Superoxide anion generation in activated neutrophils

	Formyl Met-Leu-Phe	A-23187
Control	5.2 ± 0.2	20.0 ± 1.0
Fraction II		
1 μl	n.d.	16.7 ± 0.7
10 μl	3.1 ± 0.2	12.7 ± 0.7
100 μl	2.0 ± 0.2	5.4 ± 0.6

Figures represent nmoles of O_2^- per 10^6 cells generated in 10 min. The average deviation of two measurements is given. n.d., not done.

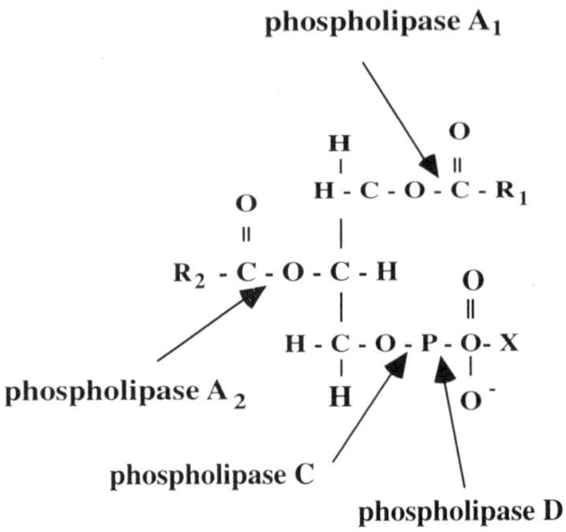

Figure 3. Bonds hydrolyzed by phospholipases.

tioned into a lower chloroform phase and an upper methanol: water phase. The phases were dried, the respective residues were dissolved in water and tested for inhibition of PAF-induced platelet aggregation. In parallel, PAF was incubated with phospholipases and partitioned in the same way. For PAF, the upper phase was inactive. The dissolved residue of the lower phase was then added to platelet rich plasma and tested as an inducer of aggregation (without addition of PAF). Phospholipase B, a mixture of phospholipases A_1 and A_2 slightly decreased PAFA activity. Phospholipase A_2 was no different with PAFA, whereas it totally abolished the activity of PAF, which is characterized by the acetyl group at the C2 position. Phospholipases C and D significantly decreased activity of the inhibitor obtained from the organic phase, as well as the activity of PAF itself. The reduction in PAFA inhibitor activity by a mixture of all four phospholipases was 59 and 72 percent for the lower and upper phases respectively. From the loss of activity on incubation with phospholipases, PAFA is identified as a phosphoglyceride.

PAFA concentration was determined by its content of phosphate (7) assuming one phosphate group per molecule. The micromolar concentration of PAFA causing 50 percent inhibition of platelet aggregation (IC_{50}) compared with that of other PAF antagonists was found to be as follows: PAFA, 0.37; WEB-2086, 0.1; BN-52051, 0.63 and CV-3988, 2.27. The order of potency is therefore WEB-2086 > PAFA > BN-5021 > CV-3988.

PAFA has been selected by evolution for its function as a potent inhibitor of platelet aggregation and of neutrophil activation. In our experience leech therapy accelerated the cure of chronic leg ulcers in patients with post-phlebitic syndrome. PAFA may become an important compound for the treatment of thromboembolic disorders and inflammation.

ACKNOWLEDGMENTS

This research was supported in part by a grant from the National Planning and Funding Committee for Biotechnology. We thank Takeda Chemical Industries, Osaka, Japan, for providing us with CV-3988; Institut Henri Beaufour, Le Plessis Robinson, France, for BN-52021 and Boehringer-Ingelheim KG, Ingelheim-am-Rhein, Germany, for WEB-2086.

REFERENCES

1. M. Orevi, M. Rigbi, E. Hy-Am., Y. Matzner and A. Eldor (1992) A potent inhibitor of platelet activating factor from the saliva of the leech *Hirudo medicinalis*. Prostaglandins. **43**:483–485.
2. M. Rigbi, H. Levy, F. Iraqi, M. Teitelbaum, M. Orevi, A. Alajoutsijärvi, A. Horovitz and R. Galun (1987). The saliva of the medicinal leech *Hirudo medicinalis* - I. Biochemical characterization of the high molecular weight fraction. Comp. Biochem. Physiol. (B) **87**:567–573.
3. M. Rigbi, C.M. Jackson and Z.S. Latallo. A specific inhibitor of bovine Factor Xa in the saliva of the leech Hirudo medicinalis. 14th International Congress of Biochemistry. July. 10–15, 1988. Prague. Abstracts, FR 037 p.53.
4. A.G. Stewart, P.N. Dubbin, T.Harris and G.J. Dusting (1990) Platelet-activating factor may act as a second messenger in the release of icosanoids and superoxide anions from leukocytes and endothelial cells. Proc. Natl. Acad. Sci. **87**:3215–3219.
5. J. Filep and E. Foldes-Filep (1988) Platelet-activating factors, neutrophil granulocyte function and BN-52021 in P. Braquet (ed.). Gingkolides-Chemistry, Biology, Pharmacology and Clinical Perspectives. J.R. Prous Science Publishers, Barcelona, Spain. pp. 151–159
6. E.G. Bligh and W. Dyer (1959) A rapid method of total lipid extraction and purification. Can. J. Biochem. Physiol. **37**:911–917.
7. J. Broekman (1989) Endogeous phospatidylinositol 4,5-biphosphate, phosphitidylinositol, and phosphatidic acid in stimulated human platelets. Methods in Enzymology. **169**:415–431.

14

PLATELET-ACTIVATING FACTOR RECEPTOR

Gene Structure and Tissue-Specific Regulation

Takao Shimizu,[1,*] Hiroyuki Mutoh,[1] and Shigeaki Kato[2]

[1]Department of Biochemistry, Faculty of Medicine
The University of Tokyo
Bunkyo-ku, Tokyo 113, Japan
[2]Department of Agricultural Chemistry, Faculty of Agriculture
Tokyo University of Agriculture
Setagaya-ku, Tokyo 156, Japan

ABSTRACT

The human platelet-activating factor receptor gene exists as a single copy on chromosome 1. Two 5′-noncoding exons (Exon 1 and 2) has distinct transcription initiation sites and promoters. These exons are alternatively spliced to a common splice acceptor site on exon 3 that contains a total coding regions. The transcript 1 is expressed ubiquitously with an emphasis of differentiated eosinophilic cell line (Eol-1), and leukocytes. On the other hand, the transcript 2 is expressed tissue-specifically. The latter is not expressed in leukocytes or brain. The transcript 1 has three tandem repeats of NF-κB, and SP-1 site, and responded to various inflammatory reagents including PAF itself, lipopolysaccharide, or phorbol ester. By northern blotting of tissue or cells with various nutritional or hormonal treatments, the PAF receptor messages are up-regulated. Estrogen increased the expression of the PAF receptor in human endometrial glandular cells, and vitamin A (retinoic acid) or thyroid hormone treatment up-regulates the PAF receptor expression only tissues with transcript 2. By various *in vivo* and *in vitro* transcriptional assays (CAT reporter assay, gel mobility shift assay), we identified estrogen responsible element, and hormone responsive element. The PAF receptor hormone responsive element is composed of three direct repeated TGACCT-like hexamer motifs with 2 and 4 bp spaces, and the two upstream and two downstream motifs were identified as response elements for RA and T_3.

* Address correspondence to Dr. T. Shimizu at the Department of Biochemistry, Faculty of Medicine, The University of Tokyo Tel 81-3-5802-2925 Fax 81-3-3813-8732.

1. INTRODUCTION

Platelet-activating factor (PAF; 1-*O*-alkyl-2-acetyl-*sn*- glycero-3-phosphocholine) is a potent pathological mediator involved in inflammatory as well as various physiological processes (1–5). PAF receptor cDNAs were cloned from various species including human, guinea-pig, rat, and mouse (6–11). The multiple signal transduction pathway were reported (12–14), which includes activation of phospholipases C, D, A_2, and mitogen-activated protein kinase cascade, and inhibition of adenylate cyclase. The activation of phospholipase A_2 produces arachidonic acid which is further converted to various types of eicosanoids. Such multiple intracellular signalings as well as a variety of eicosanoids might explain the versatile biological activities of PAF, in spite that PAF is a single compound, and PAF receptor has no subtypes. The present manuscript summarizes up-date findings of structure and regulation of expression of PAF receptor gene.

2. THE STRUCTURE OF HUMAN PAF RECEPTOR GENE

The human PAF receptor gene exists as a single copy per haploid in chromosome 1. Two human PAFR transcripts are found (PAFR transcripts 1 and 2) (15). Analysis of the genomic structure encoding the human PAFR showed that two 5′-noncoding exons (exon 1 and 2) directed by two distinct promoters (PAFR promoters 1 and 2) are alternatively spliced to a common splice acceptor site on a third exon (exon 3) that contains the total open reading frame. This yields two species of functional mRNAs (PAFR transcripts 1 and 2) (15) . These transcripts are differentially expressed in human tissues (15). The human PAFR transcript 1 is ubiquitous, and most abundant in peripheral leukocytes and a differentiated eosinophilic cell line (Eol-1 cells), while transcript 2 is located in the heart, lung, spleen and kidney, but not in brain or blood cells. Therefore, the human PAFR promoter 2 seems to contribute to the tissue-specific expression of the PAFR gene (Fig. 1).

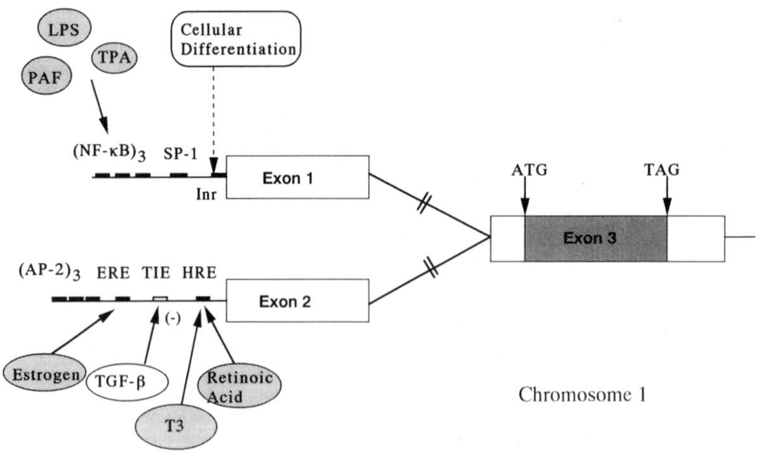

Figure 1. Structure and function of the human PAFR gene promoters. A schematic representation of the human PAFR gene structure which generates two human PAFR mRNAs (PAFR transcript 1 and 2) (15). The transcript 1 is generated by stimulation of NF-kB, while the transcript 2 expression is regulated by various hormones including estrogen, TGF-β, RA or T_3 (18–20).

3. REGULATION OF PAF RECEPTOR TRANSCRIPT 1 BY VARIOUS INFLAMMATORY REAGENTS

It was shown that the blood cells contain only transcript 1. By the stimulation of monocytes with lipopolysaccharide (LPS), the functional expression of PAF receptor is much increased, which was inhibited by actinomycin treatment (16, Ishii et al., unpublished). We found that PAF itself activates the expression of PAF receptor in human alveolar macrophages (17), thus providing a kind of a positive-feedback loop of PAF action. The PAF responsive element is most likely a three repeat of NF-κB, which is located between -898 and -785 of the transcription initiation site of transcript 1 (18). By chrolamphenicol acetyltransferase (CAT) assay using the deletion mutants of this region, the promoter is no longer responsive to treatment of PAF or phorbol ester (18). Thus, the transcript 1 is mostly involved in the regulation of PAF receptor under inflammatory and various pathological processes.

4. REGULATION OF PAF RECEPTOR TRANSCRIPT 2 BY ESTROGEN AND TGF-β

Primer extension analysis revealed that the levels of the PAF receptor transcript 2 were increased by estrogen treatment, but decreased by TGF-β treatment in JR-St stomach cancer cell line (19). Furthermore, Sato *et al.* (in this book) found that PAF receptor transcript 2 is also up-regulated by estrogen treatment in human endometrial glandular cells, as determined by Ca signaling and RT-PCR. The effect was blocked by estrogen antagonist, tamoxiferol. By CAT assay using various deletion construct, we found that a negative response element for TGF-β was mapped on the sequence from -90 bp to -81 bp, which has a consensus sequence for TIE (GNNTTGGTGA, TGF-β inhibitory element). Although consensus estrogen responsive element (AGGTCATnnnTGACCT) is not present in this promoter sequence, the entire sequence comprising two AGGTCA half motifs spaced by 153 bp conferred weak but significant estrogen responsiveness (19).

5. RETINOIC ACID AND THYROID HORMONE (T_3) ARE POTENT INDUCERS OF PAF RECEPTOR GENE

5.1 Northern Blotting Using Intact Animals

As we found some AGGTCA (or TCCAGT)-like elements in the promoter 2, next, we examined the *in vivo* effects of vitamin A (retinoic acid, RA) and thyroid hormone (T_3) on the expression of the PAF receptor gene in the heart, skin, and brain of rats under various retinoid acid and T_3 status. The levels of PAF receptor mRNAs decreased in the heart and skin of RA- and T_3-deficient rats. Oral administration of RA (100 mg/rat) to the retinol-deficient rats and intraperitoneal administration of T_3 (100 mg/rat) to the PTU-treated rats restored within 4 hours the levels of PAFR mRNAs in the heart and skin (20). Moreover, positive regulation was confirmed by the induction of the PAF receptor gene 6 hr after an excess of RA (1 mg/rat) or T_3 (500 mg/rat) was given to normal rats. In contrast, the PAF receptor gene expression in the brain which expresses only transcript 1 did not respond to RA or T_3. In sharp contrast to RA and T_3, PAF receptor gene expression was not induced 6 hr after an excess of vitamin D (1 mg/rat) given to normal rats (data not shown).

5.2 Transcriptional Assays

To delineate DNA elements of the human PAF receptor promoter 2 that mediate the response to RA or T_3, we constructed a series of deletion mutants in the human PAF receptor promoter 2. The results suggest the involvement of the consensus sequence Sp-1 between -91 bp and -68 bp in the ligand-inducibility. The sequence (-68/-43) includes three imperfect direct repeats [TGGCTT *cc* TGGCCT *cagc* TGCCCT (Box A+B+C; Fig. 2)] of the 5′-TGACCT-3′ (5′-AGGTCA-3′) motif, which is the consensus binding half-site motif for RA, thyroid hormone, and vitamin D receptor. Further analyses using various mutant constructs suggested that the sequence responsible for RA is mapped to Box A+B (two motifs with a 2 bp spacer) , and the sequence for T_3 to Box B+C (two motifs with a 4 bp spacer) . The mutations of two bases in each motif clearly impaired the responsiveness to RA and T_3 . Thus, it is most likely that the two upstream (Box A+B) and the two downstream motifs (Box B+C) act as response elements for retinoic acid and T_3, respectively. We further performed an *in vitro* DNA binding assay (gel-shift assay) with partially purified nuclear receptors to determine whether the these receptors indeed bind these sequences. Gel-shift analysis using the labeled synthetic oligonucleotides containing either Box A+B, Box B+C, and Box A+B+C again support the above notion (20). These results, together with their transcriptional activities, clarified that Box A+B is an RA responsive element (RARE) and Box B+C is a thyroid hormone responsive element (TRE) in PAF receptor promoter 2.

```
-217  CCTCTCTCCCCTGCCAGCCCAAGCCCAGCCCTTCCTCCCAGAGGGGCGCA

-167  GCATCAGAGGGGCCCAGCCCAGTATTGTGGGATGAGCTGGTGTGCCTGGC
                                                   Sp-1
-117  TGGATGTGGGTATTCCCGGTGACCTTGGTGTTGGGGCCCCGCCCCAGGAG
        Box-A    Box-B       Box-C
 -67  CTGGCTTCCTGGCCTCAGCTGCCCTGCCCACCCAGCAGAGCCCGGGGGCA
                         +1
 -17  GGCCTTTCCTAATCCTTCATCCTGCTGCTGCTCATTGGAGGGTAGAGCAG

 +34  AGTGAGTCACACACCTGTGCCTGGGCCCGTCCTCTGCCTCCTGTGCACTC

 +84  ATTACCTGCTTCCTGAGCTCCCCGAGAAGTCATCCAGGACCTCCCCGAGA

+134  AGCCGTCCAGGAAACATGCTCTCAGGGGACCCCCATCTGCCTCAGCCTCT
                                                  intron
+184  TTGTCACTGCCTGGACCATTGTCCCTGCTGTTTCTCAGgtaagaggctgg
```

Figure 2. DNA sequence of the 5′-flanking region of the PAFR transcript 2. The transcription start site is located at +1 bp (arrow). The degenerative TGACCT motifs are indicated as Boxes A, B, and C. PAF receptor-hormone responsive element which consists of Box A, B, and C is located from -67 bp to -44 bp. The consensus sequence for transcription factor Sp-1 is shown as "Sp-1". The intron is shown in lower-case letters.

6. DISCUSSION

In the series of experiments, we found that human PAFR gene expression is directed by two distinct promoters (PAFR promoters 1 and 2) to generate two transcripts (PAFR transcripts 1 and 2). The PAFR transcript 1 is ubiquitous and especially high in blood cells, whereas transcript 2 is expressed in a tissue-specific manner (15). The transcript 1 is mostly involved in inflammatory and pathological process. Although the element responsible of each stimulation is not fully clarified yet, PAF receptor mRNA is regulated by various cytokines or tumor necrosis factors (21, 22). The responses of PAF receptor gene expression to RA and T_3 were tissue-specific in rats, and indeed the RA and T_3 responsive elements were located in the human PAF receptor promoter 2. On the other hand, neither RA nor T_3 affected the transcriptional activity of the PAF receptor promoter 1. We do not know what is the advantage of the two different promoters which produce an identical coding regions. Similar alternative 5′-termini and promoters have been found in distinct exons of a mouse glucocorticoid receptor gene (23). As discussed by Schibler and Sierra (24), and Kozak (25), the transcription of a single gene from multiple promoters provides additional flexibility in the control of gene expression. Such promoters could have different cell type- and/or development-specific activities. The present study opens a new research area to elucidate the role of RA or T_3 in PAF-related events.

ACKNOWLEDGMENT

We are grateful to Prof. P. Chambon and Dr. H. Gronemeyer for the generous gift of the purified nuclear receptors, and Professor Frank K. Austen for discussion. The work was supported in part in grant-in-aid from the Ministry of Education, Culture and Science of Japan, and by grants from Yamanouchi Foundation, and Human Science Foundation.

REFERENCES

1. Braquet, P., Touqui,L., Shen,T.Y. & Vargaftig, B.B. (1987) Pharmacol. Rev. 39, 97–145.
2. Prescott,S.M., Zimmerman, G.A. & McIntyre, T.M. (1990) J. Biol. Chem. 265, 17381–17384.
3. Shimizu,T., Honda,Z., Nakamura,M., Bito, H. & Izumi, T. (1992) Biochem. Pharmacol. 44, 1001–1008.
4. Kornecki, E. & Ehrlich, Y.H. (1988) Science 240, 1792–1794.
5. Izumi, T., and Shimizu, T. (1995) Biochim. Biophys. Acta (in press)
6. Honda,Z., Nakamura,M., Miki,M., Minami, M., Watanabe, T., Seyama, Y., Okado,H., Toh,H., Ito, K., Miyamoto, T. & Shimizu,T. (1991) Nature 349, 342–346.
7. Nakamura,M., Honda,Z., Izumi,T., Sakanaka, C., Mutoh, H., Minami, M., Bito, H., Seyama, Y., Matsumoto,T., Noma, M. & Shimizu,T. (1991) J. Biol. Chem. 266, 20400–20405.
8. Sugimoto, T., Tsuchimochi, H., McGregor,C.G.A., Mutoh, H., Shimizu,T. & Kurachi, Y. (1992) Biochem. Biophys. Res. Commun. 189, 617–624.
9. Ye, R.D., Prossnitz, E.R., Zou, A.H. & Cochrane, C. G.(1991) Biochem.Biophys.Res. Commun. 180, 105–111.
10. Kunz, D., Gerard, N.P. & Gerard, C.(1992) J. Biol. Chem. 267, 9101–9106.
11. Bito, H., Honda, Z., Nakamura, M., and Shimizu, T. (1994) Eur. J. Biochem. 267, 9101–9106
12. Honda, Z., Takano, T., Gotoh, Y., Nishida, E., Ito, K., and Shimizu, T. (1994) J. Biol. Chem. 269, 2307–2315
13. Takano, T.,Honda, Z., Sakanaka, C., Izumi, T., Kameyama, K., Haga, K., Haga, T., Kurokawa, K., and Shimizu, T. (1994) J. Biol. Chem. 269, 22453–22458
14. Ferby, I. M., Waga, I., Sakanaka, C., Kume, K., and Shimizu, T. (1994) J. Biol. Chem. 269, 30485–30488

15. Mutoh, H., Bito, H., Minami, M., Nakamura,M., Honda,Z., Izumi,T., Nakata, R., Kurachi, Y., Terano, A. & Shimizu,T. (1993) FEBS Lett.. 322, 129–134.
16. Liu, H., Chao, W., and Olson, M. S. (1992) J. Biol. Chem. 267, 20811–20819
17. Shirasaki, Nishikawa, M., Adcock, I. M., Mak, J.C. W.,Sakamoto, T., Shimizu, T., and Barnes, P. J. (1994) Am. J. Respir. Cell Mol. Biol. 10, 533–537
18. Mutoh, H., Ishii, S., !zumi,T., Kato, S. & Shimizu,T. (1994) Biochem. Biophys. Res. Commun. 205, 1137–1142.
19. Mutoh, H., Kume, K., Sato, S., Kato, S. & Shimizu,T. (1994) Biochem. Biophys. Res. Commun. 205, 1130–1136.
20. Mutoh, H., Fukuda, T., Masushige, S., Sasaki, H., Shimizu, T., and Kato, S. (1995) Proc. Natl. Acad. Sci. U.S. A. (in press)
21. Aepfelbacher, M., Ziegler, H. H., Lux, I., and Weber, P.C. (1992) J. Immunol. 148, 2186–2193
22. Ouellet, S., Muller, E., and Rola-Pleszczynski, M. (1994) J. Immunol. 152, 5092–5099
23. Strahle, U., Schmidt, A., Kelsey, G., Steward, A. F., Cole, T. J., Schmid, W., and Schutz, G. (1992) Proc. Natl. Acad. Sci. U.S.A. 89, 6731–6735
24. Schibler,U. & Sierra,F. (1987) Annu. Rev. Genet. 21, 237–257.
25. Kozak, M. (1988) J. Cell Biol. 107, 1–7.

15

TRANSCRIPTIONAL REGULATION OF PLATELET-ACTIVATING FACTOR RECEPTOR GENE EXPRESSION IN LEUKOCYTES

Jong-Hwei S. Pang, Haw-Harn Yang, and Lee-Young Chau

Division of Cardiovascular Research
Institute of Biomedical Sciences
Academia Sinica, Nankang
Taipei, Taiwan, Republic of China

Platelet-activating factor (PAF) is a potent inflammatory mediator implicated in a variety of pathophysiological states[1–2]. PAF exerts a wide spectrum of biological activities via binding to specific receptors present on the surface of many types of cells, including platelets, neutrophils, monocytes, eosinophils, lymphocytes, vascular endothelial cells and smooth muscle cells. A PAF receptor cDNA was first cloned from guinea pig lung by Honda et.al.[3] Subsequently, two cDNAs for human PAF receptor were cloned from leukocytes (transcript I)[4–6] and heart (transcript II)[7], respectively. It was shown that both human transcripts contain identical coding region, but different in the 5′-noncoding sequence, indicating that two distinct promoters are involved in the transcriptional regulation of PAF receptor gene expression in various human tissues and cells[7]. Nevertheless, leukocytes have been shown to express exclusively the transcript I of the PAF receptor[7]. In attempt to understand the molecular mechanism that directs the expression of the gene encoding the PAF receptorin leukocytes, the 5′-flanking region of the human PAF receptor for transcript I was isolated and its promoter activity was characterized in myeloid and lymphoid cell lines.

RESULTS AND DISCUSSION

The transcriptional start site of the transcript I in human promonocytic U937 cells was mapped to an adenosine residue located 137 bp upstream of the ATG translation initiation site by both primer extension and 5′-RACE method[8]. The 5′-flanking region of the PAF receptor gene was isolated from a human leukocyte genomic library. Sequence analysis revealed that the 5′-flanking region lacks a typical TATA or CCAAT box. However, the sequence encompassing the transcription start site, TCTTCA_{+1}CTTCTG, shows high homology to the Inr consensus sequence, YYA_{+1}NT/AYY (Y : pyrimidine nucleo-

Platelet-Activating Factor and Related Lipid Mediators 2
edited by Nigam *et al.*, Plenum Press, New York, 1996

tide)[9–13]. A potential binding site for the universal transcriptional factor, SP1[14], is located at nucleotides -350 to -345 proximal to the initiation site. Another potential regulatory element, CCCCACCCC, commonly observed in the promoters of many myeloid -specific genes[15–17] is present at nucleotides -153 to -144. To examine whethr the 5′-flanking region of the PAF receptor gene can function as a promoter , we prepared a series of truncated fragments of the 5′-flanking sequence inserted before a promoterless luciferase reporter gene, and the promoter activity was determined by transient transfection of the receptor/luciferase fusion plasmid constructs into various cell lines. It was shown that the shortest construct, p(-44/+27)Luc, which contains the sequence spanning from -44 to +27 relative to the transcription start site, was sufficient to direct a high level of reporter gene expression in human myeloid cells (U937 and THP-1)[8] and lymphoid cells (Ramos). The activity of p(-44/+27)Luc in the nonexpressing HeLa cells was considerabley less than that in the myeloid and lymphoid cells, suggesting that the regulatory sequence within this region is important to confine cell-type specific expression of the PAF receptor gene in leukocytes.

To identify the putative nuclear factors, which intereact with the regulatory DNA sequence, we prepared the ^{32}P-labeled DNA fragment from nucleotide -44 to +27 of the PAF receptor gene and performed the gel mobility shift assay with the nuclear extracts prepared from various cells. It was shown that nuclear extract from U937 cells bound to this DNA fragment to form a DNA-protein complex, which could be specifically blocked by excess unlabeled DNA fragment from nucleotide -16 to +18 containing the Inr element[8]. Specific Inr-binding activity was also observed in the nuclear extracts from THP-1 and Ramos cells , but not from HeLa cells. Furthermore, it was noted that the Inr-protein complex formed in Ramos nuclear extract was distinct from that observed in U937 cells or THP-1 cells. To further test the functional role of the Inr element, a construct (m(-44/+23)Luc), which contains point mutations in the Inr element, was prepared and the promoter activity of the mutant was compared with that of the wild type construct. As revealed in the transfection experiments, the mutation in the Inr element, which attenuates the DNA-binding activity significantly, would cause 50%, 60%, and 65% decrease in the promoter activities in U937, THP-1 and Ramos cells, respectively. This result suggests that other regulatory element adjacent to the Inr site may act in concert to direct the optimal transcription. Subsequently, we performed the gel mobility shift assay using the upstream DNA sequence from nucleotide -44 to -18. When U937 nuclear extract was incubated with the radiolabeled upstream DNA fragment in the absence of Mg^{++}, which is the condition used in the Inr-binding assay, no DNA-protein complex formation was detected. In contrast, when the incubation was conducted in the presence of 5 mM Mg^{++}, two specific DNA-protein complexes were observed. Likewise, when the experiment was conducted with nuclear extract from THP-1 or Ramos cells in the presence of Mg^{++}, a distinct DNA-protein complex with a slower electrophoretic mobility was detected. The DNA binding activity was inhibited by excess amounts of unlabeled DNA probe, but not by other DNA sequences containing the binding sites for Sp1, AP1 and AP2, respectively, suggesting that it is sequence-specific. The binding activity to the -44/-18 DNA fragment , however, was also detected in nuclear extract from HeLa cells, indicating that it is not cell-type specific.

Taken together, these results clearly demonstrate that the DNA sequence from nucleotide -44 to +27 relative to the transcription start site is sufficient to direct the transcription of the PAF receptor gene in leukocytes through interacting with distinct nuclear factors in meyloid and lymphoid cells, respectively.

ACKNOWLEDGMENTS

Supported by grants from the Academia Sinica and the National Science Council of Taiwan.

REFERENCES

1. Braquet, P., Touqui, L., Shen, T.Y., and Vargaftig, B.B. (1987) Pharmacol. Rev. 39, 97–145
2. Venable, M.E., Zimmerman, G.A., McIntyre, T.M., and Prescott, S.M. (1993) J. Lipid Res. 34, 691–702
3. Honda, Z-I, Nakamura, M., Miki, I., Minami, M., Watanabe, T., Seyana, Y., Okado, H., Toh, H., Ito, K. Miyamota, T., and Shimizu, T. (1991) Nature 349, 342–346
4. Nakamura, M., Honda, Z., Izumi, T., Sakanaka, C., Mutoh, H., Minami, M., Bito, H., Seyama, Y., Matsumoto, T., Noma, M., and Shimizu, T. (1991) J. Biol. Chem. 266, 20400–20405
5. Ye, R.D., Prossnitz, E.R., Zou, A., and Cochrane, C.G. (1991) Biochem. Biophys. Res. Commun. 180, 105–111
6. Kunz, D., Gerard, N.P., and Gerard , C. (1992) J. Biol. Chem. 265, 4261–4265
7. Mutoh, H., Bito, H., Minami, M., Nakamura, M., Honda, Z., Izumi, T., Nakata, R., Kurachi, Y., Terano, A., and Shimizu, T. (1993) FEBS Lett. 322, 129–134
8. Pang, J.-H. S., Hung, R.-Y., Wu, C.-J., Fang, Y.-Y., and Chau, L.-Y. (1995) J. Biol. Chem. 270, 14123–14129
9. Smale, S.T., and Baltimore, D. (1989) CELL 57, 103–113
10. Smale, S.T., Schmidt, M.C., Berk, A.J., and Baltimore, D. (1990) Proc. Natl. Acad. Sci. U.S.A. 87, 4509–4513
11. O'Shea-Greenfield, A.O., and Smale, S.T. (1992) J. Biol. Chem. 267, 1391–1402
12. Javahery, R., Khachi, A., Lo, K., Zenzie-Gregory, B., and Smale, S.T. (1994) Mol. Cell. Biol. 14, 116–127
13. Du, H., Roy, A.L., and Roeder, R.G. (1993) EMBO J. 12, 501–511
14. Briggs, M.R., Kadonaga, J.T., Bell, S.P., and Tjian, R. (1986) Science 234, 47–52
15. Shapiro, L.H., Ashmun, R.A., Roberts, W.M., and Look, A.T. (1991) J. Biol. Chem. 266, 11999–12007
16. Gomolin, H.I.,Yamaguchi, Y., Paulpillai, A.V., Dvorak, L.A., Ackerman, S.J., and Tenen, D.G. (1993) Blood 82, 1868–1874
17. Shelley, C.S., Farokhzad, O.C., and Arnaout, M.A. (1993) Proc. Natl. Acad. Sci. U.S.A. 90, 5364–5368

16

PAF RECEPTOR ANCHORS *STREPTOCOCCUS PNEUMONIAE* TO ACTIVATED HUMAN ENDOTHELIAL CELLS

Diana R. Cundell,[1] Craig Gerard,[2] Ilona Idanpaan-Heikkila,[1]
Elaine I. Tuomanen,[1] and Norma P. Gerard[2]

[1]Laboratory of Molecular Infectious Diseases
Rockefeller University
New York, New York
[2]Ina Sue Perlmutter Laboratory and Department of Pediatrics
Children's Hospital
The Department of Medicine
Beth Israel and Brigham and Women's Hospitals
The Center for Blood Research and the Thorndike Laboratory of
Harvard Medical School, Boston, Massachusetts

I. INTRODUCTION

Streptococcus pneumoniae is a Gram positive bacteria that is a major cause of pneumonia, sepsis, and meningitis[1]. In contrast to the severity of invasive disease, ~40% of individuals harbor pneumococcus asymptomatically in the nasopharynx[2]. Further, it has long been recognized that the mere presence of pneumococci in the pulmonary alveolus does not infer progression to pneumonia[3,4]. These disparate courses suggest that as yet unknown elements of the encounter between host and pathogen determine the outcome. A clinical clue to the nature of these elements is the observation of the propensity of patients and experimental animals to progress to bacterial pneumonia in the context of an intercurrent upper respiratory tract viral infection[5]. Taken together with the recent observation that pneumococci have been shown to adhere in greater numbers to virally infected cells in vitro[6], we hypothesized a change from a state of simple bacterial binding to the surface of the nasopharyngeal or pulmonary epithelium to a state promoting translocation across underlying endothelial cells into the blood stream in the context of an inflammatory stimulus.

Here we describe a molecular shift to a new receptor specificity for a pathogen encountering activated versus resting human cells. Activation of cells through the chemokine or cytokine cascades, such as by thrombin, TNFα, or IL-1, enhances and mechanistically alters the adherence of pneumococci to airway epithelial and vascular endothelial cells. Only the interaction with activated cells promotes invasion. The major novel receptor on activated cells is identified as the PAF receptor and phosphorylcholine on the bacterial teichoic acid is an important cognate ligand.

II. ADHERENCE OF PNEUMOCOCCI TO ACTIVATED HUMAN VASCULAR ENDOTHELIAL CELLS AND TYPE II PNEUMOCYTES

The adherence of *S. pneumoniae* to human vein endothelial cells (EC) and type II lung epithelial cells was quantitated before and after stimulation of the eucaryotic cells with thrombin, IL-1α or TNFα (Table 1)[7]. Adherence of the encapsulated strain AII (strain 2), its unencapsulated derivative R6, and encapsulated strain SIII (serotype 3) were indistinguishable. Incubation of EC with thrombin resulted in a rapid increase in pneumococcal adherence that reached a peak at 10 min and decreased to basal levels by 60 min. Cytokine stimulation resulted in a more delayed increase in pneumococcal adherence that was maximal after 3 h incubation with TNFα and 4 h incubation with IL-1α. In contrast to endothelial cells, pneumococcal adherence to type II pneumocytes was not affected by TNFα or thrombin but increased by ~40% with IL-1α. The magnitude of the increase in adherence correlated with input bacterial concentration[7]. Stimulation of EC increased pneumococcal adherence by as much as 81% for thrombin, 190% for TNFα, and 91% for IL-1α.

III. CARBOHYDRATE SPECIFICITY OF PNEUMOCOCCAL ADHERENCE TO ACTIVATED CELLS

To identify whether the enhanced pneumococcal adherence to stimulated EC and LC was a result of expansion of existing receptors on resting cells or the appearance of a new

Table 1. Factors influencing pneumococcal adherence to endothelial cells and lung cells

	Number of pneumococci adherent to 100 host cells (% of control)			
Treatment[a]	Control	L659,989[b]	Ethanolamine[c]	TEPC-15[d]
Endothelial cells				
resting	239 ± 12	273 ± 11 (99%)	242 ± 13 (101%)	239 ± 15 (100%)
TNFα	415 ± 25	230 ± 15 (55%)	121 ± 11 (29%)	108 ± 18 (26%)
IL-1α	368 ± 18	250 ± 10 (68%)	130 ± 9 (35%)	114 ± 12 (31%)
thrombin	382 ± 12	225 ± 12 (59%)	136 ± 8 (36%)	110 ± 15 (29%)
Lung cells				
resting	273 ± 14	272 ± 8 (99%)	294 ± 14 (107%)	275 ± 18 (101%)
IL-1α	355 ± 18	239 ± 15 (67%)	135 ± 13 (38%)	108 ± 18 (26%)

[a]Endothelial cells were stimulated with TNFα (10 ng/ml, 4 h, 37°C), IL-1α (5 ng/ml, 4 h, 37°C) or thrombin (2 U/ml, 10 min, 37°C); A549 type II lung cells were stimulated with IL-1α (5 ng/ml, 4 h, 37°C). Adherence of flourescein labeled R6 pneumococci (2×10^7 cfu/ml) was assessed visually following incubation (30 min, 37°C) with the monolayers. Values are the mean ± SD of 6 experiments.

[b]Stimulated monolayers were incubated (30 min, 37°C) with PAF receptor antagonist L659,989 (1 μM), and adherence compared with cells incubated in albumin buffer alone (control).

[c]R6 pneumococci were grown in defined medium with ethanolamine replacing choline as the aminoalcohol.

[d]R6 pneumococci were incubated with the antiphosphorylcholine antibody TEPC-15 (10 min, 22°C, ascites fluid, M. Potter, NCI, Bethesda, MD), washed and co-incubated with monolayers.

class of receptor, differences in the abilities of monosaccharides or glycoconjugates to inhibit pneumococcal adherence to resting and activated cells were sought. Resting EC and LC bear two classes of receptors, GalNAcβ1–4Gal and GalNAcβ1–3Gal[8]; the combination of these carbohydrates inhibited 74–83% of pneumococcal adherence to both resting cell types. Upon thrombin or cytokine stimulation of EC, inhibition of the binding by the two carbohydrates decreased, and addition of a new sugar specificity, N-acetylglucosamine (GlcNAc), was required to inhibit the additional binding to activated EC and LC[7]. GlcNAc reduced adherence by ~50% for thrombin or cytokine activated EC or LC, in contrast to <12% of binding to resting cells. Fucose, sialic acid, glucose and galactose had no effect. This suggested that activation of eucaryotic cells not only enhanced the amount, but also changed the mechanism of pneumococcal attachment.

IV. IDENTIFICATION OF THE NOVEL LIGAND-RECEPTOR PAIR FOR ACTIVATED CELLS

Four observations suggested that the cognate ligand on the pneumococcus for the novel receptor specificity seen on activated EC and LC was the phosphorylcholine determinant of the cell wall. Purified pneumococcal cell wall (50 μg/ml) partially inhibited the enhanced pneumococcal adherence to activated EC and LC. Replacement of the choline on the pneumococcal cell wall with ethanolamine (by substitution of the aminoalcohol in the culture medium) resulted in a >60% attenuation in pneumococcal binding to thrombin-activated EC and to cytokine-activated EC and LC, but not to resting cells. Further, pre-incubation of pneumococci with choline or the anti-phosphorylcholine antibody TEPC 15, which masks the pneumococcal teichoic acid, significantly attenuated pneumococcal adherence to activated cells.

Choline is a critical determinant in the bioactivity of both the C-polysaccharide/teichoic acid on the pneumococcus and PAF[9,10]. The dereased adherence to activated EC and LC displayed by pneumococci lacking this determinant suggested the PAF receptor as a candidate receptor for pneumococcal adherence. To support this posibility, two types of studies were undertaken, Northern analysis to document the presence of PAF receptor mRNA in the activated cells and inhibition of adherence by specific PAF receptor antagonists. Northern analyses of RNA from both EC and LC indicated an mRNA signal hybridizing with the PAF receptor cDNA, indicating the presence of PAF receptor message as described previously in human lung[11]. As shown in Figure 1 the transcript was visualized in cells treated with TNFα or IL-1α, but not in control cells or cells treated with thrombin. Similar data have been reported by Shirisaki, et al[12] in studies of human lung tissue suggesting that the immediate upregulation of receptor by thrombin involves mobilization of intracellular stores in contrast to new synthesis in cytokine treated cells.

Further evidence consistent with the PAF receptor acting to anchor pneumococci, was the finding that PAF and two specific PAF receptor antagonists inhibited the enhanced pneumococcal adhesion to activated cells[7]. PAF reduced pneumococcal adherence to cytokine activated EC and LC to levels seen on resting cells. Similar results were also found for thrombin-activated EC. PAF receptor antagonists, WEB2086[11] and L659,989[10], were able to substantially abrogate the enhanced adhesion of pneumococci to cytokine-activated cells in a dose dependent manner. Similar ED50 values have been reported for WEB2086 in vitro[12] and in an animal model for L659,989[10]. Pneumococcal adherence to thrombin-stimulated EC was also inhibited by the L659,989, although WEB2086 was without significant effect. Neither receptor antagonist effected pneumococcal adherence to resting EC or LC.

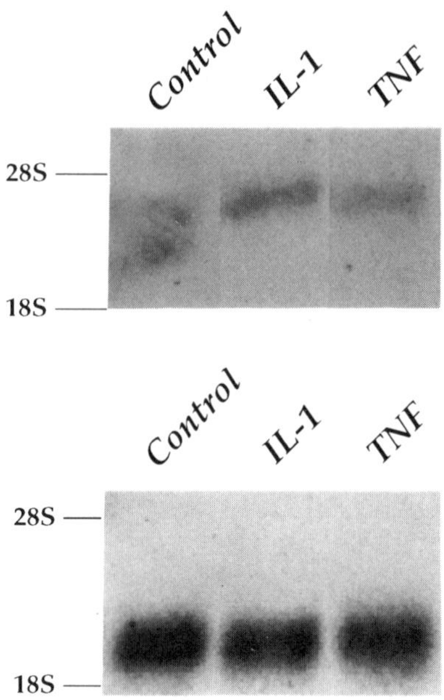

Figure 1. Northern analysis of endothelial cell PAF receptor mRNA. Poly(A^+) RNA was prepared from endothelial cell monolayers stimulated with TNFα, IL-1α, or medium alone as described for Table 1, electrophoresed on agarose formaldehyde gels, transblotted to Genescreen Plus membranes (Dupont NEN) and hybridized with [^{32}P]-labeled cDNA corresponding to the human PAF receptor coding sequence (upper panel). Following exposure, bands hybridizing with the PAF receptor specific probe are only observed for IL-1α or TNFα-treated cells. The blot was re-probed for actin (lower panel) and indicates aimilar quantities od RNA in each lane.

V. INTERACTION OF PNEUMOCOCCI WITH TRANSFECTED HUMAN PAF RECEPTORS

To determine directly if pneumococci could adhere to the PAF receptor, COS cells were transfected with the human PAF receptor cDNA modified to encode a Flag epitope[13]. The construct encoding the Flag epitope demonstrates indistinguishable pharmacology from the wild type receptor, but has the advantage of allowing confirmation of receptor expression at the cell surface. Transfected cells strongly supported pneumococcal adherence compared to untransfected controls (Table 2). Consistent with the characteristics of pneumococcal adherence to native EC and LC, adherence of pneumococci grown in ethanolamine to the PAF receptor transfected cells was >10-fold less than choline-grown bacteria. Adherence to transfected cells was inhibited by exogenous PAF and the PAF antagonists WEB2086 and L659,989. Avirulent, opaque pneumococcal colony variants adhered poorly to the transfected cells while virulent, transparent variants adhered strongly.

It was of interest to determine if the single glycosyl determinant on the human PAF receptor contributed to receptor recognition. Similar to native activated cells, pneumococcal adherence to transfected COS cells was decreased ~40% in the presence of GlcNAc. Globoside and asialo-GM2, glycoconjugates specifically inhibitory for adherence to resting cells, were inactive. Site specific mutation, so as to remove the glycosylation site of the PAF receptor, reduced pneumococcal adherence by ~75%[14]. Pharmacological and immunohistochemical analyses using the Flag epitope indicated similar reduction in cell sur-

Table 2. Pneumococcal adherence to PAF receptor-transfected COS cells[a]

	Innoculum			
Receptor transfected	10^7 cfu/ml	(%)[b]	10^8 cfu/ml	(%)
WT PAF-R[c]	286±15	(100±5%)	1495±318	(100±21%)
PAF-R dCHO	78±22	(27±8%)	448±112	(30±7%)
untx COS	16±4	(6±1%)	41±10	(3±1%)

[a]Monolayers of COS cells 64-72 after transfection were washed twice with Medium 199, incubated with fluosescein labeled R6 pneumococci (30 min at 37°C), washed and adherent bacteria counted.
[b]Percent adherent pneumococci relative to WT PAF-R.
[c]WT PAF-R, wild type human PAF receptor; PAF-R dCHO, non-glycosylated human PAF receptor mutant[14]; untx COS, untransfected COS.

face expression of the non-glycosylated mutant, suggesting additional, protein determinants on the PAF receptor are recognized by pneumococcus as well.

VI. INTERNALIZATION OF PNEUMOCOCCI FOLLOWING INTERACTION WITH PAF RECEPTORS

The PAF receptor is known to be rapidly internalized after interaction with ligand[13]. To determine if attachment to the PAF receptor may afford the bacteria a route of migration into or across endothelial cells, pneumococci were allowed to adhere to cytokine and thrombin activated EC or PAF receptor transfected COS cells and the number of internalized bacteria was detrmined by protection from killing by exogenous gentamycin. Of the $5x10^5$ bacteria initially adherent to unstimulated monolayers, 750 were recovered from lysates of resting cells treated with gentamycin for 2 h, indicating that ~0.1% of adherent bacteria entered the cells. Activation of endothelial cell monolayers with either thrombin or TNFα resulted in an increase in both pneumococcal adherence (~$1x10^6$) to the monolayer and in the entry of adherent bacteria to the cells: 2.3 and 3.7%, respectively, of the initially adherent population survived gentamycin. Pneumococcal internalization by activated endothelial cells was significantly reduced in the presence of the PAF receptor antagonist L659,989. Similar results for pneumococcal internalization were also oberved in monolayers transfected with PAF receptors.

VII. PAF RECEPTOR SIGNAL TRANSDUCTION BY PNEUMOCOCCI

Ligation of the PAF receptor by PAF results in activation of phospholipase C[15]. To determine the effect of pneumococcal ligation of the PAF receptor on signal transduction, generation of phosphatidyl inositols was determined in two settings: a) following adherence of pneumococci to PAF receptor transfected cells, and b) the ability of pneumococci to alter the PAF induced signal generated in transfected cells. Pneumococcal adherence to the PAF receptor failed to stimulate signal transduction as measured by activation of phosphatidylinositol-specific phospholipase C. The pneumococci also did not effect the PAF-induced activation of phospholipase C. The non-glycosylated PAF receptor stimulated as much phospholipase C as the native receptor [14].

VIII. SUMMARY

Streptococus pneumoniae can produce asymptomatic colonization or aggressive sepsis. We sought to differentiate the molecular mechanisms of these disparate courses. Cytokine or thrombin activation of human vascular endothelial cells and type II pneumocytes enhanced pneumococcal adherence relative to resting cells. Adherence and subsequent invasion was dramatically reduced by PAF receptor antagonists. Cells transfected with the PAF receptor gained the ability to support pneumococcal adherence. PAF or PAF receptor antagonists inhibited attachment and invasion. Adherence involved phosphorylcholine on the pneumococcal teichoic acid. Virulent pneumococci target the PAF receptor on activated human cells, a necessary step to facilitate subsequent invasion.

REFERENCES

1. Burman, LA, Norrby, R, Trollfors, B. (1985) Invasive pneumococcal infections: incidence, predisposing factors, and prognosis. Rev. Infect. Dis. 7, 133–142.
2. Austrian, R. (1986) Some aspects of the pneumococcal carrier state. J. Antimicrob. Chemother. 18, 35–45.
3. Vial, WC, Toews, GB, Pierce, AK. (1984) Early pulmonary granulocyte recruitment in response to *Streptococcus pneumoniae*. Am. Rev. Respir. Dis. 129, 87–91.
4. Hamburger, M, Robertson, O. (1940) Studies of the pathogenesis of experimental pneumococcus pneumoniae in the dog. J. Exp. Med. 72, 261–274.
5. Plotowski, M-C, Puchelle, E, Beck, G, Jacquot, J, Hannoun, C. (1986) Adherence of type I *Streptococcus pneumoniae* to tracheal epithelium of mice infected with influenza A/PR8 virus. Am. Rev. Respir. Dis. 134, 1040–1044.
6. Hakansson, A, Kidd, A, Wadell, G, Sabharwal, H, Svanborg, C. (1994) Adenovirus infection enhances in vitro adhrence of *Streptococus pneumoniae*. Infect. Immun. 62, 2707–2714.
7. Cundell, DR, Gerard, NP, Gerard, C, Idanpaan-Heikkila, I, Tuomanen, EI. (1995) *Streptococcus pneumoniae* anchors to activated human cells by the receptor for platelet-activating factor. Nature 377, 435–438.
8. Cundell, DR, Tuomanen, EI. (1994) Receptor specificity of adherence of *Streptococcus pneumoniae* to type II pneumocytes and vascular endothelial cells *in vitro*. Microb. Pathol. 17, 361–374.
9. Cabellos, C, MacIntyre, DE, Forrest, M, Burroughs, M, Prasad, S, Tuomanen, EI. (1992) Differing roles for platelet-activating factor during inflammation of the lung and subarachnoid space. The spacial case of *Streptococcus pneumoniae*. J. Clin. Invest. 90, 612–618.
10. Wissner, A, Schaub, RE, Sum, PE, Kohler, CA, Goldstein, BM. (1986) Analogues of platelet-activating factor: some modifications of the phosphorylcholine moiety. J. Med. Chem. 29, 328–333.
11. Kunz, D, Gerard, NP, Gerard, C. (1992) The human leukocyte platelet-activating factor receptor. J. Biol. Chem. 267, 9101–9106.
12. Shirasaki, H, Nishikawa, M, Adcock, IM, Mak, JC, Sakamoro, T, Shimuzu, T, Barnes, P. (1994) Expression of platelet-activating factor receptor mRNA in human and guinea pig lung. Am. J. Respir. Cell Mol. Biiol. 10, 533–537.
13. Gerard, NP, Gerard, C. (1994) Receptor-dependent internalization of platelet-activating factor. J. Immunol. 152, 793–800.
14. Rodriguez, CG, Cundell, DR, Tuomanen, EI, Kolakowski, LF, Gerard, C, Gerard, NP. (1995) The role of N-glycosylation for functional expression of the human PAF receptor: Glycosylation is required for efficient membrane trafficking. J. Biol. Chem. (in press).
15. Amatruda, TT, Gerard, NP, Gerard, C, Simon, MI. (1993) Specific interactions of chemoattractant receptors with G-proteins. J. Biol. Chem. 268, 10139–10144.

17

UP-REGULATION OF THE INTRACELLULAR Ca^{2+} SIGNALING AND mRNA EXPRESSION OF PLATELET-ACTIVATING FACTOR RECEPTOR BY ESTRADIOL IN HUMAN UTERINE ENDOMETRIAL CELLS

Sayuri Sato,[1] Kazuhiko Kume,[1] Tomoko Takan,[1] Hiroyuki Mutoh,[1]
Yuji Taketani,[2] and Takao Shimizu [1]

[1]Department of Biochemistry
[2]Department of Obstetrics and Gynecology
Faculty of Medicine
The University of Tokyo
Bunkyo, Tokyo, 113, Japan

SUMMARY

Platelet-activating factor (PAF, 1-O-alkyl-2-acetyl-sn-glycero-3-phosphocholine), a potent chemical mediator in inflammation, plays a role in reproduction. Using primary culture of human uterine endometrial cells, we investigated the effect of sex steroid hormones on the PAF-induced signal and its receptor mRNA expression. After a 24 hr treatment with estradiol, PAF increased the intracellular calcium ion ($[Ca^{2+}]_i$) in the glandular cells, but not in the stromal cells. This response was not observed in the non-treated cells, and was blocked by a PAF antagonist, WEB2086. Two types of mRNA (transcript 1 and transcript 2) occurred for PAF receptor by alternative splicing, which are under control of two distinct promoters. Using RT-PCR analysis, it was shown that both transcripts existed in endometrial cells and that estradiol alone or a combination of estradiol and progesterone induced the accumulation of transcript 2, the promoter of which responded to estrogen in our previous studies. The regulation of PAF receptor by sex steroid hormones in human uterine endometrial cells suggests that PAF is involved in the physiological process of reproduction.

1. INTRODUCTION

PAF is known to possess diverse and potent biological activities on various cells and tissues. In reproduction, PAF is reported to play roles including; ovulation (1), sperm motility (2), decidual cell reaction (3) and implantation (4). PAF is detected in the uterus and its level is hormonally controlled (5). The presence of specific binding sites for PAF in rabbit uterus was demonstrated by autoradiography after the binding of [^{3}H] PAF, which peaked on the day 6 of pregnancy (6). PAF receptor mRNA are expressed in human endometrial carcinoma cell line (7). After cloning of PAF receptor cDNA from guinea pig lung and human leukocytes (8,9), we investigated the structure of human PAF receptor gene and its transcriptional control. Two types of transcripts (transcript 1 and transcript 2) of human PAF receptor by alternative splicing were identified, consisting of two different 5′-non-coding exons (exon 1 and exon 2) and a common exon (exon 3) coding for the entire open reading frame (ORF). They are regulated by two distinct promoters in tissue-specific manner (10). Transcript 2 is expressed in limited organs including heart, lung, kidney and spleen, but not in leukocytes and brain, while transcript 1 is expressed ubiquitously among various tissues. The promoter regions of the two transcripts were analyzed for their transcriptional activation using heterozygous reporter gene assay, and the promoter for transcript 2 was found to be activated by estrogen (11). In this study, we used a primary culture of human endometrial cells to examine the regulation of PAF receptor signal and the expression of PAF receptor mRNA by sex steroid hormones, in order to know physiological relevance of PAF in reproduction.

2. MATERIALS AND METHODS

2.1 Primary Cell Culture and Hormone Treatment

Human endometrial tissues were obtained from patients undergoing hysterectomy due to histologically benign diseases and used as the source of cells. All patients had a history of regular menstrual cycle and were not given sex steroid hormones prior to the operation, and their informed consents were obtained. The specimens used in this study were in the early to mid-secretory phases. A mixture of endometrial stromal fibroblasts and epithelial cells were prepared as described previously (12). Briefly, the tissue samples collected in medium 199 (GIBCO) were minced and digested with 0.1% collagenase (Sigma). The dispersed cells were washed three times and then suspended in medium 199 supplemented with 10% fetal bovine serum (GIBCO) treated with dextran-coated charcoal, 5 μg/ml insulin (Sigma), and 1% antibiotic-antimycotic mixture (GIBCO). The cells were prepared on serum-coating microcoverglasses in 6-well culture plate for the fluorescence studies of $[Ca^{2+}]_i$ or in 10 cm culture dishes for RT-PCR studies. After the cells reached 80% confluency, the cells were treated with progesterone (10 nM), estradiol (1 nM), tamoxifen (1 μM), or their combinations for 24 hrs.

2.2 $[Ca^{2+}]_i$ Measurements

Endometrial cells grown on microcoverglasses with or without hormone treatment were loaded with 3 μM fura-2/AM (Dojindo, Kumamoto, Japan) for 60 min at 37 °C. After loading, the cells were superfused with Hepes-Tyrode's buffer , and PAF (100 nM), WEB2086 (1 μM) and ATP (100 μM) were applied. $[Ca^{2+}]_i$ was measured as the ratio of the fluorescence intensities with excitation at 340 nm and 380 nm in a fura-2/AM loaded single cell, using a fluorospectrometer model CAM230 (Jasco, Tokyo, Japan) attached to an inverted microscope (Nikon, Tokyo) as described previously (13).

2.3 Detection of Expressed PAF Receptor mRNA by RT-PCR

Total RNA were extracted from cultured endometrial cells by the acid-phenol method or ISOGEN (Nippongene, Tokyo), and then reverse-transcribed with Moloney murine leukemia virus (M-MuLV) reverse transcriptase. PCR primers were designed for two different promoters, primer L1 (5′-GGCTGGGGCCAGGACCCAGA-3′, nucleotide 104 to -85 of exon 1) or primer H1 (5′-CCTGAGCTCCCCGAGAAGTCA-3′, nucleotide -165 to -145 of exon 2) and one primer from exon 3 common to the two transcripts, primer C1 (5′-CCCGAGCACAAAGATGATGC-3′, nucleotide 87 to 68 of exon 3) as described previously (10). As control, primers A1 (5′-TTTGAGACCTTCAACACCCC-3′) and A2 (5′-CGCTCATTGCCAATGGTGAT-3′) from human ß-actin gene were used. PCR was performed for 32 thermal cycles under the following conditions : denaturation at 95 °C for 30 s, annealing at 55 °C for 30 s, elongation at 72 °C for 30 s. Amplification products were separated in agarose gels and detected by ethidium bromide staining.

3. RESULTS

3.1 PAF-Induced $[Ca^{2+}]_i$ Increase in Cultured Endometrial Glandular Cells Treated by Estradiol, but Not in Stromal Cells

As shown in Fig 1, PAF elicited an increase in $[Ca^{2+}]_i$ when the cells were pretreated by estradiol or estradiol and progesterone. The responses for PAF were very weak in the cells without hormone treatment or with progesterone treatment alone. The extent of $[Ca^{2+}]_i$ in-

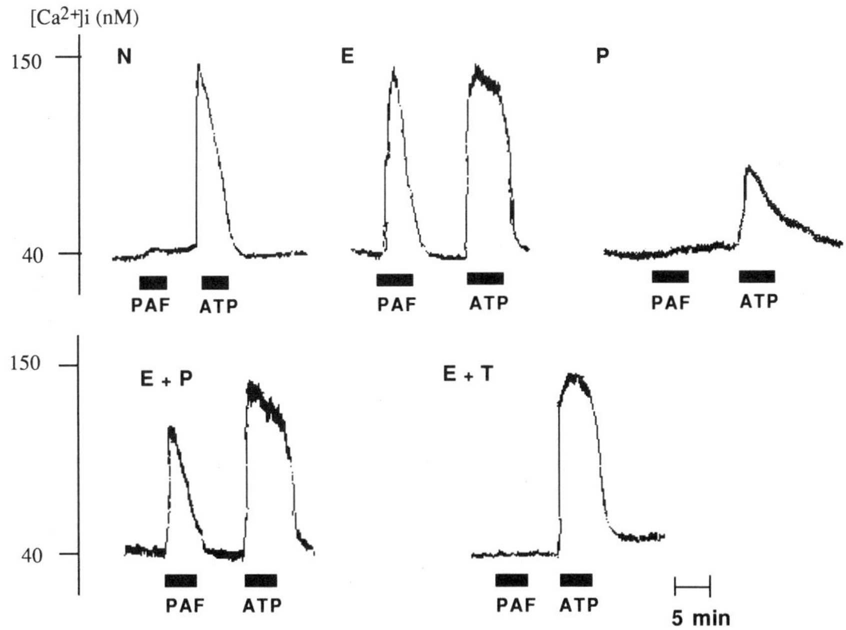

Figure 1. Effect of sex steroid hormones on PAF-elicited $[Ca^{2+}]_i$ Human endometrial cells grown on a coverglass were incubated with none (N), 1 nM estradiol (E), 10 nM progesterone (P), 1 nM estradiol and 10 nM progesterone (E+P) or 1 nM estradiol and 1 µM tamoxifen (E+T) for 24 hrs. Then the cells were loaded with 3 µM fura-2/AM, and challenged with PAF (100 nM) and ATP (100 µM), as indicated by horizontal bars. $[Ca^{2+}]_i$ of one glandular cell was measured as described in Materials and Methods.

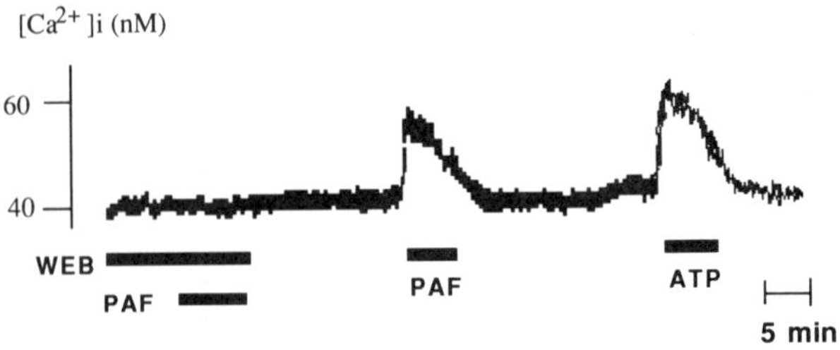

Figure 2. Inhibition of PAF-elicited $[Ca^{2+}]_i$ response by WEB2086. Cells were incubated with 1 nM estradiol as in Fig. 1, and were challenged first with WEB2086 (1 μM) and PAF (100 nM). After washing, cells were challenged with PAF alone and ATP, as indicated.

crease by PAF was comparable to that by 100 μM ATP used for a positive control, when the cells were treated by estradiol. The inducing effect of estradiol was inhibited by an estrogen antagonist, tamoxifen. PAF-elicited $[Ca^{2+}]_i$ increase was also blocked by a PAF receptor antagonist, WEB2086 (Fig.2). These responses to PAF were observed only in glandular cells, but not in stromal cells. They were readily distinguished morphologically, and further ensured by immunohistochemical staining with an antibody against cytokeratin No.19 (DAKO, Denmark), with which only glandular cells were positive (data not shown).

3.2 Expression of Both Forms of PAF Receptor Transcripts in Cultured Endometrial Cells

By RT-PCR analysis shown in Fig 3, the expression of transcript 1 (primer-pairs L1/C1; 191 bp fragment, lower band) was ubiquitously observed and was not altered by hormone treatment in all cases. In contrast, transcript 2 (primer-pairs H1/C1 ; 252 bp fragment, upper band) was accumulated after treatment by estradiol in case 1 and case 2. However, in case 3, expression of transcript 2 was detected only after treatment with both estradiol and progesterone.

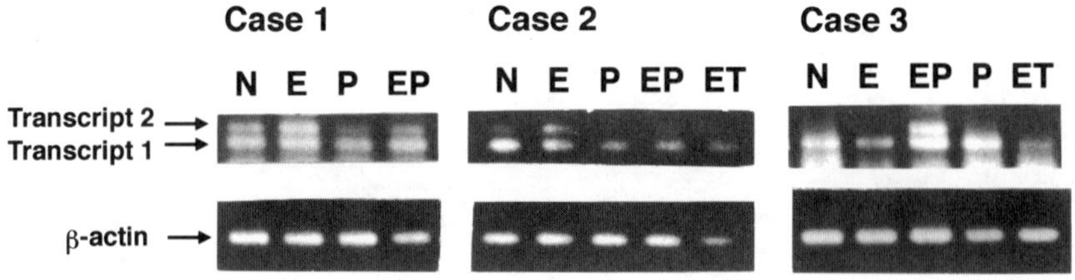

Figure 3. RT-PCR analysis of the expression of two forms of transcripts. Cells grown in 10 cm dish were treated with the indicated hormones and drug as in Fig 1. RT-PCR were performed as described in Materials and Methods, using three primers L1, H1 and C1 (upper gel) or β-actin primer set (lower gel), each for the detection of transcript 1 (191-bp fragment), transcript 2 (252-bp fragment) of human PAF receptor and human β-actin (391-bp fragment), respectively. Amplification products were detected by separation in 2 % agarose gels .

4. DISCUSSION

The presence of PAF (5) and PAF binding sites (6) in uterus were previously reported. The expression of PAF receptor mRNA and PAF-elicited $[Ca^{2+}]_i$ increase were recently described in human endometrial adenocarcinoma cell lines (7). However, in our preliminary experiments using similar cell lines, HEC1, HEC1A and HEC1B, we could not detect any effects of the sex steroid hormones on PAF receptor mRNA expression by RT-PCR. Since it often happens to a transformed cell line to lose hormone sensitivities, we used primary culture system. Primary culture of human endometrial cells is a good system for investigation of physiological functions of uterus. For example, EGF receptor was identified in these cells and found to be controlled by sex hormones (14). In this study, we investigated the existence and regulation of PAF-elicited $[Ca^{2+}]_i$ increase and PAF receptor mRNA in cultured human uterine endometrial cells by ligand-induced Ca response and RT-PCR. PAF-elicited $[Ca^{2+}]_i$ increase was found only in glandular cells, but not in stromal cells, which was augmented by treatment with estradiol, while the response was blocked by PAF antagonist. These results indicate that the response is regulated by estradiol and is mediated by the PAF receptor. By RT-PCR, two different forms of PAF receptor mRNA, namely transcript 1 and transcript 2 were detected. The expression patterns of two transcripts after hormone treatment differed depending on cases. In two cases, estradiol alone induced transcript 2, while progesterone together with estradiol were required in one case. There is no definite explanation for the discrepancy in these cases, but one possibility is that they differed in hormonal conditions due to menstrual phases before operation. The inhibitory effects of tamoxifen to the expression of transcript 2 confers that principally estradiol activates its expression. In the promoter region of transcript 2, an atypical estrogen responsive element (ERE) exists. The two half-sites of the palindromic ERE, which typically separated by 3 bp, are apart by 153 bp (from-257 bp to -93 bp). However, it responded to estrogen in a reporter gene assay (10). Our results suggest that in more physiological conditions, transcript 2 is up-regulated by estrogen. Since serum concentrations of estradiol and progesterone become elevated in the early phase of pregnancy, and PAF is reported to play roles in implantation (15), it is conceivable that transcript 2 plays the major role in this period. The fact that sex steroid hormones regulate PAF receptor-mediated signal and expression of PAF receptor mRNA in uterine endometrial cells suggests that PAF is involved in the physiological process of reproduction.

ACKNOWLEDGMENTS

We thank Drs. A. Miyauchi, Y. Ohsuga and T. Yokomizo for helpful advice. WEB 2086 is a generous gift from Boehringer Ingelheim (Ingelheim, Germany). This work was supported by a Grant-in-Aid from the Ministry of Education, Science and Culture of Japan, and by grants from Naito Foundation and Yamanouchi Foundation.

REFERENCES

1. Abisogun, A.O., Braquet, P. and Tsafriri, A. (1989) The involvement of platelet activating factor in ovulation. Science. 243, 381–383.
2. Ricker, D.D., Minhas, B.S., Kumar, R., Robertson, J.L. and Dodson, M.G. (1989) The effects of platelet activating factor on the motility of human spermatozoa. Fertil. Steril. 52, 655–658.

3. Acker, G., Braquet, P. and Mencia-Huerta, J.M. (1989) Role of platelet-activating factor (PAF) in the inhibition of the decidual reaction in the rat. J. Reprod. Fertil. 85, 623–629.
4. Spinks, N.R., Ryan, J.P. and O'Neill, C. (1990) Antagonists of embryo-derived platelet-activating factor act by inhibiting the ability of the mouse embryo to implant. J. Reprod. Fertil. 88, 241–248.
5. Alecozay, A.A., Casslen, B.G., Riehl, R.M., Deleon, F.D., Harper, M.J., Silva, M., Nouchi, T.A. and Hanahan, D.J. (1989) Platelet-activating factor in human luteal phase endometrium. Biol. Reprod. 41, 578–586.
6. Kudolo, G.B., Kasamo, M. and Harper, M.J.K. (1991) Autoradiographic localization of platelet-activating factor (PAF) binding sites in the rabbit endometrium during the peri-implantation period. Cell and Tissue Res. 265, 231–241.
7. Ahmed, A., Sage, S.O., Plevin, R., Shoaibi, M.A., Shakey, A.M. and Smith, S.K. (1994) Functional platelet-activating factor receptors linked to inositol lipid hydrolysis, calcium mobilization and tyrosine kinase activity in the human endometrial HEC-1B cell line. J. Repro. Fertil. 101,459–466.
8. Honda, Z., Nakamura, M., Miki, I., Minami, M., Watanabe, T., Seyama, Y., Okado, H., Toh, H., Ito, K., Miyamoto, T. and Shimizu, T. (1991) Cloning by functional expression of platelet-activating factor receptor from guinea-pig lung. Nature. 349, 342–346.
9. Nakamura, M., Honda, Z., Izumi, T., Sakanaka, C., Mutoh, H., Minami, M., Bito, H., Seyama, Y., Matsumoto,T., Noma, M. and Shimizu, T. (1991) Molecular cloning and expression of platelet-activating factor receptor from human leukocytes. J. Biol. Chem. 266, 20400–20405.
10. Mutoh, H., Bito, H., Minami, M., Nakamura, M., Honda, Z., Izumi, T., Nakata, R., Kurachi, Y., Terano, A. and Shimizu, T. (1993) Two different promoters direct expression of two distinct forms of mRNAs of human platelet-activating factor receptor. FEBS Lett. 322, 129–134.
11. Mutoh, H., Kume, K., Sato, S., Kato, S. and Shimizu, T. (1994) Positive and negative regulations of human platelet-activating factor tanscript 2 (tissue-type) by estrogen and TGF-β1. Biochem. Biophys. Res. Commun. 205, 1130–1136.
12. Ishihara, S., Taketani, Y. and Mizuno, M. (1988) Stimulatory action of progesterone on the synthesis of glycogen in primary cell culture of human endometrium. Asia-Oceania J. Obstet. Gynaecol. 14, 117–122.
13. Yamazawa, T., Iino, M. and Endo, M. (1992) Presence of functionally different compartments of the Ca^{2+} store in single intestinal smooth muscle cells. FEBS Lett. 301, 181–184.
14. Taketani, Y. and Mizuno, M. (1991) Evidence for direct regulation of epidermal growth factor receptors by steroid hormones in human endometrial cells. Hum. Reprod. 6, 1365–1369.
15. O'Neill, C., Gidley-Baird, A. A., Pike, I.L. and Saunders, D.M. (1987) Use of a bioassay for embryo-derived platelet-activating factor as a means of assaying quality and pregnancy potential of human embryos. Fertil. Steril. 47, 969–975.

FUNCTIONAL ROLE OF PLATELET-ACTIVATING FACTOR RECEPTOR IN SECRETORY RESPONSE IN ADRENAL CHROMAFFIN CELLS

T. Dohi, K. Morita, Y. Imai, and S. Kitayama

Department of Pharmacology
Hiroshima University School of Dentistry
Kasumi 1-2-3, Minami-ku, Hiroshima 734, Japan

1. INTRODUCTION

Platelet-Activating Factor(PAF), a potent bioactive phospholipid mediator in cell-cell communications, is involved in many physiological and also in several pathological processes. The notion that PAF might play an important role in the central nervous system(CNS), particularly in brain ischemia and traumatic injury-induced neuronal damage, was inferred from evidence that a variety of PAF receptor antagonists modified the pathological consequences of brain injury. PAF biosynthesis and degradation were detected in brain as well as its accumulation during ischemia and convulsions. In recent studies, specific biding sites[1] and PAF receptor mRNA has been found in CNS[2]. It has also been shown that PAF increases in intracellular free Ca^{2+} concentration($[Ca^{2+}]i$) in PC12 cells[3,4], NG108-15 cells[5], NCB-20[6] cells and in rat hippocampal cells[2], and stimulates the release of neurotransmitter from PC12 cells[3,4], acetylcholine(ACh) release at neuromuscular junction[7] and glutamate release from neuronal cells[8]. On the other hand, PAF inhibits the release of ACh from rat cortex and hippocampus[9]. PAF also participates in secretory functions of exocrine cells[10]. We have previously shown that PAF potentiated the stimulation-evoked $[Ca^{2+}]i$ rise and catecholamine(CA) release in adrenal chromaffin cells[11]. Thus, PAF may be involved in stimulation-secretion coupling.

2. MECHANISMS OF THE POTENTIATION BY PAF OF STIMULATION-EVOKED SECRETORY RESPONSE

PAF potentiated ACh- and excess K^+-evoked $[Ca^{2+}]i$ rise and CA release dose-dependently from a concentration as low as 1 nM in perfused dog adrenal glands and isolated cultured bovine adrenal chromaffin cells. In contrast to the effect of PAF in PC 12 cells in which PAF has been reported to cause [Ca^{2+}]i rise and neurotransmitter release, PAF by itself had no effect on [Ca^{2+}]i and CA release in adrenal chromaffin cells.

PAF receptors are functionally coupled to several transmembrane signalling mechanisms, some of which involve G proteins sensitive to pertussis toxin. Stimulation of phosphoinositide turnover and the resulting mobilization of Ca^{2+} by inositol trisphosphate from intracellular stores is one of documented mechanism of PAF action in a number of cells. However, PAF-induced potentiation was not blocked by phospholipase C inhibitors and pertussis toxin. Enhancement by PAF of ACh-evoked [Ca^{2+}]i rise and CA release seems neither due to modulation of cholinergic receptor function nor to affect some processes required for exocytosis subsequent to Ca^{2+} rise, because PAF potentiated the excess K^+-depolarization-evoked cellular responses and had no effect on the secretion due to intracellular Ca^{2+} mobilization by caffeine. Thus the potentiation of [Ca^{2+}]i transient by PAF may result in enhanced CA release. Bay K 8644, an opener of voltage-sensitive Ca^{2+} channel, induced CA release and [Ca^{2+}]i rise. PAF did not affect these responses to Bay K 8644. Thus it is not likely that PAF directly affect voltage-sensitive Ca^{2+} channels. This is in agreement with the reported results in PC12 cells in which PAF-induced [Ca^{2+}]i transient was not specifically blocked by Ca antagonists. The involvement of voltage-dependent Na^+ channel in PAF-potentiation has been suggested by the evidence that PAF failed to potentiate ACh-induced [Ca^{2+}]i rise in the presence of tetrodotoxin, a specific inhibitor of voltage-dependent Na^+ channel or in the absence of Na^+ in the medium, and amiloride, an inhibitor of membrane Na^+-Ca^{2+} exchanger, effectively blocked PAF-potentiation.

We have previously shown that cyclic AMP potentiates the stimulation-evoked [Ca^{2+}]i rise and CA release in adrenal chromaffin cells. The effect of cyclic AMP was Na^+-dependent and due to inhibition of Na^+,K^+-ATPase activity as demonstrated in membrane preparations and in intact cell preparations of chromaffin cells[12]. Thus, one of the mechanisms for cyclic AMP potentiation of these cellular events is based on the inhibition of Na^+,K^+-ATPase and the resultant increase of [Na^+]i leading to an acceleration of membrane depolarization, an increase in [Ca^{2+}]i through activation of voltage-dependent Ca^{2+} channels and Na^+-Ca^{2+} exchange mechanism. It has been also reported that PAF activates Na^+-Ca^{2+} exchange pathway in rat ileal plasmalemma[13] and in brain[14] and that PAF inhibits Na^+,K^+-ATPase of red blood cell membranes[15]. From the feature of the potentiating effect of PAF in chromaffin cells, a similar mechanism, as in the case of cyclic AMP, might be involved in PAF-potentiation, if PAF inhibits Na^+,K^+-ATPase in adrenal chromaffin cells. An examination whether PAF affects Na^+,K^+-ATPase activity revealed that PAF inhibits Na^+,K^+-ATPase in purified fragment of bovine adrenal medulla. However, the inhibitory action of PAF on the ATPase observed in membrane preparation seems not to be specific and is not mediated by the PAF receptor, because the concentrations of PAF required for the inhibition were more than micromolar range, the inhibition was dependent on protein concentration of the membrane, lyso-PAF also produced the similar inhibition of the ATPase and the inhibition of PAF and lyso-PAF were not antagonized by PAF-receptor antagonists. ^{86}Rb uptake is an indicator of K^+ transport and therefore provides a useful measurement of activity of Na^+,K^+-ATPase-mediated monovalent cation active transport system. To ascertain the effect of PAF-receptor-stimulation on Na^+,K^+-ATPase activity in intact cells, changes in ^{86}Rb uptake were examined and PAF was shown to inhibit the ouabain-sensitive ^{86}Rb uptake. Therefore, it is suggested that PAF produces inhibitory effects on Na^+,K^+-ATPase in intact chromaffin cells.

Tyrosine kinase, phospholipase A_2 and mitogen-activated protein kinase(MAPK) have been shown to couple with PAF receptors in various cells. Both tyrosine kinase inhibitors, herbimycin A and erbstatin analog(methyl 2,5-dihydroxycinnamate) and phospholipase A_2 inhibitor, p-bromophenacyl bromide effectively blocked PAF-potentiation of $[Ca^{2+}]i$ transient and CA release in adrenal chromaffin cells. MAPK was significantly activated in PAF-treated chromaffin cells. Hydrolysis of membrane phosholipid by phospholipase A_2 could interfere with Na^+,K^+-ATPase activity. Therefore, the evidence may assure to evaluate further the possibility that PAF receptors couple with tyrosine kinase in adrenal chromaffin cells and thus affect Na^+,K^+-ATPase via subsequent activation of, for example, MAPK and phospholipase A_2.

3. ANALYSIS OF PAF RECEPTOR

$[^3H]$WEB 2086 binding to the membranes from adrenal medulla was time-dependent and saturable. The binding was fully displaced by PAF but not by lyso-PAF. Scatchard analysis of the saturation-binding data indicated single binding site for $[^3H]$WEB 2086 with a dissociation constant of 76.8 nM and a binding capacity of 42 fmol/mg protein. The presence of two populations, high and low binding site in the P2 fraction of gerbil brain[16], and low affinity sites in plasma membrane fractions and high and low affinity sites in intracellular membrane fractions of rat cerebral cortex[17] were detected. Membranes of adrenal medulla, corresponding to rat brain[2], appear to have site with a lower affinity and a smaller maximum binding capacity than those reported.

Functional analyses, ligand binding studies all demonstrated the presence of PAF receptors in adrenal chromaffin cells and suggested its important role in secretory functions in the cells. To explore further the PAF receptor expression in adrenal medulla chromaffin cells, PAF receptor gene expression was examined using reverse transcriptase polymerase chain reaction (RT-PCR) technique. Primers were designed and synthesized based on the sequence of PAF receptor cDNA from guinea pig lung[18]. The specific amplification of PAF receptor-specific PCR fragment from dog adrenal mRNA as well as guinea-pig lung mRNA only in the previous first strand cDNA synthesis was detected. These results clearly demonstrate the expression of PAF receptor mRNA in dog adrenal glands, which is homologous to guinea-pig's. The PAF receptor expression in PC12 cells was also examined, when cells were treated with nerve growth factor(NGF) and/or leukemia infibitory factor(LIF) also known as cholinergic neuronal differentiation factor for 12 days. Any amplification of PAF receptor-specific fragment was not detected in control PC12 cells. However, when PC12 cells were treated with NGF, the expression of PAF receptor mRNA was clearly shown in relation to the stage of cell differentiation. Also LIF accelerated PAF receptor mRNA expression induced by NGF. These results suggest that PAF receptor mRNA message increase in accordance with differentiation of PC12 cell to neuron like feature.

4. ROLES OF PAF RECEPTOR

PAF contributes to neuronal differentiation. Recent studies suggest that PAF is a retrograde messenger of long-term potentiation[19,20]. On the other hand, pathological conditions including ischemia or convulsions, increase in $[Ca^{2+}]i$ and enhanced release of neurotransmitter are taken place, and during such pathological conditions, PAF content in brain increases. As a result, PAF may participate in the modulation of neurotransmitter release and thus pathogenesis. It may be of interest to investigate the alteration and regula-

Table 1. Some diseases associated with increased levels of PAF in plasma

Disease	PAF concentation	Ratio(fold)#	Reference
Cirrhosis of liver	1.78 ng/ml	8.9	21
Drug-induced allergic reaction	22.2 pg/ml	2.6	22
Sepsis	700 pg/ml	4.1	23
Breast cancer	1070 fmol/ml*	7.5	24
Disseminated intra-vascular coagulation syndrome	60–200 pg/ml	5–6.6	25
Neonatal necro-tizing enterocolitis	18.1 ng/ml	5.8	26
Ischemic stroke	294 pg/ml	2.1	27
Diabetes mellitus	1070 pg/ml	26.7	28

*PAF+lyso-PAF, #ratio of plasma PAF concentration in disease to that in normal.

tion of PAF-receptor expression in relation to such physiological and pathological events. Adrenal chromaffin cells are usefull model for such study.

Although PAF is normally at very low levels in blood, increased plasma levels of PAF have been detected in some diseases as shown in Table 1. In these studies, the elevated levels of plasma PAF reach several to more than ten-fold of normal subject (corresponding to 100 pM to nM ranges), and this seems to be sufficient enough to modify the secretion of adrenal chromaffin cells. PAF may play an important role in the profound cardiovascular derangements that occur in response to acute systemic anaphylaxis, several shock states, inflammatory responses and immune reactions. Hypotension is the common response to systemic administration of PAF and the effect is accompanied by large increments in the circulatory levels of CA, primarily epinepheine. Although it has been considered that PAF has no direct actions on the nervous system in periphery and PAF-induced increase in circulatory CA level in such state is due to sympathoadreno-reflex, the present results provide the possible involvement of PAF-mediated modification of CA release from adrenal chromaffin cells in periphery in the changes of circulating CA dynamics. It could be possible that the functions of adrenal chromaffin cells receives control by circular PAF, and this modification participates in the developement of complications of cardiovascular and metabolic abnormalities in *in vivo* associated with some diseases.

REFERENCES

1. Bazan, N.G., Squinto, S.P., Braquet, P., Panetta, T. and Marcheselli, V.L. (1991) Lipids 26, 1236–1242.
2. Bito, H., Nakamura, M., Honda, Z., Izumi, T., Iwatsubo, T., Seyama, Y., Ogura, A., Kudo, Y. and Simizu, T. (1992) Neuron 9, 285–294.
3. Bussolino, F., Tessari, F., Turrini, F., Braquet, P., Camussi, G., Prosdocimi, M. and Bosia, A. (1988) Am. J. Physiol. 255, C559-C565.
4. Kornecki, E. and Ehrlich, Y. H. (1988) Science 240, 1792–1794.
5. Yue, T.-Y., Gleason, M.M., Gu, J.-L. Lysko, P.G., Hallenbeck, J. and Feuerstein, G. (1991) J. Pharmacol. Exp. Ther. 257, 374–381.
6. Yue, T.-Y., Gleason, M.M., Hallenbeck, J. and Feuerstein, G. (1991) Neuroscience 41, 177–185.
7. Kornecki, E., Neel, D., Parsons, R. and Ehrlich, Y. H. (1987) J. Neurochem. 48 (Suppl.), S113.
8. Clark, G.D., Happel, L.T., Zorumski, C.F. and Bazan, N.G. (1992) Neuron 9, 1211–1216.
9. Wang, H.-Y., Yue, T.-L, Feuerstein, G. and Friedman, E. (1994) J. Neurochem. 63, 1720–1725.
10. Dohi, T., Morita, K., Kitayama, S. and Tsujimoto, A. (1991) Biochem. J. 276, 175–182.
11. Morita, K., Suemitsu, T., Uchiyama, Y., Miyasako, T. and Dohi, T. (1995) J. Lipid Mediators Cell Signalling 11, 219–230.
12. Morita, K., Dohi, T., Minami, N., Kitayama S. and Tsujimoto, A. (1991) Neurochem. Int. 19, 81–85.
13. Kester, M., Kumar, R. and Hanahan, D.J. (1986) Biochim. Biophys. Acta 888, 306–315.

14. Kumar, R., Harvey, S.A.K., Kester, M., Hanahan, D.J. and Olson, M.S. (1988) Biochim. Biophys. Acta 963, 375–383.
15. Hollister, A.S. and Kochhar, S. (1989) Clin. Res. 37, 395A.
16. Domingo, M.T., Spinnewyn, B., Chabrier, P.E. and Brauquet, P. (1988) Biochem. Biophys. Res. Commun. 151, 730–736.
17. Marcheselli, V.L., Rossowska, M.J., Domingo, M.-T., Braquet, P. and Bazan, N.G. (1990) J. Biol. Chem. 265, 9140–9145.
18. Honda, Z., Nakamura, M., Miki, I., Minami, M., Watanabe, T., Seyama, Y., Okado, H., Toh, H., Ito, K., Miyamoto, T. and Shimizu, T. (1991) Nature 349, 342–346.
19. Wieraszko, A., Li, G., Kornecki, E., Hogan, M.V. and Ehrlich, Y.H. (1993) Neuron 10, 553–557.
20. Kato, K., Clark, G.D., Bagan, N.G. and Zorumski, C.F. (1994) Nature 367, 175–179.
21. Caramelo, C., Fernandez-Gallardo, S., Santos, J.C., Inarrea, P., Sanchez-Crespo, M., Lopez-Novoa, J.M. and Hernando, L. (1987) Eur. J. Clin. Invest. 17, 7–11.
22. Lazanas, M., Demopoulos, C.A. Tourinas, S., Koussissis, S., Labrakis-Lazanas, K. and Tsarouhas, X. (1988) Arch. Dermatol. Res. 280, 124–126.
23. Diez, H., Nieto, M.L., Fernandez-Gallardo, S., Gijon, M.A. and Sanchez-Crespo, M. (1989) J. Clin. Invest. 83, 1733–1740.
24. Nigam, S., Muller, S. and Benedetto, C. (1989) J. Lipid Mediators 1, 323–328.
25. Sakaguchi, K., Masugi, F., Chen, Y.H., Inoue, M., Ogihara, T., Yamada, K. and Yamatsu, I. (1989) J. Lipid Mediators 1, 171–173.
26. Caplan, M.S., Sun, X.-M., Hsueh, W. and Hageman, J.R. (1990) J. Pediatr. 116, 960–964.
27. Satoh, K., Imaizumi, T., Yoshida, H., Hiramoto, M. and Takamatsu, S. (1992) Acta Neurol. Scand. 85, 122–127.
28. Nathan, N., Denizot, Y., Huc, M.C., Claverie, C., Laubie, B., Benveniste, J. and Arnoux, B. (1992) Diabete Metab. 18, 59–62.

19

FUNCTIONAL AND STRUCTURAL FEATURES OF PLASMA PLATELET-ACTIVATING FACTOR ACETYLHYDROLASE

Larry W. Tjoelker,[1] Chris Eberhardt,[1] Cheryl Wilder,[1] Greg Dietsch,[1] Hai Le Trong,[1] Lawrence S. Cousens,[1] Guy A. Zimmerman,[2] Thomas M. McIntyre,[2] Diana M. Stafforini,[2] Stephen M. Prescott,[2] and Patrick W. Gray[1]

[1]ICOS Corporation
Bothell, Washington 98021
[2]Program in Human Molecular Biology and Genetics
University of Utah
Salt Lake City, Utah 84112

1. INTRODUCTION

Platelet-activating factor (PAF) is a potent pro-inflammatory phospholipid that exerts its effects by binding to a receptor on target cells such as monocytes, polymorphonuclear leukocytes, platelets, and smooth muscle cells[1]. Other phospholipids that have undergone oxidative fragmentation are structurally similar to PAF, bind to its receptor, and mimic its biological properties. Although PAF synthesis is tightly regulated, the fragmented phospholipids arise via unregulated chemical oxidation.

The biological actions of PAF and oxidatively fragmented phospholipids are abolished by hydrolysis of the short acyl chain at the *sn*-2 position, a reaction catalyzed by PAF acetylhydrolase[2]. There are at least two forms of this enzyme—one intracellular and another that circulates in plasma. Although they catalyze the same reaction and have similar substrate specificity, they are distinct proteins[3]. An intracellular PAF acetylhydrolase was recently cloned from bovine brain[4]. We have focused on the plasma form of PAF acetylhydrolase, which has several interesting biochemical properties: (i) It circulates in plasma bound to high and low density lipoproteins. It can exchange between these particles in a pH-dependent manner, but requires detergent to be solubilized[5]. (ii) Although PAF acetylhydrolase is a phospholipase A_2 (PLA_2), it is almost totally specific for substrates with a short acyl chain at the *sn*-2 position. This is unusual for PLA_2s, which usually are more promiscuous in their substrate utilization. (iii) Also unlike other secreted PLA_2s, plasma PAF acetylhydrolase activity does not require calcium. (iv) The activity of the enzyme is inhibited by diisopropyl fluorophosphate (DFP), suggesting its catalytic mechanism involves an active site serine.

2. FEATURES OF RECOMBINANT PAF ACETYLHYDROLASE

Human plasma PAF acetylhydrolase was purified from detergent-solubilized low density lipoprotein particles using three chromatography steps: DEAE Sepharose, Cibacron Blue Sepharose, and copper-chelating Sepharose. A final preparative polyacrylamide gel electrophoresis purification step yielded a band that comigrated with PAF acetylhydrolase activity. This band was excised and subjected to peptide sequencing. The derived sequence was unique and was used to design degenerate PCR primers. PCR and subsequent DNA hybridization analyses identified a cDNA in a macrophage library that encoded the sequence identified by peptide sequencing. The cDNA was 1500 base pairs in length and contained an open reading frame predicted to encode a protein of 441 amino acids. Northern blot analysis using this cDNA as a probe revealed the presence of message in thymus, tonsil, brain, placenta, and macrophages. No hybridization to RNA from heart, kidney, or liver was observed[6].

To compare the biochemical features of the protein encoded by the macrophage cDNA with the known properties of plasma PAF acetylhydrolase, we expressed the cDNA in both mammalian cells and *E. coli*. Recombinant enzyme from both sources had PAF acetylhydrolase activity but since *E. coli* produced much higher levels than the mammalian expression system, we purified recombinant enzyme from *E. coli* for further characterization. To determine if the recombinant enzyme utilizes phospholipids with long-chain fatty acids at the *sn*-2 position, we chose 1-palmitoyl-2-arachidonoyl-*sn*-glycero-3-phosphocholine since this is the preferred substrate for PLA_2. We added this phospholipid to the standard enzyme activity assay and found that it did not inhibit PAF hydrolysis even at the highest concentration, which was 5-fold greater than the concentration of PAF. Another important group of substrates are the oxidatively fragmented phospholipids which, like PAF, can be hydrolyzed by PAF acetylhydrolase[2]. Incubation of the fragmented phospholipids with the recombinant enzyme resulted in their rapid degradation. In contrast, the long chain substrate (as above) was not recognized, confirming the strict specificity. Thus, recombinant PAF acetylhydrolase exhibits the same substrate selectivity as the native enzyme.

Plasma PAF acetylhydrolase has several other properties that distinguish it from other phospholipases including calcium-independence and resistance to compounds that modify sulfhydryl groups or disrupt disulfides[5]. However, it is sensitive to DFP, indicating that a serine comprises part of the active site. The recombinant enzyme exhibited identical behavior. Another unusual feature of plasma PAF acetylhydrolase is that it is tightly associated with lipoproteins, which influence its catalytic efficiency. We incubated the recombinant enzyme with human plasma (previously treated with DFP to abolish the endogenous activity) and found that it associated with low and high density lipoproteins in the same manner as the native activity. Moreover, recombinant PAF acetylhydrolase bound to artificial lipid particles; at low concentrations of particle the catalytic rate increased, but at high concentrations it was inhibited. This demonstrated that the recombinant enzyme exhibited the property of surface dilution kinetics, which is characteristic of many phospholipases including the native plasma PAF acetylhydrolase[5]. These data confirm that the cDNA clone obtained from macrophages encodes plasma PAF acetylhydrolase.

Since PAF is an extremely potent pro-inflammatory agent, we tested the effects of PAF acetylhydrolase on its actions *in vitro* and *in vivo*. The effects of PAF are mediated partly by its ability to activate neutrophils; this response includes changes in intracellular calcium, secretion of granule products, and marked morphological alterations. PAF pre-incubated with recombinant PAF acetylhydrolase, however, completely lost its ability to activate neutrophils *in vitro* as indicated by the absence of a shape change in these cells. One of the prominent *in vivo* effects of PAF is a dramatic increase in vascular permeability, which is a central component of several inflammatory states in which PAF has been implicated. We tested whether recombinant PAF acetylhydrolase could suppress this response in two animal models: (i) injection of PAF into a rat's foot pad reproducibly induced edema but systemic pretreatment with recombinant PAF acetylhydrolase reduced the amount of edema by more than 80%; (ii) PAF-induced vascular leakage into the pleural cavity could be reduced by more than 80% in rats pretreated with the recombinant enzyme. These results suggest that plasma PAF acetylhydrolase abolishes the inflammatory effects of PAF on leukocytes and the vasculature.

3. NATURE OF THE ACTIVE SITE OF PLASMA PAF ACETYLHYDROLASE

Comparison of the plasma PAF acetylhydrolase sequence with the Genbank database indicated that it is a unique protein with no conserved elements except for a Gly-Xaa-Ser-Xaa-Gly (GxSxG) motif (at Ser273) that is characteristic of lipases and esterases. This motif is significant because the central serine serves as the active site nucleophile in these enzymes[7]. The presence of this motif in PAF acetylhydrolase is consistent with observations that its activity can be blocked by the active site serine-specific inhibitor, DFP.

Plasma PAF acetylhydrolase is a PLA_2 because it hydrolyzes phospholipids at the *sn*-2 position. However, the enzyme is distinct from the many known secretory PLA_2s in several respects. The secretory PLA_2s are small (119–143 amino acids), have a high disulfide bond content (10 or 14 cysteines) and require calcium for catalysis[8]. Plasma PAF acetylhydrolase, however, is larger, has fewer cysteines (five), and functions in the absence of calcium. The known intracellular PLA_2s are larger proteins than the secreted ones and do not utilize calcium as part of the catalytic site[9]. However, the intracellular localization and the preference of these enzymes for phospholipids with arachidonate or other long chain fatty acids at the *sn*-2 position distinguish them from plasma PAF acetylhydrolase. Thus, the unique functional properties of PAF acetylhydrolase likely reflect a molecular structure that is distinct from that of other known PLA_2s. In contrast, the neutral lipases share the calcium-independence of PAF acetylhydrolase and also contain the GXSXG motif. The three-dimensional structure of several of these lipases from various organisms has been defined as containing the core α/β hydrolase fold structure characteristic of a variety of esterases and hydrolases[6,10]. The active site contains a catalytic triad comprised of the nucleophilic serine (GXSXG), an acidic residue that is usually aspartate, and an invariant histidine. In most of the lipases, the triad lies within a hydrophobic pocket that is exposed upon contact with the substrate via interfacial activation, a property shared by plasma PAF acetylhydrolase and other PLA_2s.

Plasma PAF acetylhydrolase exhibits characteristics of both the PLA_2s and neutral lipases. However, since the enzyme contains the characteristic serine esterase GXSXG motif and is inhibited by DFP, it may be mechanistically more closely related to the neutral lipases than to the classical PLA_2s. To evaluate the possibility that PAF acetylhydrolase invokes a catalytic triad mechanism of catalysis, we used site-directed mutagenesis to identify active site amino acids. Three distinct criteria were used to identify candidate catalytic residues: (i) conservation of the residues across species, (ii) sequence similarities with other lipases or serine esterases, and (iii) location of the candidate residues relative to each other in the linear sequence.

To compare sequences of plasma PAF acetylhydrolase from other species, we used low stringency hybridization with the human cDNA to obtain clones from mouse, bovine, dog, and chicken. The degree of identity between any two species ranged from 50% to 82%. Importantly, the GXSXG motif is conserved in all of the species. As described above, the only homology found between PAF acetylhydrolase and other proteins spans the GXSXG motif (Ser273). Since the active site serine of other lipases and serine esterases is contained within this motif and, since the motif is conserved in the five species we examined, Ser273 is a likely candidate for the active site nucleophile in PAF acetylhydrolase. This was supported by our observation that mutation of Ser273 to alanine completely abolished the enzyme's activity.

Identifying the position of the active site serine provided a landmark from which the positions of the aspartate and histidine components of a catalytic triad might be predicted. As a basis for such predictions, we compared the linear orientation of triad residues in lipases with known structures. These lipases all have the α/β hydrolase core structure with a catalytic triad, the components of which are arranged in the linear order (N- to C-termini) of serine, aspartate, histidine. Thus, since Ser273 appears to be the essential serine catalytic component for PAF acetylhydrolase, the aspartate and histidine would be expected to lie C-terminal to Ser273. Within this region in the human sequence are five histidines, only two of which (His351, His395) are conserved in the other species. In contrast, there are nine conserved aspartate residues. Thus, we focused first on identifying the essential histidine then on examining the aspartates found between that histidine and Ser273.

Both conserved histidines C-terminal to Ser273 were mutated to alanine. The results were clear: mutation of His351 reduced enzyme activity to undetectable levels while mutation of His395 had no impact on the enzyme's function. If PAF acetylhydrolase has the prototypical lipase structure, the active site aspartate would be expected to lie between Ser273 and His351. Within this region are four conserved or semi-conserved aspartates (Asp286 is glutamate in chicken; Asp304 is glutamate in mouse). Each was mutated to alanine and evaluated for any effect on PAF acetylhydrolase activity. Mutation of the two aspartates most distal to Ser273, Asp304 and Asp338, did not adversely affect the activity of the enzyme. In contrast, mutation of Asp296 virtually abolished the enzyme's function and the Asp286 mutation reduced activity by more than 99%. We demonstrated by using two conformation-specific monoclonal antibodies in a sandwich ELISA that changing these two aspartates to alanines changed the conformation of the enzyme. Since this conformational shift rather than a loss of an essential catalytic residue could account for the loss of enzyme activity, we made a more conservative mutation at each of these positions: aspartate to asparagine. These mutations restored the ability of the sandwich ELISA to detect both mutant proteins and also restored activity of the Asp286 mutant to within 60% of wild-type levels. In contrast, the Asp296 mutation remained inactive. These data all support the possibility that Asp296 is the acidic component of the catalytic triad. Interestingly, the distance of 23 amino acids between the active site Ser273 and Asp296 is similar to that of the pancreatic lipase family (24 amino acids).

Cloning of the plasma PAF acetylhydrolase cDNA has paved the way for a variety of functional and structural studies. We have demonstrated that the recombinant enzyme exhibits the biochemical properties of the native plasma enzyme and can block PAF-induced inflammatory events, both *in vitro* and *in vivo*. The identification of essential catalytic amino acids provides a plausible solution to the paradox that plasma PAF acetylhydrolase, while functionally a PLA_2, exhibited biochemical properties and primary sequence features that are more characteristic of the neutral lipases. We have demonstrated that the catalytic triad which defines the active sites of neutral lipases and other esterases is required for activity of the phospholipase, PAF acetylhydrolase.

REFERENCES

1. Venable, M.E., Zimmerman, G.A., McIntyre, T.M. and Prescott, S.M. (1993) *J. Lipid Res.* **34**, 691–702
2. Stremler, K.E., Stafforini, D.M., Prescott, S.M. and McIntyre, T.M. (1991) *J. Biol. Chem.* **266**, 11095–11103
3. Stafforini, D.M., Prescott, S.M., Zimmerman, G.A. and McIntyre, T.M. (1991) *Lipids* **26**, 979–985
4. Hattori, M., Adachi, H., Tsujimoto, M., Arai, H. and Inoue, K. (1994) *J. Biol. Chem.* **269**, 23150–23155
5. Stafforini, D.M., Prescott, S.M. and McIntyre, T.M. (1987) *J. Biol. Chem.* **262**, 4223–4230
6. Tjoelker, L.W., Wilder, C., Eberhardt, C., Stafforini, D.M., Dietsch, G., Schimpf, B., Hooper, S., Trong, H.L., Cousens, L.S., Zimmerman, G.A., Yamada, Y., McIntyre, T.M., Prescott, S.M. and Gray, P.W. (1995) *Nature* **374**,549–553
7. Derewenda, Z.S. (1994) *Adv. Protein Chem.* **45**, 1–52
8. Scott, D.L. and Sigler, P.B. (1994) *Adv. Protein Chem.* **45**, 53–88
9. Dennis, E.A. (1994) *J. Biol. Chem.* **269**, 13057–13060
10. Ollis, D.L., Cheah, E., Cygler, M., Dijkstra, B., Frolow, F., Franken, S.M., Harel, M., Remington, S.J., Silman, I., Schrag, J., Sussman, J.L., Verschueren, K.H.G. and Goldman, A. (1992) *Protein Eng.* **5**, 197–211

20

ACETYLATION OF SPHINGOSINE BY PAF-DEPENDENT TRANSACETYLASE

Ten-ching Lee

Medical Sciences Division
Oak Ridge Associated Universities
P.O. Box 117, Oak Ridge, Tennessee

ABSTRACT

Sphingosine induces many protein kinase C-dependent and -independent effects on biological systems. In parallel, C_2-ceramide used by investigators as an unnatural, cell permeable analog of long-chain acyl-ceramides, possesses biological activities similar with natural ceramides. We have recently characterized a membrane-associated, CoA-independent transacetylase that can transfer the acetate group from PAF to sphingosine and form C_2-ceramide. This enzyme has a strict stereochemical configuration requirement for sphingosine; only the naturally-occurring D-erythro-isomers of sphingosine accepts the acetate from PAF. Also, it has a rigid substrate specificity for sphingolipid-related analogues. Dipalmitoyl-glycerophosphocholine (-GPC) or hexadecylarachidonoyl-GPC can not transfer their long-chain acyl groups directly to sphingosine and sphingosine can not be acetylated by acetyl-CoA:lyso-PAF acetyltransferase. Results obtained from studies on pH optima, subcellular distribution, temperature sensitivities, inhibitors, tissue distributions, and expression of enzyme activities in Xenopus oocytes suggest that PAF:sphingosine transacetylase is similar, but not identical to the PAF:lysophospholipid transacetylase we have previously identified. The transacetylases function to diversify the biological responses of PAF.

1. INTRODUCTION

Our laboratory has previously demonstrated the presence of a unique membrane-associated CoA-independent PAF:lysophospholipid transacetylase in HL-60 cells that transfers the acetate group from PAF to a variety of lysophospholipids including lysoplasmalogen, acyl(alkyl)lyso-sn-glycero-3-phosphoethanolamine(-GPE), acyl-(alkyl, alk-1-enyl)lyso-sn-glycero-3-phosphocholine(-GPC), acyllyso-sn-glycero-3-phosphoserine(-GPS), acyllyso-sn-glycero-3-phosphoinositol(-GPI), alkyllyso-sn-glycero-3-phos-

Platelet-Activating Factor and Related Lipid Mediators 2
edited by Nigam *et al.*, Plenum Press, New York, 1996

Table 1. Opposing biological effects of sphingosine and N-acetylsphingosine (C_2-ceramide)

	Sphingosine	C_2-ceramide
PKC activity[8]	↓	—
Intracellular Ca^{2+} release[9]	↑	—
EGF-evoked increase in $[CA^{2+}]_i$ [10]	↑	↓
Cell growth[11,12]	↑	↓
Phospholipase D[13,14,15]	↑	↓
Phosphatidate phosphohydrolase[15,16]	↓	↑
Cell differentiation[17,18]	↓	↑

phate(-GP), and acyllyso-GP. In addition, cis-9-octadecen-1-ol can also serve as an acceptor whereas alkylglycerol, acylglycerol, or cholesterol are inactive [1]. This transacetylase has no requirement for Ca^{2+}, Mg^{2+}, or CoA and a broad pH optimum (7.0–8.0). Differences in substrate acceptor specificity, sensitivity toward phenylmethylsulfonyl fluoride, and response to temperature suggest that the transacetylase and the CoA-independent transacylase that transfers long-chain acyl moieties (for reviews see references 2 and 3) are two separate enzymes. Furthermore, results obtained from intact cells activated by stimuli suggest that the PAF:lysoplasmalogen transacetylase instead of acetyl-CoA:alk-1-enyllyso-GPE acetyltransferase is responsible for the synthesis of alk-1-enylacetyl-GPE. The calculated amount of alk-1-enylacetyl-GPE produced by the PAF:alk-1-enyllyso-GPE transacetylase is 2.8 μM, when differentiated HL-60 cells ($2x10^7$ cells) were incubated with 10^{-6} M $[^3H]$PAF and 4 μM ionophore A23187 for 15 min. Thus, this transacetylase could play an important physiological role in modifying the biological responses of PAF by generating a variety of its analogs.

Recently, we have characterized a novel and different type of transacetylase that can transfer the acetate from PAF to sphingosine forming N-acetylsphingosine (C_2-ceramide)[4]. C_2-ceramide, used widely by many investigators as an unnatural, cell permeable analog of long-chain acyl ceramides, at micromolar concentrations stimulates the activities of a mi-

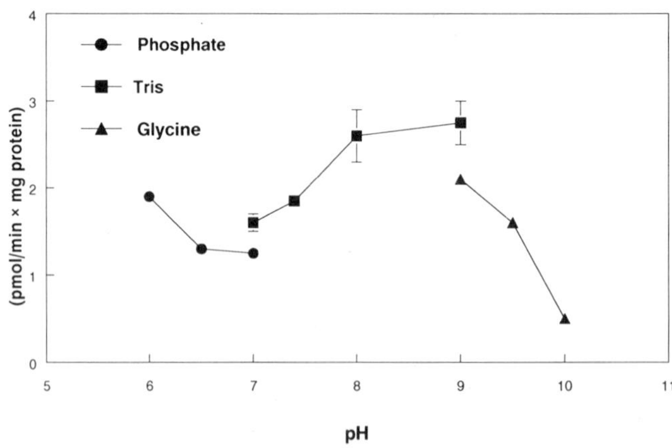

Figure 1. pH dependency of PAF:sphingosine transacetylase in the membrane fraction of undifferentiated HL-60 cells. PAF:sphingosine transacetylase was determined as previously described [4] except the pH of the incubation mixtures was varied as indicated.

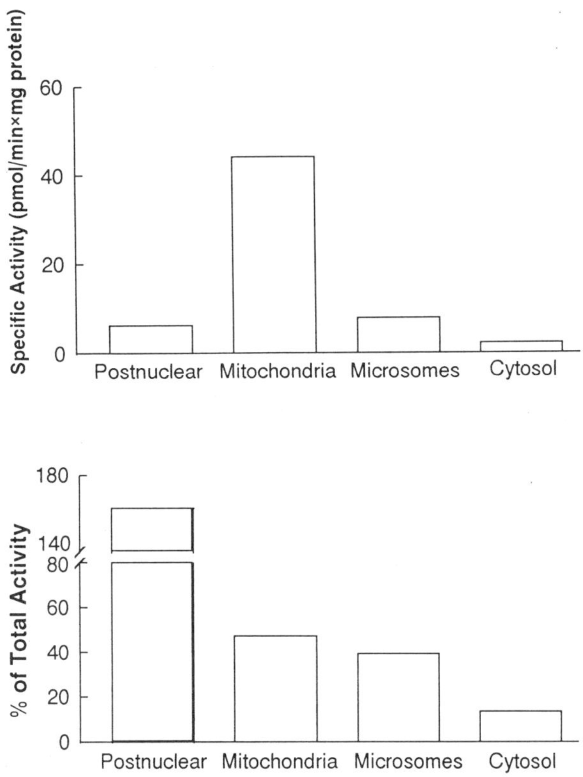

Figure 2. Subcellular distribution of PAF:sphingosine transacetylase in undifferentiated HL-60 cells. Subcellular fractions were isolated as described in the "Methods" section and the PAF:sphingosine transacetylase activities were assayed as before [4]. Values are the average of four determinations for two incubation time periods.

togen-activated protein kinase[5], a cytosolic protein phosphatase[6], and protein phosphatase 2A [7]. In addition, sphingosine and C_2-ceramide have many opposing biological effects (Table 1). Our findings that significant amounts of C_2-ceramide are generated by the PAF:sphingosine transacetylase and physiological concentrations of C_2-ceramide are detected in the undifferentiated and differentiated HL-60 cells emphasize the biological relevance of this enzyme [4]. In this report, we describe the pH optimum, subcellular distribution, and response to heat inactivation of PAF:sphingosine transacetylase, and summarize and compare some of the characteristics of this enzyme to that of PAF:lysophospholipid transacetylase.

Table 2. Comparison of the biological properties between PAF:sphingosine and PAF:lysoplasmalogen transacetylases

Similarities	Dissimilarities
Subcellular distribution	pH optimum
Response to temperature inactivation	Substrate specificity
	Tissue distribution
	Enzyme inhibitor responses
	Time-dependent expression of enzyme activity in Xenopus oocytes

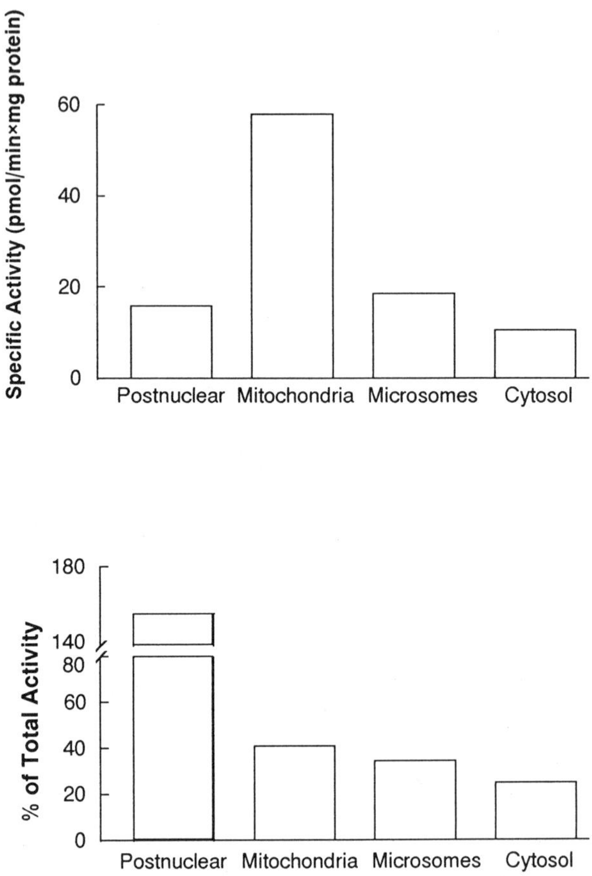

Figure 3. Subcellular distribution of PAF:sphingosine transacetylase in differentiated HL-60 cells. See legend of Figure 2 for details.

2. METHODS

Detailed methods of culturing HL-60 cells and assaying PAF:lysoplasmalogen transacetylase and PAF:sphingosine transacetylase have been described in our earlier reports [1,4]. Subcellular fractions are isolated by first homogenizing the differentiated (dimethylsulfoxide-induced) or undifferentiated HL-60 cells (1 x 10^8 cells) in 0.25 M Sucrose, 10 mM Hepes, 0.5 mM EGTA, 1 mM dithiothreitol, and 1 μg/ml leupeptin (pH 7.0) with either sonication or N_2 cavitation. The postnuclear fraction was prepared from the supernatant of homogenates centrifuged at 500 x g for 10 min. The pellet obtained from 15,000 x g, 10 min centrifugation of the postnuclear fraction was designated as the mitochondrial fraction; the mitochondrial supernatant was further centrifuged at 100,000 x g for 60 min to isolate microsomes and the cytosolic fraction.

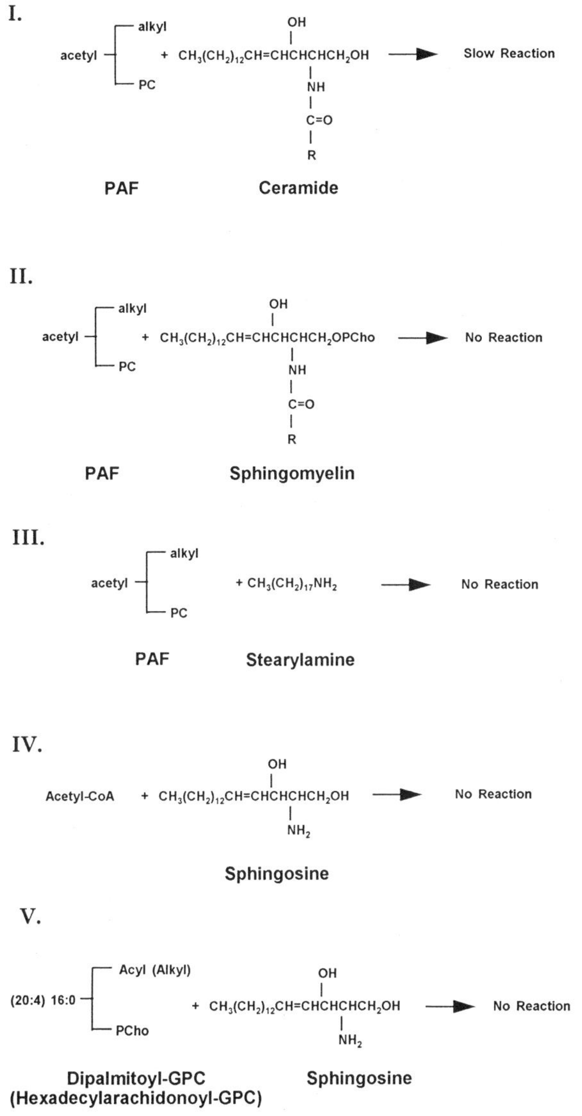

Figure 4. List of reactions in which PAF or sphingosine cannot serve as a donor or an acceptor molecule.

3. RESULTS AND DISCUSSION

The optimal pH for the PAF:sphingosine transacetylase activity occurs between 8.0–9.0 (Fig. 1). A different pH optimum (7.0–8.0) was observed for the PAF:lysoplasmalogen transacetylase [4]. The subcellular distribution of the PAF:sphingosine transacetylases in undifferentiated and differentiated HL-60 cells are shown in Figures 2 and 3, respectively. In both cell types, the PAF:sphingosine transacetylase activities are equally distributed between mitochondrial and microsomal fractions. However, the specific activ-

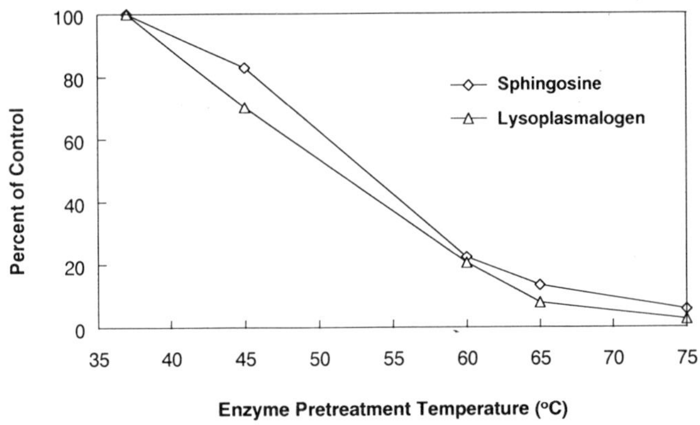

Figure 5. Temperature sensitivities of PAF:sphingosine transacetylase and PAF:lysoplasmalogen transacetylase activities in undifferentiated HL-60 cells. Membrane fractions isolated from undifferentiated HL-60 cells were preincubated at designated temperatures for 15 min before transacetylase activities were assayed according to earlier reports[1,4].

ity of the enzyme is the highest in the mitochondrial fraction. Similar subcellular distribution is also observed for PAF:lysoplasmalogen transacetylase (data not shown). In addition, 8.7% of PAF:sphingosine transacetylase is located in the nuclear fraction and the recovery of the transacetylase activities from mitochondria, microsomes, and cytosols is more than that present in the postnuclear fraction. These results suggest the possible presence of an inhibitor(s) in one of the subcellular fractions.

Vance and co-workers [19, 20] demonstrated that a membrane fraction co-isolated with mitochondria term "mitochondria-associated membrane" contains enzymes of similar or higher specific activities than in the endoplasmic reticulum for the synthesis of phospholipid, triacylglycerols, cholesterol, and cholesterol esters. Therefore, it is likely that transacetylases may distribute in this endoplasmic reticulum like mitochondria-associated membrane fraction.

PAF:sphingosine transacetylase has a narrow substrate specificity using only the naturally-occurring D-erythro-sphingosine as a substrate [4]. Figure 4 outlines several additional reactions we have tested and found that PAF or sphingosine does not participate as a substrate in these reactions. Moreover, PAF:sphingosine transacetylase displays parallel temperature sensitivity response as that of PAF:lysoplasmalogen transacetylase (Fig. 4). Table 2 summarizes and compares various enzymatic features between PAF:sphingosine transacetylase and PAF:lysoplasmalogen transacetylase that are either given in this report or elsewhere [1,4]. It appears that two closely related isoforms of transacetylases catalyze the transfer of acetate to sphingosine and lysophospholipids that generate an array of versatile lipid mediators.

ACKNOWLEDGMENT

Supported by DOE (DE-AC05-760R00033) and NIH (HL27109-14).

REFERENCES

1. Lee, T-c., Uemura, Y., and Snyder, F. (1992) J. Biol. Chem. 267, 19992–20001.
2. MacDonald, J.L.S., and Sprecher, H. (1991) Biochim. Biophys. Acta 1084, 105–121.
3. Snyder, F., Lee, T-c., and Blank, M.L. (1992) Prog. Lipid Res. 31, 65–86.
4. Lee, T-c., Ou, M-c., Shinozaki, K., Malone, B., and Snyder, F. (1995) J. Biol. Chem. (in press).
5. Raines, M.A., Kolesnick, N., and Golde, D.W. (1993) J. Biol. Chem. 268, 14572–14575.
6. Dobrowsky, R.T., and Hannun, Y.A. (1992) J. Biol. Chem. 267, 5048–5051.
7. Dobrowsky, R.T., Kamibayashi, C., Mumby, M.C., and Hannun, Y.A. (1993) J. Biol. Chem. 268, 15523–15530.
8. Hannun, Y.A., Loomis, C.R., Merrill, A.H.Jr., and Bell, R.M. (1986) J. Biol. Chem. 261, 12604–12609.
9. Ghosh, T.K., Bian, J., and Gill, D.L. (1990) Science 248, 1653–1656.
10. Hudson, P.L., Pederson, W.A., Saltsman, W.S., Liscovitch, M., MacLaughlin, D.T., Donahoe, P.K., and Blusztajn, J.K. (1994) J. Biol. Chem. 269, 21885–21890.
11. Zhang, H., Buckley, N.E., Gibson, K., and Spiegel, S. (1990) J. Biol. Chem. 265, 76–81.
12. Obeid, L.M., Linardic, C.M., Karolak, L.A., and Hannun, Y.A. (1993) Science 259, 1769–1771.
13. Lavie, Y., and Liscovitch, M. (1990) J. Biol. Chem. 265, 3868–3872.
14. Jones, M.J., and Murray, A.W. (1995) J. Biol. Chem. 270, 5007–5013.
15. Gomez-Muñoz, A., Martin, A., O'Brien, L., and Brindley, D.N. (1994) J. Biol. Chem. 269, 8937–8943.
16. Perry, D.K., Hand, W.L., Edmundson, D.E., and Lambeth, J.D. (1992) J. Immunol. 149, 2749–2758.
17. Merrill, A.H. Jr., Sereni, A.M., Stevens, V.L., Hannun, Y.A., Bell, R.M., and Kinkade, J.M. Jr. (1986) J. Biol. Chem. 261, 12610–12615.
18. Kim, M-Y., Linardic, C.M., Obeid, L.M., and Hannun, Y.A. (1991) J. Biol. Chem. 266, 484–489.
19. Vance, J.E. (1990) J. Biol. Chem. 265, 7248–7256.
20. Rusiñol, A.E., Chi, Z., Chen, M.H., and Vance, J.E. (1994) J. Biol. Chem. 269, 27494–27502.

21

PLATELET-ACTIVATING FACTOR ACETYLHYDROLASE ACTIVITY IN HUMAN FOLLICULAR FLUID

Hisashi Narahara,[1,2] Yuichiro Tanaka,[1] Yasushi Kawano,[1] Isao Miyakawa,[1] and John M. Johnston[2]

[1]Department of Obstetrics and Gynecology
Oita Medical University
1 Hasama-cho, Oita-gun, Oita 879-55, Japan
[2]Departments of Biochemistry and Obstetrics-Gynecology and
The Cecil H. and Ida Green Center for Reproductive Biology Sciences
Southwestern Medical Center
University of Texas
5323 Harry Hines Boulevard, Dallas, Texas 75235-9051

1. ABSTRACT

Platelet-activating factor (PAF) has been implicated in a number of reproductive processes ranging from ovulation to parturition. To examine the role of PAF in the human periovulatory processes, the PAF-acetylhydrolase (PAF-AH) activity was assayed in the follicular fluid (FF) obtained in conjunction with the in vitro fertilization and embryo transfer (IVF-ET) procedure and the activity related to oocyte maturation. The PAF-AH activity was also related to the concentrations of estradiol (E_2) and progesterone (P) in FF. PAF-AH activity was significantly lower in the FFs obtained from follicles of more than 20 mm in diameter. The enzyme activity was significantly lower in the FFs of patients with a successful outcome of their pregnancies. E_2 concentrations were negatively correlated with PAF-AH activities in the FFs. No correlation was found between the PAF-AH activity and concentration of P in the FF. Significantly more mature oocytes were recovered in the group who subsequently become pregnant compared to the non-pregnant group. It is suggested that PAF may be increased following follicular maturation. The increase in PAF may contribute to oocyte maturation and to the successful outcome of pregnancy following fertilization. An additional function of the increased PAF in FF may also be the stimulation of the contraction of smooth muscle in the ovary, thereby assisting the extrusion of the oocyte cumulus cell mass and signaling the completion of ovulation.

Platelet-Activating Factor and Related Lipid Mediators 2
edited by Nigam *et al.*, Plenum Press, New York, 1996

2. INTRODUCTION

The gonadotropin surge stimulates a large variety of events which ultimately leads to the release of the oocyte and the cumulus mass produced by the granulosa cells. Following the luteinizing hormone (LH) surge or the administration of human chorionic gonadotropin (hCG), the concentrations of estradiol (E_2) and progesterone (P) continue to increase in the ovarian follicle. This process is enhanced by release of various proteolytic enzymes, such as plasmin and collagenase, resulting in digestion of collagen in the follicular wall and increasing its distensibility. The proteolytic enzymes are activated in an orderly sequence. The granulosa and theca cells produce plasminogen activator in response to the gonadotropin surge, which, in turn, causes an increase in intrafollicular concentrations of plasmin. Plasmin and other proteases stimulate collagenase activity resulting in the degradation of the connective tissue of the follicular wall.[1]

It has been suggested that the ovulatory process involves changes which are common to the inflammation process.[2] Platelet-activating factor (PAF) is one of the most potent proinflammatory mediators described and has been implicated in the pathophysiology of a number of diseases including asthma, endotoxin shock, ischemic diseases, diabetes mellitus, and cardiovascular disease.[3–6] PAF has also been shown to be involved in a number of processes associated with reproduction including ovulation,[7–11] sperm motility,[12] implantation,[13] fetal lung maturation, and the initiation and maintenance of parturition.[14]

The first suggestion that PAF may be involved in ovulation was that of Abisogun et al.[7] who observed that the local administration of a PAF receptor antagonist inhibited follicle rupture in rats. Espey et al.[8] subsequently reported a decline of the PAF concentration in rat ovarian tissue within several hours following administration of human chorionic gonadotropin (hCG). It has been shown that PAF is secreted from cultured ovine folliclar cells and that the PAF secretion is increased after the endogenous preovulatory surge of luteinizing hormone (LH).[9] It has also been shown that PAF is present in human follicular fluid (FF).[10] Further support for a role of PAF in ovulation has been recently provided by Li et al.[11] who demonstrated that PAF is involved in the ovulatory process of rats and that PAF action is independent from the known effects of prostaglandins (PG). Based on these findings, it is suggested that the increased quantities of PAF as well as PG are present in the preovulatory follicular fluid following the gonadotropin surge.

Inactivation of PAF is catalyzed by the enzyme PAF-acetylhydrolase (PAF-AH), which removes the acetyl group from the *sn*-2 position and converts PAF into the biologically inactive lyso-PAF.[15] It has been suggested that this reaction is one of the mechanisms by which the PAF concentration is tightly regulated.[5] The properties of PAF-AH and its potential role in human disease have recently been reviewed.[6] We and others have reported that human peripheral blood derived macrophages,[6] human decidual macrophages[16] and phorbol ester-stimulated HL-60 cells[17] secrete PAF-AH activity of the plasma type. We have previously reported that the treatment of rats with either dexamethasone or medroxyprogesterone causes an increase in plasma PAF-AH activity, whereas estrogen treatment results in a decrease in the activity of the plasma enzyme.[18,19] We have also suggested that the hormonal regulation of plasma PAF-AH activity in the rat may be, in part, mediated by the secretion of PAF-AH by macrophages.[20]

3. PAF-AH IN HUMAN FF

It has recently been reported that the PAF-AH activity is present in human follicular fluid.[21] The reported half-life of PAF was estimated to be 7–12 minutes in plasma and approximately 2 h in FF. In order to elucidate further the role of PAF in periovulatory processes of the human, we have assayed the PAF-AH activity in the FF obtained in conjunction with the *in-vitro* fertilization and embryo transfer (IVF-ET) procedure.

In Figure 1 is illustrated the PAF-AH activity in follicular fluids obtained from various follicles in relation to the size of the follicule. The samples were obtained from patients participating in the IVF-ET program at Oita Medical University Hospital. A unit of PAF-AH activity is defined as 1 nmol of acetate release per hour at 37C. The enzyme activity was assayed according to the method of Miwa et al..[22] As shown in the Figure, a significantly lower activity was found in follicles of more than 20 mm in diameter. It is suggested that the lower PAF-AH activity may result in an elevated PAF concentrations in the mature follicles.

Among thirty patients with tubal infertility, a successful pregnancy was achieved in eight women (pregnancy rate per patients = 26.7%). As shown in Table 1, the average age of the pregnant group was 31.3 ± 1.7 (mean ± standard deviation) and was not different from the non-pregnant group (32.9 ± 1.5). No difference was observed between pregnant and non-pregnant groups with respect to the number of oocytes retrieved (7.3 ± 1.8 vs. 6.1 ± 1.5, respectively), the rate of oocytes fertilized (62.7% and 58.2%, respectively), or the number of embryos transferred (3.2 ± 0.3 vs. 3.5 ± 0.4, respectively). When oocyte maturation was evaluated, a statistically significant greater number of mature ($P < 0.01$) and smaller number of immature ($P < 0.01$) oocytes were retrieved from the women who become pregnant compared to the non-pregnant group.

Human FFs aspirated from preovulatory follicles contained PAF-acetylhydrolase activity ranging from 20.2 units/ml to 89.9 units/ml. The enzyme activity (mean ± standard error) was significantly lower ($P < 0.01$) in the women with a successful pregnancy outcome (29.7 ± 8.7 units/ml) than in the non-pregnant patients (58.6 ± 19.9 units/ml) (Figure 2).

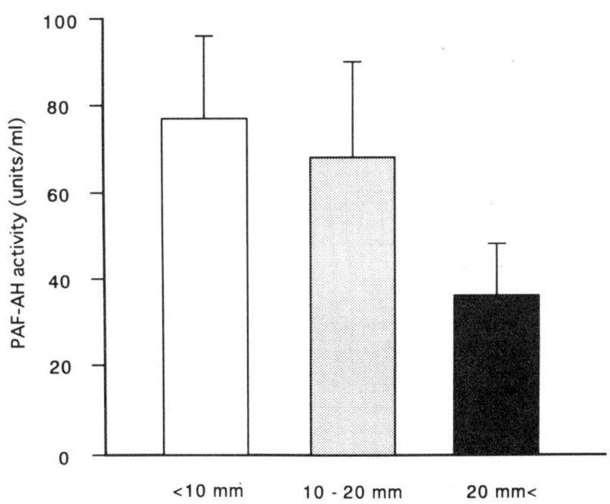

Figure 1. PAF-AH activity in various follicles in relation to the size.

Table 1. *In vitro* fertilization outcome of patients with tubal infertility

	Outcome	
	Pregnant	Non-pregnant
No. of patients	8	22
Age (years)	31.3 ± 1.7	32.9 ± 1.5
No. of oocytes retrieved	7.3 ± 1.8	6.1 +6.1 + 1.5
No. of mature oocytes	4.3 ± 1.2[a]	2.5 ± 0.8
No. of intermediate oocytes	2.2 ± 0.6	2.4 ± 0.5
No. of immature oocytes	0.8 ± 0.3[a]	1.2 ± 0.3
No. of oocytes fertilized (%)	62.7	58.2
No. of embryos transfered	3.6 ± 0.8	3.2 ± 0.4

[a]Values are means ± SD. Means in the pregnant group were significantly different from those in the non-pregnant group ($P < 0.01$).

Based on these observations, it is suggested that the increased PAF-AH activity in the immature follicles may lead to a decreased PAF in the follicles. The PAF-AH activity is decreased following follicular maturation, resulting in the increase in PAF. The increase in PAF may contribute to oocyte maturation and contribute to the successful outcome of the pregnancy following fertilization.

Although the origin of the PAF-AH in FF is yet to be defined, there are at least two possibilities: the enzyme in FF might be exudated from plasma or might be secreted by macrophages which are known to be present in human FF.[23]

4. POSSIBLE REGULATORY FACTORS OF PAF-AH IN HUMAN FF

It has been reported that oophorectomy of the rat causes a decrease in the uterine PAF concentration by one third and that the PAF level returns to a normal value following the subcutaneous administration of E_2.[24] As discussed, the plasma activity of PAF-AH is hormonally regulated.[18,19] Therefore, we assessed the relationship between the enzyme ac-

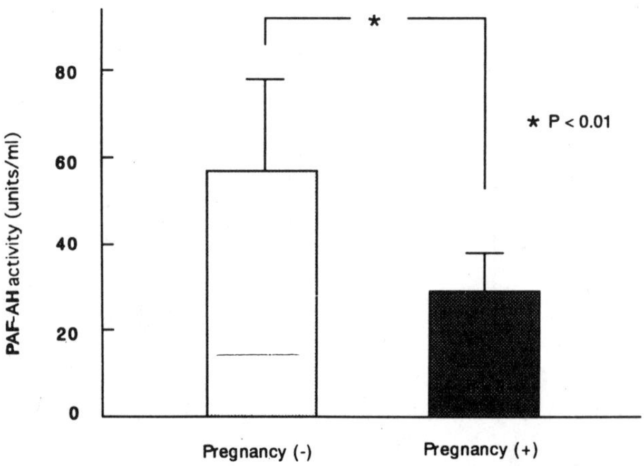

Figure 2. PAF-AH activity in the follicular fluid of pregnant (n=8) and non-pregnant (n=22) groups.

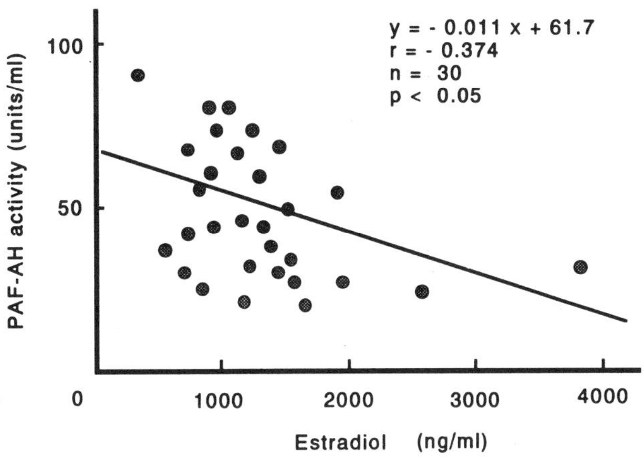

Figure 3. Correlation between estradiol levels and PAF-AH activity in the follicular fluid.

tivity in follicular fluid and the ovarian steroid hormones, E_2 and P concentration in the follicular fluid. As shown in Figure 3, E_2 levels in FFs were negatively correlated with PAF-AH activity in FFs of patients ($n = 30$, $r = -0.374$, $P < 0.05$). In contrast, the P concentrations did not show any correlation with PAF-AH activity in follicular fluids (data not shown).

The finding that PAF-AH activities in FFs were negatively correlated with E_2 levels in the FFs may be an indication of the regulatory role for E_2 in the metabolism of PAF in FF. It has been reported that PAF induces P secretion and morphological changes in cultured luteinizing granulosa cells.[25] In the present report, however, no correlation was observed between the levels of P in FFs and the PAF-AH activity. This apparent discrepancy may be due to the fact that PAF-AH is not affected by P, although PAF affects P in FF.

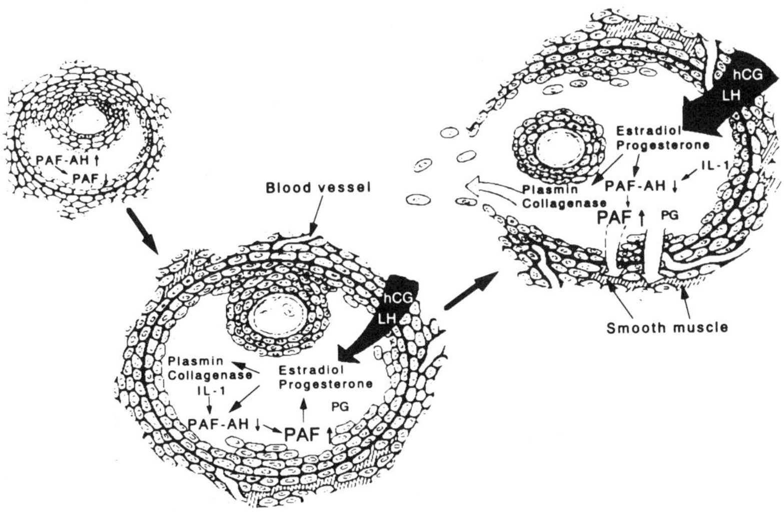

Figure 4. The role of PAF and PAF-AH in ovulation.

Alternatively, our samples may reflect an earlier stage of luteinization than that previously reported.

It has been suggested that a cytokine, interleukin-1 (IL-1), is involved in mammalian ovulation.[26] IL-1 has been shown to stimulate PAF production by monocytes and neutrophils.[4] We have recently reported that IL-1 inhibited the PAF-AH secretion by decidual macrophages[27] and peripheral blood-derived macrophages (Narahara and Johnston, unpublished observation). Based on these observations it is suggested that PAF metabolism is modulated by IL-1 in follicular tissue: the cytokine might increase the PAF concentration not only by the stimulation of the PAF production but also by the inhibition of PAF-AH secretion.

5. THE ROLE OF PAF AND PAF-AH IN OVULATION

It is suggested that the induction of ovulation by gonadotropins such as LH and hCG stimulate estradiol and progesterone production in the follicle which leads to an increase in plasmin and collagenase activities. PAF as well as PG are known to be increased in the preovulatory follicles.

We have demonstrated that PAF-AH activity is present in the follicular fluid and that an increased PAF-AH activity may account for the decreased PAF present in the small immature follicles. The PAF-AH activity is decreased following follicular maturation and this decrease may be due, in part, to the increased E_2 and/or IL-1, which favors the accumulation of PAF in the follicle. The increase in PAF may contribute to oocyte maturation and possibly to the successful outcome of pregnancy following fertilization. The increase in PAF in the follicular fluid may also stimulate progesterone production, resulting in an increase of the proteolytic enzymes plasmin and collagenase. The increased PAF in the follicular fluid may also lead to the contraction of smooth muscle in the ovary, thereby assisting the extrusion of the oocyte cumulus cell mass, signaling the completion of ovulation.

ACKNOWLEDGMENTS

This work was supported, in part, by NIH grant HD11149, The Robert A. Welch Foundation, Houston, TX. HN was a recipient of a fellowship from the Chilton Foundation.

REFERENCES

1. L. Speroff, R.H. Glass, and N.G. Kase, Regulation of the menstrual cycle, in: Clinical Gynecologic Endocrinology and Infertility, L. Speroff, R.H. Glass, N.G. Kase, eds., 4th edition (1989) pp91.
2. L.L. Espey, Ovulation as an inflammatory reaction: a hypothesis, Biol. Reprod. 22:73 (1980).
3. F Snyder, Platelet-activating factor and related acetylated lipids as potent biologically active cellular mechanisms, Am. J. Physiol.259:C697 (1990).
4. P. Braquet, L, Touqui, T.Y. Shen, and B.B. Vargaftig, Perspectives in platelet-activating factor research, Parmacol. Rev.39:97 (1987)
5. S.M. Prescott, G.A. Zimmerman, and T.M. McIntyre, Platelet-activating factor, J. Biol. Chem. 265:17381 (1990);.

6. M.R. Elstad, D.M. Stafforini, K.E. Stremler, G.A. Zimmermen, T.M. McIntyre, and S.M. Prescott, The potential role of the PAF acetylhydrolase in human disease, in. Platelet-Activating Factor and Disease, K. Saito, D.J. Hanahan, eds. International Medical Publishers, Tokyo, Japan: 69 (1989).
7. A.O. Abisogun, P. Braquet, and A, Tsafriri, The involvement of platelet-activating factor in ovulation, Science 243:381 (1989).
8. L.L. Espey, N. Tanaka, D.S. Woodard, M.J.K. Harper, and H. Okamura. Decrease in ovarian platelet-activating factor during ovulation in the gonadotropin-primed immature rat, Biol. Reprod. 41:104 (1989).
9. B.M. Alexander, E.A. Van Kirk, and W.J. Murdoch, Secretion of platelet-activating factor by periovulatory ovine follicles, Life Sci 47:865 (1990).
10. M.J. Amiel, J. Testart, and J. Benveniste, Platelet-activating factor-acether is a component of human follicular fluid Fertil Steril 56:62 (1991).
11. X.-M. Li, N. Sagawa, Y. Ihara, A. Okagaki, H. Hasegawa, K. Inamori, H. Itoh, T. Mori, C. Ban, The involvement of platelet-activating factor in thrombocytopenia and follicular rupture during gonadotropin-induced superovulation in immature rats, Endocrinology 128:3132 (1991).
12. B.S. Minhas, Platelet-activating factor treatment of human spermatozoa enhances fertilization potential, Am. J. Obstet. Gynecol. 168:1314 (1993).
13. C. O'Neill, PAF and the establishment of pregnancy, in: Platelet-activating factor and human disease, P.C. Barnes, C.P. Page, P.M. Henson, eds... Blackwell, Oxford. 282–296 (1990).
14. H. Narahara, R.A. Frenkel, and J.M. Johnston, The role of PAF in reproductive biology, in Advances in Molecular Biology-Lipobiology, R. Gross, ed.. JAI Press Inc., Greenwich, CT, in press. .
15. R.S. Farr, M.L. Wardlow, C.P. Cox, K.E. Meng, and D.E. Greene, Human serum acid-labile factor in an acyl-hydrolase that inactivates platelet-activating factor, Fed. Proc. 42:3120 (1983).
16. H. Narahara, Y. Nishioka, and J.M. Johnston, Secretion of platelet-activating factor acetylhydrolase by human decidual macrophages, J. Endocrinol. Metab. 77:1258–62 (1993).
17. H. Narahara, R. Frenkel, and J.M. Johnston, Secretion of platelet-activating factor acetylhydrolase (PAF-AH) following phorbol ester-stimulated differentiation of HL-60 cells, Arch. Biochem. Biophys. 301:275 (1993).
18. S. Miyaura, N. Maki, W. Byrd, and J.M. Johnston, The hormonal regulation of platelet-activating factor acetylhydrolase activity in plasma, Lipids 26:1015 (1991).
19. K. Yasuda, and J.M. Johnston, The hormonal regulation of platelet-activating factor acetylhydrolase in the rat, Endocrinology 130:708 (1992).
20. K. Yasuda, H. Eguchi, H.Narahara, and J.M. Johnston, Platelet-activating factor: Its regulation in parturition. in: Eicosanoids and Other Bioactive Lipids in Cancer, Inflammation and Radiation Injury, S. Nigam, K.V. Honn, L.J. Marnett, T.L. Walden, eds., Kluwer Academic Publishers, Boston, 727 (1993).
21. N. Lepage, P. Miron, R. Hammings, K.D. Roberts, and J. Langlais, Distribution of lysophaspholipids and metabolism of platelet-activating factor in human follicular and peritoneal fluids J. Reprod. Fertil. 98:349 (1993).
22. M. Miwa, T. Miyake, T. Yamanaka, J. Sugatani, Y. Suzuki, S. Sakata, Y. Araki, and M. Matsumoto, Characterization of serum platelet-activating factor (PAF) acetylhydrolase. Correlation between deficiency of serum PAF acetylhydrolase and respiratory symptoms in asthmatic children J. Clin. Invest. 82:1983 (1988).
23. J.A. Loukides, R.A. Loy, R. Edwards, J. Hong, I. Visintin, and M.L. Polan, Human follicular fluid contain tissue macrophages, J. Clin. Endocrinol. Metab. 71:1363 (1990).
24. R. Nakayama, K. Yasuda, T. Okumura, and K. Saito, Effect of 17ß-estradiol on PAF and prostaglandin levels in oophorectomized rat uterus Biochem. Biophys. Acta 1085:235 (1991).
25. J. Rabinovici, and M.J. Angle, Platelet-activating factor induces progesterone secretion and changes in morphological appearance on luteinizing granulosa cells *in vitro*,. Fertil. Steril. 55:1106 (1991).
26. Y. Takehara, A.M. Dharmarajan, G. Kaufman, E.E. Wallach, Effect of interleukin-1ß on ovulation in the in-vitro perfused rabbit ovary, Endocrinology 134:1788 (1994).
27. H. Narahara, and J.M. Johnston, Effects of endotoxins and cytokines on the secretion of platelet-activating factor-acetylhydrolase by human decidual macrophages, Am. J. Obstet. Gynecol. 169;531 (1993).

22

THE PREVENTION OF NECROTIZING ENTEROCOLITIS

The Role of Platelet-Activating Factor Acetylhydrolase

Kouei Muguruma,[1] Masayuki Furukawa,[1] Larry W. Tjoelker,[2] Greg Dietsch,[2] Patrick W. Gray,[2] Biren Zhao,[1] and John M. Johnston[1]

[1]Departments of Biochemistry, Obstetrics-Gynecology and
The Cecil H. and Ida Green Center for Reproductive Biology Sciences
The University of Texas Southwestern Medical Center at Dallas
5323 Harry Hines Boulevard, Dallas, Texas 75235–9051
[2]ICOS Corporation, 22021
20th Avenue S.E., Bothell, Washington 98021

1. INTRODUCTION

A disease similar, if not identical, to necrotizing enterocolitis (NEC) was first described in the last century in a 2 day-old infant who died of inflammation and perforation of the ileum.[1] The incidence of NEC, a disease linked to prematurity, has increased recently due to innovative treatments of premature infants, particularly surfactant replacement. It has been estimated that premature infants suffering from this disease may have a mortality rate as high as 50%. Minimal basic information is known concerning the mechanism(s) involved in the development of NEC. Numerous hypotheses have been invoked as causative or modifying factors of this disease. These include hypothermia, hypoxia, infectious agents, ischemia, various feeding regimes,[2] oxygen free radical generation, etc..

2. ROLE OF PAF IN NECROTIZING ENTEROCOLITIS DEVELOPMENT

It is well established that PAF is one of the most potent lipid mediators described, being active at concentrations as low as 10^{-12} M.[3] This is especially true in relation to its pro-inflammatory action. The first report relating PAF to the pathogenesis of NEC in animals was that of Gonzales-Crussi and Hsueh.[4] Subsequently, this same group of investigators reported that the plasma PAF concentration was elevated and the PAF-acetylhydrolase (PAF-AH) activity was lower in patients with NEC.[5] We[6,7] and others[8] have emphasized the importance of PAF-AH in the regulation of the PAF concentration and had observed that the plasma PAF-AH was low in newborns and increased with age.[6]

Several years ago, Whittle and colleagues[9] established a role for PAF as a potent ulcerogen and noted that this autacoid was a likely candidate as a pathological mediator of gastrointestinal disorders associated with shock and inflammation. Subsequently, Wallace and collaborators further established the role of PAF in the pathogenesis of gastric ulcer.[10]

Several years ago, Hsueh and colleagues[4,11] described a rat model for NEC in which PAF, tumor necrosis factor, or endotoxins were injected. This group also reported that the administration of a PAF receptor blocker prevented the development of NEC in a rat model.[12] In a similar study,[13] we observed that within 2 hr following the injection of PAF into the descending aorta, the serosal surface became diffusely discolored and a full-thickness necrosis and hemorrhage of the mucosa of the duodenum, jejunum, and ileum. The pathology was identical to that seen in patients with NEC.[13]

We originally established that rats injected with dexamethasone injection causes an increase in the plasma PAF-AH activity.[14,15] In order to further examine the role of increased plasma PAF-AH activity on the NEC development, rats were treated with dexamethasone for three days.[13] The plasma PAF-AH activity increased by the injection of dexamethasone from 55.2 ± 7.1 nmol x min^{-1} x ml^{-1} plasma (mean ± SD) on day 1 to 117.5 ± 7.9 nmol x min^{-1} x ml^{-1} plasma by day 4. The rats were then injected with PAF into the descending aorta. The small intestine in the dexamethasone treated and control animals were normal in all respects as judged by both gross and microscopic examination. All of the PAF-treated rats, however, developed NEC.

The incidence of NEC is known to be decreased in newborns who nurse.[16] We, therefore, determined the PAF-AH in human milk.[17] PAF-AH activity was demonstrated in human milk. It was established that the activity was not due to other known lipases in milk. We also determined the PAF-AH activity in milk obtained from various species. The specific activity of PAF-AH was similar in milk samples obtained from the human, pig, goat, and rat. The PAF-AH activity of cow's milk, however, contained little if any PAF-AH activity. In two species, human and pig, the colosterum had a PAF-AH activity that was 2–3 times higher than milk from that species.

Elstad *et al.*[18] reported that PAF-AH of the plasma type secreted by human peripheral macrophages. We reported that the enzyme is also secreted by rat alveolar,[19] and decidua,[20] macrophages, and phorbol ester treated HL-60 cells.[21] Macrophages are known to be present in human milk.[22] Macrophages from human milk were isolated and it was reported that PAF-AH was secreted by these cells. The milk enzyme was shown to be the plasma isozyme.

3. MECHANISMS OF THE BENEFICIAL EFFECTS OF GLUCOCORTICOIDS AND BREAST FEEDING ON THE PREVENTION OF NEC

The mechanism by which PAF-AH may alter the development of NEC is not completely established. The observed increase of the plasma PAF-AH activity caused by dexamethasone may contribute to the inactivation of PAF. Corticosteroids are widely used in the treatment of inflammation. We have suggested[14,15] that one action of glucocorticoids was to increase the activity of PAF-AH and, in turn, decrease the concentration of PAF.

Several groups of investigators have reported that the pretreatment with corticosteroids resulted in a significant decrease in the incidence of NEC.[23,24] It is suggested that this decreased incidence of NEC under these conditions is due, in part, to the elevation of PAF-AH activity with its associated decrease in PAF.

The beneficial effects of breast feeding compared to formula feeding in the prevention of NEC in the human are well established.[16] The presence of PAF-AH in human milk may contribute to its protective effect. In a canine model system for NEC, it was reported that raw cow's milk was not a satisfactory substitute for dog milk in the nursing pup.[25] The absence of PAF-AH in bovine milk may contribute to this difference. The observation that the PAF-AH of human milk is resistant to exposure to a low pH may also be important in the prevention of PAF accumulation in the small intestine.[17] It has also been reported that the presence of the white cells of milk may be beneficial in the prevention of this disease.[26] The secretion of PAF-AH activity by human milk macrophages may explain, in part, the beneficial effects of white cells.

4. UTILIZATION OF THE HUMAN RECOMBINANT PAF-AH IN THE PREVENTION OF NEC

Recently the molecular cloning and characterization of human plasma PAF-acetylhydrolase has been reported.[27] The recombinant enzyme has a substrate specificity and lipoprotein specificity which was similar to the native enzyme. Recombinant human plasma PAF-acetylhydrolase has been shown to block PAF-mediated inflammatory reactions *in vitro*.[27]

We have utilized the recombinant PAF-AH to determine whether this enzyme could be utilized to prevent NEC development in the rat model. We observed a severe discoloration and hemorrhage of both the serosal and mucosal surfaces of the entire small intestine 2 hr following the administration of PAF (0.2 μg/100 gm) into abdominal aorta as we have previously described.[13] The presence of NEC was confirmed by microscopic examination.

When animals were pretreated with the recombinant PAF-AH 15 minutes prior to the abdominal PAF injection, a normal small bowel was observed as judged by both gross and microscopic examination and the small intestine was indistinguishable from the control animals. The animals injected with PAF into the descending aorta developed a disease that was indistinguishable from NEC. In subsequent studies, we established that the increase of the plasma activity which was approximately 2–3 times the normal plasma activity was completely protective. This protective activity was similar to that observed when dexamethasone was utilized to increase the plasma PAF-AH activity. Recombinant PAF-AH is known to bind to the plasma lipoprotein fraction.[27] It was demonstrated that the protective effect was similar to that predicted from the half-life of the lipoprotein. The recombinant PAF-AH must be administered shortly before the time of the PAF injection since treatment with PAF-AH 15 min after the PAF injection was not protective.

In the present studies, we report the further development of an animal model for necrotizing enterocolitis. This lipid mediator has proven to be one of the most potent lipid pro-inflammatory agents thus far described that is involved in the development of this disease. Employing the basic concepts that we have described for the regulation of the plasma enzyme that inactivates PAF we have suggested a mechanism for the protective effect of glucocorticoids and human milk both of which have been recognized for several years. In order to establish that the primary protective effect of glucocorticoids was due to its known effect to increase the plasma PAF-AH activity,[15] we report that the administration of the recombinant PAF-AH will prevent NEC development in a similar manor to that previously described for dexamethasone. We have previously suggested that the anti-inflammatory properties of glucocorticoids may be explained, in part, due to the elevation of the PAF-AH activity.[14,15] The beneficial effects of the steroid and nursing provides an explanation as to why they result in a decreased incidence in the disease.

Based on the reported investigation it is clear that PAF occupies a central role in the development of NEC. Although numerous factors have been reported to be associated with an increased incidence of NEC, most of these can be linked to an increased PAF production. As previously described, the intestine is known to be more susceptible to asphyxia during feeding[2] and increased systemic PAF formation has been reported to be associated with enteral feeding[28] We have reported that the endotoxin, lipopolysaccharide, will inhibit PAF-AH release by various macrophages.[29] Thus, a mechanism for the increase in PAF concentration is suggested for a bacterial component. In addition to endotoxins, the activation of the inflammatory response resulting in the increase in various cytokines such as tumor necrosis factor α and the interleukins may further increase PAF production since these cytokines also inhibit PAF-AH release.[29] It is also recognized that prematurity is a predisposing factor for NEC development. In an attempt to explain this observation, we have assayed the critical enzymes involved in PAF biosynthesis by both the remodeling pathway and *de novo* pathways (Muguruma and Johnston, unpublished observations). It was found that the newborn intestine had a higher capacity than for PAF biosynthesis than any other newborn or adult tissues. Thus, a possible explanation as to why the intestine of the newborn may be more susceptible to NEC is due to its capacity to synthesize this potent pro-inflammatory mediator.

The reported observations that the incidence of NEC can be reduced by increasing the plasma PAF-AH activity by glucocorticoid treatment or recombinant enzyme administration in conjunction with the observation in relation to infant feeding bacterial endotoxins and breast feeding provide convincing evidence for a role of PAF in NEC. The possible inclusion of the recombinant enzyme in newborn formula provide a mechanism for the treatment of this disease.

REFERENCES

1. A. Genersich, Bauchfellentzündung beim Neugebornen in Folge von Perforation des Ileums, *Virch. Arch. Path. Anat.* 126:485 (1891).
2. J.S. Szabo, S.R. Mayfield, W. Oh, and B.S. Stonestreet, Postprandial gastrointestinal blood flow and oxygen consumption: Effects of hypoxemia in neonatal piglets, *Pediatr. Res.* 27:93 (1987)
3. F. Snyder, Platelet-activating factor and related acetylated lipids as potent biologically active cellular mediators, *Am. J. Physiol.* 259:C697 (1990).
4. F. Gonzalez-Crussi and W. Hsueh, Experimental model of ischemic bowel necrosis. The role of platelet-activating factor and endotoxin, *Am. J. Pathol.* 112:127 (1983).
5. M.S. Caplan, X.-M. Sun, W. Hsueh, J.R. Hageman, Role of platelet activating factor and tumor necrosis factor-alpha in neonatal necrotizing enterocolitis, *J. Pediatr.* 116:960 (1990).
6. N. Maki, D.R. Hoffman, and J.M. Johnston, Platelet-activating factor acetylhydrolase activity in maternal, fetal, and newborn rabbit plasma during pregnancy and lactation, *Proc. Natl. Acad. Sci. USA* 85:728 (1988).
7. J.M. Johnston, The function of PAF in the communication between fetal and maternal compartments during parturition, in: *Platelet-Activating Factor and Diseases*, K. Saito, and D.J. Hanahan, eds., International Medical Pubs, Tokyo, p. 129 (1989).
8. S.M. Prescott, G.A. Zimmerman, and T.M. McIntyre, Platelet-activating factor, *J. Biol. Chem.* 262:4125 (1990).

9. B.J.R. Whittle, and J.V. Esplugues, PAF and the gastro-intestinal tract, in: *Platelet-Activating Factor and Human Disease*, P.J. Barnes, C.P. page, and P.M. Henson, eds., Blackwell Scientific Pub., Oxford, p 198 (1989).
10. J.L. Wallace, Lipid mediators of inflammation in gastric ulcer, *Am. J. Physiol.* 258:G1 (1990).
11. X.-m. Sun, W. Hsueh, Bowel necrosis induced by tumor necrosis factor in rats is mediated by platelet-activating factor, *J. Clin. Invest.* 81:1328 (1988).
12. W.E.I. Hsueh, F. Gonzalez-Crussi, J.L. Arroyave, R.C. Anderson, M.L. Lee, W.J. Houlihan, Platelet activating factor-induced ischemic bowel necrosis: The effect of PAF antagonists, *Eur. J. Pharmacol.* 123:79 (1986).
13. Furukawa, E.L. Lee, and J.M. Johnston, Platelet-activating factor-induced ischemic bowel necrosis: The effect of platelet-activating factor acetylhydrolase, *Pediatr. Res.* 34:237 (1993).
14. S. Miyaura, N. Maki, W. Byrd, and J.M. Johnston, The hormonal regulation of platelet-activating factor acetylhydroalse activity in plasma, *Lipids* 26:1015 (1991).
15. K. Yasuda, and J.M. Johnston: The hormonal regulation of platelet-activating factor acetylhydrolase in the rat, *Endocrinology* 130:708 (1992).
16. A. Lucas, T.J. Cole, Breast milk and neonatal necrotizing enterocolitis, *Lancet* 336:1519 (1990).
17. M. Furukawwa, H. Narahara, K.Yasuda, and J.M. Johnston, Presence of platelet-activating factor-acetylhydrolase in milk, *J. Lipid Res.* 34:1603 (1993).
18. M.R. Elstad, D.M. Stafforini, T.M. McIntyre, S.M. Prescott, and G.A. Zimmerman, Platelet-activating factor-acetylhydrolase increases during macrophage differentiation. A novel mechanism that regulates acumulation of platelet-activating factor, *J. Biol. Chem.* 264:8467 (1989).
19. K. Yasuda, H. Eguchi, H. Narahara, and J.M. Johnston, Platelet-activating factor: Its regulation in parturition, in: *Eicosanoids and Other Bioactive Lipids in Cancer, Inflammation and Radiation Injury*, S. Nigam, K.V. Honn, L.J. Marnett, T.L. Walden, eds, Kluwer Academic Pubs, Boston, p 727 (1993).
20. H. Narahara, Y. Nishioka, J.M. Johnston, Secretion of platelet-activating factor acetylhydrolase y human decidual macrophages, *J. Clin. Endocrinol. Metab.* 77:1258 (1993).
21. H. Narahara, R.A. Frenkel, and J.M. Johnston, Secretion of platelet-activating factor-acetylhydrolase (PAF-AH) following phorbol ester-stimulated differentiation of HL-60 cells, *Arch. Biochem. Biophy.* 301:275 (1993).
22. F.R. Balkwill and N. Hogg, Characterization of human breast milk macrophages cytostatic for human cell lines, *J. Immunol.* 123:1451 (1979).
23. C.R. Bauer, J.C. Morrison, W.K. Poole, S.B. Korones, J.J. Boehm, H. Rigatto, and R.D. Zachman, A decreased incidence of necrotizing enterocolitis after prenatal glucocorticoid therapy, *Pediatrics* 73:682 (1984).
24. E. Halac, J. Halac, E.F. Bégué, J.M. Casanas, D.R. Indiveri, J.F. Petit, M.J. Figueroa, J.M. Olmas, L.A. Rodriguez, R.J. Obregón, M.V. Martinez, D.A. Grinblat, and H.O. Vilarrodona, Prenatal and postnatal corticosteroid therapy to prevent neonatal necrotizing enterocolitis: A controlled trial, *J. Pediatr.* 117:132 (1990).
25. D.A. Park, G. B. Bulkley, D.N. Granger, Role of oxygen-derived free radicals in digestive tract disease. *Surgery* 94:45 (1983).
26. J. Pitt, B. Barlow, and W.C. Heird, Protection against experimental necrotizing enterocolitis by maternal milk. I. Role of milk leukocytes, *Pediatr. Res* 11:906 (1977).
27. L.W. Tjoelker, C. Wilder, C. Eberhardt, D.M. Stafforini, G. Dietsch, B. Schimpf, S. Hooper, H.L. Trong, L.S. Cousens, G.A. Zimmerman, Y. Yamada, T.M. McIntyre, S.M. Prescott, P.W. Gray, Anti-inflammatory properties of a platelet-activating factor acetylhydrolase, *Nature* 374:549 (1995).
28. W. MacKendrick, N. Hill, W. Hsueh, *et al.*, Increase in plasma platelet activating factor levels in entrally fed preterm infants. *Biol. Neonate* 64:89 (1993).
29. H. Narahara, Y. Nishioka, J.M. Johnston, Secretion of platelet-activating factor acetylhydrolase by human decidual macrophages, *J. Clin. Endocrinol. Metab.* 77:1258 (1993).

23

IMPLICATION OF PAF AND ACETYLHYDROLASE (PAF-AH) ACTIVITY IN PERIODONTAL DISEASE

George Baltas,[1] Helen Kotsifaki,[2] Smaragdi Antonopoulou,[3]
Anthoula Kipioti,[1] and Constantinos A. Demopoulos[3]

[1]Department of Periodontology
[2]Department of Physiology
[3]Department of Chemistry
University of Athens
Athens, Greece

1. INTRODUCTION

Platelet-activating factor (PAF) is a bioactive phospholipid produced by activated inflammatory cells such as polymorphonuclear leukocytes, macrophages, platelets and endothelial cells[1]. Previous research showed that PAF in vivo, increases vascular permeability and infiltration of PMNs [2]. Recent studies, however, have implicated PAF with virtually all inflammatory and immune responses [3]. So, PAF influences either directly or indirectly many organ and tissue functions. In PMNs PAF induces chemotaxis, aggregation, adherence, lysosomal enzymes release and the production of leukotrienes and free radicals [4, 5]. In monocytes macrophages, low concentrations of PAF cause not only chemotaxis and aggregation but also potentiate, the IL-1, TNF and superoxide production [6, 7]. Also PAF induces inhibition of lymphocyte proliferation and IL-2 production was observed [8].

Besides PAF, other molecules as oxidatively-fragmented phospholipids, have PAF-like activity and act through PAF receptors [9].

The inactivation of PAF into the biologically inactive lyso-PAF is catalysed by PAF-acetylhydrolase. This enzyme was detected in many tissues and cells as well as in biologically fluids and apparently play a major role in the regulation of PAF levels [10].

Periodontal disease is a bacterial infection. Bacteria or their products and components induce the initial infiltration of inflammatory cells, such as PMNs, lymphocytes, and monocytes. The activation of inflammatory cells has as result the production and release of mediators. Moreover, microbial components, mainly lipopolysaccharides (LPS), activate macrophages to synthesize and secret cytokines (IL-1, TNF-a), prostaglandins and hydrolytic enzymes. These cytokines, having catabolic activities, are concidered the causative agent of periodontal tissue breakdown [11, 12].

Platelet-Activating Factor and Related Lipid Mediators 2
edited by Nigam *et al.*, Plenum Press, New York, 1996

To the best of our knowledge, the relationship between PAF and periodontal disease has not been examined so far thoroughly. In a very interesting study, PAF levels in mixed saliva of edentulous subjects, were estimated in comparison to that of dentate ones. PAF was not detected in the saliva of 60% of the edentulous and when present, the levels were significantly less than in dentate subjects. However the presence of PAF-AH has not been detected since PAF activity in saliva was unchanged after proper incubation [13].

The presence of substance(s) in certain samples of saliva which inhibited platelet aggregation induced by PAF as well as by arachidonic acid, has been referred [14]. Positive correlation between PAF levels in saliva and the severity of periodontal disease was demonstrated [15]. Also the levels of PAF produced in inflammed gingival were higher than in normal ones but the activity of PAF-AH didn't show any significant difference [16].

This study was designed to estimate PAF and PAF-AH levels in gingival crevicular fluid (GCF) (an exudation, mainly inflammatory) in comparison to clinical parameters of periodontal disease, a) during experimental gingivitis and b) before and after conventional periodontal treatment.

2. EXPERIMENTAL PROCEDURES

2.1 GCF Samples

The crevicular fluid's samples were collected by using paper strips that were inserted into the crevice of the six maxilla's anterior teeth and left for 60 sec. All samples were kept into ice.

PAF-AH was extracted from paper strips with phosphate buffer PH 7.5. Then the paper strips were transferred to another tube and PAF-like molecule(s) was extracted according to Bligh- Dyer method [17].

2.2 HPLC Purification

Bioactive phospholipid was further purified on a cation exchange column SS 10μm Partisil 25cm x 4.6mm I.D., PXS 10/25 SCX and the biological properties of the active fraction were studied [18]. HPLC was performed on a dual pump Jasko model 880-PU HPLC, supplied with a 330μL loop Rheodyne (P/N 7125–047) injector. A Jasko model 875 UV spectrophotometer was used as detector (210nm). The spectrophotometer was connected to a Hewlett-Packard model HP-3396A integrator-plotter.

2.3 PAF-Bioassay

PAF was estimated by bioassay studies based on washed rabbit platelets and PAF levels were expressed as fmoles synthetic PAF per 100 μg protein [19]. Experiments with specific inhibitors (creatine phosphate (CP) 0.7 mM /creatine phosphate kinase (CPK) 13 units per ml, indomethacin 10 μM and BN 52021 0.1 mM), were also performed. These inhibitors are added to washed rabbit platelets 1 min prior to the addition of the examined sample into the aggregometer cuvette. This experiment was carried out according to Lazanas *et al* [20].

2.4 Treatment with Acetylhydrolase

In a test tube prewarmed at 37 °C, phosphate buffer (0.08 M, pH 7.5) was added along with human serum acetylhydrolase, purified by Dr. D. Stafforini and the examined sample in 2.5 mg bovine serum albumin (BSA) per ml saline. The enzymic system is incubated at 37 °C and at different times aliquots were taken and tested for their ability to induce washed rabbit platelet aggregation.

2.5 PAF-Acetylhydrolase Assay

The estimation of PAF-AH was based on its capacity to degrade [^{3}H] PAF and PAF-AH levels were expressed as fmoles degraded [^{3}H] PAF per 100 mg protein after 45 min incubation [21].

2.6 Experimental Gingivitis

In plaque-free volunteers with inflammation free gingiva plaque begun to accumulate when all oral hygene ceased for 21 days. Then plaque was removed and oral hygiene repeated. As result gingiva returned to clinical heathy status [22].

3. RESULTS

The biologically active substance extracted from GCF induced washed rabbit platelet aggregation in a dose-dependent manner, with PAF-like aggregation tracing. This active molecule was not co-migrated with PAF in HPLC but it eluted within the area of lysophosphadidylcholine. Furthermore, this molecule desensitized platelets against itself and PAF but not against thrombin. The aggregation induced by this PAF-like molecule was inhibited by BN52021. No inhibition was observed by indomethacin and CP/CPK. This PAF-like molecule was inactivated by human plasma PAF-AH in a slower rate than that of PAF. Also, inhibition of synthetic [^{3}H]-PAF with GCF revealed the existence of PAF-AH activity.

Furthermore, the levels of PAF-like activity and PAF-AH were estimated in GCF in 11 volunteers during the course of experimental gingivitis. The crude extract of GCF was used for the estimation of PAF-like activity. From our mesurments increase in the level of PAF-like activity was detected at the begining of plaque

accumulation (2nd day) and after 11 days. The levels of PAF-AH increased in a statistically significant maner ($p<0.1$) at the end of experimental gingivitis. But during the course a negative correlation ($p<0.05$) was detected between PAF-like and PAF-AH activity (fig.1).

In addition PAF-AH activity was detected in dental plaque received from these volunteers. In 21th day of experiment gingivitis the PAF-AH activity was 230 fmols degraded [^{3}H]PAF per 100μg protein of GCF. Also, there is evidence for the presence of PAF-activity as well as inhibitor(s) in dental plaque. Furthermore, PAF-like activity and PAF-AH levels were investigated in GCF from 12 patients with chronic adult periodontitis pre and post treatment. The success of treatment was based on the measurments of clinical parameters, plaque index (PL I), bleeding on probing (BOP) and probing depth (PD). They showed a statistically significant decrease post treatment ($p<0.001$) (fig.2). Estima-

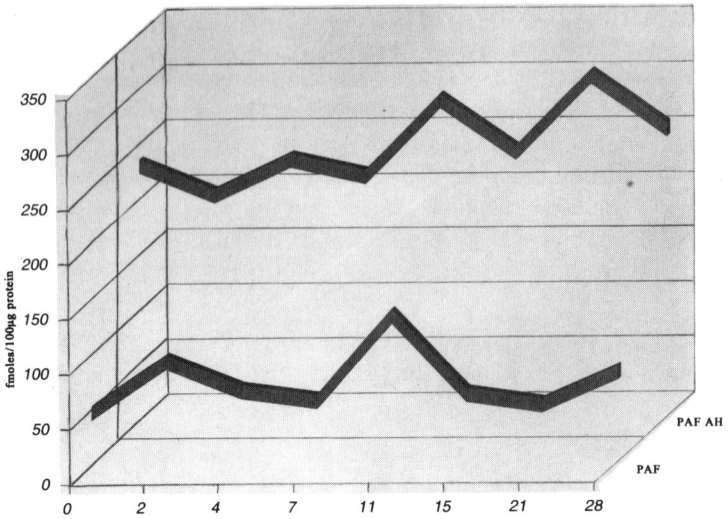

Figure 1. Correlation of PAF and PAF-AH levels in crevicular fluid of 11 volunteers during the course of experimental gingivitis.

tion of PAF-like activity did not show any statistically significant difference pre and post treatment.

Concerning the PAF-AH activity, a decrease was noticed after the treatment. The mean value before treatment was 525.5 ± 162.1 and after treatment was 391.2 ± 110.3 fmoles degraded [^{3}H]PAF per 100µg protein ($p<0.1$) (fig.3).

As well as in experimental gingivitis and in adult periodontitis a negative correlation ($p<0.01$) between PAF-like activity and PAF-AH activity was observed (table 1).

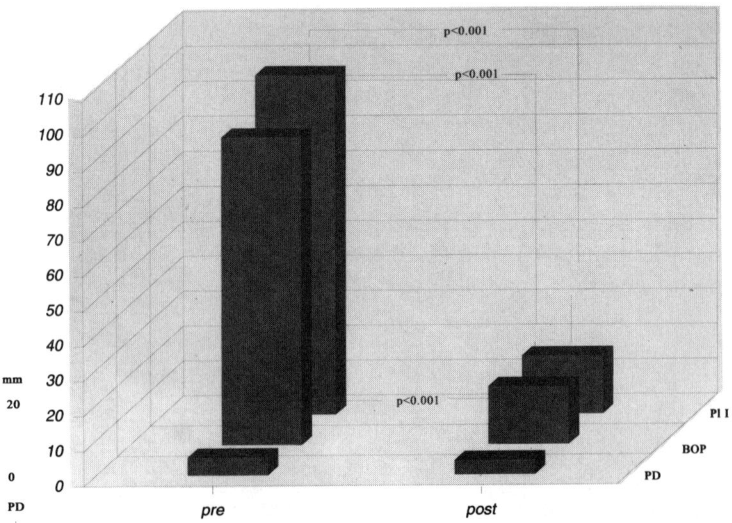

Figure 2. Clinical parameters before and after treatment.

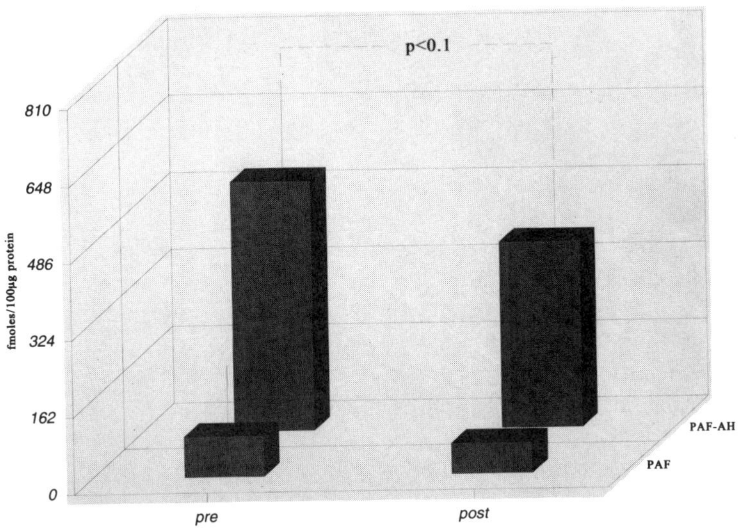

Figure 3. PAF and PAF-AH levels in crevicular fluid of 12 patients with chronic adult periodontitis in pre and post treatment.

The PAF-AH activity in patients with chronic adult periodontitis was two times more than that of healthy volunteers (fig.4).

To exclude contamination of GCF with saliva, PAF-AH activity in whole saliva and PAF activity in crude extracts was measured in parallel. In spite of other studies PAF-AH was detected in saliva. Moreover a decrease in PAF-AH activity observed after the treatment (fig.5).

4. DISCUSSION

In this study the presence in GCF of PAF-like activity and PAF-AH activity was demonstrated. The extracted molecule(s) has PAF-like activity, inducing rabbit platelet aggregation in a dose dependent manner with PAF-aggregation tracing. This substance has not the typical structure of PAF because it doesn't co-migrate with PAF in HPLC. It is eluted within the area of lyso-phosphatidylcholine. Also, the human plasma PAF-AH inactivates this substance with a slower rate than that of PAF. However this molecule acts through PAF receptors because it desensitizes platelets against itself and PAF but not against thrombin. Moreover the aggregation induced by this substance is inhibited by PAF inhibitor BN52021 but not by indomethacin and CP/CPK.

Table 1. Correlation between PAF and PAF-AH levels in adult periodontitis pre and post treatment. ↓: indicates a decrease; ↑: indicates an increase

Number of patients	6.00	3.00	3.00
PAF	↑	↓	↓
PAF-AH	↓	↑	↓

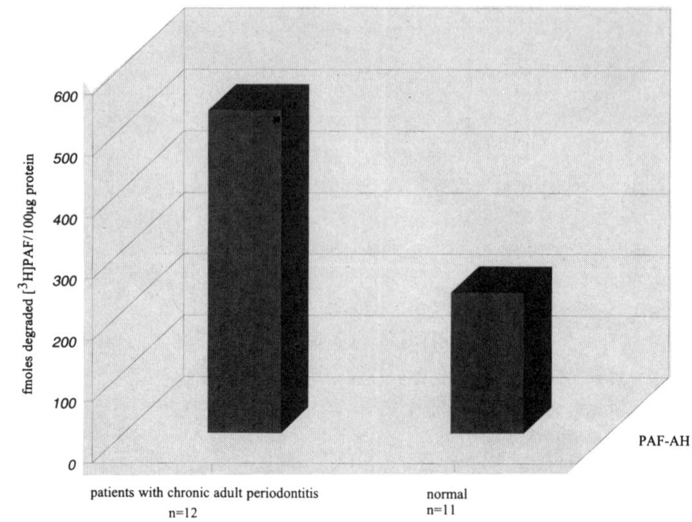

Figure 4. PAF-AH activity in GCF of normal volunteers and in patients with chronic adult periodontitis.

In experimental gingivitis, components of microbial plaque have the capacity to induce the initial infiltration of inflammatory cells, including PMNs, monocytes and lymphocytes[11, 12]. This inflammatory

cells activation may be resulted in the release of PAF analogue. Moreover the inflammatory state may be led to oxidation and /or modification of cellular phospholipids to compounts with PAF-like activity[9].

The transient increase of PAF-like activity at 2nd and 11th days after the start of experimental gingivitis may be due to PAF-AH activity. Also the increase of PAF levels 11 days after the start of the course may by caused by a second burst of stimulus.

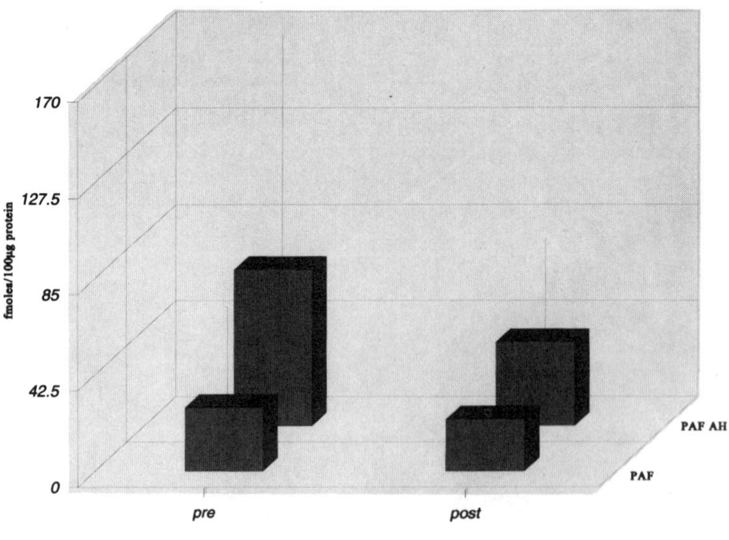

Figure 5. PAF and PAF-AH levels in saliva of 12 patients with chronic adult periodontitis pre and post treatment.

The activity of PAF-AH increased in a statistically significant manner at the end of experimental gingivitis and during the course a negative correlation between the levels of PAF-like activity and PAF-AH activity was observed. Previous studies showed that negative correlation between PAF and PAF-AH was observed only in macrophages[23]. Furthermore in periodontal disease, there is strong activation of periodontal macrophages by bacterial LPS. So we could suppose that the origin of PAF-AH is found mainly to periodontal macrophages. This needs further investigation. Negative correlation between PAF and PAF-AH was also observed in chronic adult periodontitis but PAF levels did not show any statistically significant difference.

In chronic adult periodontitis prolonged bacteria activation and the different bacterial flora have resulted in increased PAF-AH activity compared to those of experimental gingivitis. As it was expected, PAF-AH activity is decreased post-treatment. This may be due to the absence of the essential stimulus of periodontal macrophages, the LPS.

Finally, besides the above findings, PAF-AH, PAF-inhibitors as well as PAF presence is reported for the first time in dental plaque, as well as the presence of PAF-AH in saliva.

REFERENCES

1. Pinckard RN, Ludwing JC, McManus LM. (1988) In Platelet-activating factors; Editors Gallin JI, Goldstein IM, Snyderman R.; pp. 139; Raven Press Ltd., N.Y.
2. Humphrey D.M., Hanahan D.J. and Pinckard R.N. (1982). Lab. Invest, 47: 227.
3. Braquet P. and Rola-Pleszcynski M. (1987) Immunology today, 8: 345.
4. Shaw J.O., Pinckard R.N., Ferrigni K.S., McManus L.M. and Hanahan D.J. (1981). J. Immun., 127: 1250.
5. Chilton F.H., O'Flaherty J.T., Walsh C.E., Thomas M.J., Wyklie R.L., DeChatelet L.R. and Waite B.M. (1982) J. Biol. Chem., 257: 5402.
6. Yasaka T., Boxer L.A. and Baehner R.L (1982). J. Immun., 128: 1939.
7. Pignol B., Henane S., Mencia-Huerta J.-M., Rola-Pleszynsky M. and Braquet P.(1987). Prostsglandins, 33: 931.
8. Rola-Pleszcynsky M., Pignol B., Pouliot C. and Braquet O. (1987). Biochem. Biophys. Res. Commun., 142: 754.
9. Zimmerman GA, Prescott S.M., and McIntyre I.M. (1995) J. Nutr., 125: 1661S.
10. Blank M.L., Lee T. C., Fitzegrald V. and Snyder F. (1981). J.Biol. Chem., 256: 175.
11. Gecno R.J. (1992). J. Periodontololy, 63: 338.
12. Pege R. (1991). J. Periodontol Res., 26: 230.
13. McManus LM., Marze B.T., and Schiess A.V. (1990). J. Peridont. Res., 25: 347.
14. Smal MA and Baldo BA (1991). Lipids, 26: 1144.
15. McDonnell HT (1993). J. Periodontology, 64: 1109.
16. Noguchi K., Morita I. and Murota S., (1989). Arch Oral Biol, 34: 37.
17. Bligh E. G., Dyer W. J. (1959) Can. J. Biochem. Physiol., 37: 911.
18. Andrikopoulos N. K., Demopoulos C. A., Siafaka-Kapadai A. (1986) J. Chromatogr., 363: 412.
19. Demopoulos C. A., Pinckard R. N., Hanahan D. J. (1979) J. Biol. Chem., 254: 9355.
20. Lazanas M., Demopoulos C. A., Tournis S., Koussissis S., Labrakis-Lazanas K., Tsarouxas X. (1988). Arch. Dermatol. Res., 280: 124.
21. Pinckard RN and Ludwing JC (1986) Fed Proc, 45: 856.
22. Loe H., Theilade E., Proglum-Jencen S. (1965). J. Periodontology, 36: 177.
23. Elstad M.R., Stafforini D.M., McLntyre T.M., Prescott S.M. and Zimmerman G.A. (1989). J.Biol. Chem., 264: 8467.

24

STIMULATION OF NF-κB ACTIVATION AND GENE EXPRESSION BY PLATELET-ACTIVATING FACTOR

Richard D. Ye,* Vladimir V. Kravchenko, Zhixing Pan, and Lili Feng

Department of Immunology
The Scripps Research Institute
La Jolla, California 92037

1. ABSTRACT

PAF stimulation of NF-κB activation and transcription of immediate-early genes have been investigated in peripheral blood mononuclear cells and in transfected Chinese hamster ovary cells expressing the cloned PAF receptor. These studies identified a G protein-coupled pathway for PAF induction of gene expression and transcription factor activation, which differs from the mechanisms employed by other immediate-early gene inducers. Potential significance of PAF induced NF-κB activation is discussed.

2. INTRODUCTION

Intercellular communication and signaling are mediated largely by molecules that are either secreted by the cells or attached to the cell surface. The lipid mediator platelet-activating factor (PAF; 1-*O*-alkyl-2-acetyl-*sn*-glycero-3-phosphocholine) is capable of acting in both the paracrine and juxtacrine modes. In its secreted form, PAF has been shown to stimulate the activation of platelet and leukocyte functions, to induce vasoconstriction, and to promote glycogen synthesis, among other functions (1–3). As a cell surface associated molecule, PAF is intimately involved in the interaction between endothelial cells and leukocytes (3). The latter allows a prolonged exposure of the target cell to this potent mediator, raising the possibility that PAF may actively participate in the regulation of gene transcription in such pathological conditions as inflammation and atherogenesis.

In this brief report, we present data derived from our experiments that substantiate the above hypothesis. Our results indicate that PAF is capable of stimulating the activation

* Send correspondence to Dr. Richard D. Ye.

of NF-κB, a transcription activator that induces the expression of many immediate-early genes during inflammation and under other conditions (4). PAF-stimulated NF-κB activation occurs at nanomolar concentrations, requires the cell surface G protein-coupled PAF receptor, and leads to the expression of proinflammatory cytokine and growth factor genes. These preliminary data are likely to inspire further investigations of PAF with respect to its role in transcription regulation.

3. RESULTS

3.1 PAF Stimulates NF-κB Activation in Transfected Cells

Stable expression of the cloned human PAF receptor (5) in the Chinese hamster ovary cells resulted in high affinity binding of PAF and agonist specific calcium mobilization (not shown). Previous studies by others also indicated functional expression of the PAF receptor in the same cells (6). We used this transfected cell line (CHO-PAFR) in our initial experiments to examine whether PAF, a proinflammatory mediator, is capable of activating the transcription factor NF-κB. Nuclear extracts were prepared from CHO-PAFR and vector-transfected CHO cells after stimulation with PAF or with the control NF-κB inducer TNFα, and the κB DNA binding activity was examined by electrophoretic mobility shift assays (EMSA). Exposure of both CHO-PAFR and vector-transfected cells to TNFα resulted in a marked increase in the κB binding activity. In comparison, PAF significantly induced a κB binding activity in CHO-PAFR cells, but only weakly in vector-transfected cells (Fig. 1A). This response required a PAF concentration of approximately

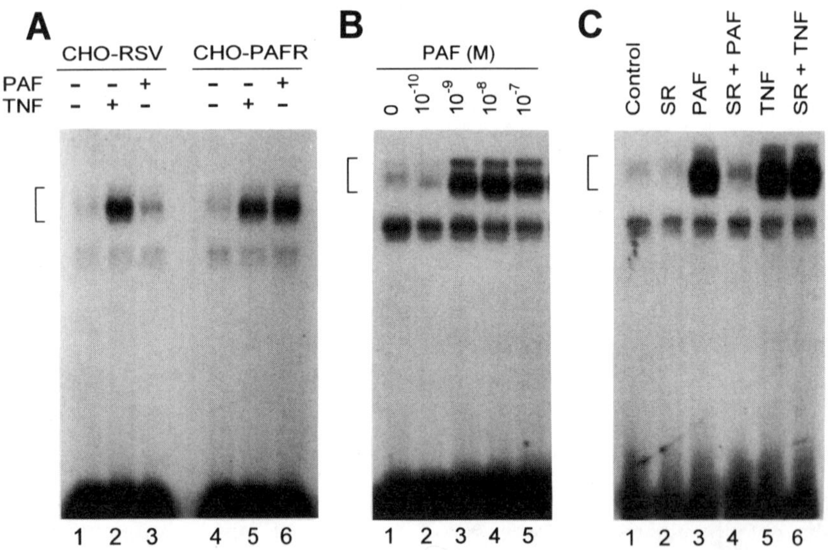

Figure 1. PAF induced κB DNA binding activity in transfected CHO cells expressing the PAF receptor. A. Comparison between vector transfected cells (CHO-RSV) and PAF receptor cDNA transfected cells (CHO-PAFR). B. Dose response of PAF induced κB binding activity in CHO-PAFR cells. C. The inhibitory effects of the PAF antagonist SR 27417 (SR) on κB binding activity. In all of the above experiments, the cells were stimulated for 40 min with PAF (10–100 nM or as indicated) or with TNFα (40 ng/ml). SR 27417 was used at a concentration of 10 nM.

1 nM; however, further increase in PAF concentration had no additional effect on the activation of NF-κB (Fig. 1B). The PAF antagonist SR 27417 blocked PAF-induced κB binding activity but had no effect on the same activity induced by TNFα (Fig.1C). These results indicate that PAF-induced κB binding activity requires the transfected PAF receptor.

The specificity of the observed DNA-protein interaction was examined by adding unlabeled κB oligonucleotide in competition assays. Unlabeled probe at 10-fold and 100-fold molar excess successfully competed with the labeled probe, whereas a 100-fold molar excess of a point mutated oligonucleotide had no effect (7).

We next investigated whether the increased κB activity correlates with gene transcription. Recent studies revealed a κB-like site (GGAAATTCCC) in the murine IκBα promoter, which is stimulated by the p65 subunit of NF-κB in transfected cells (8). We co-transfected the CHO-PAFR cells with two CAT constructs driven by the κB site of the IκBα promoter (8) to determine whether PAF could induce the IκBα promoter-directed CAT gene expression in CHO-PAFR cells. As illustrated in Fig. 2, both PAF and TNFα (as a positive control) greatly increased the amount of CAT activity produced in transiently transfected CHO-PAFR cells by the wild-type IκBα promoter-CAT construct, but not in transfectants containing the construct with a mutant κB site (CGAAATTAAT). Thus, PAF-induced κB binding activity leads to gene transcription activity.

3.2 Biochemical Properties of PAF-Induced NF-κB Activation

NF-κB, in its original form, is defined as a heterodimer consisting of the 50 kDa (p50, NFKB1) and 65 kDa (p65, RelA) protein subunits. Additional proteins have recently been identified that belong to the NF-κB/Rel family (4,9). We performed gel supershift assays to determine whether the PAF-induced DNA-protein complexes contain p50 and p65, components of the prototypic NF-κB. Using nuclear extracts from PAF-stimulated cells, the anti-p50 and anti-p65 antibodies induced a small by significant shift of the

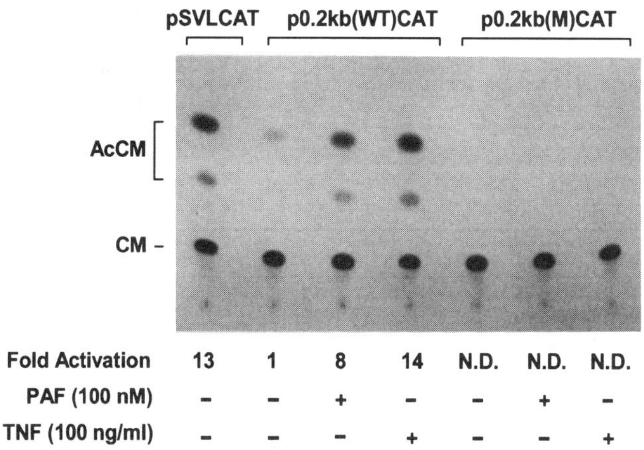

Figure 2. PAF and TNFα induced IκBα promoter-directed CAT gene expression in transiently transfected CHO-PAFR cells. CHO-PAFR cells were transfected with 5 μg of the IκBα promoter-CAT reporter plasmids (specified at the top) or 5 μg of the pSVLCAT plasmid, together with 1 μg of pCMVβ plasmid. CAT activity was determined in these cells after stimulation of PAF or TNFα. The results shown here are representative of 3 CAT assays. N.D., not detectable.

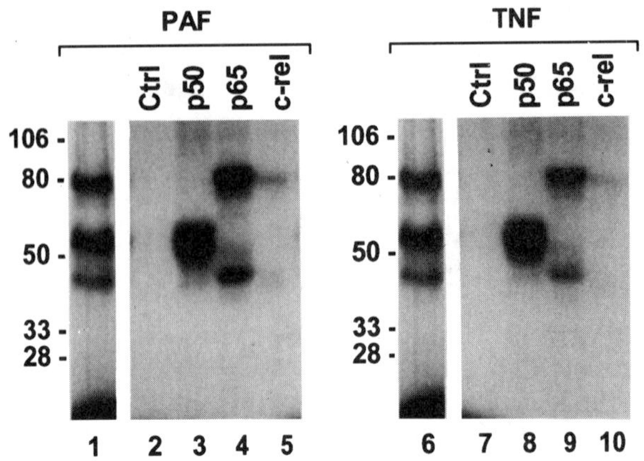

Figure 3. Identification of p50 and p65 in the DNA-protein complexes induced by PAF. [^{32}P]-labeled probe, containing 5′-bromo-2′ deoxyuridine 5′-triphosphate, was incubated with nuclear extracts from PAF- and TNF-stimulated CHO-PAFR cells (lanes 1 and 6, respectively). DNA-Protein complexes in EMSA gel were UV cross-linked, excised, eluted, and resolved by SDS-PAGE. In separate experiments, UV cross-linked samples were immunoprecipitated with non-specific (Ctrl, shown in lanes 2 and 7) or specific antibodies against 3 members of the NFκB/Rel family of proteins as noted at the top of each lane and analyzed by SDS-PAGE.

DNA-protein complexes (7). To further examine the involvement of p50 and p65, complexes formed with a photoreactive κB probe were UV cross-linked in native EMSA gel, eluted, and analyzed by SDS-PAGE. This resulted in three DNA-protein complexes, with relative molecular weight of approximately 80, 60, and 45 kDa, respectively (Fig. 3, lanes 1 and 6). To determine whether p50 and p65 were present in the DNA-protein complexes, UV cross-linked samples were immunoprecipitated with antibodies against p50, p65 and c-Rel. SDS-PAGE analysis of the samples indicated recognition of the 60 kDa adduct by the anti-p50 Ab; whereas the anti-p65 Ab immunoprecipitated a major species of 80 kDa and a minor species of 45 kDa, respectively (lanes 3–5). The sizes of these adducts are similar to those reported with cells from other species. The 45 kDa minor species is likely to be the proteolytic product of p65 (10). Similar patterns of immunoprecipitation were observed in samples stimulated with TNFα (Fig. 3, lanes 8–10). Thus, these results suggest that PAF-induced κB binding activity consists of the p50/p65 heterodimer.

It has been known from previous studies that NF-κB activation is the result of IκB degradation and nuclear translocation of the p50/p65 proteins (11,12). Through an autocrine regulatory mechanism, NF-κB stimulates the transcription of IκB, which then inhibits the action of NF-κB (12). We examined whether PAF induced NF-κB follows the same time course. PAF induced a rapid and sustained NF-κB activation, observed within 15 min and lasted for at least 2.5 hours (Fig. 4, upper panel). The appearance of the DNA-protein complex coincided with the disappearance of IκBα, as detected by immunoblotting with an anti-IκBα antibody (Fig. 4, middle panel). This loss of IκBα was followed by its synthesis at 60 min, and by 150 min IκBα reached the prestimulation level. Results from our experiments indicate that the messenger RNA for IκBα was upregulated by nearly 2.5-fold following PAF stimulation for 60 min (Fig. 4, bottom panel), consistent with findings by others using different ligand and cell systems (8,13,14).

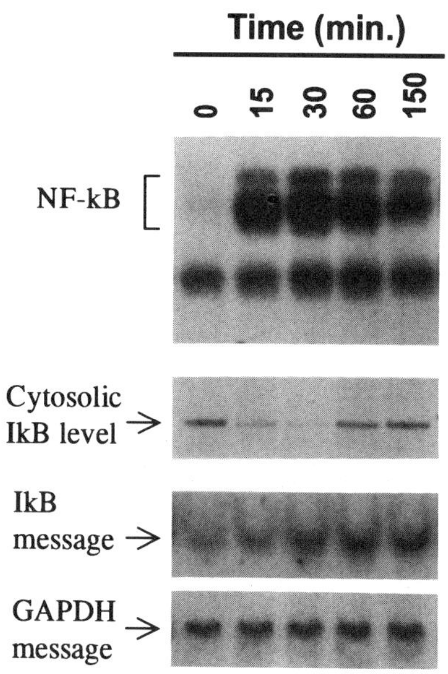

Figure 4. Time-dependent stimulation of the κB binding activity and degradation of IκBα. Shown are simultaneous measurements of NF-κB activation by EMSA (top panel), cytosolic IκB content by western blot (second panel), and the message for IκB by Northern blot (third panel), as a function of time after stimulation with 10 nM PAF. GAPDH was used as a control for Northern blot.

3.3 PAF-Induced NF-κB Activation in Other Cells and the Physiological Relevance

We have also examined PAF-induced NF-κB in other cells that express endogenous PAF receptors. In the murine P388D_1 macrophage and the human ASK.0 B cell lines, PAF induced NF-κB activation similar to that observed in the transfected CHO-PAFR cells. This response, too, was inhibited by the PAF receptor antagonist SR 27417 (7).

In human peripheral blood monocytes, PAF-stimulated NF-κB activation correlated with PAF-induced upregulation of the heparin-binding EGF-like growth factor (HB-EGF) and IL-1β messenger RNA levels with respect to PAF doses and time course (Fig. 5). Although the transcription of IL-1β gene could be stimulated by NF-κB, it was not known that HB-EGF gene transcription could be regulated by NF-κB prior to these experiments. To establish a causal relationship, we examined the effect of pyrrolidine dithiocarbamate (PDTC) on both the gene expression and NF-κB activation in PAF-stimulated monocytes. PDTC is an antioxidant inhibitor of NF-κB, which blocks IκB phosphorylation and degradation (15). The result, shown in Fig. 6, indicated that PDTC reduced the formation of the NF-κB•DNA complex in gel shift assays, at the same time abolished PAF-induced HB-EGF mRNA in monocytes. Thus, we conclude that NF-κB activation is responsible for the transcription activation of the HB-EGF gene. This notion is supported by a recent finding of two κB sites in the promoter region of the mouse HB-EGF gene (M. Klagsbrun, personal communication). Furthermore, we have studied PAF-induced HB-EGF expression in the rat glomerular epithelial cells (GEC) and found an increase in the secretion of HB-EGF by these cells after PAF stimulation (16).

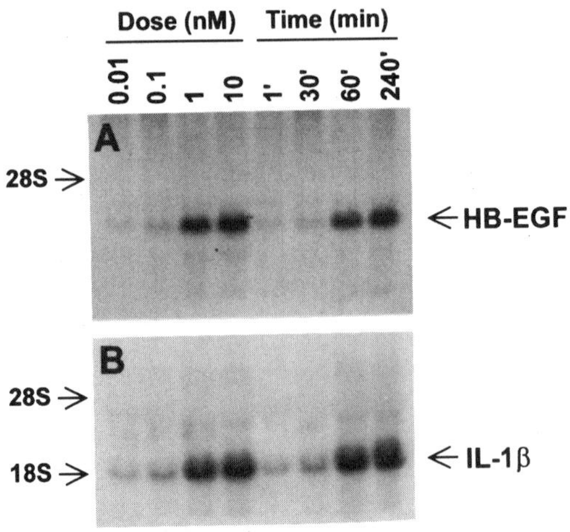

Figure 5. Accumulation of cytoplasmic RNA in PAF stimulated monocytes. Total RNA from monocytes was prepared after stimulation with PAF (100 nM) for various times or for 1 h at different concentrations. Twenty micrograms of the total RNA was analyzed by Northern blot and probed with cDNA for HB-EGF (A) or IL-1β (B).

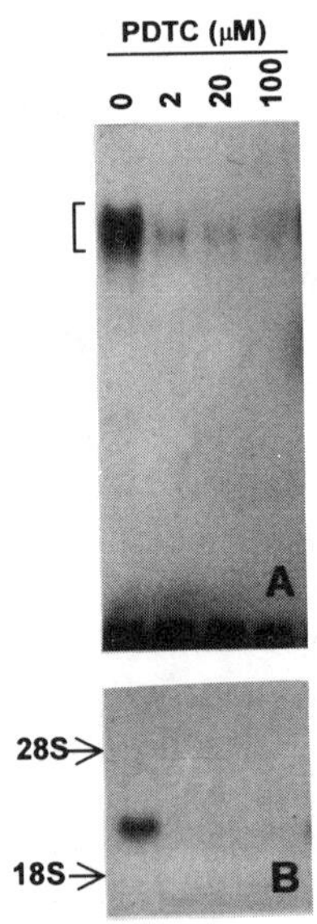

Figure 6. Inhibition of PAF-induced κB binding activity and HB-EGF gene expression by PDTC. A. PDTC inhibits PAF-induced κB binding activity. Nuclear protein extracts were prepared from monocytes preincubated for 40 min with PDTC at the indicated concentrations, followed by PAF stimulation (100 nM) for 60 min. The autoradiograph of a representative EMSA gel is shown. B. PDTC inhibits PAF-induced HB-EGF gene expression. Cells pretreated with PDTC, as indicated above, were stimulated with PAF (100 nM) for 60 min. Total cellular RNA was extracted for Northern blot. The autoradiograph of a Northern blot is shown with positions of the ribosomal RNAs indicated.

3.4 The Signal Transduction Pathway Leading to PAF-Stimulated NF-κB Activation

NF-κB activation is believed to be the consequence of IκB phosphorylation and degradation that leads to the release and nuclear translocation of the p50/p65 heterodimer (12). The NF-κB activation induced by PAF is no exception, as shown by our data from the CHO-PAFR experiments. Although the kinase(s) that phosphorylate IκB *in vivo* has not been identified to date, it is known that several signaling molecules are involved in the activation of NF-κB. These include oxygen radicals, ceramide, and a number of candidate protein kinases.

In an attempt to delineate the signaling pathway for PAF-induced NF-κB activation, we discovered some differences between this G protein-coupled receptor mediated pathway and the NF-κB response stimulated by other agonists such as TNFα. First, PAF-induced NF-κB activation requires the coupling of the PAF receptor to G proteins, as bacterial toxins that block this coupling can inhibit NF-κB activation. However, the exact G protein that is involved in this response seems to vary in different cell types. For example, in peripheral blood monocytes, PAF-induced NF-κB activation and gene transcription is subject to blockade by pertussis toxin (17); whereas in the CHO cells cholera toxin appears to inhibit NF-κB activation induced by PAF (7). This difference is probably due to the fact that PAF receptor can couple to more than one G protein. We do not know at this time why in the CHO cells PAF receptor preferentially couples to a cholera toxin sensitive pathway; in the few G protein-coupled receptors we have examined (e.g., the bradykinin receptor) they are almost invariably coupled to pertussis toxin sensitive pathways for NF-κB activation. Secondly, certain protein kinase inhibitors including herbimycin A and staurosporine can block PAF-induced NF-κB activation and gene transcription, but have no inhibitory effect on TNFα-induced NF-κB activation (Fig. 7A). Thirdly, two phosphatidyl inositol-3 kinase (PI3K) inhibitors selectively blocked PAF-induced NF-κB activa-

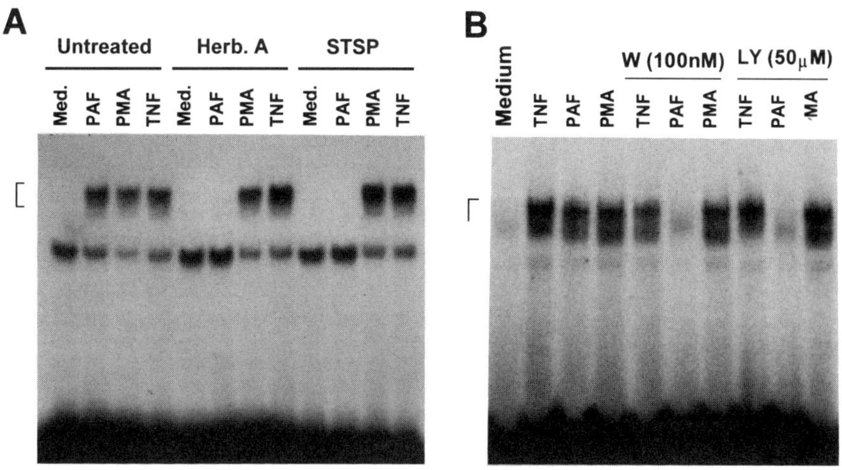

Figure 7. Selective inhibition of PAF induced κB binding activity by protein kinase inhibitors. A. Autoradiograph of EMSA showing the differential effects of herbimycin A (Herb. A, 1 μM) and staurosporine (STSP, 200 nM) on κB binding activity induced by PAF (100 nM), PMA (100 nM) and TNFα (40 ng/ml). Mononuclear cells were treated with the inhibitors for 30 min followed by agonist stimulation for 60 min. B. Similar to A, but mononuclear cells were treated with wortmannin or LY294002 for 5 min before agonist stimulation.

tion but did not affect the NF-κB activation induced by either TNFα or phorbol esters (Fig. 7B). The last two findings suggest a signaling pathway consisting the PI3K and protein tyrosine phosphorylation for PAF induction of NF-κB activation.

4. CONCLUSION

The finding that PAF stimulates NF-κB activation extends our current knowledge of this pluripotent lipid mediator. It also suggest a possible mechanism for the role that PAF plays in the course of inflammation and other pathological conditions. Our results complement recent reports that PAF induces the expression of c-fos and promotes DNA synthesis, and together suggest an important function of this molecule in gene expression and cell growth. Most recently, it has been reported that conjugation of P-selection potentiates the function of PAF in gene transcription (18). Therefore, PAF stimulation of transcription activation may be undertaken in the context of cell-cell interactions. The interplay of PAF and other molecules in transcription regulation is a topic that requires further investigation. Future studies will be directed to the relevance of PAF-stimulated transcription activation in various disease models, utilizing the bountiful PAF receptor antagonists developed during the past decade. These investigations are likely to result in a better understanding of the newly discovered functions of PAF and related mediators.

ACKNOWLEDGMENTS

This work was done during the tenure of an Established Investigatorship (to R.D.Y.) from the American Heart Association, and was supported in part by USPHS grants GM46572 and AI33503. Figures 1–4 were reproduced from Kravchenko et al (1995), and Figures 5–6, from Pan et al (1995), with permissions from the Journal of Biological Chemistry.

REFERENCES

1. Hanahan, D. J. (1986) Platelet activating factor: A biologically active phosphoglyceride. *Ann. Rev. Biochem.* **55**, 483–509
2. Snyder, F. (1990) Platelet-activating factor and related acetylated lipids as potent biologically active cellular mediators. *Am. J. Physiol. Cell Physiol.* **259**, C697-C708
3. Zimmerman, G. A. , Prescott, S. M. , and McIntyre, T. M. (1992) Platelet-activating factor: a fluid-phase and cell-associated mediator of inflammation. In *Inflammation: Basic principles and clinical correlates* (Gallin, J. I. , Goldstein, I. M. , and Snyderman, R. , eds) pp. 149–176, Reven Press, New York
4. Baeuerle, P. A. and Henkel, T. (1994) Function and activation of NF-κB in the immune system. *Annu. Rev. Immunol.* **32**, 141–179
5. Ye, R.D. , Prossnitz, E. R. , Zou, A., and Cochrane, C.G. (1991) Characterization of a human cDNA that encodes a functional receptor for platelet activating factor. *Biochem. Biophys. Res. Commun.* **180**, 105–111
6. Honda, Z. ; Takano, T. ; Gotoh, Y. ; Nishida, E. ; Ito, K. and Shimizu, T. (1994) Transfected platelet-activating factor receptor activates mitogen-activated protein (MAP) Kinase and MAP kinase Kinase in Chinese hamster ovary cells. *J. Biol. Chem.* **269**, 2307–2315
7. Kravchenko, V. V. ; Pan, Z. ; Han, J. ; Herbert, J. M. ; Ulevitch, R. J. and Ye, R. D. (1995) Platelet-activating factor induces NF-kappa B activation through a G protein-coupled pathway. *J Biol Chem* **25**, 14928–14934
8. Chiao, P. J. ; Miyamoto, S. and Verma, I. M. (1994) Autoregulation of IκBα activity. *Proc. Natl. Acad. Sci. U. S. A.* **91**, 28–32

9. Liou, H. C. and Baltimore, D. (1993) Regulation of the NF-κB/rel transcription factor and IκB inhibitor system. *Cur. Biol.* **5**, 477–487
10. Oeth, P. A. ; Parry, G. C. ; Kunsch, C. ; Nantermet, P. ; Rosen, C. A. and Mackman, N. (1994) Lipopolysaccharide induction of tissue factor gene expression in monocytic cells is mediated by binding of c-Rel/p65 heterodimers to a kappa B-like site. *Mol Cell Biol* **14**, 3772–3781
11. Baeuerle, P. A. and Baltimore, D. (1988) IκB: A specific inhibitor of the NF-κB transcription factor. *Science.* **242**, 540–546
12. Finco, T. S. and Baldwin, A. S.,Jr.(1995) Mechanistic aspects of NF-kappa B regulation: The emerging role of phosphorylation and proteolysis. *Immunity* **3**, 263–272
13. Sun, S. C. ; Ganchi, P. A. ; Ballard, D. W. and Greene, W. C. (1993) NF-κB controls expression of inhibitor IκBα: Evidence for an inducible autoregulatory pathway. *Science.* **259**, 1912–1915
14. Brown, K. ; Park, S. ; Kanno, T. ; Franzoso, G. and Siebenlist, U. (1993) Mutual regulation of the transcriptional activator NF-κB and its inhibitor, IκB-α. *Proc. Natl. Acad. Sci. U. S. A.* **90**, 2532–2536
15. Schreck, R. ; Meier, B. ; Männel, D. N. ; Dröge, W. and Baeuerle, P. A. (1992) Dithiocarbamates as potent inhibitors of nuclear factor κB activation in intact cells. *J. Exp. Med.* **175**, 1181–1194
16. Feng, L. ; Pan, Z. ; Xia, Y. ; Wilson, C. B. and Ye, R. D. (1994) Platelet-activating factor induces expression of the growth factor, heparin-binding EGF-like growth factor in glomerular epithelial cells. *J. Am. Soc. Nephrol.* **5**, 679(Abstract)
17. Pan, Z. ; Kravchenko, V. V. and Ye, R. D. (1995) Platelet-activating factor stimulates transcription of the heparin-binding epidermal growth factor-like growth factor in monocyte. Correlation with an increased kappa B binding activity. *J. Biol. Chem* **270**, 7787–7790
18. Weyrich, A. S. ; McIntyre, T. M. ; McEver, R. P. ; Prescott, S. M. and Zimmerman, G. A. (1995) Monocyte tethering by P-selectin regulates monocyte chemotactic protein-1 and tumor necrosis factor-alpha secretion. *J. Clin. Invest.* **95**, 2297–2303

25

TYROSINE KINASE ACTIVATION BY PAF LEADS TO DOWNSTREAM GENE EXPRESSION

Shivendra D. Shukla

Department of Pharmacology
School of Medicine
University of Missouri
Columbia, Missouri 65212

Developments in the past decade have highlighted the complexity of signal transduction pathways activated by PAF receptor (PAF-R) (Shukla, 1992; Shimizu *et al.*, 1992). Molecular details and identities of individual components involved in these signalling steps have yet to be resolved. In a broader perspective, PAF receptor activates two major pathways (a) mediated via phospholipases (of types C, D, and A_2) and (b) protein kinases (i.e., PKC, tyrosine kinase, MAP kinase). Further, Ca^{2+} mobilization (intracellular release, influx) also play important roles (James-Kracke *et al.*, 1994). How these multiple pathways lead to expression of genes induced by PAF is the topic of this article.

PAF-R, pp60$^{c\text{-}src}$ AND PLC NEXUS IN PLATELETS

Tyrosine kinase inhibitors (genistein, erbstatin, 3–5 dihydroxycinnamate, tyrphostin) inhibited PAF-induced aggregation. Genistein inhibited PAF-induced PLC activation and protein phosphorylation (Shukla, 1992). Interestingly, in platelets, genistein (250 μM) activated MAP kinase by 2–3 fold (Shukla *et al.*, 1995). Thus, elevated MAPK inhibited PAF-induced platelet aggregation. Although PAF modestly activates MAPK in platelets (Samiei *et al.*, 1993), this could be part of a feedback inhibitory mechanism to restrain platelets from aggregation. Further, PAF caused phosphorylation of pp60$^{c\text{-}src}$ (independent of extra- or intracellular Ca^{2+}) increased its intrinsic tyrosine kinase activity (Zhu and Shukla, 1993) and caused its translocation from cytoplasm to the membrane fraction of platelets (see Dhar and Shukla, 1994). In a further development, it has been demonstrated that introduction of pp60$^{v\text{-}src}$ monoclonal antibody into rabbit platelets blocks PAF activated responses (e.g., aggregation, PLC-mediated IP_3 production) in a specific manner (Dhar and Shukla, 1994). It is proposed that pp60$^{v\text{-}src}$ mAb interfered with the interaction among PAF-R, pp60$^{c\text{-}src}$ and PLC and thus inhibited downstream responses of PAF. Thus, pp60$^{c\text{-}src}$ modulates PAF-R linked PLC activity. At this stage, the role of pertussis toxin (PT) sensitive G-protein remains to be ascertained in this process.

Platelet-Activating Factor and Related Lipid Mediators 2
edited by Nigam *et al.*, Plenum Press, New York, 1996

TYROSINE KINASE-MEDIATED GENE EXPRESSION BY PAF IN A431 CELLS

PAF activates c-fos and TIS1 genes in human epidermoid A431 cells. Since these cells have well-defined tyrosine kinase and PLC- γ1 pathways, they were used in studies with PAF. In A431 cells, PAF activated IP_3 production was greater than EGF, but the phosphorylation of PLC- γ1 was lower than that seen with EGF (Thurston *et al.,* 1993). Interestingly, pertussis toxin only partially abolished the PAF-stimulated IP_3 production, whereas genistein totally eliminated this response. Therefore, a dual mechanism for PLC activation by PAF exists in A431 cells in which tyrosine kinase is common to both a membrane-associated PT-sensitive G-protein and a G-protein independent mechanisms. Involvement of such a dual mechanism in c-fos and TIS1 expression by PAF has also been observed. Treatment of A431 cells with PT partially inhibited the c-fos expression whereas genistein totally inhibited this response. It is noteworthy that PAF stimulated tyrosine phosphorylation of several proteins in A431 cells including $pp60^{c\text{-}src}$ (Thurston *et al.*, 1993). Both PKC and tyrosine kinase play roles in these gene expressions (Tripathi *et al.*, 1992). It was further demonstrated that activation of tyrosine kinase precedes the PKC activation in the signal transduction cascade leading to gene expression (Tripathi *et al.*, 1992). Expression of different genes by PAF in a variety of cells (see Bazan *et al.,* 1993) is predicted to be mediated through the early activation of tyrosine kinases. It is relevant here to note that a recent report has demonstrated PAF activation of NF- κB through a G-protein coupled pathway and may have potential significance in PAF-induced gene expressions (Kravchenko *et al.*, 1995). Several interlinked signalling pathways activated by PAF can thus lead to transcriptional activation of genes. Some of the pathways known to be activated by PAF and likely to be involved in gene expression are schematically presented in Figure 1.

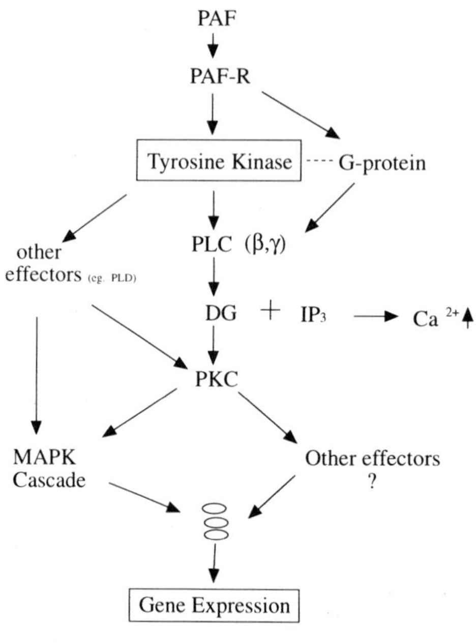

Figure 1. Activation of tyrosine kinase by PAF triggers the downstream signalling cascade for gene expressions. DG, diaglyceride; G-protein, GTP binding protein; IP_3, inositol trisphosphate; MAPK, mitogen activated protein kinase; PAF, platelet activating factor; PAF-R, platelet activating factor receptor; PKC, protein kinase C; PLC, phospholipase C; PLD, phospholipase D

ACKNOWLEDGMENTS

The work described was supported by National Institutes of Health grant DK35170 and by a Research Career Development Award DK01782. Typing assistance by Ms. Judy Richey is gratefully acknowledged.

REFERENCES

Bazan, N. G. and Doucet, J. P. Platelet activating factor and intracellular signalling pathways that modulate gene expression, in *Platelet Activating Factor Receptor: Signal Mechanisms and Molecular Biology*, Edited by S. D. Shukla, CRC Press, Boca Raton, 1993, pp. 137–146.

Dhar, A. and Shukla, S. D. Electro-transjection of pp60 $^{v\text{-}src}$ antibody inhibits activation of phospholipase C in platelets: A new mechanism for PAF responses. J. Biol. Chem. 269:9123–9127, 1994.

James-Kracke, M., Sexe, R. and Shukla, S. D. Picomolar PAF mobilizes Ca to change platelet shape without activation PLC or PKC: Simultaneous fluorometric measurement of $[Ca]_i$ and aggregation. J. Pharm. Exp. Ther. 271:824–831, 1994.

Kravchenko, V.V., Pan, Z., Han, J., Herbert, J., Ulevitch, R.J., and Ye, R.D. Platelet activating factor induces NF-κB activation through a G-protein-coupled pathway. J. Biol. Chem. 270:14928–14934, 1995.

Samiei, M., Sanghera, J.S. and Pelech, S.L. Activation of myelin basic protein and S6 peptide kinases in phorbol ester- and PAF-treated sheep platelets. Biochim. Biophys. Acta 1176:287–298, 1993.

Shimizu, T., Honda, Z., Nakamura, M., Bito, H. And Izumi, T. Platelet activating factor receptor and signal transduction. Biochem. Pharm. 44:1001–1008, 1992.

Shukla, S. D. Platelet activating factor receptor and signal transduction mechanisms. FASEB J. 6:2296–2301, 1992.

Shukla, S.D., Dhar, A. and Kansra, S. Tyrosine kinase and MAP kinase in PAF responses in platelets. Thromb. and Hemostasis (in press), 1995.

Thurston, Jr., A. W., Rhee, S. G. and Shukla, S. D. Role of G-protein and tyrosine kinase in PAF activation of phospholipase C in A431 cells: Proposal for dual mechanisms. J. Pharmacol. Exp. Ther. 266:1106–1112, 1993.

Tripathi, Y. B., Lim, R. W., Fernandez-Gallardo, S., Kandala, J., Guntaka, R. V. and Shukla, S. D. Involvement of tyrosine kinase and protein kinase C in platelet activating factor induced c-fos gene expression in A431 cells. Biochem. J. 286:527–533, 1992.

Zhu, C. Y. and Shukla, S. D. Increased tyrosine kinase activity in pp60 $^{c\text{-}src}$ immunoprecipitate from PAF-stimulated human platelets: *In vitro* phosphorylation of a synthetic peptide. Life Sci. 53:175–183, 1993.

HETEROCYCLIC ALKYLPHOSPHOLIPIDS WITH AN IMPROVED THERAPEUTIC RANGE

P. Hilgard, J. Stekar, T. Klenner, G. Nössner, B. Kutscher, and J. Engel

ASTA Medica AG
Experimental Cancer Reseach Department
Weismüllerstr. 45, D-60314 Frankfurt am Main, Germany

Hexadecylphosphocholine (INN: Miltefosine) is the prototype of a new class of antitumor phospholipids. Its specificially formulated solution (Miltex®) is an approved drug in Germany and has become a valuable tool in the treatment of cutaneous breast cancer and other malignant lesions (1). The systemic treatment with an oral miltefosine formulation was, however, relatively toxic to the GI-tract. Therefore, the drug could not be adequately dosed and in initial clinical studies the overall response rates remained low (2).

It was the aim of our analogue research to identify alkylphosphocholine derivatives with an improved therapeutic index. Since some of the side effects observed in experimental animals and humans were reminiscent of parasympathomimetic effects, the hypothesis was put forward that one of miltefosine's major metabolites, phosphocholine, might be involved in the toxicity of the compound (3). The chemical similarity of phosphocholine with acetylcholine - one of the most potent parasympathicomimetic agents - is indeed striking (Figure 1). Therefore, novel miltefosine congeners were synthesized in which the choline moiety was replaced by heterocyclic nitrogen compounds. Among the first structures identified, D-20133, in which a piperidine was attached to the alkylchain, showed outstanding properties. In comparison to miltefosine this compound was on a molecular basis generally more active in vitro and equally active in various in vivo systems. In addition, the acute oral toxicity (LD_{50}) was considerably reduced. The emetic potential of D-20133 tested in ferrets, was distinctly less pronounced than that of miltefosine. The pharmacology of this compound was previously described in great detail (4).

The above findings suggested, that the substitution of the choline moiety by a heterocyclic structure such a piperidine was indeed a feasible and effective way to circumvent GI-tract toxicity of alkylphosphocholines. In the hope that additional molecular alterations would further improve the pharmacological and toxicological properties, more novel alkylphospholipids were synthesized along these lines. Finally, D-21266 was shown to fulfill most of the required characteristics (Figure 2). The in vitro efficacy of D-21266,

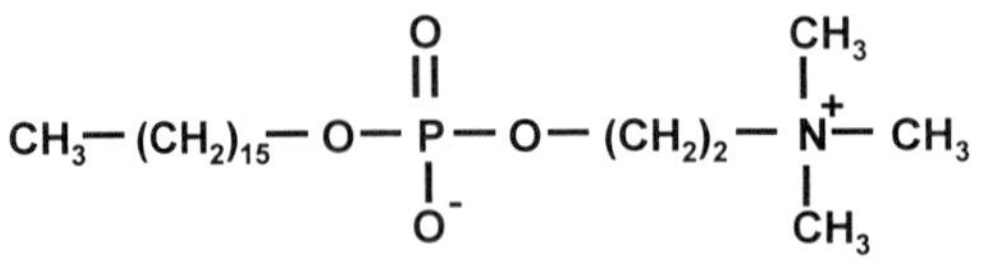

Hexadecylphosphocholine (INN: Miltefosine)

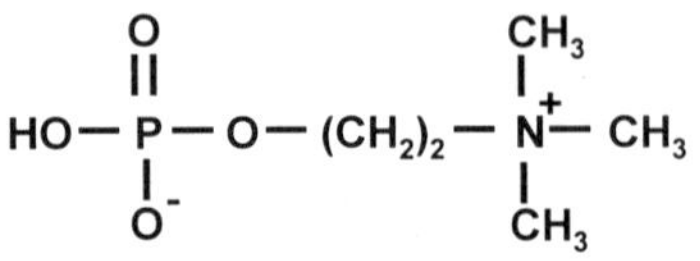

Phosphocholine

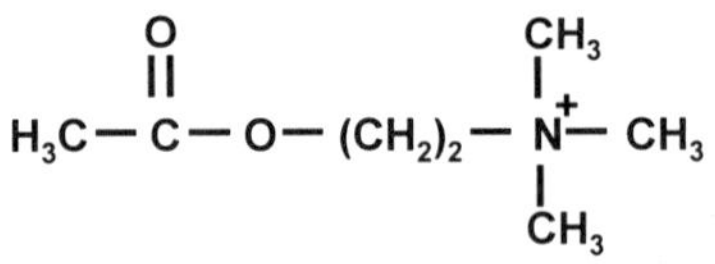

Acetylcholine

Figure 1. Chemical structures of hexadecylphosphocholine, its major metabolite phosphocholine, and acetylcholine.

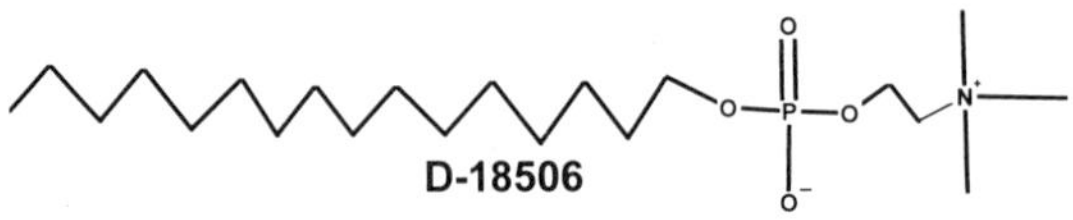

D-18506

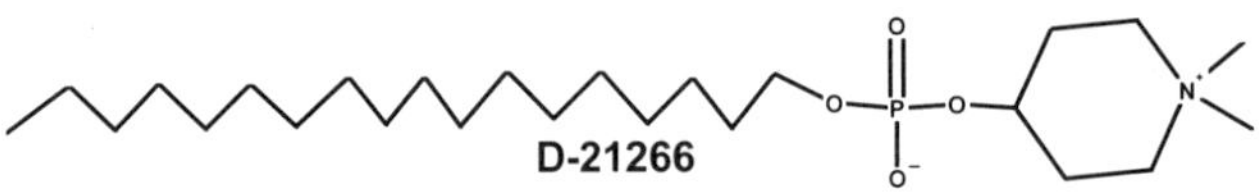

D-21266

Figure 2. Chemical structures of hexadecylphosphocholine (D-18506) and D-21266.

Table 1. Growth inhibitory activities of miltefosine (D-18506), D-20133 and D-21266 against selected human and murine cell lines

		$IC50_{50}$ (µg/ml)		
Cell line		D-18506	D-20133	D-21266
Mouth	KB	1.6	1.3	1.0
Larynx	HEp-2	6.3	4.8	3.4
Lung	PC-1	5.6	3.6	3.2
	LC-1sq	7.4	4.4	4.3
	LU65A	2.1	0.9	0.8
Prostate	LNCaP	5.5	3.8	2.7
Colon	LS180	5.8	8.0	6.7
	DLD-1	5.5	4.5	4.9
	HCT-8	6.0		5.8
Liver	SK-HEP-1	4.2		4.3
Pancrea	MIA-PaCa-2	6.1		4.1

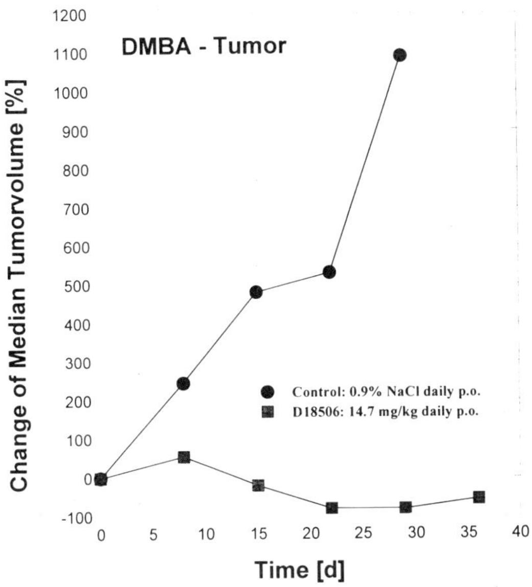

Figure 3. Growth curves of control and miltefosine (D-18506)-treated DMBA tumors in rats.

in comparison to miltefosine and D-20133, is summarized in table 1. Obviously, both new compounds were more active than the parent compound in the human cell lines tested.

The activity of D-21266 against dimethylbenzantrathene (DMBA) induced rat tumors was superior to that of miltefosine (Figure 3, 4). The reduction of body weight as a global indicator of gastrointestinal toxicity, however, was considerably less pronounced than with miltefosine (Figure 5, 6).

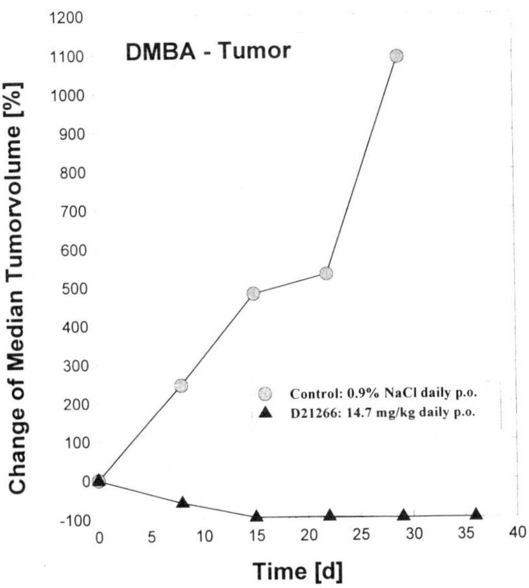

Figure 4. Growth curves of control and D-21266-treated DMBA tumors in rats.

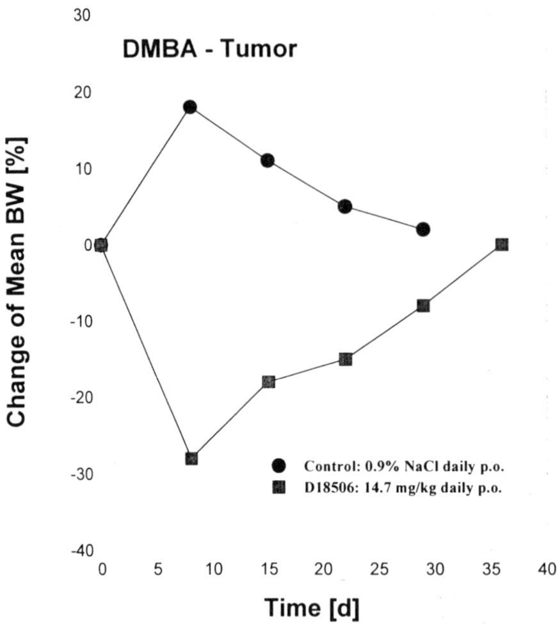

Figure 5. Change of body weight in DMBA tumor bearing rats following saline or miltefosine (D-18506) treatment.

Studies into the schedule dependency of D-21266 have revealed, that an initially high dose followed by low oral maintenance doses gives the best therapeutic results in DMBA-tumor bearing animals with extremely good GI-tract tolerance (Figure 7).

Interestingly, even very low daily doses without an initial loading dose exert significant therapeutic effects, however dependent on the dose, it takes some time until this becomes evident (Figure 8). If such a response pattern would also occur in clinical trials, a completely

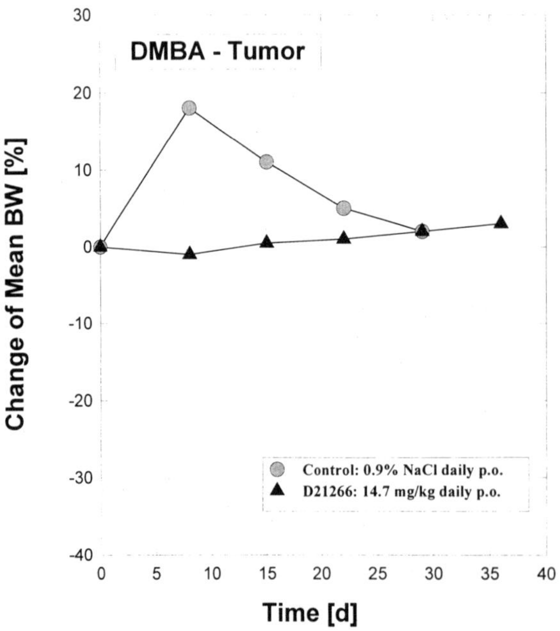

Figure 6. Change of body weight in DMBA tumor bearing rats following saline or D-21266 treatment.

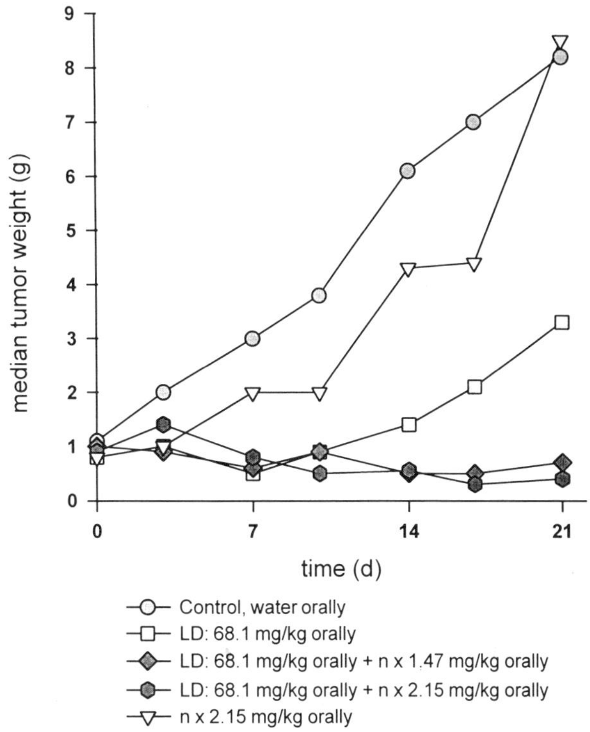

Figure 7. D-21266 in the DMBA-induced mammary carcinoma (Loading Dose + Maintenance Dose).

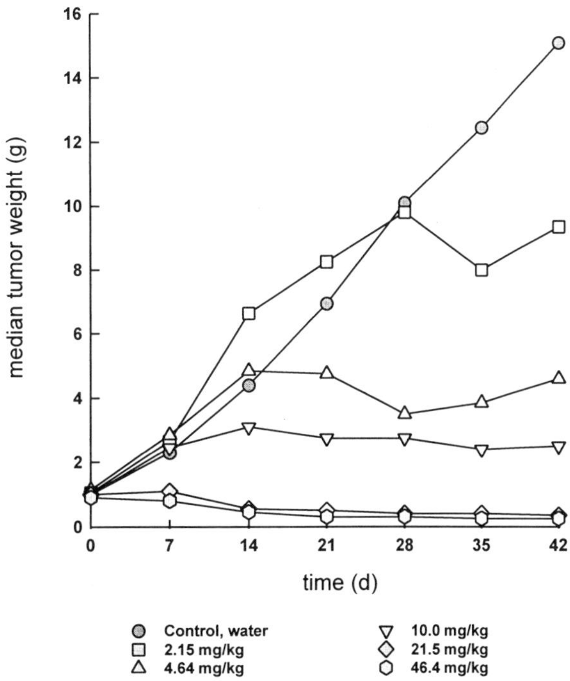

Figure 8. D-21266 at different doses in the DMBA-induced mammary carcinoma-25 x orally.

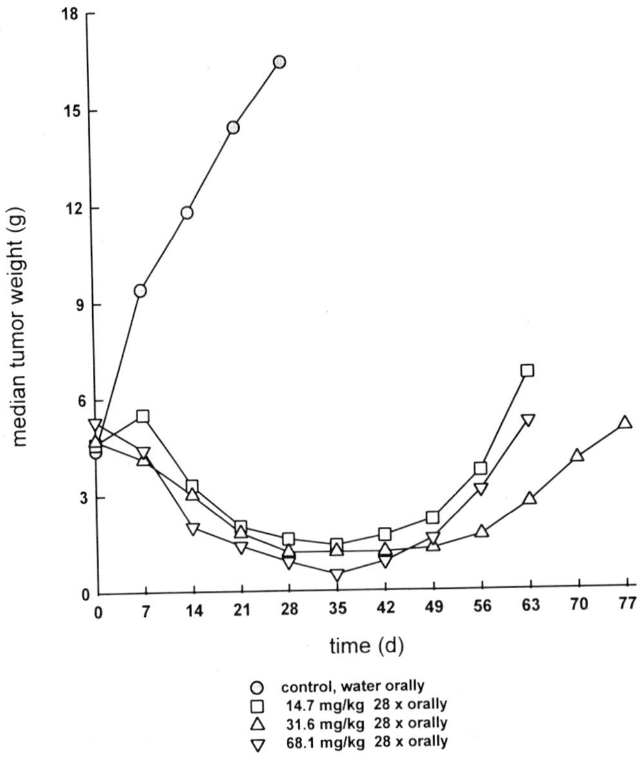

Figure 9. Antineoplastic effect of D-21266 given orally on 28 consecutive days to rats bearing large DMBA-induced mammary carcinomas.

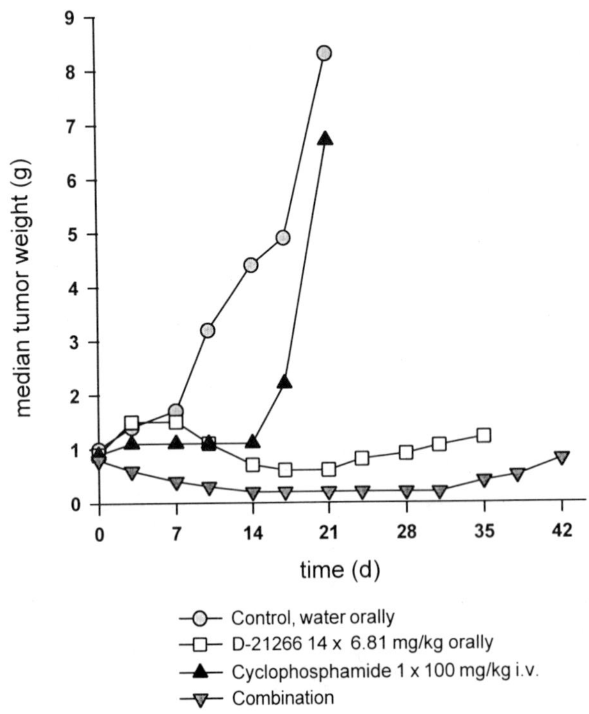

Figure 10. Cyclophosphamide in combination with D-21266 in the DMBA-induced mammary carcinoma.

new study concept is required to discover such an activity, which might become apparent only after a longer initial therapy. Considering the fact that for the patient long-term cancer control it is more important than short lived complete remissions, this therapeutic approach might open up new avenues, paticularly in the treatment of slowly growing tumors.

As shown in figure 9, even large DMBA-tumors respond to daily treatment with D-21266, whereby the dose-response-relationsship is not very pronounced in the doses tested. However, it is important to note, that the responses are long-lasting, suggesting, that even intermittent treatment cycles might be a feasible approach for dose-finding studies in the clinic.

Combination therapies have been performed with cyclophosphamide (Figure 10), adriamycin and cisplatin. There was some evidence of additive or synergistic effects with cyclophosphamide, confirming previous combination therapy data with miltefosine (5). It is worth mentioning that there was no evidence of additive toxicities with any of the tested combinations.

Based on the assumption that the cumulative weight loss of rats following miltefosine treatment reflects the consequence of the GI-tract toxicity encountered in clinical studies, D-21266 should have a significantly superior therapeutic index. This is further supported by the fact that the weight loss after miltefosine is preceeded by a major decrease in food up-take, a clinical phenomenon regularly encountered as loss of appetite in the phase II-clinical- trials with the compound. In additional experimental studies carried out in our laboratories, D-21266 did not reduce food-intake over a wide dose range. In addition to the better tolerance of D-21266 itself superior schedules of administration may further improve the tolerance of the treatment. Since there was some clinical evidence in the phase II-studies with miltefosine, that parts of the GI-tract toxicity were mediated by central nervous reactions, we are currently conducting experiments in order to identify centrally active pharmacological principles which might counteract this effect.

The mode of action of D-21266 is likely to be similar to that of other alkylphosphocholines. It appears that these compounds mediate biochemical signals (6) which subsequently activate appropriate genes to either induced differentiation or apoptosis or both (Figure 11). If this is the case also in vivo, long-term treatment becomes mandatory and rapid tumor responses may not be expected.

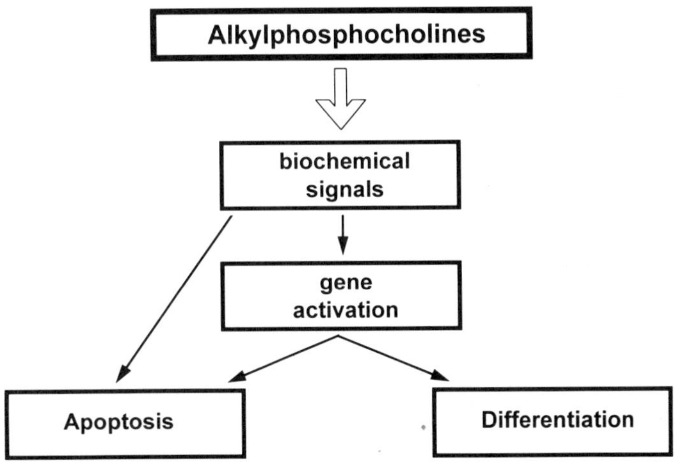

Figure 11. Proposed mode of action of alkylphosphocholines, including D-21266.

This work was partially supported by a grant from the Ministry of Education, Science, Research and Technology (BMBF), Bonn-Bad Godesberg.

REFERENCES

1. Burk, K., David, M., Junge, K., Sindermann, H. Overview on the clinical development of miltefosine solution (Miltex®) for the treatment of cutaneous breast cancer. Drugs of Today 30, (Suppl. B): 59–72, 1994
2. Verweij, J., Gandia, D., Planting, A.S.Th., Stoter, G., Armand, J.P. Phase II study of oral miltefosine in patients with squamous cell head and neck cancer. Eur. J. Cancer 29A, 778–779, 1993
3. Stekar, J., Hilgard, P., Klenner, T., Noessner, G., Schumacher, W. A second generation of alkylphospholipids with high antineoplastic activity. Proc. Amer. Ass. Cancer Res. 34: 335 (Abstr. 1996), 1993
4. Stekar, J., Hilgard, P., Voegeli, R., Maurer, R., Engel, J., Kutscher, B., Noessner, G., Schumacher, W. Antineoplastic activity and tolerability of a novel heterocyclic alkylphospholipid, D-20133. Cancer Chemother. Pharmacol. 32: 437–444, 1993
5. Spruss, T., Bernhardt, G., Reile, H., Schönenberger, H., Engel, J. Combination therapy with platinum complexes and hexadecylphosphocholine of MXT mouse mammary adenocarcinomas. (Abstr. TH10), J. Cancer Res. Clin. Oncol. 115:55, 1989
6. Beckers, T., Voegeli, R. Molecular targets of miltefosine. Drugs of Today 30 (Suppl. B): 5–12, 1994

27

SYSTEMIC ADMINISTRATION OF ALKYLPHOSPHOCHOLINES

Erucylphosphocholine and Liposomal Hexadecylphosphocholine

P. Kaufmann-Kolle,[1] M. R. Berger,[2] C. Unger,[3] and H. Eibl[1]

[1]Max-Planck-Institut für biophysikalische Chemie
Am Fassberg, Göttingen, Germany
[2]DKFZ, Institut für Toxikologie
Im Neuenheimer Feld 280, Heidelberg, Germany
[3]Medizin. Universitätsklinik
Robert-Koch-Str. 40, Göttingen, Germany

The class of alkylphosphocholines[1], best represented by hexadecylphosphocholine (HePC), shares common structural and physical properties with the lysolecithins, yet their biological activities vastly differ[2–6]. An understanding of this behaviour arises from the fact that while alkylphosphocholines, lysolecithins and (ether)-lysolecithins as well incorporate into cell membranes, alkylphosphocholines are metabolically more stable under physiological conditions. We demonstrated that this stability-induced concentration could be used to initiate an antineoplasic response in the cells of methylnitroso urea (MNU) induced mammary carcinomas of Sprague-Dawley (S-D) rats[7,8]. In the same model, lysolecithins and (ether)-lysolecithins were inactivated by degradation or acylation. Based on this evidence, clinical investigations on the scope of HePC's application began in the late nineteen eighties[9]. Success emerged in 1992, with the approval of its use, under the tradename of Miltex®, for the topical treatment of skin tumors which originated from the breast. Although, approved for topical usage, the oral administration of HePC has been limited by the appearance of nausea and vomiting (at dosages below the effective) and thrombophlebitis following i.v. administration, respectively. Liposomes, bilayer spheres of phospholipids, provide an unique tool to deliver toxic species, such as HePC, to their cellular targets. Currently, our efforts focus on gaining an understanding of the effects that modification of HePC has on its ability to be delivered in a liposomal fashion (i.e., HePC-SUVs). Wherein, the crucial factors must support an increased tolerance for oral and intravenous administration. Beyond, we provide a detailed account of the effects of elongating the carbon chain (i.e., from C_{16} to C_{22}) and introducing a cis-alkene at the ω-9 position. This was accomplished by changing the hexadecyl to the erucyl chain.

Platelet-Activating Factor and Related Lipid Mediators 2
edited by Nigam *et al.*, Plenum Press, New York, 1996

Table 1. Ratios of ErPC / HePC concentrations in tissues after oral application

Tumor	Spleen	Lung	Small intestine	Brain	Serum	Liver	Kidney
1.3	1.2	0.7	0.7	0.7	0.4	0.3	0.1

The S-D rats bore MNU induced mammary carcinomas. Choosen oral doses (255 μmol HePC/kg/week, 600 μmol ErPC/kg/week) show same efficacy in reduction of the tumor volume and the number of tumors after two weeks treatment[13].

In vivo models were conducted by feeding healthy Wistar rats 4 weeks with HePC (250 μmol/kg/week). The HePC animals displayed approximately 10 percent reduction in body weight, slightly ruffled skin, macroscopic gastric and intestinal damage. The main injuries found in this population were located in the duodenum, which is one of the primary sites for HePC absorption[10]. In contrast, these anormalities were not be detected in a population treated with the same dosage of liposomal HePC[11].

The pharmacokinetics of HePC and liposomal HePC were studied after a single administration[11]. Healthy Wistar rats were presented with a single intravenous injection of HePC-SUVs. Elimination occured in a biphasic mode with a very rapid initial phase ($t_{½}$ = 0.5 h) and a much slower terminal phase ($t_{½}$ = 16 hrs). Elimination upon oral application of HePC ($t_{½}$ = 48 hrs) and HePC-SUVs ($t_{½}$ = 120 hrs) was sustained for longer periods of time and was much more pronounced when introduced in liposomes. Surprisingly, oral AUC (area under the curve) serum values for HePC as well as for HePC-SUVs were much higher than AUC serum values after intravenous injection of HePC-SUVs. Indicating, intravenously applied HePC-SUVs are not destroyed in the bloodstream but are eliminated by macrophages (resulting in a lower AUC value and shorter half-life). In contrast after oral administration of HePC and HePC-SUVs, we observed a slow but almost complete absorption. The resorbed HePC was bound to serum albumin[10] leading to higher AUC values and extended half-lives[11]. After intravenous injection of HePC-SUVs, the highest concentrations were found in serum, liver and duodenum. With the except for small intestine, peak concentrations were measured 24 hrs after oral application. The peak concentrations for orally applied HePC were higher than of the orally applied HePC-SUVs. Furthermore, this reduction was not accompanied by a change in biodistribution.

The antineoplasic activity was studied in S-D rats bearing autochthonous, MNU induced mammary carcinomas. The oral dose of ErPC needed to regress tumors was much higher than that found for oral HePC[12]. In spite of that higher doses of ErPC were presented orally, we detected lower concentrations in the tissues[12]. In contrast to HePC, orally administered ErPC was resorbed only to a low extent[13]. Nevertheless, the ErPC concentrations in tumors were higher than for HePC (Table 1). Therapy was continued for a period of 5 weeks. The subjects were observed for total period comprised 10 weeks. The median tumor volume was markedly reduced during intravenous treatment with liposomal HePC without mortality (Table 2). The intravenously delivered optimal dosage of liposomal HePC (196 μmol/kg/week) was markedly lower than the oral administered HePC dose (368 μmol/kg/week) at a comparable toxicity. A liposomal i.v. applied dose (196 μmol/kg/week) caused significant inhibition of tumor growth without detectable toxicity. The double liposomal HePC dose (392 μmol/kg/week) led to an even more extensive tumor reduction (Table 2). However, the higher liposomal i.v. dose was poorly tolerated as indicated by decreased median body weight and increasing mortality. ErPC was far superior when delivered in intravenous therapy. Intravenous administration of ErPC (360 μmol/kg/week) resulted in dramatic tumor regression without toxicity or mortality (Table 2).

Table 2. Antineoplastic efficacy and toxicity

Treatment	Weekly dose [μmol/kg]	Weekly dose [mg/kg]	T/C [%]	BWC [%]	Mortality [%]
No	—	—	100	+14	0
HePC (p.o.)	172	70	30	+11	17
HePC (p.o.)	245	100	16	+8	0
HePC (p.o.)	368	150	7	+3	0
no	—	—	100	+4	15
HePC-SUVs (i.v.)	196	80	9	+7	0
HePC-SUVs (i.v.)	392	160	3	−7	20
no	—	—	100	+6	50
ErPC (i.v.)	90	44	31	+6	30
ErPC (i.v.)	180	88	6	+6	20
ErPC (i.v.)	360	176	0.01	+7	0

The MNU induced mammary carcinoma bearing S-D rats were treated with HePC (p.o.), HePC-SUVs (i.v.) or ErPC (i.v.) for 5 weeks. Evaluation was done at week 6. An oral pulse dose once a week was similarly effective but better tolerated than an equimolar dose administered five times a week[8]. The HePC-SUVs were given twice weekly. The median tumor volume of each group was expressed in percent of untreated control (T/C, %). The body weight change (BWC, %) is the mean weight difference at the end of the treatment in percent of initial body weight. The HePC-SUVs consisted of HePC/Chol/DPPG in the molar ratio 4/5/1. Note, that treated animals have different control groups.

To summarize, chemical or galenic modification of alkylphosphocholines can be used to reduce unwanted side effects by altering the physico-chemical behaviour, as well as resorption and biodistribution. In MNU induced tumors, intravenously injected erucylphosphocholine was the most potent antineoplastic drug, which showed remarkable enrichment in tumor tissue. Hexadecylphosphocholine can be given orally as liposomes to reduce side effects. Liposomal HePC can be repeatedly injected intravenously without thrombophlebitis. It is interesting to note that besides antineoplastic efficacy, HePC also shows significant activity against leishmaniasis[14].

ABBREVIATIONS

Area under the curve (AUC); cholesterol (Chol); erucylphosphocholine (ErPC); hexadecylphosphocholine (HePC); liposomal HePC (HePC-SUVs); intravenous (i.v.); methylnitroso urea (MNU); peroral (p.o.); 1.2-dipalmitoyl-*sn*-glycero-3-phosphoglycerol (DPPG); Sprague Dawley rat (S-D rat); small unilamellar vesicles (SUVs).

REFERENCES

1. Eibl H, Unger C (1990) Hexadecylphosphocholine: A new and selective antitumor drug. Cancer Treat Rev 17: 233–242
2. Berkovic D, Fleer E.A.M, Eibl H, Unger C (1992) Effects of hexadecylphosphocholine on cellular function. In: Eibl H, Hilgard P, Unger C (eds): Progress in experimental tumor research. New drugs in cancer therapy. Vol 34, S. Karger, Basel, pp 59–68

3. Fleer EAM, Unger C, Kim DJ, Eibl H (1987) Metabolism of ether phospholipids and analogs in neoplastic cells. Lipids 22: 856–861
4. Fleer EAM (1992) Biochemistry and possible clinical applications of etherlipids, especially of hexadecylphosphocholine (HePC). Dtsch Zschr Onkol 24: 10–18
5. Überall F, Oberhuber H, Marly K, Zaknun J, Demuth L, Grunicke HH (1991) Hexadecylphosphocholine inhibits inositol phosphate formation and protein kinase C activity. Cancer Res 51: 807–812
6. Vehmeyer K, Liersch T, Eibl H, Unger C (1992) Hexadecylphosphocholine amplifies the effect of granulocyte colony-stimulating factor on differentiating hemapoietic progenitor cells. In: Eibl H, Hilgard P, Unger C (eds): Progress in experimental tumor research. New drugs in cancer therapy. Vol 34, S. Karger, Basel, pp 69–76
7. Muschiol C, Berger MR, Schuler B, Scherf HR, Garzon FT, Zeller WJ, Unger C, Eibl H, Schmähl D (1987) Alkylphosphocholines: toxicity and anticancer properties. Lipids 22: 930–934
8. Berger MR, Yanapirut P, Reinhardt M, Klenner T, Scherf HR, Schmeiser HH, Eibl H (1992) Antitumor activity of alkylphosphocholines and analouges in methylnitrosourea-induced rat mammary carcinomas. In: Eibl H, Hilgard P, Unger C (eds): Progress in experimental tumor research. New drugs in cancer therapy. Vol 34, S. Karger, Basel, pp 98–115
9. Unger C, Sindermann H, Peukert M, Hilgard P, Engel J (1992) Hexadecylphosphocholine in the topical treatment of skin metastases in breast cancer patients. In: Eibl H, Hilgard P, Unger C (eds): Progress in experimental tumor research. New drugs in cancer therapy. Vol 34, S. Karger, Basel, pp 153–159
10. Kötting J, Marschner N, Neumüller W, Unger C, Eibl H (1992) Hexadecylphosphocholine and octadecylmethyl-glycero-3-phosphocholine: a comparison of hemolytic activity, serum binding and tissue distribution. In: Eibl H, Hilgard P, Unger C (eds): Progress in experimental tumor research. New drugs in cancer therapy. Vol 34, S. Karger, Basel, pp 131–142
11. Kaufmann-Kolle P, Drevs J, Berger MR, Kötting J, Marschner N, Unger C, Eibl H (1994) Pharmacokinetic behaviour and antineoplastic activity of liposomal hexadecylphosphocholine. Cancer Chemother Pharmacol 34: 393–398
12. Eibl H, Kaufmann-Kolle P (1995) Medical application of synthetic phospholipids as liposomes and drugs. J Liposome Res 5: 131–148
13. Kötting J, Berger MR, Unger C, Eibl H (1992) Alkylphosphocholines: Influence of structural variation on biodistribution at antineoplastically active concentrations. Cancer Chemother Pharmacol 30: 105–112
14. Kuhlencord A, Maniera T, Eibl H, Unger C (1992) Hexadecylphosphocholine: Oral treatment of visceral leishmaniasis in Mice. Antimicr Agents Chemother 36: 1630–1634

ARACHIDONATE–PHOSPHOLIPID REMODELING AND CELL PROLIFERATION

Floyd H. Chilton,[1,2] Marc E. Surette,[1] and James D. Winkler[3]

[1]Department of Medicine (Pulmonary and Critical Care Medicine)
[2]Department of Biochemistry
Bowman Gray School of Medicine of Wake Forest University
Winston-Salem, North Carolina 27103
[3]Division of Pharmacology
Smithkline Beecham Pharmaceuticals
King of Prussia, Pennsylvania 19406

I. INTRODUCTION

It has been recognized since the mid-1980's that arachidonic acid is taken up and remodeled between as many as 20 different glycerolipid molecular species in most cell types [1]. However, it has been difficult to understand why it is necessary for mammalian cells to maintain so many different arachidonate-containing glycerolipids and the ramifications of remodeling arachidonate between different glycerolipid molecular species. It is known that arachidonate-phospholipid remodeling occurs at a relative "slow" rate in most resting cells but this rate can be accelerated many fold during priming or stimulation of cells [2, 3]. Recently, we have begun to utilize newly-discovered inhibitors of the enzyme, CoA-independent transacylase (CoA-IT), which is thought to remodel arachidonate into the phospholipid subclass 1-alkyl-2-arachidonoyl-GPC and ethanolamine-containing phospholipids. Using these inhibitors of remodeling, it has been discovered that rapid remodeling of arachidonate is probably necessary to support continued arachidonic acid release and PAF generation which is initiated by phospholipase A_2.

There is an extensive literature on the effects of alkyl-lyso-phospholipids on cell proliferation and cancer. One of the most studied compounds is 1-0-octadecyl-2-0-methyl-sn-glycero-3-phosphocholine (ET-18-0-CH_3). This compound, along with other alkyl-lyso-phospholipids has been shown to decrease cell proliferation of leukemia, carcinoma and sarcoma cell lines [4, 5, 6] . This effect on proliferation appears to be selective, in that these compounds do not block proliferation of non-cancerous cells, such as normal bone marrow cells, neutrophils and skin fibroblasts. The mechanism of action of these anti-proliferative alkyl-lyso-phospholipids remains elusive. We have recently discovered that Et-18–0-CH_3 is a very potent inhibitor of CoA-IT. These findings raise a fundamental

Platelet-Activating Factor and Related Lipid Mediators 2
edited by Nigam *et al.*, Plenum Press, New York, 1996

question as to the role of arachidonate-phospholipid remodeling in cell proliferation. The current study addresses this important question.

II. MATERIALS AND METHODS

a. Compounds: Et-18-0-CH_3 was obtained from Biomol. SK&F 98625 (diethyl 7-(3,4,5-triphenyl-2-oxo-2,3-dihydro-imidazole-1-yl) heptane-phosphonate) was made by the Department of Medicial Chemistry, Smithkline Beecham. SK&F 45905 is (2-[2-[3-(4-chloro-3-trifuoromethylphenyl) ureido]-4-trifluoro methyl phenoxy]-4,5-dichlorobenzene sulfonic acid) was obtained from Bader Chemical Company.

b. Cell Proliferation and Apoptosis: The functional status of cells were measured in two ways. First cell numbers and viability were assessed utilizing the trypan blue stain. The second method, measurement of DNA fragmentation, determined not only the amount of cell death, but also provided insight into mechanism. When cells undergo programmed cell death or apoptosis, but not necrosis cellular DNA is broken into oligo-nucleosomal fragments which can be discerned on 1% agarose gels.

c. CoA-IT Activity: CoA-IT activity was measured in microsomes from U937 cells as previously described[7].

d. Extraction and Analysis of Lipids: Experiments were terminated by washing cells (2x) by centrifugation with ice cold buffer followed by extraction of lipids by the method of Bligh and Dyer. Fatty acids in a fraction of this extract were analyzed after methanolic base hydrolysis by negative ion chemical ionization GC/MS (NICI-GC/MS). Glycerolipid classes were isolated from another fraction by normal phase HPLC and fatty acids analyzed following methanolic base hydrolysis utilizing NICI-GC/MS. Phosphatidylethanolamine (PE) and phospatidylcholine (PC) fractions were further separated into 1-acyl, 1-alkyl and 1-alk-1-enyl subclasses and fatty acids analyzed follow methanolic base hydrolysis utilizing NICI-GC/MS.

e. Incubation of Cells with Labeled Lipids: In experiments where cells were labeled with [^{3}H] AA, cells were washed and resuspended at 20x10^6 cells per ml. [^{3}H]-AA was added and cells were incubated for 15 minutes at 37^{0}C. The cells were then washed and resuspended in culture media and were immediately incubated at 37^{0}C with CoA-IT inhibitors.

III. RESULTS AND DISCUSSION

The findings that Et-18-0-CH_3 is a potent inhibitor of CoA-IT raised the question of whether structurally different CoA-IT inhibitors would influence proliferation in a manner comparable to Et-18-0-CH_3. Two other CoA-IT inhibitors, SK&F 98625 and SK&F 45905 blocked increases in HL-60 cell numbers in a comparable fashion to Et-18-0-CH_3. All inhibitors caused a delayed loss of cell viability between 24 and 48 hours.

A major mechanism by which ET-18-O-CH_3 blocks cell proliferation has been the induction of programmed cell death or apoptosis [8]. The next group of experiments were performed to determine if CoA-IT inhibitors would instigate apoptosis. Apoptosis was determined in these studies by examining internucleosomal DNA fragmentation patterns before and after treatment with CoA-IT inhibitors. Table 1 shows the results from these experiments. All three inhibitors caused DNA "ladder" formation. DNA ladder formation

Table 1. Degree of DNA "ladder" formation

Vehicle	–
Et-18-O-CH_3 (6 μM)	+++
SK&F 98625 (50 μM)	++++
SK&F 45905 (50 μM)	++

depended on the dose of the inhibitor and occur within 3 hr and was maximal 24 hr after treatment with CoA-IT inhibitors.

Subsequent studies were designed to elucidate how blocking arachidonate-phospholipid remodeling relates to apoptosis. Since CoA-IT inhibitors had never been provided long-term to cells in culture, initial studies examined the influence of SK&F 98625, SK&F 45905 and Et-18-0-CH_3 on the remodeling of labeled arachidonate between phospholipid subclasses. HL-60 cells were pulse labeled 15 min, washed and then placed back in culture with CoA-IT inhibitors for 24 and 48 hr. When cells were placed back in culture, there was a time-dependent movement of labeled arachidonate from PC to ethanolamine-containing molecular species with PE containing 57% of labeled arachidonate after 48 hr. SK&F 98625, SK&F 45905 and Et-18-0-CH_3 all blocked the movement of labeled arachidonate from PC to PE.

Although the aforementioned studies demonstrated that CoA-IT inhibitors could block the remodeling of newly-incorporated arachidonate between phospholipids, it was not apparent that this treatment would have an impact on mole quantities of endogenous arachidonate in phospholipid classes and subclasses. All three CoA-IT inhibitors caused a similar dose-dependent loss of arachidonate from PE with a concomitant increase in arachidonate formed in PC. In contrast to PE and PC, the cellular arachidonate content of PI remained unchanged.

Experiments were next performed to determine whether the changes in distribution of PC and PE were reflected in all subclasses or were the result of losses or gains in specific subclasses. All inhibitors caused a marked reduction in the quantity of arachidonate in 1-acyl-2-arachidonoyl-GPE and 1-alk-1-enyl-2-arachidonoyl-GPE. In contrast to PE subclasses, the arachidonate content of 1-acyl linked PC increased following treatment with all three inhibitors. There was no significant change in the arachidonate content of other PC subclasses.

Since the redistribution of cellular arachidonate and more specifically the loss of arachidonate from PE were major changes induced by CoA-IT inhibitors, supplementation experiments were performed in an attempt to reverse the inhibition of cell proliferation. Specifically, HL-60 cells were cultured with exogenous arachidonic acid (50 μM) or ethanolamine (100 μM) or both together with CoA-IT inhibitors. It was reasoned that if the loss of in arachidonate in PE was blocking the ability HL-60 cells to mobilize arachidonic acid during the cell cycle, supplementation with exogenous arachidonic acid would overcome this potential effect. In the case of ethanolamine supplementation, this was attempted to increase the overall PE content in the cell and overcome the loss of arachidonate-containing PE. The inhibitory effect of SK&F 45905, SK&F 98625 or Et-18-O-CH_3 on cell proliferation were unaffected by supplementing cells with either arachidonic acid or ethanolamine or both.

Taken together, these data suggest that CoA-IT inhibitors and more specifically a disruption in the ratio of arachidonate in phospholipids represent an important signal leading to cellular events such as apoptosis. Currently, it is not understood how this alteration

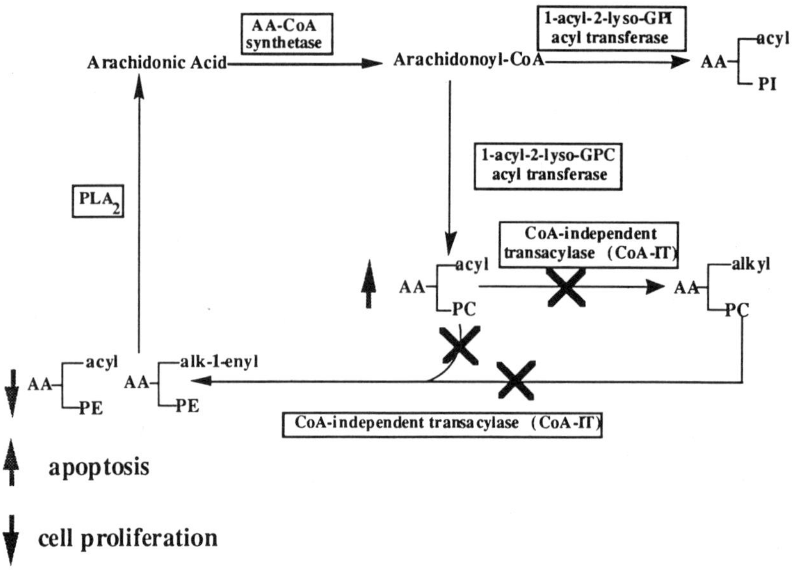

Figure 1. Proposed relationship between Arachidonate–Phospholipid remodeling and Apoptosis.

generates a signal. In light of the data presented here, the scheme in figure 1 summarizes the role of CoA-IT in arachidonate remodeling within phospholipids and the consequences of blocking this remodeling on cell proliferation and death.

REFERENCES

1. Chilton, F.H. and Murphy, R.C. (1986) J. Biol. Chem. 261, 7771
2. Fonteh, A.N. and Chilton, F.H. (1992) J. Immunol. 148, 1784
3. Winkler, J.D., Sung, C.M., Huang, L. and Chilton, F.H. (1994) Biochim Biophys Acta 1215, 133
4. Berdel, W.E., Bausert, W.R., Fink, U., Rastetter, J. and Munder, P.G. (1981) Anticancer Res 1(6), 345
5. Modolell, M., Andreesen, R., Pahlke, W., Brugger, U. and Munder, P.G. (1979) Cancer Res 39, 4681
6. Hoffman, D.R., Hajdu, J. and Snyder, F. (1984) Blood 63, 545
7. Winkler, J.D., Sung, C.M., Bennett, C.F. and Chilton, F.H. (1991) Biochim Biophys Acta 1081, 339
8. Diomede, L., Colotta, F., Piovani, B., Re, F., Modest, E.J. and Salmona, M. (1993) Int J Cancer 53, 124

PROTEIN KINASE C INHIBITION BY ET-18-OCH_3 AND RELATED ANALOGS

A Target for Cancer Chemotherapy

Susan B. Pauig and Larry W. Daniel

Department of Biochemistry
Bowman Gray School of Medicine
Wake Forest University
Medical Center Boulevard
Winston-Salem, North Carolina 27157-1016

1. TARGETING PKC SIGNALLING PATHWAYS IN CANCER CHEMOTHERAPY

Protein kinase C (PKC), is an important component of signal transduction mechanisms involving many physiological and pharmacological agonists[1]. While many isoforms are known[2,3], those isoforms which are responsive to the phorbol ester, TPA[4], and the lipid second messenger, diacylglycerol (DAG)[5] are the most completely characterized. PKC activation is a necessary component for TPA induced differentiation of the human acute myelogenous leukemia cell line, HL-60[6–9]. However, PKC activation does not appear to be required for tumor necrosis factor α (TNFα) induced differentiation[10]. PKC may be involved in differentiation induced by vitamin D_3 and all-trans retinoic acid, since both cause upregulation of PKC isozyme levels[11,12]. Differentiation induction therapy provides an alternative therapeutic approach for patients with acute myeloid leukemia (AML) who are either unsuitable for or unreponsive to conventional cytotoxic chemotherapy[13,14].

HL-60 cells are useful for studying the signalling pathway(s) leading to programmed cell death (apoptosis). TNFα and ceramide induce apoptosis[15]. TPA and cell permeable DAG (DiC_8) inhibit the induction of apoptosis by the anti-neoplastic agent, cytosine arabinoside (ara-C)[15,16]. Similarly, downregulation of PKC with bryostatin I results in enhanced apoptosis by ara-C in HL-60 cells[17]. Many pharmacological inhibitors of PKC induce apoptosis, and it has been suggested that basal PKC activity could render cells less sensitive to apoptotic inducing agents[18]. Therefore, activation of PKC is antagonistic to induction of apoptosis in myeloid leukemia cells. Thus inhibition of PKC activity may provide an additional approach to enhancing the efficacy of cytotoxic chemotherapeutic agents in the treatment of leukemia.

Platelet-Activating Factor and Related Lipid Mediators 2
edited by Nigam *et al.*, Plenum Press, New York, 1996

2. ANTINEOPLASTIC ACTIVITY OF ET-18-OCH$_3$ AND OTHER RELATED ETHER LIPID ANALOGS

2.1 Background

Recently, new strategies for anti-cancer drug development have begun to focus on cellular targets other than DNA replication and mitosis. One group of chemotherapeutic agents are synthetic analogs of naturally occurring ether lipids. These include 1-*O*-octadecyl-2-*O*-methyl-*rac*-glycero-3-phosphocholine (ET-18-OCH$_3$) and sulfur-linked compounds including 1-*S*-octadecyl- analogue (BM 41.440)[19,20] (Figure 1). ET-18-OCH$_3$ and related analogues selectively inhibit the growth of neoplastic cells *in vitro* and *in vivo* and have been demonstrated to induce granulocytic differentation[21,22] and apoptosis of myeloid leukemic cells[23,24]. Both ET-18-OCH$_3$ and BM 41.440 have been used in clinical trials for several neoplastic diseases, and promising strategies include bone marrow purging[25].

Uptake of these ether lipids and membrane composition are some of the known factors which confer sensitivity of leukemic cell types to these agents[26–30]. The mechanism of growth inhibition is not due to a direct detergent effect since ET-18-OCH$_3$ is cytostatic at non-lytic concentrations[31].

2.2 PKC Inhibition

PKC appears to be important in the mechanism of action of ether lipids since its activity is inhibited both in cell-free assays and intact cell systems[32,33]. ET-18-OCH$_3$ is a competitive inhibitor with respect to phosphatidylserine binding to the regulatory domain of PKC[32]. The mechanism of action of BM41.440 apppears identical to ET-18-OCH$_3$ because similar biochemical effects are induced in cells by these ether lipids[34,35].

Further evidence supporting PKC as an important target for ET-18-OCH$_3$ and BM 41.440 comes from studies with TPA treatment of HL-60 cells. ET-18-OCH$_3$ antagonizes a number of biochemical effects of PKC stimulation by TPA[33,36,37] including transcription factor activation[38]. An end result is inhibition of TPA induced monocytic differentiation by ET-18-OCH$_3$ in HL-60 cells[39].

3. ET-18-OCH$_3$ AND BM 41.440 ENHANCE Ara-C INDUCED APOPTOSIS IN HL-60 CELLS

3.1 Cytosine Arabinoside

Ara-C, a DNA synthesis inhibitor, has been the most effective single agent against acute myelogenous leukemias[40]. However, aquired resistance to ara-C in relapsed leukemias remains a major obstacle to cure. Further, high dose ara-C therapy regimens result in side effects such as myelosuppression and neurocytotoxcity. Therefore, a clear need exists for improving the induction and maintenance of complete remissions by araC. While high concentrations of ara-C *in vitro* and *in vivo* induce apoptosis in leukemic cells, low concentrations can induce differentiation. Recently, ara-C has been used in combination with retinoic acid in differentiation induction therapy, with some success[41–44].

Ara-C has been demonstrated to induce the formation of lipid second messengers, ceramide and diglyceride[45]; transcription factor activation[45,46]; and activation of a protein

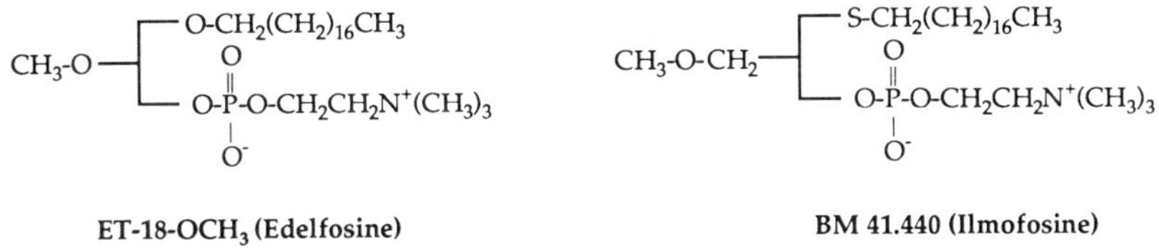

Figure 1. Structures of representative ether-linked lipids. ET-18-OCH_3 (1-*O*-octadecyl-2-*O*-methyl-*rac*-glycero-3-phosphocholine); BM 41.440

kinase[46,47]. This protein kinase has been found to be similar in some respects to a Ca^{2+}-independent protein kinase C[47].

3.2 Growth Inhibition and Monocytic Differentiation in Response to Ara-C and Ether Lipids

To investigate the effect of BM 41.440 and ara-C on leukemic cell growth, HL-60 cells were treated with either 0.1 μM or 10 μM ara-C in the presence or absence of BM 41.440 (0.05–2.0 μM). Low dose (0.1 μM) ara-C is growth inhibitory over 48 hours in HL-60 cells (data not shown). By 24 hours, 10 μM ara-C induces approximately a 40% decrease in viable cells compared with untreated controls (Figure 2). Co-incubation with ara-C (0.1 or 10 μM) and BM 41.440 (or ET-18-OCH_3, data not shown) (0.05 μM to 2.0 μM) results in a dose dependent decrease in cell viability compared to either 0.1 μM and 10 μM ara-C treatment. At 48 hours, treatment of HL-60 cells with BM 41.440, results in a 33% decrease in viability only at the highest concentration tested, the other concentrations (0.05–1.0 μM) do not significantly affect HL-60 cell viability.

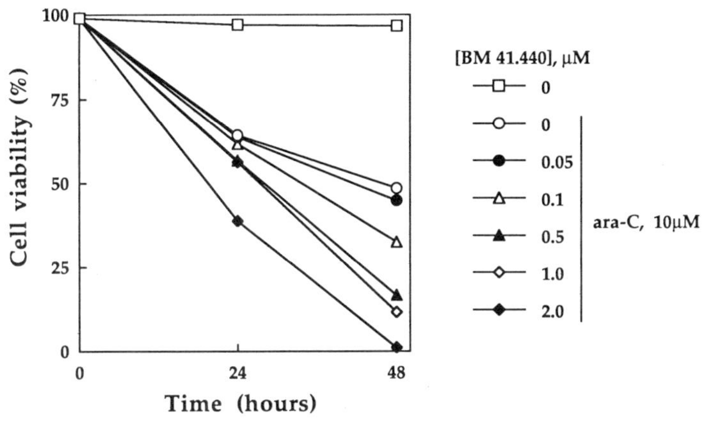

Figure 2. Dose-dependent decrease in cell viability by BM 41.440 in ara-C treated HL-60 cells. HL-60 cells (2 x 10^5 cells/mL; RPMI, 10% FBS)) were exposed to 10 μM ara-C and the indicated BM 41.440 concentrations (0–2.0 μM) for up to 48 hours. Cell viability was determined by the ability to exclude the vital dye, trypan blue. Viability due to ether lipid treatment only was greater than 90% at 24 hrs; at 48 hours, 2 μM ether lipid alone caused significant decrease in cell viability to 67% of control. Vehicle control (0.1% ethanol) was without effect. The data are representative of five separate experiments. o, control, no ara-C or BM 41.440; the other symbols represent ara-C (10 μM) with the indicated concentrations of BM 41.440.

Ara-C induces monocytic differentiation of HL-60 cells, but the role of PKC is unclear. We next examined the effect of ether lipids on ara-C induced monocytic differentiation. HL-60 cells were treated with 0.1 μM ara-C in the absence or presence of ET-18-OCH_3 (0.5–2.0 μM) for up to 6 days. Ara-C induces approximately 10–18% monocytes in the surviving population, as determined by the presence of non-specific esterase (NSE) activity and morphological changes typical of monocytes (data not shown). Simultaneous treatment of cells with ET-18-OCH_3 and ara-C leads to an increase in the percent of NSE-positive cells. However, the viable cell population decreases more during the combination treatment compared with the viable cell numbers obtained during ara-C treatment. Thus, the total number of monocytes decreases during the combination treatment compared with ara-C treatment.

3.3 Apoptosis in HL-60 Cells in Response to Ether Lipids and Ara-C

The above data suggest that ET-18-OCH_3 and BM 41.440 are having more of an effect on cell viability in ara-C treated cells; perhaps reducing cell numbers prior to a committment to the differentiation pathway. To examine this possibility, we tested the effects of these ether lipids on the induction of apoptosis by ara-C in HL-60 cells. Ara-C induced apoptosis was determined by measuring DNA fragmentation into multiples of 180–200 base pairs. The ether lipids enhanced DNA fragmentation in ara-C (10 μM) treated cells. Figure 3 shows the results from cells treated for 8 hours with BM 41.440 (0–2.0 μM) in the presence of 10 μM ara-C. BM 41.440 and ET-18-OCH_3 did not induce DNA fragmentation in the time period examined. By 12 hours, treatment of cells with a low dose of ara-C (0.1 μM) induced a low level of DNA fragmentation (Figure 4). A subcytotoxic concentration of BM 41.440 (0.5 μM) enhanced this effect.

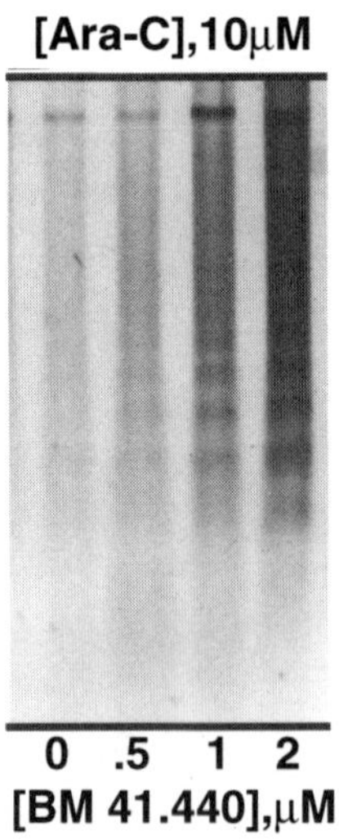

Figure 3. Dose-dependent enhancement of apoptosis by BM 41.440 in ara-C treated HL-60 cells. HL-60 cells (2.5 x 10^5 cells/mL) were treated for 8 hours with 10 μM ara-C and the indicated concentrations of BM 41.440 (0–2.0 μM). Total genomic DNA was isolated by phenol/chloroform extraction and ethanol precipitation and equal DNA was loaded into the wells of a 1.5% agarose gel and electrophoresed. The gel was stained with ethidium bromide and DNA fragments were visualized under ultraviolet light. Treatment with BM 41.440 (0.5–2.0 μM) alone did not induce the DNA laddering pattern by 8 hours of exposure. The results are representative of 2 separate experiments.

4. PERSPECTIVES

While PKC appears to be involved in ara-C-induced differentiation, it is antagonistic to ara-C-induced apoptosis. Ether lipids inhibit differentiation and enhance apoptotic pathways via PKC inhibition. Our results suggest that a combination of ether lipids with DNA synthesis inhibitors may be therapeutically beneficial. We found that differentiation inducing concentrations of ara-C in combination with subcytotoxic levels of ether lipid can enhance leukemic cell death during treatment periods similar to those used in clinical high dose ara-C therapy. These data indicate that ara-C stimulates opposing signal transduction pathways and that inhibiting PKC enhances the induction of apoptosis by ara-C.

Conjugates of ara-C and ether lipids have been synthesized and tested in various cell lines[48]. It has not been established whether these conjugates are the active form or if they act as prodrugs. The promising preclinical results observed with these conjugates may be due to PKC inhibition by the ether lipid moiety. In addition, PKC inhibitors may be useful in combination with Adriamycin, since resistance to this drug may be due to increased PKC levels[49]. Further studies on the mechanism of these ether lipids may lead to development of more effective compounds and combination regimens.

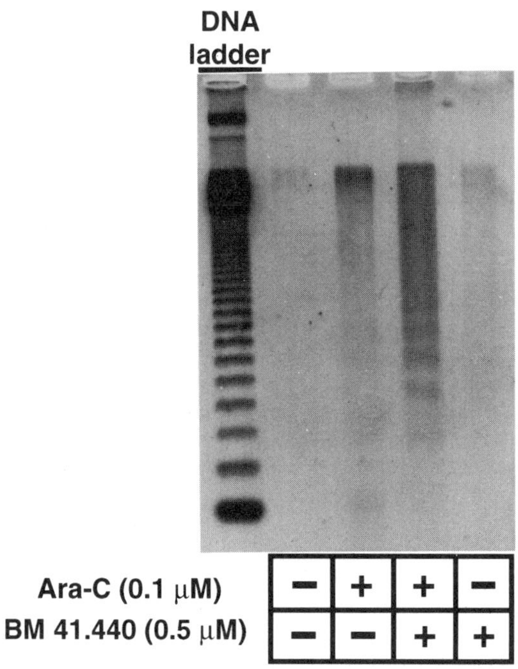

Figure 4. BM 41.440 enhances low dose ara-C induced apoptosis in HL-60 cells. HL-60 cells were treated for 12 hours with 0.1 μM ara-C in the absence or presence of 0.5 μM BM 41.440. DNA fragmentation was determined as in Figure 3.

REFERENCES

1. Blumberg, P.M., Aces, G., Areces, L.B., Kazanietz, M.G., Lewin, N.E., and Szallasi, Z. Protein kinase C in signal transduction and carcinogenesis. In: *Receptor-Mediated Biological Processes: Implications for Evaluating Carcinogenesis*, pp. 3–19, Wiley-Liss, Inc.. 1994.
2. Kikkawa, U., Kishimoto, A., and Nishizuka, Y. The protein kinase C family: heterogeneity and its implications. *Ann. Rev. Biochem.*, *58*: 31–44, 1989.
3. Parker, P.J., Kour, G., Marais, R.M., Mitchell, F., Pears, C., Schaap, D., Stabel, S., and Webster, C. Protein kinase C-α family affair. *Mol. Cell. Endo.*, *65*: 1–11, 1989.
4. Castagna, M., Takai, Y., Kaibuchi, K., Sano, K., Kikkawa, U., and Nishizuka, Y. Direct activation of calcium-activated, phospholipid-dependent protein kinase by tumour-promoting phorbol esters. *J. Biol. Chem.*, *257*: 7847–7851, 1982.
5. Kishimoto, A., Takai, Y., Mori, T., Kikkawa, U., and Nishizuka, Y. Activation of calcium and phospholipid-dependent protein kinase by diacylglycerol, its possible relation to phosphatidylinositol turnover. *J. Biol. Chem.*, *255*: 2273–2276, 1980.
6. Kiss, Z., Deli, E., and Kuo, J.F. Temporal changes in intracellular distribution of protein kinase C during differentiation of human leukemia HL-60 cells induced by phorbol ester. *FEBS Letters*, *231*: 41–46, 1988.
7. Nishikawa, M., Komada, F., Uemura, Y., Hidaka, H., and Shirakawa, S. Decreased expression of type II protein kinase C in HL-60 variant cells resistant to induction of cell differentiation by phorbol diester. *Cancer Research*, *50*: 621–626, 1990.
8. Nakaki, T., Mita, S., Yamamoto, S., Nakadate, T., and Kato, R. Inhibition by palmitoylcarnitine of adhesion and morphological changes in HL-60 cells induced by 12-*O*-tetradecanoylphorbol-13-acetate. *Cancer Research*, *44*: 1908–1912, 1984.
9. Macfarlane, D.E. and Manzel, L. Activation of β-isozyme of protein kinase C (PKCβ) is necessary and sufficient for phorbol ester-induced differentiation of HL-60 promyelocytes. *J. Biol. Chem.*, *269*: 4327–4331, 1994.
10. Chedid, M., Yoza, B.K., Brooks, J.W., and Mizel, S.B. Activation of AP-1 by IL-1 and phorbol esters in T cells. Role of protein kinase A and protein phosphatases. *J. Immunology*, *147*: 867–873, 1991.
11. Obeid, L.M., Okazaki, T., Karolak, L.A., and Hannun, Y.A. Transcriptional regulation of protein kinase C by 1,25-dihydroxyvitamin D_3 in HL-60 cells. *J. Biol. Chem.*, *265*: 2370–2374, 1990.
12. Makowski, M., Ballester, R., Cayre, V., and Rosen, O.M. Immunochemical evidence that three protein kinase C isozymes increase in abundance during HL-60 differentiation induced by dimethyl sulfoxide and retinoic acid. *J. Biol. Chem.*, *263*: 3402–3410, 1988.
13. Hassan, H.T. Differentiation induction of therapy of acute myelogenous leukemias. *Haematologia*, *21*: 141–150, 1988.
14. Hassan, H.T. Differentiation induction therapy: An alternative for the treatment of elderly patients with acute myelogenous leukemias. *J. Clin. Exp. Gerontology,*, *10*: 63–73, 1988.
15. Obeid, L.M., Linardic, C.M., Karolak, L.A., and Hannun, Y.A. Programmed cell death induced by ceramide. *Science*, *259*: 1769–1771, 1993.
16. Jarvis, W.D., Fornari, F.A., Browning, J.L., Gewirtz, D.A., Kolesnick, R.N., and Grant, S. Attenuation of ceramide-induced apoptosis by diglyceride in human myeloid leukemia cells. *J. Biol. Chem.*, *269*: 31685–31692, 1994.
17. Jarvis, W.D., Povirk, L.F., Turner, A.J., Traylor, R.S., Gewirtz, D.A., Pettit, G.R., and Grant, S. Effects of bryostatin 1 and other pharmacological activators of protein kinase C on 1-[β-D-arabinofuranosyl]cytosine-induced apoptosis in HL-60 human promyelocytic leukemia cells. *Biochemical Pharmacology*, *47*: 839–852, 1994.
18. Jarvis, W.D., Turner, A.J., Povirk, L.F., Traylor, R.S., and Grant, S. Induction of apoptotic DNA fragmentation and cell death in HL-60 human promyelocytic leukemia cells by pharmacological inhibitors of protein kinase C. *Cancer Research*, *54*: 1707–1714, 1994.
19. Berdel, G.E. Ether lipids and derivatives as investigational anticancer drugs. *Onkologie*, *13*: 245–250, 1990.
20. Berdel, W.E. Membrane-interactive lipids as experimental anticancer drugs. *British Journal of Cancer*, *64*: 208–211, 1991.
21. Honma, Y., Kasukabe, T., Hozumi, M., Tsushima, S., and Nomura, H. Induction of differentiation of cultured human and mouse myeloid leukemia cells by alkyl-lysophospholipids. *Cancer Research*, *41*: 3211–3216, 1981.

22. Rogers, M.A., Samples, L., Chabot, M.C., Marasco, C.J., Piantadosi, C., and Daniel, L.W. Leukemic cell differentiation by ether-linked lipids: studies on quarternary ammonium alkylglycerols. *Proceedings of the American Association for Cancer Research, 31*: 410, 1990.
23. Diomede, L., Colotta, F., Piovani, B., Re, F., Modest, E.J., and Salmona, M. Induction of apoptosis in human leukemic cells by the ether lipid 1-*O*-octadecyl-2-methyl-*rac*-glycero-3-phosphocholine: a possible basis for its selective action. *Internationl Journal of Cancer, 53*: 124–130, 1993.
24. Diomede, L., Piovani, B., Re, F., Principe, P., Colotta, F., and Modest, E.J. The induction of apoptosis is a common feature of the cytotoxic action of ether-linked glycerophospholipids in human leukemic cells. *International Journal of Cancer, 57*: 645–649, 1994.
25. Vogler, W.R., Berdel, W.E., Olson, A.C., Winton, E.F., Heffner, L.T., and Gordon, D.S. Autologous bone marrow transplantation in acute leukemia with marrow purged with alkyl-lysophospholipid. *Blood, 80*: 1423–1429, 1992.
26. Hoffman, D.R., Hoffman, L.H., and Snyder, F. Cytotoxicity and metabolism of alkyl phospholipid analogues in neoplastic cells. *Cancer Research, 46*: 5803–5809, 1986.
27. Wilcox, R.W., Wykle, R.L., Schmitt, J.D., and Daniel, L.W. The degradation of platelet activating factor and related lipids: susceptibility to phospholipases C and D. *Lipids, 22*: 800–807, 1987.
28. Daniel, L.W., Small, G.W., and Strum, J.C. Characterization of cells sensitive and resistant to ET-18-OCH_3. *Proceedings of the American Association for Cancer Research, 31*: 412, 1990.
29. Chabot, M.C., Wykle, R.L., Modest, E.J., and Daniel, L.W. Correlation of ether lipid content of human leukemia cell lines and their susceptibility to 1-*O*-octadecyl-2-*O*-methyl-*rac*-glycero-3-phosphocholine. *Cancer Research, 49*: 4441–4445, 1989.
30. Diomede, L., Bizzi, A., Magistrelli, A., Modest, E.J., Salmona, M., and Noseda, A. Role of cell cholesterol in modulating antineoplastic ether lipid uptake, membrane effects, and cytotoxicity. *International Journal of Cancer, 46*: 341–346, 1990.
31. Daniel, L.W. Ether lipids in experimental cancer chemotherapy. In: J.A. Hickman and T.R. Tritton (eds.), *Cancer Chemotherapy*, pp. 146–178, Oxford: Blackwell Scientific Publications. 1993.
32. Helfman, D.M., Barnes, K.C., Kinkade, J.M.Jr, Vogler, W.R., Shoji, M., and Kuo, J.F. Phospholipid-sensitive Ca^{2+}-dependent protein phosphorylation system in various types of leukemic cells from human patients and in human leukemic cell lines HL60 and K562, and its inhibition by alkyl-lysophospholipid. *Cancer Research, 43*: 2955–2961, 1983.
33. Parker, J., Daniel, L.W., and Waite, M. Evidence of protein kinase C invovlement in phorbol diester-stimulated arachidonic acid release and prostaglandin synthesis. *J. Biol. Chem., 262*: 5385–5393, 1987.
34. Shoji, M., Raynor, R.L., Berdel, W.E., Vogler, W.R., and Kuo, J.F. Effects of thioether phospholipid BM41.440 on protein kinase C and phorbol ester-induced differentiation of human leukemia HL-60 and KG-1 cells. *Cancer Research, 48*: 6669–6673, 1988.
35. Berdel, W.E., Fromm, M., Fink, U., Pahlke, W., Bicker, U., Reichert, A., and Rastetter, J. Cytotoxicity of thioether-lysophospholipids in leukemia and tumours of human origin. *Cancer Research, 43*: 5538–5543, 1983.
36. Daniel, L.W., Etkin, L.A., Morrison, B.T., Parker, J., Morris-Natschke, S., Surles, J.R., and Piantadosi, C. Ether lipids inhibit the effects of phorbol diester tumor promoters. *Lipids, 22*: 851–855, 1987.
37. Daniel, L.W. Protein kinase C inhibition by alkyl-linked lipids. In: J.J. Kabara (ed.), *The Pharmacological Effects of Lipids*, pp. 90–96, Illinois: The American Oil Chemists Society. 1990.
38. Rogers, M.A., Small, G.W., Samples, L., and Daniel, L.W. Phorbol diester stimulated HL-60 cell differentiation is associated with the activation of specific DNA binding proteins. *Proceedings of the American Association for Cancer Research, 32*: 295, 1991.
39. Kiss, Z., Deli, E., Vogler, W.R., and Kuo, J.F. Antileukemic agent alkyl-lysophospholipid regulates phosphorylation of distinct proteins in HL-60 and K562 cells and differentiation of HL-60 cells promoted by phorbol ester. *Biochemical Biophysical Research Communications, 142*: 661–666, 1987.
40. Frei, E., Bickers, J.N., Hewitt, M., Leary, W.V., and Talley, R.W. Dose schedule and antitumor studies of arabinosyl cytosine. *Cancer Research, 29*: 1325–1332, 1969.
41. Hassan, H.T. and Rees, J.K. Triple Combination of Retinoic acid + low concentration of cytosine arabinoside + hexamethylene bisacetamide induces differentiation of human AML blasts in primary culture. *Hematological Oncology, 7*: 429–440, 1989.
42. Hassan, H.T. and Rees, J.K. Low concentrations of cytosine arabinoside, 6-thioguanine, actinomycin-D and aclacinomycin A stimulate the differentiation of normal human marrow myeloid progenitor cells. *Medical Oncology & Tumor Pharamcotherapy, 6*: 213–217, 1989.
43. Hino, K. and Nakamaki, T. Differentiation therapy for myelodysplastic syndrome. *Japanese Journal of Clinical Hematology, 34*: 283–288, 1993.

44. Brach, M.A., Mertelsmann, R.H., and Herrmann, F. Modulation of cytotoxicity and differentiation-inducing potential of arabinofuranosylcytosine in myeloid leukemia cells by hematopoietic cytokines. *Cancer Investigation*, *11*: 198–211, 1993.
45. Strum, J.C., Small, G.W., Pauig, S.B., and Daniel, L.W. 1-β-D-arabinofuranosylcytosine stimulates ceramide and diglyceride formation in HL-60 cells. *J. Biol. Chem.*, *269*: 15493–15497, 1994.
46. Kharbanda, S., Datta, R., and Kufe, D. Regulation of c-jun gene expression in HL-60 leukemia cells by 1-β-D-arabinofuranosylcytosine. *Biochemistry*, *30*: 7947–7952, 1991.
47. Kharbanda, S., Emoto, Y., Kisaki, H., Saleem, A., and Kufe, D. 1-beta D-arabinofuranosylcytosine activates serine/threonine protein kinases and c-jun gene expression in phorbol ester-resistant myeloid leukemia cells. *Molecular Pharmacology*, *46*: 67–72, 1994.
48. Ueda, T., Imamura, S., and Kawai, Y. Successful treatment of myelodysplastic syndrome with 1-β-D-arabinofuranosylcytosine-5′-stearyl phosphate. *Leukemia Research*, *14*: 1067, 1990.
49. Aquino, A., Hartman, K.D., Knode, M.C., Grant, S., Huang, K.P., Niu, C.H., and Glazer, R.I. Role of protein kinase C in phosphorylation of vinculin in adriamycin resistant HL-60 leukemia cells. *Cancer Research*, *48*: 3324–3329, 1988.

30

INFLUENCE OF HEXADECYLPHOSPHOCHOLINE (MILTEFOSINE) ON CYTOKINE SYNTHESIS AND BIOLOGICAL RESPONSES

T. Klenner,[1] T. Beckers,[1] K. Nooter,[2] and H. Holtmann[3]

[1]ASTA Medica AG
Weismüllerstr. 45, 60314 Frankfurt am Main, Germany
[2]Department of Medical Oncology
University Hospital Rotterdam
Dr. Molewaterplein 40, 3015 GD Rotterdam, The Netherlands
[3]Institute for Molecular Pharmacology
Medical School
30623 Hannover

1. INTRODUCTION

Hexadecylphosphocholine, INN: Miltefosine (Figure 1), the lead compound of the alkylphosphocholines, has shown its antineoplastic activity in various in-vitro and in-vivo tumor models (Hilgard et al. 1993). After oral administration to patients miltefosine induced a remarkable leucocytosis (Pronk et al. 1994). This suggested a stimulatory effect of the compound on the hematopoietic system. To investigate this phenomenon, mouse and rat bone marrow was stimulated with either recombinant mouse/rat IL-3 or mouse GM-CSF with or without the addition of miltefosine. Additionally, Northern blot analysis was performed on human myeloid leukemic cell lines for the IL-3 and GM-CSF receptor genes and for TNFα mRNA on peripheral blood cells stimulated with concanavalin A (Con A) or lipopolysaccharid (LPS) when they were exposed to Miltefosine.

2. MATERIAL AND METHODS

The mouse colony forming unit assay has been published previously by Nooter et al. (1994). In short, mouse nucleated cells were cultured in-vitro serum-free in semi-solid methyl cellulose in the presence of 0, 0.1, 0.3 or 1.0 μg/ml miltefosine and different concen-

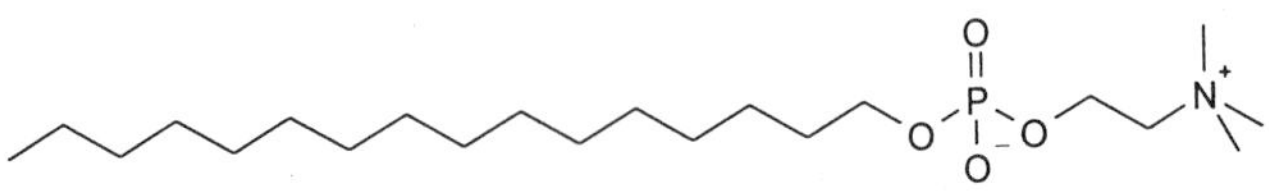

Figure 1. Formula of Hexadecylphosphocholine (HPC, Miltefosine, MIL).

trations (25 U, 400 U) of the hematopoietic growth factor Il-3 or GM-CSF. The incubation time in this experiment was 24 hours.

The rat colony forming unit assay was performed in a slightly modified form in comparison to the mouse study. Fresh femoral bone marrow suspensions were collected from healthy rats, pooled and suspended in enriched α-300 MEM culture medium, supplemented with L-glutamine, α-Mercapto ethanol, Penicillin/Streptomycin and 20% (v:v) horse serum and containing 0.3% methylcellulose. This semi-solid culture medium containing the rat bone marrow was plated in 35 mm petri dishes in triplicates. Filtration-sterilized miltefosine was added to the cultures in the final concentrations 0, 0.01, 0.03, 0.1, and 0.3 μg/ml. The rat bone marrow was stimulated with either recombinant rat IL-3 or mouse GM-CSF. Expression of glycosylated rat IL-3 was accomplished by transfection of COS cells with pILR-1, an SV-40-origin based eukaryotic expression vector containing the rat genomic IL-3 genes. After transfection of this construct in COS cells, which resulted in transient secretion of rat IL-3 protein in the medium, the supernatant was harvested and used as source of recombinant rat IL-3 (COS medium). Different concentrations of COS medium were used with miltefosine to stimulate the IL-3 responsive rat bone marrow progenitor cells. The recombinant mouse GM-CSF (Behringwerke AG, Marburg, Germany) was used as stock solution containing 1 x 10^7 U/mg and diluted in culture medium with 10% FCS. The dose levels used were 250 U (high), and 10U (low). Results are expressed as colonies/10^5 cells to allow direct comparison of the effect.

The method of Northern blot analysis of hematopoietic growth factor expression is described in detail by Beckers et al. (1994). In short, the human myeloid cell line KG1 and its phorbol ester (TPA) resistant counterpart KG1a were cultivated at 5x10^5 cells/ml density (time 0h) for 0h, 24h, 48h, 72h and 96h in medium containing miltefosine or TPA in concentrations shown in Fig. 5. The cell viability usually was between 75% and 96% after 72h. The cells were harvested and after isolation of total RNA, Northern-Blot analysis was performed using standard techniques. 10mg total RNA (for IL-3Rα poly(A)+ RNA from 100mg total RNA) were electrophoretically separated and transferred to Hybond N+ membranes (Amersham). Random-primed DNA fragments from the corresponding cDNA clones were used as ^{32}P-labelled probes. After washing hybridized membranes under stringent conditions, autoradiography was performed using Kodak X-OmatAR films with enhance screens. Usually after 1–7 days incubation time at a temperature of -80°C films were developed.

Mononuclear cells were isolated from human peripheral blood by centrifugation on Ficoll-Hypaque. The mononuclear cells (PBMC) enriched on the interface were collected, washed and suspended at 3x10^6 cells per ml in RPMI 1640 medium containing 10% (v:v) fetal calf serum. The human fibroblastoid cell line SV-80 was grown in DMEM containing 5% (v:v) FCS. Both media were supplemented with L-glutamine and antibiotics.

To determine the TNF-bioactivity in supernatants from PBMC, serial dilutions of the supernatants were added to cultures of the TNF-sensitive cell line SV-80, together with cycloheximide (50 μg/ml). Following incubation for 16 hours, viability of the cultured cells was determined by the neutral-red dye uptake method (Holtmann et al. 1988).

The results are expressed in TNF units per ml, as calculated on the basis of the cytotoxic activity of a standard TNF preparation.

The cDNA probe for human TNF-α was generated by reverse transcription-PCR using mRNA from TPA-stimulated U-937 cells.

For the Northern Blot analysis of TNF-mRNA the total RNA of the cells was isolated and electrophoresed in 1% agarose gels (6.8 % (v/v) formaldehyde), transferred and cross-linked to nylon membranes (Hybond N, Amersham Buchler, Braunschweig, FRG). Prehybridisation and hybridisation with the radiolabeled cDNA probe, prepared by random priming (DNA labeling kit, Bethesda Research Laboratories, Gaitersburgh, MD), as well as washing of the membranes were carried out essentially as described by Winzen et al. (1993). Autoradiography was performed at -80°C for 20 h using Kodak X-omatAR films and intensifying screens.

3. RESULTS

In the presence of IL-3 or GM-CSF, miltefosine induced a significant increase in the number of bone marrow colonies in soft agar in both species. In the absence of IL-3 or GM-CSF miltefosine was not stimulatory. In detail, it can be seen that IL-3 and GM-CSF have a stimulatory effect on the progenitor cells in the mouse system and that the addition of miltefosine results in a stimulation in the suboptimal concentration only (Fig. 2).

In contrast, the rat system is equally stimulated by the growth factors alone and is further stimulated by miltefosine at an optimal concentration of the respective growth factors used (Fig. 3). However, rat bone marrow is more sensitive to miltefosine, because at 0.1 μg/ml colony numbers are reduced.

In-vivo experiments in healthy rats showed that miltefosine was also active in reducing the duration of suppression of granulocytes after cytotoxic treatment with cyclophosphamide. Additionally, an "overshoot" after the suppression was observed, similar as has been reported after growth-factor application (Figure 4).

The most striking data, however, is the observation that miltefosine can protect the progenitor cells when it is given for one week before the cytotoxic agent (Figure 4, day 0–6). Again, the effect of miltefosine itself can be observed according to the application time and the reduction of nadir time and the "overshoot" of the granulocytes.

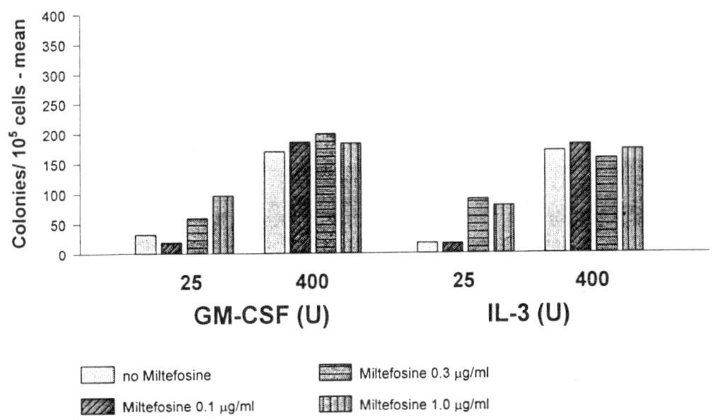

Figure 2. In-vitro colony formation of purified progenitor cells of mouse bone marrow after stimulation with mouse-GM-CSF or mouse-IL-3.

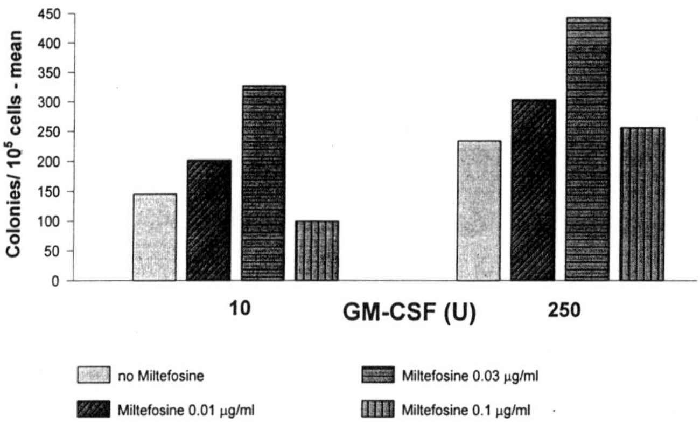

Figure 3. In-vitro colony formation of rat bone marrow after stimulation with mouse-GM-CSF.

The effect of miltefosine on the mRNA level of different cytokine receptors was analysed using the Northern Blot technique.

Human myeloid cell lines were cultivated for different periods in medium containing miltefosine (HPC) before isolation of total RNA. For comparison and as a positive control, cells were treated with 10nM TPA. After electrophoretic separation and transfer of RNA to nylon membranes, ^{32}P-labelled cDNAs were used as probes to monitor the steady state level of the corresponding cytokine receptor mRNAs.

As shown in Fig. 5, there is a significant increase in IL-3Rα and βc mRNA levels in KG1 cells treated with 75µM HPC or 10nM TPA. The effects in KG1a cells are less pronounced. The GM-CSF-R α transcript was not detectable. In contrast, there is no effect on

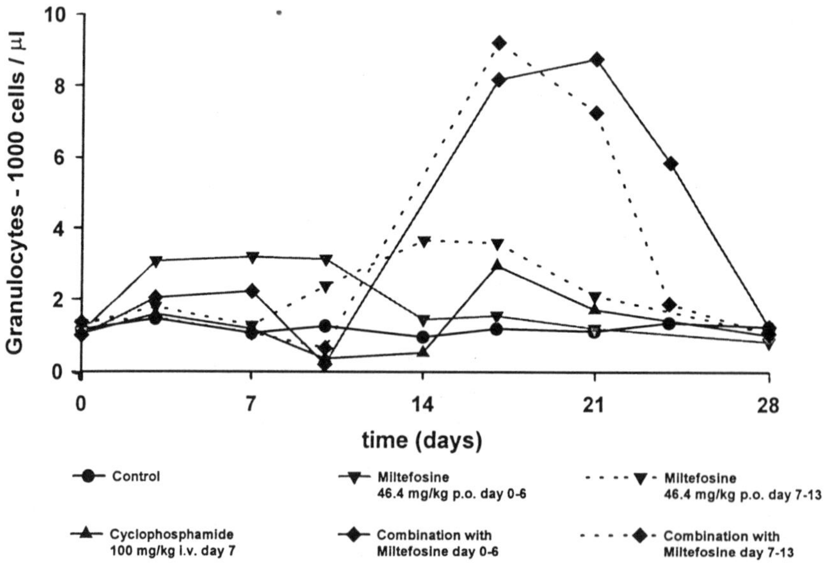

Figure 4. Granulocyte stimulation by cyclophosphamide in combination with multiple doses of miltefosine in female SD-rats.

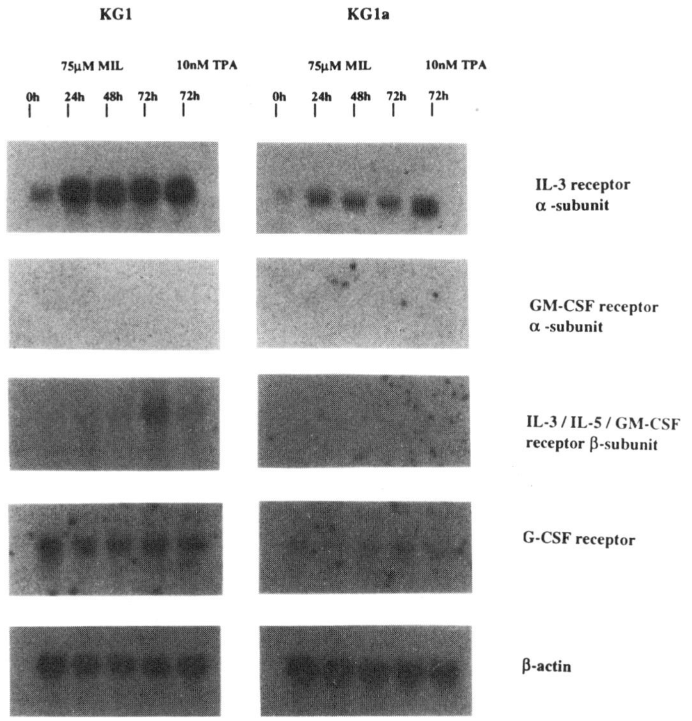

Figure 5. Human myeloid leukemia cells were incubated without or with 75 µM miltefosine (MIL) for the 24, 48, or 72 hours in comparison to incubation with 10 nM TPA.

the G-CSF-R level in the cell lines treated with miltefosine. As a control the β-actin mRNA level is not changed during the incubation time.

Miltefosine was also found to influence the formation of TNFα in peripheral blood cells stimulated with Con A or LPS. TNF bioactivity was induced by miltefosine alone to variable extents in different assays. Furthermore, miltefosine markedly increased TNF activity in culture stimulated with Con A (Fig. 6) or LPS (data not shown), particularly at late time points; 16 and 24 h).

Miltefosine does not affect the basal level of mRNA for TNF or GM-CSF by itself. However, it causes a prolongation of the increase in both mRNA species in response to Con A (Fig. 7).

4. DISCUSSION

From the results on the stimulation of colony growth in bone marrow cells and leukemia cells it can be derived that miltefosine has a differentiation inducing effect on hematopoietic cells as it has on tumor cells (Berger et al. 1992, Harlemann 1994). In humans the receptors for IL-3 and GM-CSF are heterodimeric receptor complexes with different α-subunits but one common β-subunit (βc). The β-subunit seems to be essential for signal transduction after ligand binding (Kitamura et al. 1991). In the human leukemia cells the α-subunit is predominantly stimulated but to a lesser extent also the β-subunit. Therefore,

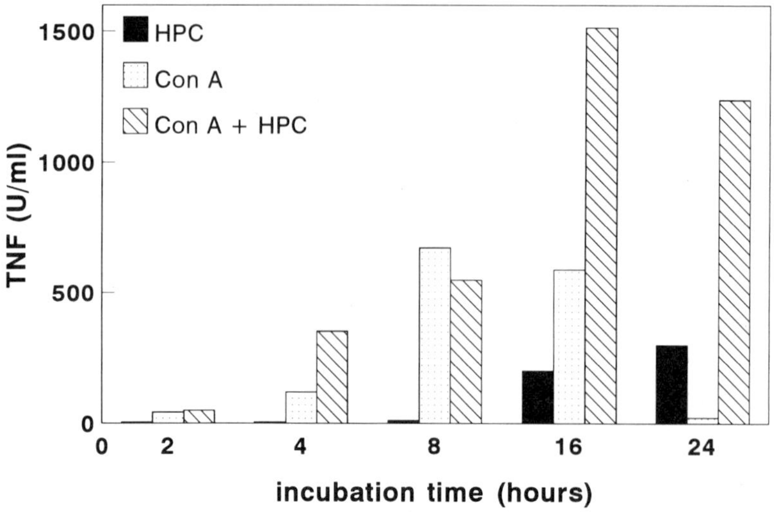

Figure 6. Human PBMC were incubated without or with 10 μM miltefosine (HPC) or 10 μg/ml Con A for the indicated times. Thereafter, the supernatants were assayed for TNF bioactivity (expressed in U/ml).

the mechanism of the effect on blood cells seems to be related to the interaction of miltefosine with the GM-CSF and/or IL-3 receptor and/or their subunits. It could be possible that miltefosine induces the cells to increase the number of functional receptors on the cell membrane. Likewise, it could enhance to affinity and/or sensitivity of the receptor for the respective growth factor. It is obvious that the influence of miltefosine on cytokine/receptor systems is not restricted to one type, but observed for others, e.g. the TNFα/TNF re-

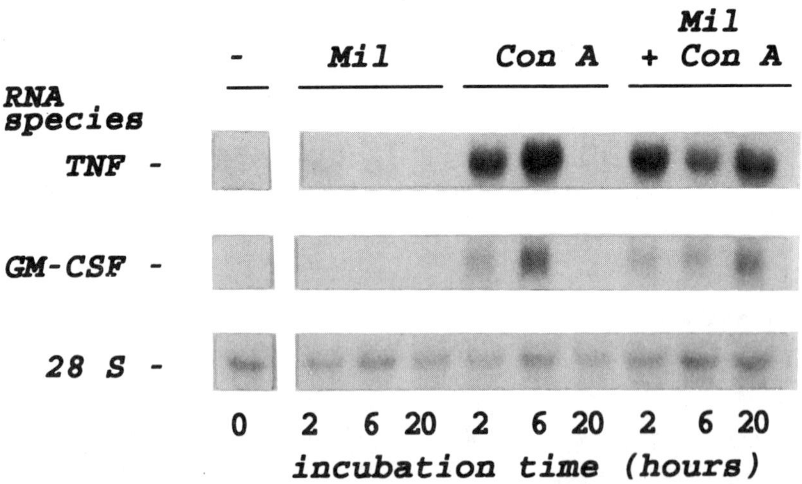

Figure 7. Human PBMC were incubated without or with 10 μM miltefosine (Mil) or 10 μg/ml Con A for the indicated times, followed by isolation of the RNA and Northern blot analysis of TNF-α and GM-CSF mRNAs.

ceptors system. The different concentrations used in the two assays are related to the EC_{50} found in previous experiments (data not shown). A similar activity on cytokines has also been shown for Ilmofosine an alkyllysophospholipid containing a glycerol backbone (Baier et al. 1995). The differentiation inducing effect could be responsible on one hand for the antineoplastic activity and on the other hand for the stimulatory effect dependent on the receptor pattern of the respective cell type. For the clinic, it appears to be possible to use the hematopoietic effect in addition to the antineoplastic effect of miltefosine or to find an alkylphospholipid which displays only the stimulatory effect and hopefully also a reduced adverse reaction pattern.

In conclusion, miltefosine has a growth stimulatory effect on bone marrow cells, possibly by increasing expression of cytokine receptors. This pharmacological property may be useful in conjunction with myelosuppressive chemotherapy.

ACKNOWLEDGMENTS

The cDNA probe for GM-CSF was kindly provided by Prof. K. Welte, Medical School, Hannover, Germany.

(Supported by The German Ministry of Research and Technology (BMBF), Bonn-Bad Godesberg).

REFERENCES

Baier JE, Brauckmann-Berger JM, Neumann HA (1995) Modulatorische Eigenschaft des Alkyllysophospholipids Ilmofosin auf Zytokine. Tumordiagn. u. Ther. 16, 88–93

Beckers T, Voegeli R, Hilgard P (1994) Molecular and cellular effects of Hexadecylphosphocholine (Miltefosine) in human myeloid leukemic cell lines. Eur. J. Cancer 30A (14), 2143–2150

Berger MR, Yanapirut P, Reinhardt M, Klenner T, Scherf HR, Schmeiser HH, Eibl H (1992) Antitumor activity of Alkylphosphocholines and analogues in methylnitrosourea-induced rat mammary carcinomas. In: Eibl H, Hilgard P, Unger C (eds): Alkylphosphocholines: New drugs in Cancer Therapy. Prog Exp Tumor Res. Basel, Karger, Vol. 34, 98–115

Harlemann J (1994) Tumor differentiation in-vivo with Miltefosine. In: Drugs of Today, Clinical aspects of miltefosine and ist topical formulation Miltex® Ed: Hilgard P, Engel J, Vol. 30, Suppl. B, 43–46

Hilgard P, Klenner T, Stekar J, Unger C (1993) Alkylphosphocholines: A new class of membrane-active anticancer agents. Cancer Chemother Pharmacol 32, 90–95

Holtmann H, Hahn T, Wallach D (1988) Immunobiol. 177, 7–22

Kitamura T, Sato N, Arai K, Miyajima A. Expression cloning of the human IL-3 receptor cDNA reveals a shared β-subunit for the human IL-3 and GM-CSF receptors. Cell 1991, **66**, 1165–1174

Nooter K, Stoter G, Verweij J (1994) Induction of hematopoietic differentiation by the alkylphosphocholine Miltefosine. In: Drugs of Today, Clinical aspects of miltefosine and ist topical formulation Miltex® Ed: Hilgard P, Engel J, Vol. 30, Suppl. B, 31–42

Pronk LC, Planting ASTh, Oosterom R, Drogendijk TE, Stoter G, Verweij J (1994) Increases in leucocyte and platelet counts induced by the alkylphospholipid hexadecylphosphocholine. Eur. J. Cancer, 30A (7), 1019–22

Überall F, Oberhuber H, Maly K, Zaknun J, Demuth L, Grunicke HH (1991) Hexadecylphosphocholine inhibits inositol phosphate formation and proteinkinase C activity. Cancer Res. 51, 807–812

Winzen R, Wallach D, Kemper O, Resch K, Holtmann H (1993) J. Immunol. 150, 4346–4353

31

EFFECT OF SM-12502 ON DISSEMINATED INTRAVASCULAR COAGULATION (DIC) IN TUMOR-BEARING RATS

Shigeaki Morooka and Yasuhiro Natsume

Sumitomo Pharmaceuticals Research Center
1-98, Kasugade-naka-3, Konohana, Osaka 554, Japan

SUMMARY

To investigate the role of PAF in tumor-associated disseminated intravascular coagulation (DIC), we have established the tumor-induced DIC model in rats and examined the effect of PAF-antagonist as compared with commonly used anti-DIC drugs. Four days after the intraperitoneal inoculation of rat ascites hepatoma AH-130, DIC-like hematological parameter changes such as decrease in platelet count, prolongation of PTT and increase in FDP were observed. In addition, serum levels of GOT and GPT were increased, indicating that liver injury was provoked. PAF-antagonist SM-12502 inhibited the development of DIC (PT, PTT and FDP) and liver injury. These results indicate that PAF plays an important role in tumor-associated DIC as well as accompanying organ failure.

1. INTRODUCTION

We examined the effect of PAF-antagonist SM-12502 on disseminated intravascular coagulation (DIC) in tumor-bearing rats. DIC is a pathological syndrome manifesting repeated hemorrhage and formation of micro-thrombi associated with sepsis, leukemia or malignant tumor. The excess activation of coagulation-fibrinolytic system results in the consumption of coagulation factors. Figure 1 schematically shows the underlining possible molecular mechanism involved in the pathogenesis of DIC elucidated to date. Endotoxin (ET), otherwise lipopolysaccharide (LPS), an outer membrane component of gram negative bacteria, is a causative substance for both DIC and sepsis. The role of ET in DIC as well as septic shock has been extensively explored in various animal models by many investigators including ourselves. ET activates monocytes/macrophages to induce the production of tissue factor (TF) and cytokines such as tumor necrosis factor (TNF), interleukin-1 (IL-1) and IL-6. ET directly or indirectly acts on endothelial cells (EC) via soluble mediators to induce the production of TF and plasminogen activator inhibitor-1

Platelet-Activating Factor and Related Lipid Mediators 2
edited by Nigam *et al.*, Plenum Press, New York, 1996

(PAI-1), expression of cell adhesion molecules as well as suppression of tissue-plasminogen activator (t-PA) and thrombomodulin (TM). These cellular responses cause the loss of anticoagulant activity of EC. ET also activates neutrophils to produce reactive oxygen and to release granulous enzymes such as elastase. These factors also injure EC that increases vascular permeability and disturbs the coagulation-fibrinolytic system. The role of PAF in ET-induced cellular activation has been examined by many researchers. Macrophage/monocytes, PMN and EC are known to produce PAF and other lipid mediators such as leukotriene B4 (LTB4) in response to ET. Kuipers et al. recently reported that PAF-antagonist TCV-309 inhibited the production of TNF, IL-1 and IL-6 in vitro and in chimpanzees in vivo[1]. The production of TF was also inhibited by TCV-309[2]. We have reported that ET-induced DIC-like hematological parameter changes were inhibited by the treatment with PAF-antagonist SM-12502 in rats and rabbits[3]. It has been reported that administration of PAF itself also provokes DIC-like changes[4]. These results indicate that PAF plays an important role in ET-induced DIC. In contrast to ET-induced DIC, DIC induced by leukemia/tumor or obstetrical injury is not well understood, since appropriate animal models were not available. TF is known as a pathogenic molecule for these diseases since TF is not only generated from macrophage in response to ET but also released from tumor cells or detected in amniotic fluid, and activates intravascular coagulation system by activating factor X to Xa. However, role of TF in the pathogenesis of DIC in leukemia, tumor and obstetrical injury has not been elucidated. Furthermore, the involvement of PAF in tumor- or obstetrical injury-induced DIC is not well understood. Several animal models have been developed for DIC. One of them, ET-induced model is commonly used for study of sepsis-related DIC. In this study, we have newly developed a tumor-induced DIC model in rats, and examined the role of PAF using our PAF-antagonist SM-12502.

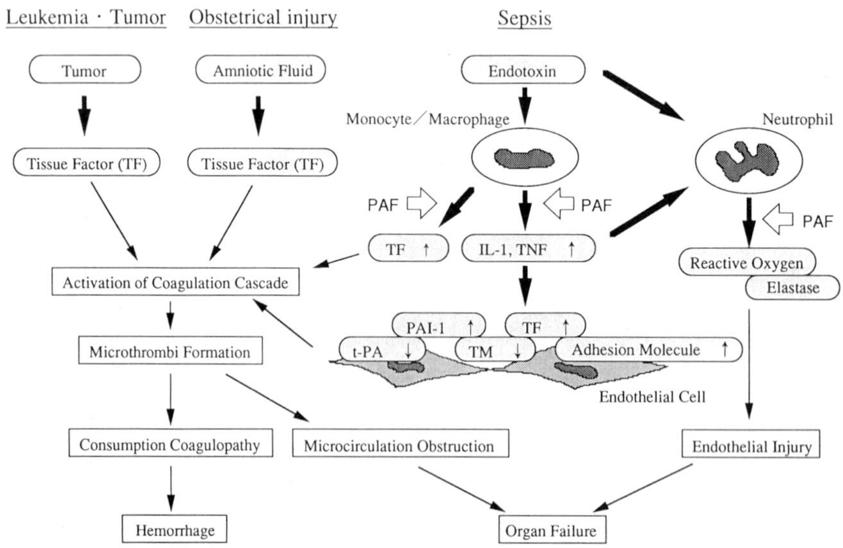

Figure 1. Pathophysilopgy of disseminated intravascular coagulation.

2. MATERIALS AND METHODS

2.1 Materials

Male Wistar rats weighing 200–300 g were used. Rat ascites hepatoma AH-130 was generously provided from Dr. Baccino, Torino University, Italy. SM-12502 was synthesized in our laboratory (Fig. 2). Gabexate mesilate (FOY) was purchased from Ono Pharmaceutical.

2.2 Experimental Condition

AH-130 (3×10^7 cell/rat) was intraperitoneally inoculated. On days 3, 4, 5 and 7, peripheral blood was withdrawn from abdominal aorta under light anesthesia with ether and hematological and hepatological parameters were measured. The number of platelet and white blood cell (WBC) were determined using automatic blood cell counter. Plasma fibrinogen, prothrombin time (PT) and partial thromboplastin time (PTT) were measured using automatic blood coagulation analyzer (Sysmex CA-5000, Toa Medical Electric Co., Ltd., Kobe, Japan). Fibrin/fibrinogen degradation products (FDP) was determined by FDPL™ test (Teikoku Hormone Mfg. Co., Tokyo, Japan). Liver function parameters GOT and GPT were measured by a commercial kit (Transaminase CII-test Wako, Wako Pure Chemical Industries, Osaka, Japan).

2.3 Drug Treatment

SM-12502 (50 mg/kg) was intravenously administered once a day for 4 days. Heparin (500 units/kg) or FOY (30 mg/kg) was subcuteneously injected twice a day for 4 days. On day 4, blood sample was collected.

3. RESULTS

3.1 Time Course of DIC Parameter Change

Time course of hematological parameter changes were studied.

On the day 3, peripheral platelet count significantly decreased from the normal value of 107.6 ± 6.1 to 73.5 ± 11.7×10^7/ml. On days 4 and 5, it was further decreased. On day 7, it still remained significantly low. PT and PTT were gradually prolonged from normal values of 15.1 ± 0.1 and 17.8 ± 0.4 sec respectively. PT value was significantly high on day 7 (22.3 ± 3.7 sec), PTT value was prolonged after day 4 (35.9 ± 6 sec). FDP level was increased, and after day 4, it was significantly high (18.0 ± 3.4 μg/ml). Fibrinogen

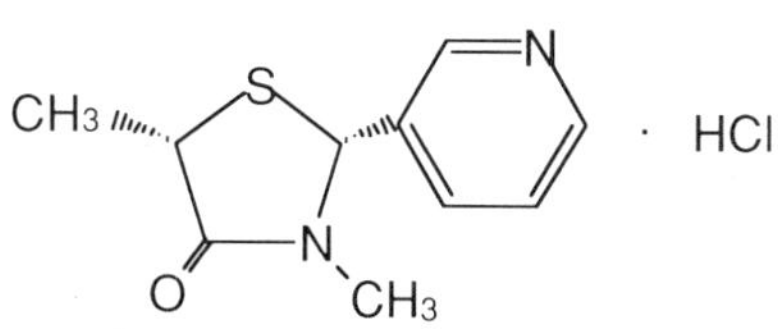

Figure 2. Chemical structure of SM-12502.

level was increased significantly from day 3, which is contrast to ET-induced DIC in which fibrinogen level decreases. In this time course study, WBC number was increased on days 3 and 7. Although there are some inconsistency with ET-induced models, tumor-bearing rats showed manifestation of DIC.

3.2 Time Course of GOT and GPT

Serum level of GOT was increased from day 3 and became significantly high on days 4, 5 and 7 (482.1 ± 107.9, 567.4 ± 42.0 and 482.5 ± 95.1 Karmen units, respectively). Similar profile was observed for GPT and on days 4 and 5, the values were significantly high (95.9 ± 15.6 and 89.4 ± 11.5, respectively). On day 7, the GPT level decreased.

3.3 Effects of Heparin and FOY on DIC Parameters

The treatment with heparin improved the change in platelet number and FDP level (26.5 ± 3.2 to 39.5 ± 2.4×10^7/ml and 77.0 ± 9.2 to 22.9 ± 4.4 μg/ml).

Heparin further increased the fibrinogen level. On the other hand, FOY was slightly but significantly effective only on FDP level (77.0 ± 9.2 to 53.0 ± 4.7 μg/ml). These results suggest that currently used anti-DIC drugs tested here are effective, and that tumor-bearing rats is a suitable animal model of tumor-induced DIC.

3.4 Effect of SM-12502 on DIC Parameters

Administration of PAF-receptor antagonist SM-12502 improved the change of DIC parameters. The prolongation of PT and PTT and the increase in FDP were significantly inhibited (16.6 ± 0.5 to 14.9 ± 0.4 sec, 36.8 ± 1.2 to 30.1 ± 2.0 sec and 37.1 ± 4.7 to 11.8± 3.1 μg/ml, respectively). Fibrinogen level was further increased by the treatment of SM-12502. However, platelet and WBC number were unaffected by SM-12502. Since rodent platelet has been known to devoid of PAF-receptor, it seems conceivable that the decrease in platelet number in this model is not mediated by PAF. Ineffectiveness of SM-12502 on thrombocytopenia also support this. The effect of SM-12502 on tumor-induced DIC is nearly identical on ET-induced DIC.

3.5 Effects of SM-12502, Heparin, and FOY on GOT and GPT

SM-12502 significantly suppressed both GOT and GPT increases (711.3 ± 89.9 to 351.8 ± 74.4 and 145.2 ± 15.8 to 98.5 ± 12.3 Karmen unit), suggesting that PAF plays a role in liver failure. On the contrary, neither heparin nor FOY affected the levels of GOT and GPT.

4. DISCUSSION

Inoculation of rat hepatoma AH-130 into rat peritoneal cavity induced the changes of hematological parameters, which are similar to the changes observed in ET-induced model. The difference is that fibrinogen level was increased in tumor-induced DIC model, while it was decreased in ET-induced DIC. The possible cause of increased level of fibrinogen may be due to the high hepatic activity in response to liver damage. Another difference is that the time course of development of DIC. In tumor-bearing rats, DIC was developed 4 days after the inoculation of tumor cells, whereas ET induces DIC in rats 2 to 4 hrs after the injection. Both heparin and FOY, which are clinically used for the treatment of DIC were effective on some parameters indicating that tumor-bearing rats is a suitable animal model of tumor-induced DIC. PAF-antagonist SM-12502 also ameriolated the DIC parameters, suggesting that PAF plays a role in the development of tumor-induced DIC, and that SM-12502 is a clinically useful drug. The increase in the serum levels of GOT and GPT indicates that liver injury was induced by tumor progression. It is known that organ failure is commonly associates with DIC, which sometimes proceeds into multiorgan failure (MOF). The liver injury observed in this study seems to present a suitable animal model for MOF. It appeared interesting that SM-12502 inhibited the increases in GOT and GPT, while neither heparin nor FOY was effective. These results indicate that SM-12502 may be useful not only on tumor-induced DIC but also liver injury-associated with DIC.

REFERENCES

1. B. Kuipers et al., J. Immunol. 152; 2438–2446 (1994).
2. M. Kawamura et al., Thromb. Res. 70; 281–293 (1993).
3. Y. Natsume et al., Arzneim.-Forsch./Drug Res. 44(II); 1208–1213 (1994).
4. Y. Imura et al., Life Sci. 39; 111–117 (1986).

PLATELET-ACTIVATING FACTOR IS AN EFFECTOR OF RAPID REACTIONS AND AN INDUCTOR OF LATE RESPONSES IN IMMUNE-MEDIATED INJURY

M. Sanchez Crespo, A. Alonso, Y. Bayon, and M. C. Garcia Rodriguez

Instituto de Biología y Genética Molecular
Facultad de Medicina
47005-Valladolid, Spain

1. BACKGROUND

The reversed passive Arthus reaction (RPA) was applied to the peritoneal cavity of rats to perform the present study. In this model, antibody is locally injected, and the cognate antigen is intravenously administered. The analysis of the sequence of events that occur in the RPA reaction includes formation of immune complexes in the microvessel wall, activation of the complement cascade, migration and adherence of polymorphonuclear leukocytes (PMN) to the endothelial cells, and release of PAF from PMN which acts on endothelial cells to cause leakage[1–3]. Recent studies have emphasized the involvement of nitric oxide (NO) in the production of tissue injury in a model of alveolitis and dermal vasculitis analogous to RPA[4,5], but the actual role of NO in immune-mediated injury has not been ascertained as yet. The purpose of this study has been to assess the effect of blocking PAF receptors on events of the RPA that occur at different times after antigen challenge.

2. RESULTS

2.1 Protein-Rich Plasma Extravasation and Accumulation of PMN in the Peritoneal Cavity of Rats in RPA

Induction of RPA produced a marked increase of the accumulation of Evans blue dye (EB) in the peritoneal cavity of rats. This was already apparent at 15 min, but maximal accumulation occurred at 30–60 min (solid bars in Fig. 1). When the rats were pretreated with an i.v. dose of 5 mg/kg of UR-12460 (Uriach S.A., Barcelona, Spain) 10 min

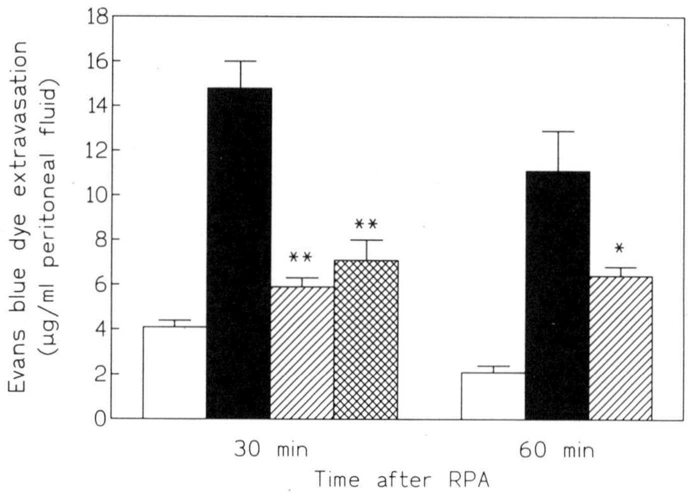

Figure 1. Extravasation of EB in the peritoneal cavity during RPA and effect of PAF-receptor antagonists. The peritoneal exudate was obtained at the times indicated after local injection of anti-ovalbumin antibody (negative control, open bars), local antibody and intravenous antigen (RPA animals, solid bars), or local antibody and intravenous antigen preceded by i.p. injection of 5 mg/kg UR-12460 10 min prior to challenge (striped bars), or 5 mg/kg BB-823 (checkered columns). Results represent mean±s.e.mean of 7 to 9 animals in each group. *P<0.05, **P<0.01. Taken from ref. 10, with permission.

prior to the induction of RPA, a reduction of extravasation to levels similar to those detected in control animals was observed at 30 min, however, the inhibition was less marked when rats were sacrificed 60 minutes after challenge. To ascertain whether the effect of UR-12460 was a consequence of its purported PAF receptor antagonist effect, experiments were also performed with BB-823 (British Bio-technology Ltd., Oxford, U.K.), a PAF receptor antagonist with a chemical structure unrelated to UR-12460. As indicated by a checkered bar in Fig. 1, BB-823 showed a similar protective effect on extravasation. Treatment with the L-arginine analogues L-NOARG and L-NMA alone or in combination at doses as high as 100 mg/kg, i.p., prior to challenge showed no protection. NO-generating compounds molsidomine and isosorbide dinitrate at doses of 15 mg/kg in combination with 200 mg/kg L-arginine did not influence extravasation. Assay of myeloperoxidase activity as an index of PMN accumulation in the peritoneal cavity showed that infiltration by PMN was a delayed event as compared to plasma leakage, with a scarce infiltration of PMN up to 1 hour, and a significant increase at 4 hours. Treatment with UR-12460 did not influence the number of infiltrating PMN.

2.2 Nitrite Production and iNOS Expression in RPA

Adherent peritoneal cells from RPA rats showed an enhanced *ex vivo* production of NO as compared to cells obtained from control animals (Fig. 2). This enhanced production occurred after 8 hours in culture, with a linear increase up to 18 hours and a less prominent increase up to 28 hours. Both this pattern of NO production and its magnitude suggested the involvement of the inducible form of NO synthase. Treatment of rats prior to the induction of RPA with either UR-12460 or BB-823 showed no enhancement of nitrite production as compared to controls (Fig. 2). Since these compounds possess different

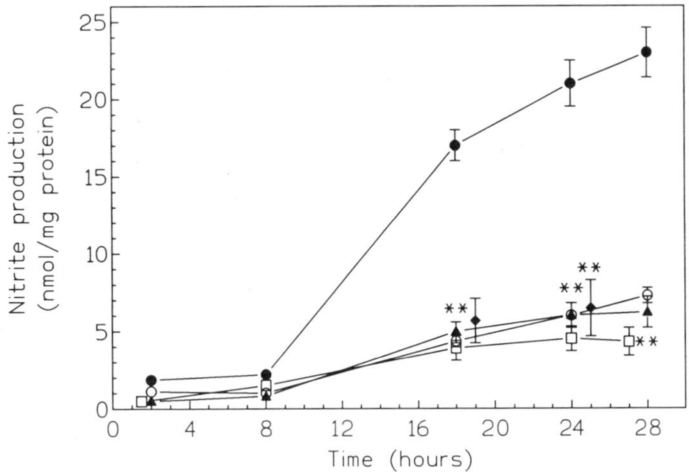

Figure 2. Production of NO by adherent peritoneal cells in culture after induction of RPA. Peritoneal cells were collected by washing the peritoneum, followed by centrifugation and resuspension in 10 ml of medium. At that time cells from each rat were distributed in aliquots and maintained at 37°C to allow the collection of adherent cells. NO_2^- production was assayed in duplicate samples at the times indicated. Open squares indicate control animals. Open circles indicate rats treated with i.p. antibody alone. Closed circles indicate animals treated with i.p. antibody and i.v. antigen. Closed triangles indicate RPA animals pretreated with 5 mg/kg UR-12460, i.p., 10 min before elicitation of RPA. Closed diamonds at 18 and 24 hours indicate rats receiving 5 mg/kg BB-823, under similar conditions to UR-12460-treated rats. Data represent mean ± s.e.mean of 9 to 12 animals in each group. **Indicates P>0.001. Taken from ref. 10, with permission.

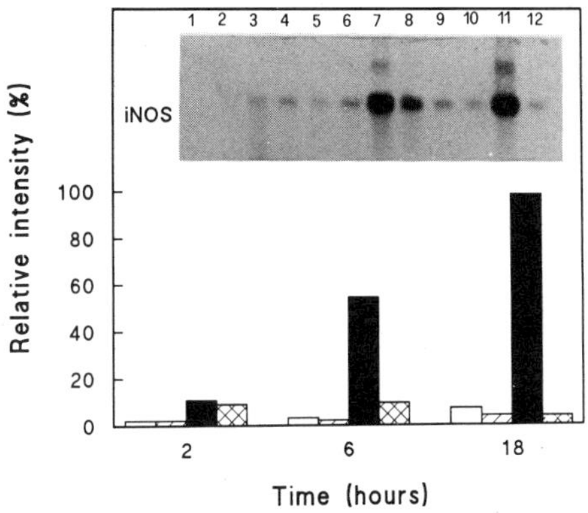

Figure 3. iNOS mRNA levels in RPA. Rats were injected with saline (lanes 1, 5 & 9), antibody (lanes 2, 6 & 10), or antibody and antigen in the absence (lanes 3, 7 & 11) or in the presence of 5 mg/kg of UR-12460 (lanes 4, 8 & 12). Adherent peritoneal cells were cultured *ex vivo* for 2 (lanes 1–4), 6 (lanes 5–8), and 18 hours (lane 9–12), and used for the analysis of iNOS mRNA using a specific probe. Quantitative analyses of mRNA levels were obtained by densitometry of the bands after normalization for the ß-actin content, and are shown in the bar diagram. Taken from ref. 10 with permission.

chemical structures, but elicit the same pharmacological effect on the PAF receptor, the results strongly suggest a role for PAF on the induction of NO synthase in RPA. In agreement with the production of NO by adherent cells after RPA, the levels of iNOS mRNA increased 14-fold after 18 hours of culture in animals challenged with antibody plus antigen. This increase could be observed at six hours since the initiation of the culture, but was most prominent at 18 hours. The appearance of iNOS mRNA was inhibited more than 90% by pretreatment of the animals with UR-12460 (Fig. 3).

2.3 PAF Enhances Nitrite Production by Control Peritoneal Cells and Abrogates the Response after RPA

Experiments with adherent cells from control rats showed that the production of nitrite can be elicited by PAF with a bell-shaped pattern of dose-dependency (open squares in Fig. 4). This was abrogated by the inclusion of UR-12460 in the medium (open triangles in Fig. 4), suggesting again that PAF might induce the production of NO via the inducible form of NOS by signaling through its receptor. When adherent peritoneal cells obtained one hour after RPA were incubated with different concentrations of PAF, as shown by closed circles in Fig. 4, a dose-dependent inhibition of nitrite production was observed. The time-dependency of this effect was studied with PAF at the dose of 10 nM.

3. DISCUSSION

The data herein presented indicate a role for PAF in the RPA model of rat immune complex peritonitis, in keeping with earlier findings in the rabbit skin model, and in rat alveolitis. The release of PAF is an early event, and can be most likely explained by the

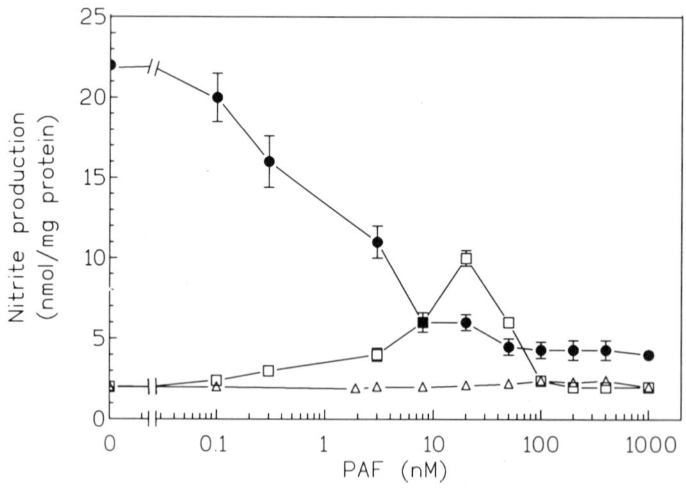

Figure 4. Effect of PAF on the production of nitrite by adherent peritoneal cells after RPA: Dose-response. Closed circles represent rats treated to induce RPA. Open squares indicate rats receiving saline solution alone. Open triangles indicate peritoneal cells from rats treated with saline solution incubated *in vitro* with different concentrations of PAF and 1 μM UR-12460. The measurement of nitrite was carried out after 15 hours in culture. Taken from ref. 10 with permission.

stimulation of PMN by immune complexes and/or complement-derived components in the microvasculature. Endothelial cells are one of the targets for PAF, as judged from the early appearance of protein-rich plasma leakage. Peritoneal macrophages are another target for PAF in RPA as judged from the effect of both PAF and PAF-receptor antagonists on nitrite production under a wide array of experimental conditions including *ex vivo* experiments with adherent peritoneal cells from RPA rats and *in vitro* studies with PAF and other agonists. The contribution of infiltrating PMN to the increased production of NO by peritoneal cells in *ex vivo* experiments seems unlikely, since the cell population was collected before infiltration of PMN had occurred, and also because the assay of nitrite production was carried out with adherent cells. In addition, the production of NO by PMN has been a matter of debate. Since iNOS induction is a delayed event involving nuclear signaling, this should be analyzed in the light of recent reports linking PAF receptor signaling and gene expression. Thus, PAF stimulates transcription of genes possessing the Fos-Jun/AP-1 responsive sequence[6], and up regulates kB binding activity[7]. There has recently been shown that the level of iNOS mRNA is controlled by at least two different signalling cascades which may act synergistically, one involves cyclic AMP and the other one is triggered by cytokines such as IL-1ß[8]. This raises the question of the possible relationship of PAF-triggered iNOS induction to any of these pathways. The biological role for NO synthase induction in RPA might be linked to either the induction of tissue damage or to the reparative phase of inflammation. Initial studies in rat immune complex alveolitis linked NO production to generation of peroxynitrite and to extravasation of protein-rich plasma, even though a diminished recruitment of PMN was not observed[4,5]. Our data differ from these results since we have not observed any effect on the extent of plasma extravasation by either inhibition of NO production with pharmacological tools or enhancement of NO production by NO-inducers. These discrepancies could be due to differences in either experimental conditions or properties of the tissues studied, which differ in both cellular components and vascularity. In keeping with the difficulty to assign a damaging role to NO production in tissue injury there are a number of reports emphasizing its protective role in hypoxia-induced intestinal injury, most probably by producing inhibition of NF-kB. Finally, the possibility that PAF could also play a significant role in the production of NO in other models of tissue injury is strongly suggested by the finding of significant attenuation of the induction of NO synthase produced by bacterial LPS in anesthetized rats treated with the PAF receptor antagonist WEB 2086[9].

4. REFERENCES

1. Hellewell, P.G. & Williams, T.J. 1986. *J. Immunol.* **137**,302–307
2. Warren, J.S., Mandel, D.M., Johnson, K.J. & Ward P.A. 1989. *J. Clin. Invest.* **83**,669–679
3. Tavares de Lima, W., Sirois, P. & Jancar, S. 1992. *Eur. J. Pharmacol.* **213**,63–7
4. Mulligan, M.S., Hevel, J.M. Marletta, M.A. & Ward, P.A. 1991. *Proc. Natl. Acad. Sci. USA.* **88**,6338–6342
5. Mulligan, M.S., Moncada, S. & Ward, P.A. 1992. *Br. J. Pharmacol.* **197**,1159–1162
6. Squinto, S.P., Block, A.L., Braquet, P. & Bazan, N.G. 1989. *J. Neurosci. Res.* **24**,558–566
7. Pan, Z., Kravchenko,, V.V., & Ye, R.D. 1995. *J. Biol. Chem.* **270**,7787–7790
8. Kunz, D., Mühl, H., Walker, G., & Pfeilschifter J. 1994. *Proc. Natl. Acad. Sci. USA.* **91**,5387–5391
9. Szabo, C., Wu, C.C., Mitchell, J.A., Gross, S.S., Thiemermann, C. & Vane, J.R. 1993. *Cir. Res.* **73**,991–999
10. Steil, A.A., García Rodríguez, M.C., Alonso, A., Sánchez Crespo, M., & Boscá, L. 1995. *Br. J. Pharmacol.* **114**,895–901

33

ETHER LIPID METABOLISM, GPI ANCHOR BIOSYNTHESIS, AND SIGNAL TRANSDUCTION ARE PUTATIVE TARGETS FOR ANTI-LEISHMANIAL ALKYL PHOSPHOLIPID ANALOGUES

H. Lux,[1] D. T. Hart,[1*] P. J. Parker,[2] and T. Klenner[3]

[1]King's College London
United Kingdom
[2]Imperial Cancer Research Fund
London, United Kingdom
[3]ASTA Medica AG
Frankfurt am Main, Germany

INTRODUCTION

Leishmaniasis is a tropical and subtropical disease with up to 12 million sufferers and is the cause of considerable morbidity and mortality (WHO-TDR 1990). The causative agent is *Leishmania* a genus of parasitic protozoa and member of the *Trypanosomatidae*. *Leishmania* has a digenic life cycle and exists as a flagellated and motile promastigote stage in the digestive tract of the sandfly vector and as an aflagellated and non-motile intracellular amastigote stage, which survive and multiply in macrophages of the vertebrate host.

Anti-leishmanial chemotherapy is heavily dependent upon the antimonial drugs Pentostam and Glucantime which possess good activity against *Leishmania* but are cytotoxic to the host and this limits their usefulness. New and effective anti-leishmanial drugs are urgently required and novel aspects of leishmanial metabolism are considered as potential targets for rational drug design (eg Hart *et al* 1989).

Leishmania have been found to contain high levels of ether lipids (Beach *et al* 1979 and Herrmann & Gercken 1980) and to use these alkyl phospholipids as plasma membrane

* Correspondence to; Dr D. T. Hart, Infection and Immunity Research Group, Divison of Life Sciences, King's College London, Campden Hill Road, London W8 7AH. UK. Telephone 44–171–333–4286; FAX 44–171–333–4500; Email d,hart@hazel.cc.kcl.ac.uk

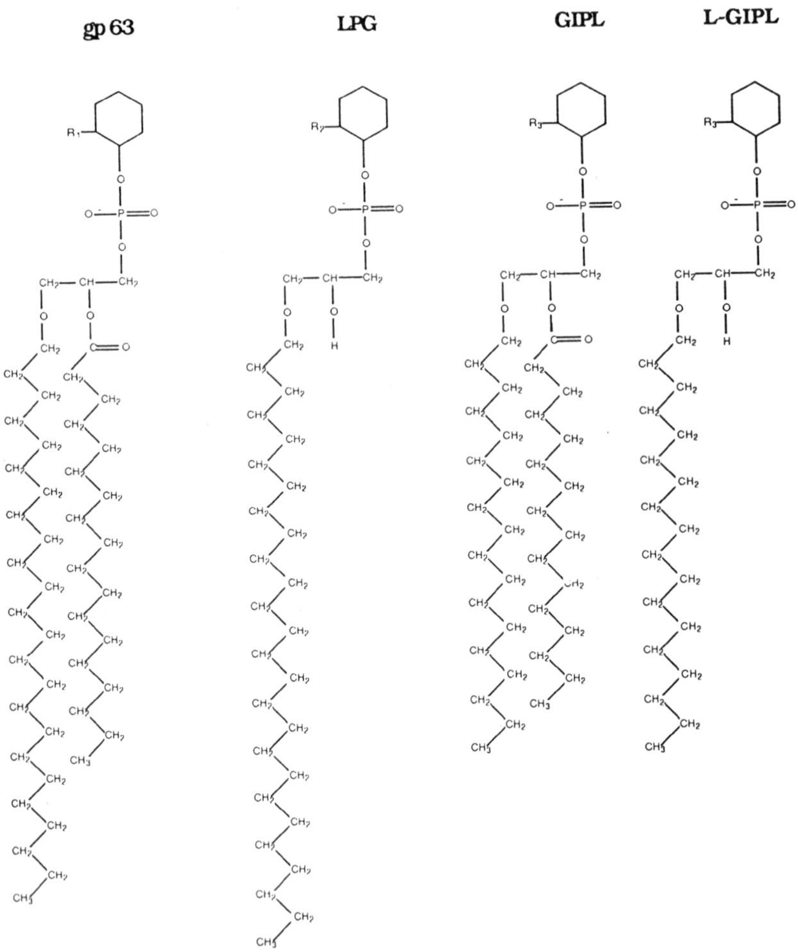

Figure 1. Lipid moities of GPI-anchored glycoprotein (gp63) and glycolipids (LPG & GIPL) from *Leishmania.* gp63, surface zince metalloproteinase (Schneider et al., 1990); LPG, lipophosphogycan (Ilg et al., 1991 and Orlandi & Turco, 1987) and GIPL (& Lyso-GIPL), glycoinositol phospholipid (McConville and Blackwell, 1991).

anchor moieties (see Figure 1). Indeed, these alkyl phospholipids anchor abundant and essential glycolipids (eg LPG & GIPL) and glycoproteins (eg gp63). The structure and function of these surface glycocalyx has recently been comprehensively reviewed by McConville and Ferguson (1993).

For example, the alkyl-lyso anchor moieties of LPG (see Figure 1) have been suggested to inhibit macrophage protein kinase C (PKC) and thus perturb signal transduction and the respiratory burst (Descoteaux *et al* 1991), while the alkyl-acyl and alkyl-lyso moieties of GIPL's, shown in Figure 1, have been suggested to block nitric oxide production in the macrophage host (Proudfoot *et al* 1995). Perturbation of signal transduction in the macrophage is therefore key to the survival of *Leishmania,* and consequently the metabolism of alkyl-acyl phospholipids in *Leishmania* is a prime target for anti-leishmanial drug design.

Alkyl phospholipids have been used as targets for anti-cancer drug design and analogues, such as Miltefosine and Edelfosine (see Figure 2). They have been shown to pos-

2-LPC L-PAF ET 18 OCH_3 ET PA HePC

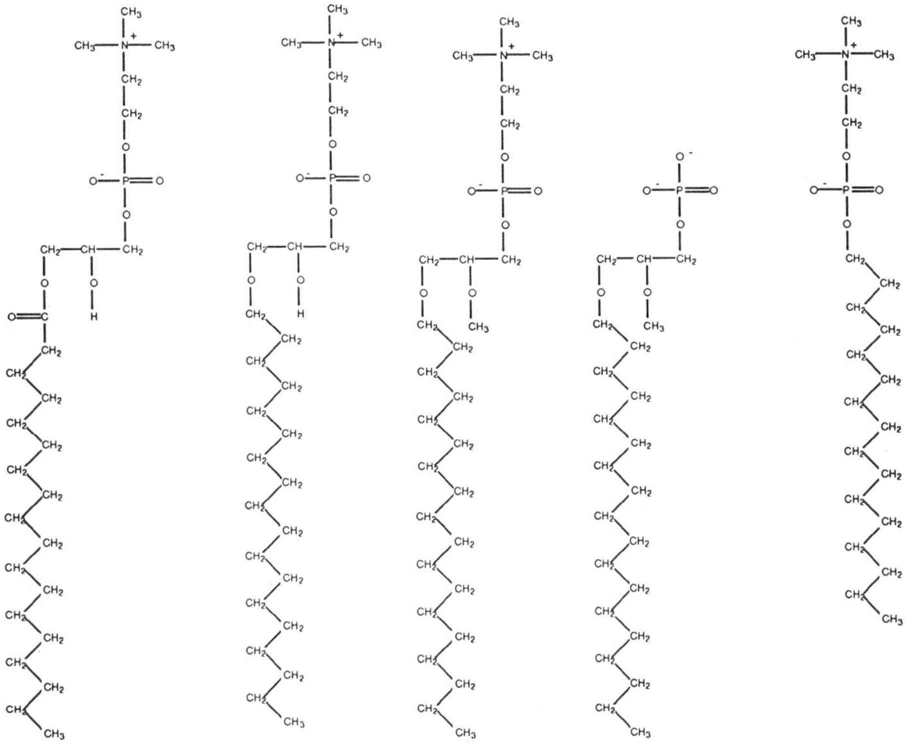

Figure 2. Structures of naturally occuring ester and ether lysophospholipids (2-LPC and L-PAF) and three synthetic analogues (ET 18 OCH_3, EPTA, and HePC). 2-LPC, L-a-lysophosphatidylcholine (C18:0); L-PAF, 1-*O*-octadecyl-rac-glycero-3-phosphophorycholine (C18:0); ET 18 OCH_3, 1-*O*-octadecyl-2-*O*-methyl-rac-glycerophosphorylcholine (Edelfosine); ETPA, 1-*O*-octadecyl-2-*O*-methyl-rac-glycerophosphatic acid and HePC, hexadecylphosphocholine (Miltefosine).

sess *in vitro* and *in vivo* anti-neoplastic activity (eg Arnold *et al* 1978, Berdel *et al* 1987, Hilgard & Klenner 1994 and Lohmeyer & Bittman 1994).

Neoplastic disease is characterised by an increase in cellular proliferation relative to the rate of differentiation. Therefore tumour cells can be considered as having impaired or inhibited differentiation and antagonists of neoplasticity, such as alkyl phospholipid analogues, exert their mechanism of action by inducing differentiation. To date the exact molecular mechanism for inhibition of proliferation and induction of differentiation has not been elucidated, however, several putative targets have been described.

Alkyl phospholipid analogues have been shown to accumulate in the plasma and organellular membranes of cancer cells and to perturb several aspects of metabolism (reviewed in Grunicke & Uberall 1992, Hilgard *et al* 1993 and Lohmeyer & Bittman 1994). Firstly, integral membrane proteins such as growth factor receptors (eg EGF) and translocator enzymes (eg Na^+/K^+-ATPase) have been suggested to be perturbed. Secondly, membrane associated enzymes involved in signal transduction (eg PLC and PKC) and ester phospholipid biosynthesis (eg phosphocholine cytidylyltransferase) would seem to be inhibited.

Most importantly alkyl phospholipid analogues have also been suggested to be a novel class of anti-leishmanial drug (Achterberg & Gercken 1987, Croft *et al* 1987, Kuhlencord *et al* 1992, Croft *et al* 1993 and Hart *et al* 1995). Here we address the leishmanicidal activity of ether lipid analogues and consider putative mechanisms of action.

MATERIALS AND METHODS

Leishmania m. mexicana MNYC/BZ/62/M379 promastigotes were grown in a modified SDM growth media (SDM 93) at 26°C (Lux & Hart, manuscript in preparation). Preparation of homogenates and cell fractionation by differential centrifugation were performed essentially as described by Steiger *et al* (1980) and Hart & Opperdoes (1984). Acyl-CoA - acylation reactions were measured at 412 nm in a 10 mM Tris-HCl buffer pH 8.0 at 26°C essentially as described by Stobart and Stymne (1990). Plasma membrane protein kinase assay was studied using histone 2a as substrate as described by Das *et al.* (1986). Inositolphosphates were isolated and separated as described by Rigley and Hicks (1991).

RESULTS

The cytotoxicity of related ester and ether phospholipids (see Figure 2) towards *Leishmania mexicana* promastigotes was determined after 24 and 48 hours (Table 1). The acyl phospholipid, 2-LPC, and the alkyl counterpart, L-PAF, were compared for leishmanicidal action and only the ether lipid was found to be significantly effective. Indeed, the synthetic ether lipid analogues ET 18 OCH_3 (Edelfosine) and HePC (Miltefosine), designed on alkyl phos-

Table 1. LD_{50} of ether lipid analogues and anti-leishmanial drugs against *Leishmania mexicana* promastigotes

	LD_{50}*			
	24 hrs		48 hrs	
Compound	μg/ml	μM	μg/ml	μM
Pentostam	>100	—	>100	—
Pentamidine	1.4	—	0.4	—
Aminosidine	7.4	—	5.2	—
2-LPC	>16	>31	>16	>31
L-PAF	13	27	14	28
ET 18 OCH_3	1.6	3.0	1.5	2.8
ETPA	>58	>128	>58	>128
HePC	1.2	2.9	1.3	3.2
DH-10	1.3	na	1.2	na
DH-11	1.2	na	1.1	na
DH-12	0.9	na	0.8	na
DH-13	0.7	na	0.7	na
DH-14	>128	na	>128	na
DH-15	1.3	na	1.4	na
DH-16	1.5	na	1.5	na
DH-17	1.3	na	1.5	na
DH-18	2.0	na	1.6	na

*Taken from Lux, Hart and Klenner (manuscript in preparation).
na: not applicable.

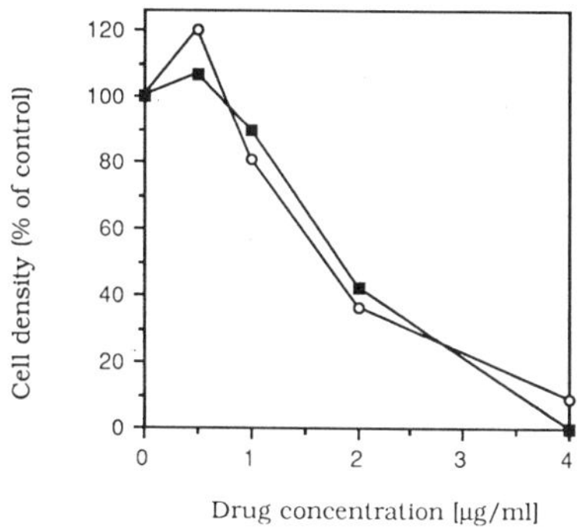

Figure 3. Effect of HePC (Miltefosine) on the proliferation of *L. m. mexicana* promastigotes. Promastigotes were seeded at 5 × 10^5 cells/ml and cell density determined after 24 hours (O) and 48 hours (■).

pholipid structures, were found to be even more effective than the anti-leishmanial drugs Pentostam and Aminosidine. Only Pentamidine had comparable efficancy to ether lipid analogues and the most active alkyl phospholipid analogues were found to be members of a novel series of structurally related analogues (DH-series) supplied by ASTA Medica. Interestingly, for these ether lipid analogues the choline head group of alkyl phospholipid analogues (see Figure 2) would seem to be important in leishmanicidal activity since ETPA was found to be more than 25 fold less active than Edelfosine (Table 1).

At low concentrations of ether lipid analogue we have observed rather interesting stimulations of promastigote proliferation of up to 20% (Figure 3), while at higher levels of drug anti-proliferative activity is seen. Somewhat similar concentration specific stimulations and inhibitions have been observed in cancer cells. The anti-neoplastic activity seen in cancer cells would seem to be comparable with the anti-leishmanial action of Miltefosine and Edelfosine since both anti-proliferative and putative pro-differentiation activity have been observed (Figure 4).

a

b

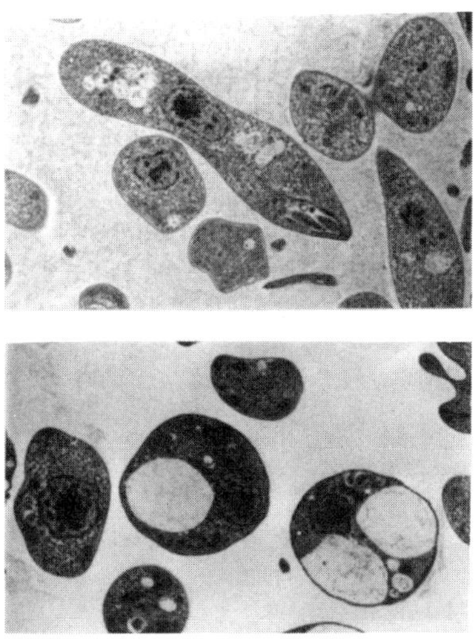

Figure 4. The effect of ET 18 OCH3 (edelfosine) on *L. m. mexicana* promastigotes. (a) Control promastigotes in late log phase of growth. (b) Promastigotes treated with 2μg/ml for 12hrs.

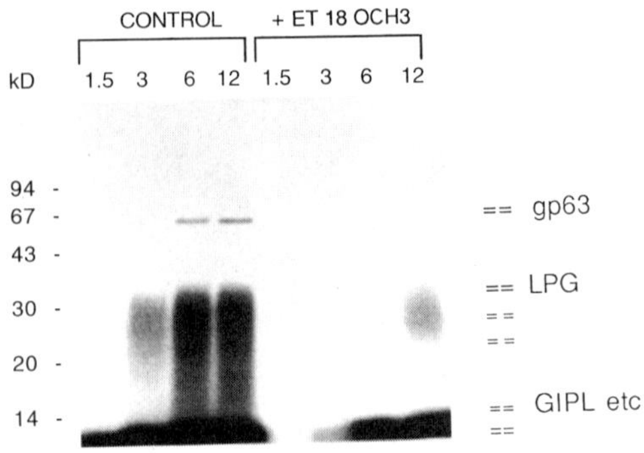

Figure 5. Effect of ET 18 OCH3 (Edelfosine) on [^{3}H] inositol incorporation into GPI anchored molecules of *L. m. mexicana* promastigotes. Each track is loaded with 20 μg of protein for both control and treated with 2 μg/ml of ET 18 OCH_3 (Edelfosine).

In Figure 4 a & b it can be seen that upon treatment of *Leishmania mexicana* promastigotes with Edelfosine the plasma and organellular membranes are disrupted and aflagellated "amastigote-like" morphologies are observed. From such preliminary morphological studies it is tempting to speculate that differentiation from promastigote to amastigote is being observed, however, more stringent markers for differentiation need to be studied.

In Figure 5 it can be seen that the biosynthesis of leishmanial GPI anchors for both LPG and GIPL classes of glycolipid and the glycoprotein gp63 can be followed by incorporation of radiolabelled inositol and separation on SDS-PAGE. In contrast promastigotes treated with Edelfosine (2μg/ml) showed complete inhibition of the GPI anchors for gp63 and >90% inhibition for LPG anchor biosynthesis, while only moderate perturbation (<30%) of GIPL anchors was observed. Since it has been suggested that GIPL's may well serve as precursors for LPG and gp63 anchors it could be concluded that Edelfosine was inhibiting alkyl or acyl transferases involved in the remodelling of the alkyl-lyso and alkyl-acyl membrane anchors of LPG and gp63 respectively.

We have therefore studied the effects of alkyl phospholipid analogues on leishmanial acyl and alkyl phospholipid acyltransferases (Figure 6). Miltefosine and Edelfosine

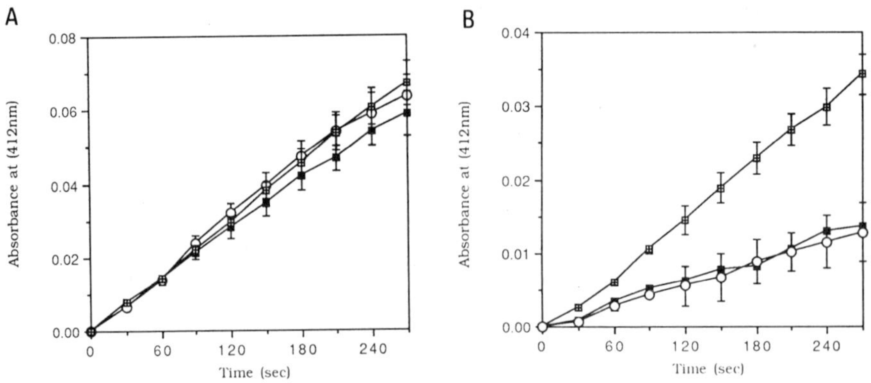

Figure 6. Effect of HePC (Miltefosine) and ET 18 OCH3 (Edelfosine) on (A), 1-acyl-glycerophosphocholine (2-LPC); and (B), 1-alkyl-glycerophosphocholine (L-PAF): acyl-CoA acyltransferase. Acylation reactions for *Leishmania mexicana mexicana* homogenates were measured at 412 nm in a 10 mM Tris-HCl buffer pH 8.0 at 26°C, essentially as described by Stobart & Stymme (1990). Control (⊞), HePC (■), and ET 18 OCH_3(O) at equimolar concentrations to 2-LPC and L-PAF.

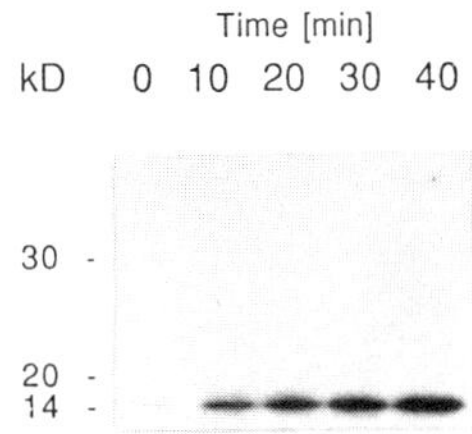

Figure 7. Protein kinase activity with intact promastigotes of *L. m. mexicana* using histone 2a as substrate. 10^6 promastigotes were incubated in 100 μl assay buffer containing 25 mM HEPES buffer, 10 mM NaF, 10 mM $MgCl_2$, 0.3 mM histories, 1 mM [γ^{32}P] ATP, 0.9% NaCl and 5 mM glucose (after Das et al., 1986).

had only moderate inhibition of 1-acyl-glycerophosphocholine (2-LPC); acyl-CoA acyltransferase activity (Figure 6A), a key enzyme in acyl glycerolipid metabolism. In contrast, both ether lipid analogues were found to inhibit 1-alkyl-glycerophosphocholine (L-PAF); acyl-CoA acyltransferase by more then 60% (Figure 6B). Thus the ether lipid remodelling enzyme is more sensitive to alkyl phospholipid analogues than the ester phospholipid counterpart. Interestingly, preliminary results (data not shown) suggests that although Miltefosine and Edelfosine inhibit the biosynthesis of leishmanial acyl phospholipids (<40%) the major lipid pathway to be effected was in fact alkyl phospholipids biosynthesis (>80%).

Since Miltefosine and Edelfosine have been shown to accumulate in the membranes of cancer cells and to perturb membrane enzymes involved in transport and signal transduction we have examined leishmanial plasma membrane enzymes. *Leishmania* have been reported to possess a novel plasma membrane protein kinase (Das *et al* 1986 and Hermoso *et al* 1991) which may be involved in the phosphorylation of host proteins and has been shown to phosphorylate complement proteins. We therefore studied the effect of ether lipid analogues on this leishmanial plasma membrane protein kinase activity (Figure 7 & 8). The leishmanial ectokinase activity was assayed using histone 2a as substrate and as can be seen in Figure 7 the phosphorylation of this 14.5 KDa protein was found to be linear with time for at least 40 mins. The stimulatory effect on the activity of this protein kinase by both the ester lipid control, 2-LPC, and the ether lipid analogue, Edelfosine, was found to be concentration dependent (Figure 8).

We have also noted interesting effects of ether lipid analogues on membrane PLC and PKC activities in *Leishmania*. For example, Figure 9 shows that Edelfosine perturbs the metabolism of key inositolphosphate intermediates (IP_2, IP_3 & IP_4). We have observed that IP_2 decreases with increasing Edelfosine concentration while IP_3 and IP_4 increase concomitantly. It is interesting to note that IP_3 is produced by the hydrolysis of PIP_2 by PLC

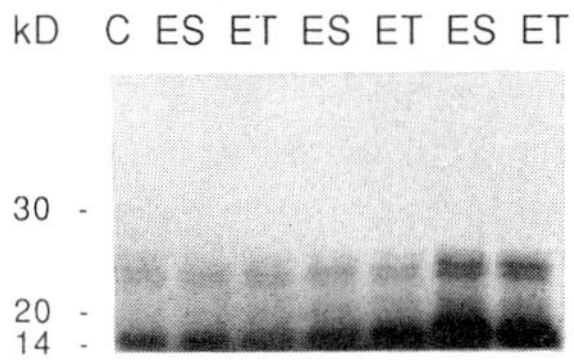

Figure 8. Effect of 2-LPC and ET 18 OCH_3 (Edelfosine) on *L. m. mexicana* promastigote plasma membrane protein kinase activity. 2 × 10^5 promastigotes were preincubated for 30 minutes in 20 μl assay buffer containing 0.4 or 40 μM 2-LPC (ES) or ET 18 OCH_3 (ET) respectively. The reaction was initiated by the addition of 2 μCi [γ-^{32}P] ATP in 5 μl assay buffer and stopped after 40 mins by adding 25 μl of treatment buffer (after Das et al. 1986).

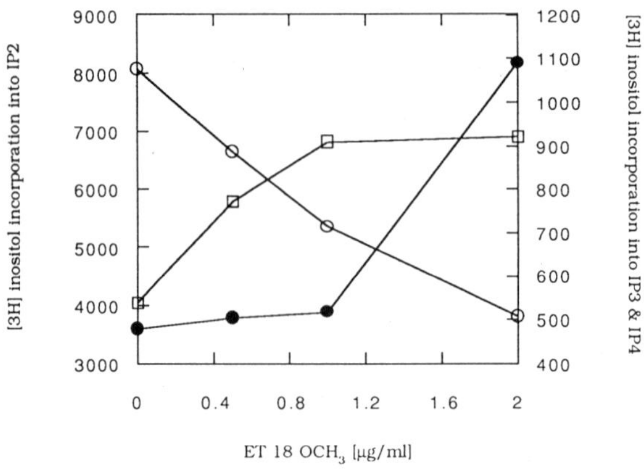

Figure 9. Effect of ET 18 OCH_3 (Edelfosine) on the incorporation of [3H] inositol into inositolphosphates in *L. m. mexicana* promastigotes. Promastigotes were incubated at 26°C for 12 hours with indicated concentrations of ET 18 OCH_3. Extraction and isolation of inositol phosphates (after Rigley and Hicks 1991) IP_2(o), IP_3(•), IP_4(■).

and is involved in Ca^{2+} mobilization and the translocation and priming of PKC. We are presently investigating in detail the effects of alkyl phospholipid analogues on specific leishmanial isoforms of PLC and PKC.

Indeed, using antibodies to specific PKC isoforms we have identified three PKC related isoforms in *Leishmania mexicana* (Table 2). Interestingly, we are able to detect PKC α, δ, and possible ζ related isoforms in *Leishmania,* while in the macrophage host PKC α, ε, and ζ isoforms were found.

Table 2. PKC related polypeptides in *Leishmania mexicana* promastigotes and mammalian counterparts

	Molecular mass kDa				
PKC isoform	*Leishmania mexicana*	J774 macrophage	Mammalian control		Tissue (s)[a]
Classical PKC					
α	79 ± 2	79 ± 2	79 ± 2	(81) [b]	Universal
β_1	nd	nd	77 ± 1	(79) [b]	Several
β_2	—	—	—	—	Many
	nd	nd	89	(80) [b]	Brain only
Novel PKC					
δ	74 ± 2	nd	77	(77)[c]	Universal
ε	nd	92 ± 2	92 ± 2	(89)[d]	Brain and others
η	nd	nd	nd	(80)[e]	Lung, skin, heart
θ	nd	nd	nd	(79)[f]	Skeletal muscle
Atypical PKC					
ζ	68 ± 2	72	72 ± 2	(80)[g]	Universal
λ	—	—	—	(74)[h]	Ovary, testis, etc.

PKC related antigens were detected by Western blot analysis using antisera raised against the C-termini of specific PKC isozymes. Comparisons were made between *Leishmania* and J774 macrophages. Mouse brain PKC isoforms were used as a control, except PKC δ, η and θ, where rat brain, lung, and skeletal muscle were used respectively. Published values for mammalian PKC's are shown in brackets with references. a, Nishizuka 1992; b, Marais and Parker 1989; c, Oliver and Parker 1991; d, Schaap *et al* 1989; e, Chida *et al* 1994; f, Osada *et al* 1992; g, Ways *et al* 1992; h, Akimoto *et al* 1994. nd - not detected.

DISCUSSION

From both our studies and those of others (Achterberg & Gercken 1987, Croft *et al* 1987, Kuhlencord *et al* 1992, Croft *et al* 1993 and Hart *et al* 1995) the consensus data suggest that ether lipid analogues possess considerable anti-leishmanial activity. In addition to leishmanicidal activity these compounds are also reported to be possible immuno modulating agents and activate host macrophages (eg Ngwenya *et al.* 1991, Schreiber *et al* 1994).

The considerable advantage in developing alkyl phospholipid analogues as anti-leishmanial drugs is that much of the pharmacokinetic and toxicity data in humans which is costly to obtain are available since many of these compounds are presently being developed as anti-tumour drugs (eg Hilgard *et al* 1993 and Lohmeyer & Bittman 1994). In addition, the topical application and promising oral effectiveness of these drugs make them particularly suitable for the treatment of cutaneous and visceral leishmaniasis respectively.

The mechanism of anti-leishmanial action of ether lipid analogues remains to be unequivocally elucidated, however, we postulate that perturbation of ether lipid metabolism and GPI anchor biosynthesis may well be involved. Similarly, the perturbation of leishmanial membrane signal transduction may also be implicated since we have detected perturbation of PLC and inositolphosphate metabolism which would in turn affect PKC activation and function.

Both ester and ether lipids were found to stimulate leishmanial plasma membrane protein kinase activity, however, the mechanism of stimulation and functional significance remain to be elucidated. Two attractive possibilities are that the specific substrate for this ectokinase is an integral membrane protein or is an alkyl-acyl GPI anchored glycoprotein, thus, the combination of ester or ether phospholipids and histone mimic the natural substrate. It is also possible, however, that this kinase has exofacial regulatory domains for phospholipids and is therefore activated in the presence of exogenous lipids.

The presence of leishmanial antigens related to one isoform in each of the three subclasses of PKC is provocative. Most mammalian cells express at least one representative of each PKC subclass (eg Mahoney & Huang 1994) and it is presumed that these confer on cells distinct functions. The presence of these antigens in *Leishmania* may indicate that a similar set of functions are controlled in an equivalent manner in this important genus of parasitic protozoan.

Future studies will concentrate upon the characterisation of the role of PLC and PKC isoforms in leishmanial and macrophage signal transduction processes. The perturbation of signal transduction in *Leishmania* and stimulation of macrophage activation are promising anti-leishmanial targets. Indeed, the elucidation of the anti-leishmanial mode of action of ether lipid analogues could be used in the design and development of new and more effective drugs.

REFERENCES

Achterberg, V. and Gercken, G. (1987). Cytotoxicity of ester and ether lysophospholipids on *Leishmania donovani* promastigotes. *Molecular and Biochemical Parasitology*, **23**, 117–122.

Akimoto, A., Mizuno, K., Osada, S., Hirai, S., Tanuma, S., Suzuki, K. and Shigeo, O. (1994). A new member of the third class in the protein kinase C family, PKC λ, expressed dominantly in an undifferentiated mouse embryonal carcinoma cell line and also in many tissues and cells. *The Journal of Biological Chemistry*, **269**, 12677–12683.

Arnold, B., Reuter, R. and Weltzien, H. U. (1978). Distribution and metabolism of synthetic alkyl analogues of lysophosphatidylcholine in mice. *Biochemica and Biophysica Acta*, **530**, 47–55

Beach, D. H., Holz, G. G. and Anekwe, G. E. (1979). Lipids of *Leishmania* promastigotes. *Journal of Parasitology*, **65**, 203–216

Berdel, W. E., Fink, U. and Rastetter, J. (1987). Clinical phase I pilot study of the alkyl lysophospholipid derivative ET-18-OCH_3. *Lipids*, **22**, 967–969

Chida, K., Sagara, H., Suzuki, Y., Murakami, A., Osada, S., Ohno, S., Hirosawa, K. and Kuroki T. (1994). The η isoform of protein kinase C is localized on rough endoplasmatic reticulum. *Molecular and Cellular Biology*, **14**, 3782–3790.

Croft, S. L., Neal, R. A., Pendergast, W. and Chan, J. H. (1987). The activity of alkyl phosphorylcholines and related derivatives against *Leishmania donovani*. *Biochemical Pharmacology*, **36**, 2633–2636.

Croft, S. L., Neal, R. A., Thornton, E. A. and Herrmann, D. B. J. (1993). Antileishmanial activity of the ether phospholipid ilmofosine. *Transactions of the Royal Society of Tropical Medcine and Hygiene*, **87**, 217–219.

Das, S., Saha, A. K., Mukhopadhyay, N. K. and Glew, R. H. (1986). A cyclic nucleotide-independent protein kinase in *Leishmania donovani*. *Biochem. J.*, **240**, 641–649.

Descoteaux, A., Turco, S. J., Sacks, D. L. and Matlashewski, G. (1991). *Leishmania donovani* lipophosphogycan selectively inhibits signal transduction in macrophages. *Journal of Immunology*, **146**, 2747–2753.

Grunicke, H. H. and Uberall, F. (1992). Protein kinase C modulation. Seminars in Cancer Biology **3** , 351–360.

Hart, D. T., Langridge, A., Barlow, D. and Sutton B. J. (1989). Antiparasitic drug design. *Parasitology Today*, **5**, 117–120.

Hart, D. T. and Opperdoes, F. R. (1984) The occurrence of glycosomes (Microbodies) in the promastigote stage of four major *Leishmnaia* species. *Molecular and Biochemical Parasitology*, **13**, 159–172.

Hart, D. T., Bhatti, M., Lux, H. and Klenner, T. (1995). Ether lipid analogues: A novel class of anti-leishmanial drug. *Trans. Roy. Soc. Trop. Med. and Hyg.* (in press)

Hermoso, T., Fishelson, Z., Becker, S. L. Hirschberg, K. and Jaffe, C. L. (1991). Leishmanial protein kinases phosphorylate components of the complement system. *EMBO*, **10**, 4061–4067.

Herrman, H. and Gerken, G. (1980). Incorporation of [1-^{14}C]octadecanol into the lipids of *Leishmania donovani*. *Lipids*, **15**, 179–186.

Hilgard, P., Klenner, T., Stekar, J. and Unger C. (1993). Alkylphosphocholines: a new class of membraneactive anticancer agents. *Cancer Chemotherapy and Pharmacology*. **32**, 90–95

Hilgard, P. and Klenner, T. (1994). Experimental pharmacology of Miltex and its constituents. *Drugs of Today*, **30**, 13–20

Ilg, T., Etges, R. and Overath, P. McConville, M. J., Thomas-Oates, J., Thomas, J., Homans, S. W. and Ferguson, M. A. J.(1991). Structure of *Leishmania mexicana* lipophosphoglycan. *Journal of Biological Chemistry*, **267**, 6834 -6810

Kuhlencord, A.; Maniera, T.; Eibel, H. and Unger, C. (1992). Hexadecylphosphocholine: Oral treatment of visceral Leishmaniasis in mice. *Antimicrobial Agents and Chemotherapy*, **36**, 1630–1634.

Lohmeyer, L. and Bittman R. (1994). Antitumor ether lipids and alkylphosphocholines, *Drugs of the Future*, **19**, 1021–1037.

Mahoney, C. W. and Huang, K.-P. (1994). Molecular and catalytic properties of protein kinase C. In Protein kinase C, edited by J. F. Kuo, Oxford University Press.

Marais, R. M. and Parker, P. J. (1989). Purification and characterisation of bovine brain protein kinase C isotypes α, β and γ. *European Journal of Biochemistry*, **182**, 129–137.

McConville, M. J. and Blackwell J. M.(1991). Developmental changes in the glycosylated phosphatidylinositols of *Leishmania donovani*. *Journal of Biological Chemistry*, **266**, 15170–15179

McConville, M. J. and Ferguson, M. A. J. (1993). The structure, biosynthesis and funktion of glycosylated phosphatidylinositols in the parasitic protozoa and higher eukaryotes. *Biochemical Journal*, **294**, 305–324.

Ngwenya, B., Fiavey, N. P. and Mogashoa, M. M.(1991). Activation of peritoneal macrophages by orally administered ether analoges of lysophospholipids. *P. S. E. B. M.*, **197**, 91–97.

Nishizuka, Y. (1992). Intracellular Signaling by hydrolysis of phospholipids and activation of protein kinase C. *Science*, **258**, 607–614.

Olivier, A. R. and Parker P. J. (1991). Expression and characterization of protein kinase C-δ. *European Journal of Biochemistry*, **200**, 805–810.

Orlandi, P. A. Jr. and Turco, S. J.(1987). Structure of the lipid moiety of the *Leishmania donovani* lipophosphoglycan. *Journal of Biological Chemistry*, **262**, 10384–10391

Osada, S., Mizuno, K., Saido, T. C., Suzuki, K., Kuroki, T. and Ohno, S. (1992). A new member of the protein kinase C family, nPKCθ, predominantly expressed in skeletal muscle. *Molecular and Cellular Biology*, **12**, 3930–3938

Proudfoot, L., O' Donnell, C. A. and Liew, F. Y. (1995). Glycoinositolphospholipids of Leishmania major inhibit nitric oxide synthesis and reduce leishmanicidal activity in murine macrophages. *Eur. J. Immunol.*, **25**, 745–750.

Rigley and Hicks (1991). In Cytokines: A practical approach, edited by Balkwill, F. R., Oxford University Press.

Schaap, D., Parker, P.J., Bristol, A., Kriz, R. and Knopf J. (1989). Unique substrate specificity and regulatory properties of PKC- ε: a rationale for diversity. *FEBS Letters*, **243**, 351–357.

Schneider, P., Ferguson, M. A. J., McConville, M.J., Mehlert, A., Homans, S. W. and Bordier, C (1990). Structure of the glycosyl-phosphatidylinositol membrane anchor of the *Leishmania major* promastigote surface protease. *Journal of Biological Chemistry*, **265**, 16955–16964

Schreiber, B. M., Layne, M. D. and Modest, E. J. (1994). Superoxide production by macrophages stimulated *in vivo* with synthetic ether lipids. *Lipids*, **29**, 237–242.

Steiger, R. F., Opperdoes, F. R. and Bontemps, J.(1980). Subcellular fractionation of *Trypanosoma brucei* bloodstream forms with special reference to hydrolases. *European Journal of Biochemistry*, **105**, 163–175

Stobart,A. K. and Stymne S. (1990). Triacylglycerol biosynthesis. *Methods in Plant Biochemistry*, **4**, 19–46.

Ways, D. K., Cook, P. P., Webster, C. and Parker P. J. (1992). Effect of Phorbol ester on PKC-ζ. *The Journal of Biological Chemistry*, **267**, 4799–4805.

WHO–TDR (1990). Leishmaniasis. In Tropical Diseases 1990, World Health Organisation, Geneva, Switzerland.

34

PAF RESPONSE TO INFLAMMATORY AND NON-INFLAMMATORY STIMULI

Kunihiko Saito

Department of Medical Chemistry*
Kansai Medical University
Osaka, Japan

1. INTRODUCTION

PAF is now well known as a chemical mediator, but it has being found widely in normal cells and tissues including higher and lower animals and plants. Biomedically, therefore, PAF has biphasic functions. One is as a chemical mediator which should be removed from the pathological region principally. The other function is an autacoid activity that is related to physiological reactions or homeostasis. The difference between these two functions might be based on the *in situ* quantity rather than quality.

In this paper, PAF in ischemic anterior segment of rabbit eye as an example of chemical mediator, and PAF in male reproductive organs of normal guinea pig and in yeast as examples of autacoid are described.

2. PAF AND PAF-ACETYL HYDROLASE ACTIVITY IN ISCHEMIC ANTERIOR SEGMENT OF RABBIT EYE

PAF was accumulated in the cornea with alkali injury (1). We investigated the relationship between anterior segment ischemia and PAF and PAF-AH activity (2). The ischemia was made by cauterization of bilateral long posterior ciliary arteries. It sometomes occurrs after retinal operation and corneal edema and neovascularization are chracteristic. The influenced iris, cornea and aqueous humor were used for analyses of PAF and PAF-AH activity. PAF was purified by TLC and HPLC and its platelet aggregation activity was inhibited with CV 6209 or WEB 2086 at a level of $1x10^{-6}$ M.

Mass spectrometry was used for identification of PAF and its molecular species (3). For assay of PAF-AH activity, sn-2 tritiated acetyl PAF was used as the substrate. They were assayed periodically for 5 days after operation.

* Prof. emer.

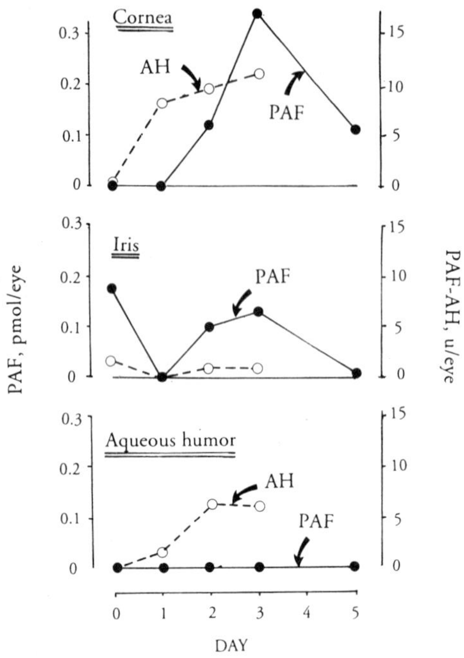

Figure 1. Change in PAF level and PAF-AH activity in cornea, iris and aqueous humor after induced anterior segment ischemia. [Cited from Ito et al. (2), with permission].

As shown in Fig 1, neither PAF nor PAF-AH activity are detected in normal cornea but by the operation PAF-AH responses prior to PAF, phenomenally at least, and increases from day1 to day 3, whereas PAF is detected later on day 2 and reached maximally on day 3. In iris, PAF and PAF-AH activity are detected in normal tissue but their level do not increase by ischemia. No PAF is detectable in aqueous humor before and after the operation, but PAF-AH activiity begins to incease after operation.

These findings indicate that the ischemic stimulus induced PAF-AH and PAF from these tissues and if PAF-AH activity is too strong PAF is no more detectable and *vice versa*.

3. PAF IN REPRODUCTION

3.1 PAF in Male Reproductive Organs

Generally reproductive organs are hormon sensitive. Following to our previous paper (4), which described the relationship between uterus PAF and estradiol, PAF in male reproductive organs of guinea pig and its response to testosterone and gonadotropin administered were described (5).

PAF levels of male reproductive organs of Hartley guinea pig and Wistar rats are shown in Fig. 2. These normal animals are not given any exogenous stimumus before experiment, so the levels mean endogenous or basal PAF contents in the organs. These values, respectively, are about one fifth and one thousandth compared to those of rat uterus or heart and stimulated human neutrophils with A 23187 (3). Seminal vesicle, different from other reproductive organs listed, has no contaminant sperm and was used for the next experiment.

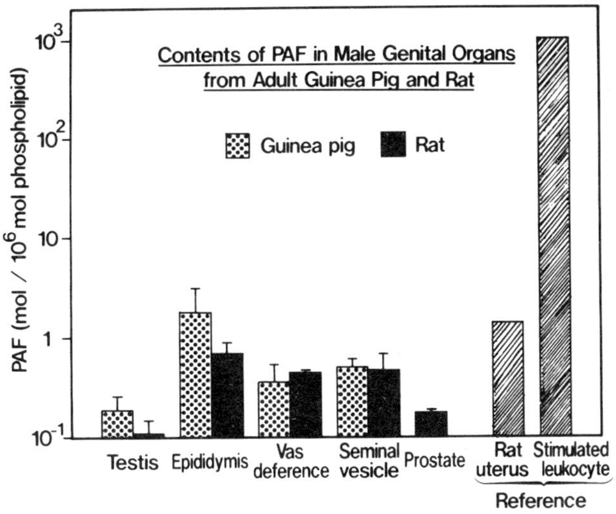

Figure 2. PAF level in male reproductive organs of normal guinea pig and rat.

Fig. 3 shows that PAF contents in seminal vesicle of guinea pig and in rat uerus decreased by castration and recovered by administration of testosteron and estradiol, respectively. The changes in PAF level between these two different sex organs are quite similar, indicating the endogenous PAF level is under hormonal control. Human chorionic gonadotropin (100IU) elevated transiently the PAF level in seminal vesicle of normal guinea pig to about 150%.

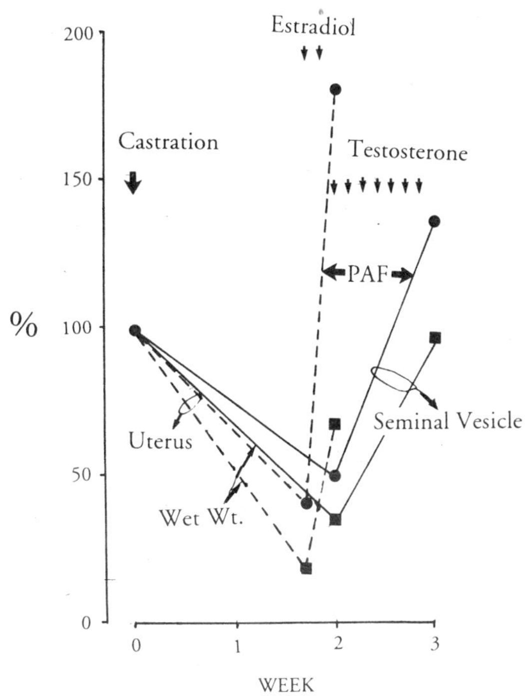

Figure 3. Change in PAF level of rat uterus and guinea pig seminal vessicle by castration and administration of estradiol and testosterone, respectively.

3.2 PAF in Yeast, *Sacchromyces cerevisiae*

Recently the evidence for the presence of endogenous PAF in lower animals and presumably in some plants is accumulating, for example, *Tetrahymena pyriformis*(6), *Dictyostelium discoideum* (7), *Incilaria bilineata*, slug (8), *Saccharomyces cerevisiae* (9) and *Cucurbita pepo L*, zucchini (10).

We screened 30 yeast strains (11 genera) for PAF production and found it mainly in the strains of *Saccharomyces* genus. The PAF molecular species of *S. cerevisiae* were 16:0, 16:1 and 18:0 PAFs, and 16:0, 16:1,18:0 and 18:1 acylPAF, among of which 16:1 and 18:0 PAFs were very minute and as seen in normal rat uterus, stomach or heart (3), acylPAF was much more abundant than PAF in the yeast as well. PAF formation in yeast cells increased at the middle stationary phase of growth and stimulated with A23187 (9).

4. DISCUSSION

As seen in cornea ischemic stimulus produced harmful amount of PAF and also its destructive enzyme, PAF-AH, and phenomenally PAF-AH activity was detected before PAF was found. It might be interesting teleologically.

The amount of endogenous PAF is about 2 moles or less per 1 million phospholipids of mother tissues and is controled by hormone as seen here and by autonomic nervous system (11) to carry out the physiological reactions. The PAF in S.cerevisiae might br involved in cell growth as will be published elsewhere.

REFERENCES

1. Bazan, H.P., Reddy, S.T. and Lin, N. (1991) Exp. Eye Res. 52, 481-491.
2. Ito, Y., Sugatani, J., Miki, H., Uyama, M. and Saito, K. (1993) Exp. Eye Res., 57, 107-115.
3. Saito, K. (1992) in: Mass Spectrometry: Clinical and Biomedical Applications, 1, edited by D. M. Desiderio, Plenum Press, New York, 319-343.
4. Nakayama, R., Yasuda K. , Okumura T. and Saito K. , (1991) Biochim. Biophys. Acta, 1085, 235-240.
5. Muguruma, K., Komatz, Y., Ikeda, M., Sugimoto T. and Saito K. (1993) Biol. Reprod., 48, 386-392.
6. Lekka, M., Tselepis, A.D. and Tsoukatos, D.(1986) FEBS Lett. 208, 52-55
7. Bussolino, F., Sordano, C.,Benfenati, E. and Bozzaro, S. (1991) Eur. J. Biochem., 196, 609-615.
8. Sugiura, T., Ojima, T, Fukuda, T., Satouchi, K., Saito, K. and Waku, K. (1991) J. Lipid Res., 32, 1795-1803.
9. Nakayama, R., Kumagai H. and Saito, K.(1994) Biochim. Biophys. Acta 1199, 137-142.
10. Scherer, G.F.E., Martiny-Baron, G. and Stoffel, B. (1988) Planta 175, 241-253.
11. Sugatani,J., Miwa, M., Fujimura, K. and Saito, K. (1993) Biochem. Pharmacol.,46,37-45.

35

ROLE OF PLATELET-ACTIVATING FACTOR IN SKELETAL MUSCLE ISCHEMIA-REPERFUSION INJURY

Donald Silver, Animesh Dhar, Milton Slocum, John G. Adams, Jr., and Shivendra Shukla

Department of Surgery
University Hospital
One Hospital Drive, Columbia, Missouri 65212

1. INTRODUCTION

Blood flow can be restored to limbs after 1 to 2 hours of ischemia with the expectation that the limbs will survive and be functional. Adverse changes may occur during reperfusion of skeletal muscle which has been ischemic for 3 hours or more. These changes, ischemia-reperfusion injury (IRI), can include endothelial cell swelling, increased microvascular permeability, leukocyte endothelial adhesion, vascular thrombosis, altered vascular resistance, no reflow phenomenon, altered monocyte/neutrophil function and even death of the ischemic reperfused cells.

Platelet-activating factor (PAF), is a potent lipid mediator which could have an important role in the genesis of IRI. Studies of the role of PAF in skeletal muscle IRI are reported.

2. MATERIALS AND METHODS

The gracilis muscles were isolated to their major vascular pedicles in 25–30kg conditioned female mongrel dogs. At an appropriate time, the muscle was rendered ischemic by application of a nontraumatic vascular clip to the artery. Electrocardiographic and rectal probe thermometer tracings were recorded continuously and the body core temperature was maintained between 37.5°C and 38.0°C. All animals received normal saline intravenously at 8ml/kg/hr. Arterial and venous cannulas were placed in the femoral vessels and positioned in the gracilis artery and vein for blood sampling and injections.

After ischemia and reperfusion, and/or infusions of test agents, the cannulas were removed and the wounds closed. At the completion of each study: the animals were anesthetized and the gracilis muscles reexposed; cannulas were replaced for blood sampling and pressure measurements; the gracilis muscles were harvested; and the animals euthanized.

Blood samples were collected in duplicate in glass test tubes containing 10μL of 0.5mol/L of ethylenediamine tetraacetic acid (final concentration 5mmol/L). Lipids were extracted and the samples were assayed for PAF with the PAF [^{3}H] scintillation proximity assay system (Amersham, Arlington Heights, IL).

The harvested muscles were weighed, divided into seven equal sections and incubated with nicotinamide adenine dinucleotide-nitroblue tetrazolium dye for 20 minutes at room temperature (25°C). Both sides of each muscle section were traced for total area and nonstained area (nonviable) using planimetry (Easydij, Geocomp; Golden, CO). The percentage of necrosis of each muscle was calculated.

3. RESULTS

3.1 Experiment #1

Group 1 gracilis muscles (N=9) were subjected to 5 hours of ischemia and 20 hours of reperfusion. The opposite muscle was isolated but not subjected to ischemia. Control blood samples were obtained. After 5 hours of ischemia, additional blood samples were obtained at 1, 5, 10, 15, and 30 minutes . Final blood samples were obtained after 20 hours of reperfusion, at which time the muscles were harvested and assayed for necrosis. Group 2 gracilis muscles (N=3) were isolated but not subjected to ischemia.

The ischemic gracilis muscles produced more PAF than did the nonischemic muscles at each time of reperfusion. The peak PAF production by the ischemic muscles occurred at 10 minutes of reperfusion, when the PAF values from the ischemic muscles (41.5*p*mole/ml) were significantly greater than that (2.5*p*mole/ml) from the nonischemic muscles ($p=0.0007$). A statistically significant increase of PAF (28.3*p*mole/ml) also occurred at 15 minutes of reperfusion ($p=0.0222$). PAF levels in venous blood from the nonischemic Group 1 muscles were compared to PAF from Group 2 muscles to evaluate the effect of surgery on PAF production. The PAF levels were higher at every study time, except at 10 minutes, in blood from control muscles in Group 1.

After 20 hours of reperfusion, the percent necrosis of the ischemic muscle was 77.13% ± 21.61%, whereas no necrosis was observed in the control muscles.

3.2 Experiment #2

Group 1 muscles (N=9) were subjected to 5 hours of ischemia. Group 2 muscles (N=7) were infused, while ischemic for 20- 40 minutes, with 10^{-4}M or 10^{-5}M PAF. Group 3 muscles (N=5) were infused, while ischemic for 20–40 minutes, with 10^{-4}M *lyso*-PAF. The opposite gracilis muscle served as a control in each dog. The ischemia and infusions were followed with 20 hours of reperfusion. Blood samples were obtained before and at 1, 5, 10, 30 and 60 minutes and 20 hours of reperfusion. The muscles were harvested after 20 hours of reperfusion.

PAF production was increased in all groups during reperfusion, although a lesser increase occurred in Group 3. PAF was significantly increased in the ischemic muscles, compared to the control muscles, of Group 1 at 10, 15 and 30 minutes; of Group 2 at 5, 10, 15, 30 and 120 minutes; and of Group 3 at 15 minutes and 20 hours. The extent of muscle necrosis was significantly greater ($p<0.05$) in the ischemic muscles in Groups 1 and 2 than their respective controls; no significant muscle necrosis occurred in Group 3.[1]

3.3 Experiment #3

Group 1 muscles (N=10) were subjected to 5 hours of ischemia. After 5 hours of ischemia, Group 2 muscles (N=6) received 15mg/kg of pentoxifylline in 100ml of saline as a bolus 10 minutes prior to the reperfusion. The Group 3 muscles (N=6), after 5 hours of ischemia, received 25mg/kg of pentoxifylline 10 minutes prior to reperfusion. Blood samples were obtained before and at 1, 5, 10, 30 and 60 minutes and 20 hours of reperfusion. The muscles were harvested after 20 hours of reperfusion and assayed for necrosis.[2]

The heart rate increased during the first two hours of reperfusion when the animals received pentoxifylline, the blood pressure was not affected. PAF was decreased, not significantly, at 10, 15 and 30 minutes in Group 2 as compared with Group 1 ($p=0.066$, $p=0.320$, $p=0.230$, respectively). PAF levels in Group 3 were decreased at all times of reperfusion when compared with Group 1. At 10 minutes of reperfusion, PAF was decreased to 4.89 ± 5.16*p*mole/ml in Group 3 compared with 41.39 ± 9.19*p*mole/ml in Group 1 ($p<0.05$).

The extent of muscle necrosis in Group 2 was less than that observed in Group 1; however, this difference did not attain significance ($p=0.08$). There was a significant reduction in muscle necrosis in Group 3 muscles (44.55% ± 21.47%) compared with Group 1 muscles (77.26% ± 20.38%) ($p<0.05$).

3.4 Experiment #4

Group 1 muscles (N=5) were not treated. Group 2 muscles (N=5) received systemic infusions of 10mg/kg of WEB-2086 immediately prior to reperfusion. Group 3 muscles (N=7) received 200mg of WEB-2086 into their gracilis arteries immediately prior to reperfusion. Blood samples were obtained before and at 1, 5, 10, 30 and 60 minutes and 20 hours of reperfusion, when the muscles were harvested. PAF in the venous effluent from the ischemic muscles was significantly ($p=0.0029$) increased at every collection when compared to the control muscle PAF effluents. The greatest increase (18 ± 7*p*mole) occurred at 15 minutes. The systemic and intra-arterial injections of WEB-2086 resulted in higher concentrations of PAF in the gracilis venous effluent (Group 1: 8 ± 2*p*mole/ml; Group 2: 17 ± 5*p*mole/ml; Group 3: 23 ± 4*p*mole/ml) at 10 minutes. There was significantly less muscle necrosis in the WEB-treated groups. The percent of muscle necrosis was 86.6% ± 3.2% in the control group, and 44.3% ± 14.5% in the intra-arterial WEB-treated group ($p=0.056$). This study demonstrates that it is possible to attenuate ischemia reperfusion-induced skeletal muscle necrosis with WEB-2086.

4. DISCUSSION

The studies reported indicate that PAF is produced by ischemic skeletal muscle during periods of reperfusion. The PAF concentrations were increased above the pre-ischemic levels at every time of study of reperfusion, up to 1 hour, with the peak elevation at 10–15 minutes. The increased PAF concentrations in the ischemia-reperfusion muscles correspond to the edema and necrosis of the ischemic muscles. The infusion of PAF into ischemic skeletal muscles subjected to brief periods (20–40 minutes) of ischemia resulted in changes similar to that seen after 5 hours of ischemia and 20 hours of reperfusion. Similar infusions with *lyso*-PAF were associated with limited (not significant) muscle changes. These studies confirm that PAF has an important role in the injury associated with ischemia-reperfusion.

Pentoxifylline has been shown to inhibit platelet aggregation,[3] block granulocyte response to PAF,[4] and decrease the metabolic derangements associated with IRI in the kidney, liver and brain.[5,6] We found that the administration of pentoxifylline 15mg/kg as a systemic bolus 10 minutes before the reperfusion of ischemic skeletal muscle, resulted in no significant decrease in PAF levels at 10, 15 and 30 minutes of reperfusion. However, the administration of 25mg/kg of pentoxifylline resulted in significant decreases of PAF during all times of reperfusion with a significant decrease ($p<0.05$) after 10 minutes. In addition, the extent of muscle necrosis was also significantly decreased in the muscles receiving the higher dose of pentoxifylline. The beneficial effects of pentoxifylline could relate to the inhibition of granulocyte function by pentoxifylline,[4] hemoreologic effects on the platelets, reductions in plasma viscosity, increases of fibrinolytic activity,[7] and/or increased prostacyclin synthesis.[8,9]

PAF enhances the binding of neutrophils to endothelial cells.[10] It has been shown that a PAF receptor antagonist, WEB-2086 (Boehringer-Ingelheim, Ridgefield, CT), will block leukocyte adhesion to endothelial cells during ischemia reperfusion. The infusion of WEB-2086 intravenously into the gracilis artery immediately prior to reperfusion resulted in a significant decrease in muscle necrosis. The levels of PAF in the venous effluents, from the ischemic muscles treated with WEB when compared to untreated ischemic reperfused muscles, were significantly increased at 10 and 15 minutes. This increase in PAF was expected since WEB-2086 interferes with PAF binding to receptors, thus making more PAF available for assay.

5. SUMMARY

These studies indicate that PAF, a known stimulator of aggregation, secretory and/or contractile activity of platelets, neutrophils, smooth muscle and other cells, is produced by skeletal muscle during IRI. The maximum increase occurs at 10 to 15 minutes and continues for at least an hour. Infusions of PAF into skeletal muscle subjected to 20 minutes of ischemia and 20 hours of reperfusion resulted in tissue necrosis similar to that produced by 5 hours of ischemia and 20 hours of reperfusion. 25mg/kg of pentoxifylline, infused immediately prior to reperfusion of ischemic skeletal muscle, decreased PAF production and muscle necrosis. It was also demonstrated that skeletal muscle necrosis can be reduced by infusion of PAF antagonists (WEB-2086) into the muscle immediately prior to reperfusion. Thus, PAF has an important role in skeletal muscle IRI and additional studies of PAF inhibition during IRI are warranted.

REFERENCES

1. Adams J, Dhar A, Kikta M, et al. Role of platelet-activating factor in skeletal muscle ischemia reperfusion injury. Surg Forum:XLV;395–397, 1994.
2. Adams J, Dhar A, Shukla S, Silver D. Effect of pentoxifylline on tissue injury and platelet-activating factor production during ischemia-reperfusion injury. JVS:21(5);742–749, 1995.
3. Stefanovich V, Jarvis P, Grigoleit HG. The effect of pentoxifylline on the 3',5'-cyclic AMP-system in bovine platelets. Int J Biochem:8;359–364, 1977.

4. Hammerschmidt D, Kotasek D, McCarthy T, et al.Pentoxifylline inhibits granulocyte and platelet function, including granulocyte priming by platelet activating factor. J Lab Clin Med: 112;254–263, 1988.
5. Berens K, Luke D. Pentoxifylline in the isolated perfused rat kidney. Transplantation:49; 876–879, 1990.
6. Bluhm R, Molnar J, Cohen M. The effect of pentoxifylline on the energy metabolism of ischemic gerbil brain. Clin Neuropharmacol:8;280–285, 1985.
7. Satewachin I, Iljin V, Schestakov V, et al. The use of pentoxifylline in a combined regimen for the prevention of re-thrombosis in major blood vessels. Pharmatherapeutica:2 Suppl 1;109–117, 1978.
8. Weithmann K. Reduced platelet aggregation by effects of pentoxifylline on vascular prostacyclin isomerase and platelet cyclic amp. Gen Pharmac:14;161–162, 1983.
9. Schror K, Matzky R, Kahlen T, Darius H. The release of prostacyclin (PGI2) by pentoxyfylline from human and animal vascular tissue and its implications for vascular and antiplatelet activities. Thromb Haemost:46;272, 1981.
10. Lorant D, Patel K, McIntyre T, et al. Coexpression of GMP-140 and PAF by endothelium stimulated by histamine or thrombin: a juxtacrine system for adhesion and activation of neutrophils. J Cell Biology:115;223–234, 1991.

PLATELET-ACTIVATING FACTOR

A Possible Role in the Modulation of the Vasomotor Tone and Blood Pressure

Rodrigo A. B. Lopes-Martins, Claudia V. Araújo, Vanessa Estato, Sheila Moreira, Renato S. B. Cordeiro, and Eduardo V. Tibiriçá

Departamento de Fisiologia e Farmacodinâmica
Instituto Oswaldo Cruz
FIOCRUZ, C. P. 926
21045-900, Rio de Janeiro, Rio de Janeiro, Brasil

INTRODUCTION

It is well known that the vascular smooth muscle tone is regulated not only by the activity of the sympathetic nervous system, but also by the release of vasoactive factors from the endothelium including nitric oxide (NO), prostacyclin, thromboxane A_2 and endothelin (1; 2; 3) that can be induced by both chemical and mechanical stimuli (4). In addition, several studies have reported endothelium-independent vasodilator responses to an unknown endogenous mediator, in rabbit and rat resistance arteries (5;6). PAF is a potent phospholipid mediator released by various cell types including platelets, leukocytes, macrophages, endothelial cells (7) and cultured rat vascular smooth muscle cells (8). This mediator has many biological actions already described (9). Several reports have demonstrated that PAF (< 1 μg kg^{-1}, i.v.) induce severe cardiovascular alterations, including a decrease in arterial blood pressure (10; 11), a direct negative chronotropic effect (12; 13) and an increase in vascular permeability (14). The hypotensive effect of PAF has been attributed mainly to the dilation of resistance vessels (15;16), thus suggesting a role for this mediator in the modulation of vascular smooth muscle tone.

Here, we provide evidence that PAF is involved in a negative feedback mechanism which modulates vascular tone, and hence blood pressure, in the anaesthetized rabbit.

METHODS

Animals and Haemodynamic Measurements

New Zealand white rabbits were anaesthetized with sodium pentobarbitone (40 mg kg^{-1}) administered via a marginal ear vein, immobilized with pancuronium bromide (1 mg

Platelet-Activating Factor and Related Lipid Mediators 2
edited by Nigam *et al.*, Plenum Press, New York, 1996

kg^{-1}, i.v.) and artificially ventilated with room air (tidal volume 10 ml kg^{-1}; stroke rate 25 min^{-1}). The right femoral vein was cannulated for i.v. injections. The instantaneous arterial pressure was continuously monitored through a catheter placed in the abdominal aorta via the right femoral artery and connected to a Hewlett Packard quartz transducer (1290 A), which in turn was connected to a 4-channel pressure processor and recorder (Hewlett Packard 7754 system with 8805B amplifier). Systolic (SAP) and diastolic (DAP) arterial pressures were obtained directly from the recordings and mean arterial pressure (MAP) was calculated as the diastolic pressure plus one-third of the pulse pressure. The heart rate (HR) was determined by counting the blood pressure waves at a high recorder speed.

In some rabbits an electromagnetic flow probe was placed around the ascending aorta and connected to a Skalar blood flow meter model MDL 1401 and cardiac output (ml min^{-1}) was continuously recorded. Systemic vascular resistance was calculated as the quotient of the mean arterial pressure and the cardiac output multiplied by a conversion factor (80) and expressed in dyn sec cm^{-5}.

The animals were then allowed to stabilize for at least 15 min (control period). Before any physiological or pharmacological manipulation was carried out, the mean of three determinations for each cardiovascular parameter was calculated from recordings obtained at 5 min intervals. These values were considered as the basal haemodynamic values.

Experimental Design

Experiment 1-Carotid Artery Occlusion. The carotid arteries were dissected and clamped for 10 minutes. This classical procedure results in systemic vasoconstriction and important increases in systemic vascular resistance, and consequently in arterial pressure.

Separate groups of rabbits received intravenous injections of saline solution (control group), PAF receptor antagonists (WEB 2086, 2.0, 5.0 and 10.0 mg kg^{-1}; BN 52021, 0.3, 1.0 and 3.0 mg kg^{-1}) or the phospholipase A_2 inhibitor mepacrine (5 mg kg^{-1}), 5 min prior to carotid artery clamping. At the end of the experiments, a hypotensive dose of PAF (3.0 μg kg^{-1}) was intravenously injected, in order to confirm the effectiveness of PAF receptor blockade; a dose which had previously been shown to induce cardiovascular collapse in this experimental model (20).

Experiment 2-Desensitization to the PAF-Induced Hypotensive Response. A group of rabbits was treated for seven days with a single dose of PAF (3.0 μg kg^{-1}; i.v.). On the day after the last injection (eighth day), the surgical procedures described above were performed, and PAF was injected at the same dose in order to monitor the hypotensive response following desensitization. When the haemodynamic parameters had returned to the pre-injection levels, the carotid arteries were clamped for 10 min and the resulting hypertensive response recorded.

Experiment 3-Effects of PAF-Antagonists and Indomethacin on the Norepinephrine-Induced Pressor Response. The rabbits were prepared as described earlier and the dose-response relationship for the norepinephrine (1.0–30.0 μg kg^{-1}, i.v.)-evoked changes on the haemodynamic parameters was analysed before and after the administration of the PAF-receptor antagonist WEB 2086 (5 mg kg^{-1}). In control experiments, a separate group of animals received an intravenous injection of saline solution between the two norepinephrine dose-response curves.

This same experimental design was used to test the effects of indomethacin (3.0 mg kg^{-1}) injection on the norepinephrine-induced cardiovascular changes.

Drugs

Norepinephrine, indomethacin, and platelet-activating factor (1-O-hexadecyl-2-acetyl-sn-glyceryl-3-phosphorylcholine) dissolved and further diluted in saline containing 0.25 % (w/v) bovine serum albumin, were purchased from Sigma Chemical Co (St. Louis, MI, U.S.A.). Pentobarbitone sodium was purchased from Sanofi Santé Nutrition Animale (Toulouse, France). Pancuronium bromide (Pavulon®) was supplied by Organon Tecknica (Fresnes, France) and WEB 2086 3–4-(2-chlorphenyl-)-9-methyl-6H-thieno-3,2-f-1,2,4-triazolo-4,3-α-1,4-diazepin-2-yl-(4-morpholinyl)-l-propanone was from Boehringer Ingelheim (Germany). BN 52021 was kindly provided by Henri Beaufour Laboratories (Paris, France) and mepacrine was from Rhône-Poulenc Santé (Vitry-sur-Seine, France). All drugs were dissolved in 0.9 % saline solution , and administered in a fixed volume of 500 μl.

Statistical Analyses

The results are expressed as the mean ± s.e.mean for each group. The effects of the different treatments on baseline haemodynamics over time were analysed by repeated measure analysis of variance (ANOVA) while comparisons between the control group (saline-injected) and treated groups were performed using one way ANOVA. When an overall difference was detected by ANOVA, the Newman-Keuls test was used to localize the statistically significant differences. Differences with P values of less than 0.05 were considered significant. All calculations were performed using a commercially available statistical package (Graphpad Instat, Graphpad Software, University of London, UK).

RESULTS

Experiment 1

Effects of the PAF-Receptor Antagonists WEB 2086 and BN 52021 and the Phospholipase A_2 Inhibitor Mepacrine on the Cardiovascular Response Elicited by Bilateral Carotid Artery Occlusion. In the control group, carotid artery occlusion increased mean arterial pressure (MAP) by 26 ± 3.4 % (n=9, P<0.05) above the baseline value. The PAF antagonist WEB 2086 potentiated the increase in MAP induced by carotid occlusion (26 ± 3.4 % before vs. 34.4 ± 4.5 %, 51 ± 9 %, 44 ± 4.4 %; n=4–6, P<0.05, after 2.0, 5.0 and 10.0 mg of WEB 2086 per kg^{-1}, respectively). A similar response was observed with another PAF antagonist BN 52021 (26 ± 3.4 % before vs. 49.4 ± 1.8 %, 37.6 ± 2.3 %, 43.8 ± 8.2 %; n=5, P<0.05; following BN 52021 doses of 0.3, 1.0 and 3.0 mg kg^{-1}, respectively; Figure 1).

The phospholipase A_2 inhibitor mepacrine (quinacrine; 5 mg kg^{-1} i.v.) also potentiated the increases in arterial blood pressure induced by carotid clamping (from 26 ± 3.4 % to 47.5 ± 6.3 %; n=5; P<0.05). This effect was particularly pronounced in the diastolic arterial pressure where the levels attained were very similar to those observed in the group treated with the PAF receptor antagonists (Table 1).

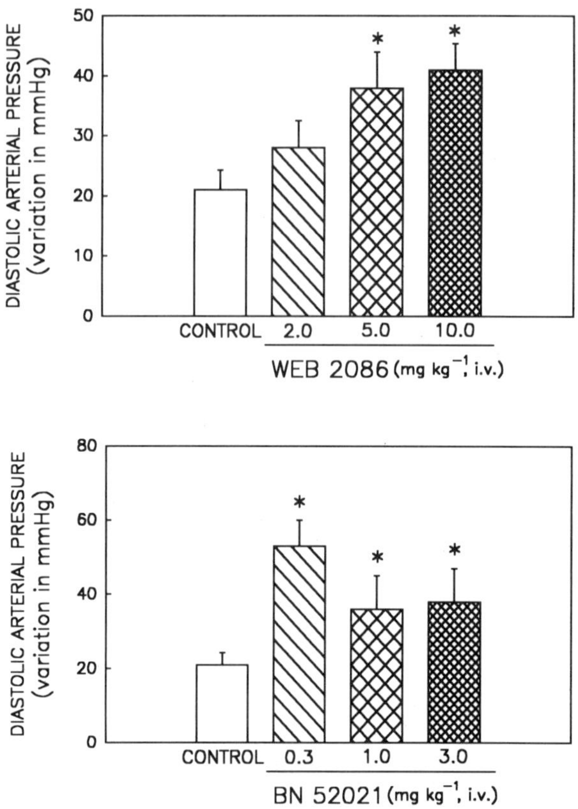

Figure 1. Effects of the pre-treatment with increasing doses of the PAF receptor antagonists WEB 2086 and BN 52021 on the hypertensive response elicited by bilateral carotid artery occlusion in pentobarbitone anaesthetized rabbits. CONTROL: saline-injected animals. Data are represented as mean ± s.e. mean of 5–9 experiments. * $P < 0.05$: significant difference between values observed in the control and treated groups.

Experiment 2

Desensitization of PAF-Induced Hypotensive Response. The chronic administration of PAF (3.0 μg kg^{-1}) reduced not only the maximum hypotensive effect (from 50.7 ± 5.3% to 29.1 ± 3.7%; n=12, P<0.05) but also diminished the average time required for the mean arterial blood pressure to fall (Figure 2). In this same group of animals, the increases in blood pressure resulting from bilateral carotid occlusion were significantly potentiated (from 26 ± 3.4 % to 57.8 ± 10.4 %; n=12, P<0.05. Figure 2).

Experiment 3

Effects of PAF Antagonists on the Norepinephrine-Induced Haemodynamic Responses. Compared to the control group which received saline solution before the second administration of norepinephrine (1.0–30.0 μg kg^{-1}), WEB 2086 (5 mg kg^{-1}, i.v.) potentiated the haemodynamic responses to this catecholamine, particularly the systemic vascular resistance (Figure 3). In the control group, the two consecutive dose-response curves to increasing doses of norepinephrine yielded similar haemodynamic responses.

Table 1. Basal and peak values for the cardiovascular parameters recorded during bilateral carotid artery occlusion in pentobarbitone anaesthetized and artificially ventilated rabbits after the injection of saline solution (control group) or of phospholipase A2 inhibitor (mepacrine)

	Control group		Mepacrine (5 mg kg^{-1}, i.v.)	
	B	O	B	O
MAP (mmHg)	92 ± 5	117 ± 8	89 ± 5	131 ± 2
SAP (mmHg)	124 ± 6	152 ± 6	117 ± 5	165 ± 7
DAP (mmHg)	78 ± 5	99 ± 8	75 ± 1.8	114 ± 3

Data represent the mean ± s.e. mean of 5–9 experiments.
*$P < 0.05$, statistically significant when compared to the hypertensive response obtained in the control (saline) group. MAP, mean arterial pressure; SAP, systolic arterial pressure; DAP, diastolic arterial pressure; B, basal values; O, carotid artery occlusion.

Indomethacin (3 mg kg^{-1}, i.v.) did not significantly alter the haemodynamic responses to norepinephrine, when compared to the control animals (n=6, $P<0.05$; Figure 3).

DISCUSSION

Platelet-activating factor (PAF), a mediator whose biological actions extend well beyond its initially discovered aggregating effect on platelets (17), has been reported to be involved in several pathophysiological events (9). In the present work, we employed two different experimental models to investigate the hypothesis that PAF could be involved in the control of vasomotor tone and blood pressure.

In the rabbits treated with PAF receptor antagonists, carotid occlusion increased the arterial pressure by two-fold compared to the control group (saline alone). The greatest potentiation was observed in the diastolic pressure, which is directly related to systemic vascular resistance. We have used two chemically dissimilar PAF receptor antagonists, BN 52021, a naturally occurring antagonist (ginkgolide B;18) and WEB 2086, a specific synthetic thienotriazolo-diazepine antagonist devoid of sedative action (19), in order to rule out the possibility of a compound-specific potentiating effect.

The inhibition of PAF biosynthesis by the phospholipase A_2 (PLA_2) inhibitor mepacrine (21; 22) led to a potentiation similar to that obtained using the PAF receptor antagonists. PLA_2 is known to be a key enzyme in the remodeling pathway for the synthesis of PAF since it regulates the deacylation of membrane-bound 1-*O*-alkyl-2-acyl-*sn*-glycero-3-phosphocoline to lyso-PAF, which is then acetylated by a specific acetyltransferase (23). It was also recently reported that inhibition of PLA_2 with a specific blocker of the enzyme, BMS-181162, significantly reduces PAF biosynthesis in isolated human polymorphonuclear leukocytes (24).

Furthermore, vasoactive mediators such as angiotensin II (25) and isoprenaline (26) are known to activate mepacrine-sensitive intracelular PLA_2 and hence influence phospholipid turnover and degradation. The release of PAF via PLA_2 activation in response to carotid occlusion observed in this study may therefore represent a complementary negative feed-back mechanism for the local control of blood pressure and flow.

The desensitization of PAF cardiovascular receptors significantly reduced the hypotensive response to PAF and also the time required for the blood pressure to return to normal levels. In contrast, the hypertensive effect induced by carotid artery occlusion in these animals was significantly potentiated and presented a profile similar to that observed in animals treated with PAF antagonists.

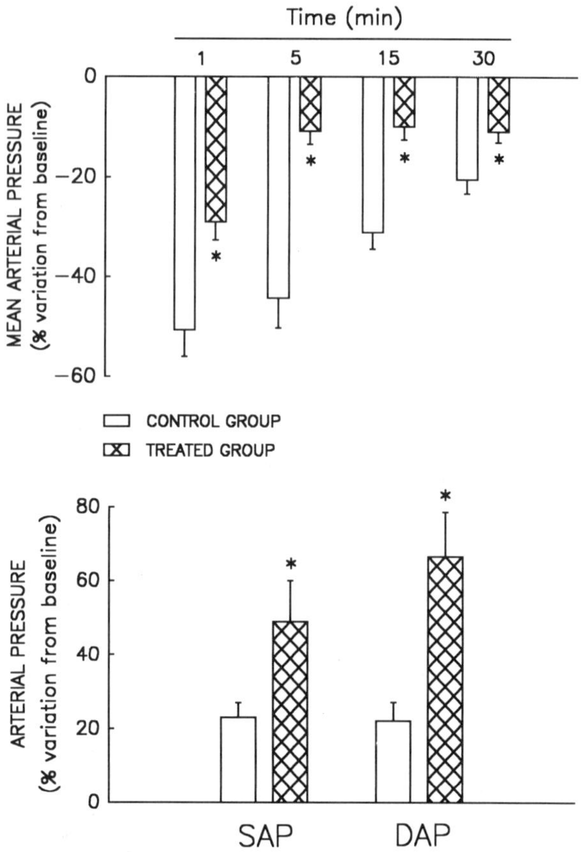

Figure 2. Average time course (upper panel) of blood pressure changes induced by the i.v. administration of PAF (3 μg/kg) to pentobarbitone anaesthetized and artificially ventilated rabbits after 7 daily injections of saline solution (control group) or of PAF (3 μg/kg) (treated group); maximum hypertensive effects (lower panel) elicited by bilateral carotid artery occlusion in these two groups of animals. Each set of points represents the mean ± s.e. mean of 8–12 experiments. * $P < 0.05$: significantly different from the effects observed in the control group. SAP, systolic arterial pressure; DAP, diastolic arterial pressure.

The increase in systemic vascular resistance produced by a high dose of norepinephrine (30.0 μg kg^{-1}, i.v.), was potentiated in animals pretreated with WEB 2086. This potentiation in the presence of the antagonist most likely occurred because the increase in vascular tone probably activated a sequence of reactions involving the release of PAF, which is this situation is not able to counteract the pressor activity of norepinephrine. These observations also suggest that a certain level of vasoconstriction must be achieved in order to activate this mechanism, since the elevation in systemic vascular resistance induced by a lower dose of norepinephrine (10 μg kg^{-1}), was not affected by this treatment. Another possible explanation for the foregoing results is that the model used (carotid occlusion) differs from the norepinephrine model in that other vasoconstrictors may be released in the blood stream corroborating to the activation of the negative feed-back mechanism.

Vasodilator prostaglandins such as prostacyclin do not appear to be important in modulating the response to norepinephrine since indomethacin had no effect on the resulting vasoconstriction.

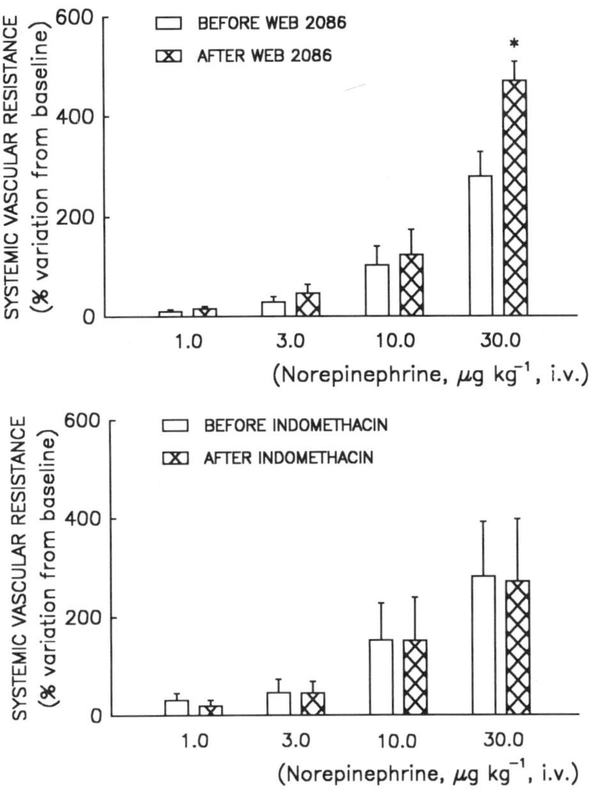

Figure 3. Effects of the intravenous administration of the PAF receptor antagonist WEB 2086 (5 mg/kg) (upper panel) or of the cyclooxygenase inhibitor indomethacin (3 mg/kg) (lower panel), on the hemodynamic response elicited by intravenous injection of increasing doses of norepinephrine in pentobarbitone anaesthetized and artificially ventilated rabbits (open columns, control curve; cross-hatched columns, after treatment). Data are represented as mean ± s.e. mean of 7 experiments. $^*P < 0.05$: significantly different when compared to control curve.

The present findings suggest that the release of PAF elicited by important increases in vascular tone may constitute a negative feedback mechanism designed to regulate vascular smooth muscle tone. *In vitro* studies are now required in order to elucidate the mechanism of action and cellular sources of PAF involved in this phenomenon.

ACKNOWLEDGMENTS

This work was supported by grants form the CNPq (Conselho Nacional de Desenvolvimento Científico e Tecnológico) and FAPERJ (Fundação de Amparo à Pesquisa do Rio de Janeiro).

REFERENCES

1. Furchgott, R.F. (1983). Role of endothelium in response of vascular smooth muscle. *Circ. Res.*, **53**, 557–573.
2. Vane, J. (1993). Control of the circulation by endothelial mediators. *Int. Arch. Allergy Immunol.*, **101**, 333–345.

3. Bevan, J.A. & Henrion, D. (1994). Pharmacological implications of the flow-dependence of vascular smooth muscle tone. *Annu. Rev. Pharmacol. Toxicol.*, **34**, 173–190.
4. Kuo, L.I.H., Chilian, W.H., Davis, M.J. (1991). Interaction of pressure- and flow-induced responses in porcine coronary resistance vessels. *Am. J. Physiol.*, **260**, H1706–H1715.
5. Gaw, A.J. & Bevan, J.A. (1993). Flow-induced relaxation of the rabbit middle cerebral artery is composed of both endothelium-dependent and -independent components. *Stroke*, **24**, 105–110.
6. Rees, D.D, Cellek, S., Palmer, R.M.J. & Moncada, S. (1990). Dexamethasone prevents the induction by endotoxin of a nitric oxide synthase and the effects on vascular tone. An insight into endotoxin shock. *Biochem. Biophys. Res. Commun.*, **173**, 541–547.
7. Cordeiro, R.S.B., Martins, M.a. & Silva, P.M.R. (1988). Proinflammatory activity of platelet-activating factor: pharmacological modulation and cellular involvement. *Prog. Biochem. Pharmacol.*, **22**, 156–167.
8. Tomlinson, P.R., Croft, K., Harris, T. & Stewart, A.G. (1994). Platelet-activating factor biosynthesis in rat vascular smooth muscle cells. *J. Vasc. Res.*, **31**, 144–152.
9. Braquet, P., Touqui, L., Shen, T.Y. & Vargaftig, B.B. (1987). Perspectives in platelet-activating factor research. *Pharmacol. Rev.*, **39**, 97–145.
10. Caillard, C.G., Mondot, S., Zundel, J.L. & Julou, L. (1982). Hypotensive activity of PAF-acether in rats. *Agents Actions* , **12**, 725–730.
11. Felix, S.B., Baumann, G., Raschke, P., Maus, C. & Berdel, W.E. (1990). Cardiovascular reactions and respiratory events during platelet activating factor-induced shock. *Basic Res. Cardiol.*, **85**, 217–226.
12. Sybertz, E.J., Watkins, R.W., Baum, T., Pula, K. & Rivelli, M. (1985). Cardiac, coronary and peripheral vascular effects of acetyl glyceryl ether phosphoryl choline in the anesthetized dog. *J. Pharmacol. Exp. Ther.*, **232**, 156–162.
13. Robertson, D.a., Genovese, a. & Levi, R. (1987). Negative inotropic effect of platelet-activating factor in human myocardium : A pharmacological study. *J. Pharmacol. Exp. Ther.* **243:** 834–839.
14. Macmanus, L.M., Hanahan, D.J., Demopoulos, C.a. & Pinckard, R.N. (1980). Pathobiology of the intravenous infusion of acetyl glyceryl ether phosphorylcholine (AGEPC), a synthetic platelet-activating factor (PAF), in the rabbit. *J. Immunol.*, **124**, 2919–2924.
15. Blank, M.L., Snyder, F., Byers, L.W., Brooks, B., Muirhead, E.E. (1979). Antihypertensive activity of an alkyl ether analog of phosphatidylcholine. *Biochem. Biophys. Res. Commun.*, **90:** 1194–1200.
16. Yamanaka, S., Katsuyuki, M., Yukimura, T., Okumura, M., Yamamoto, K. (1992). Putative mechanisms of hypotensive action of platelet-activating factor in dogs. *Circ. Res.* **70:** 893–901.
17. Benveniste, J., Henson, P.M. & Cochrane, C.J. (1972). Leukocyte-dependent histamine release from rabbit platelets: The role of IgE basophils and a platelet-activating factor. J. Exp. Med., **136**, 1356–1377.
18. Braquet, P. & Hosford, D. (1991). Ethnopharmacology and the development of natural PAF antagonists as therapeutic agents. *Ethnopharmacol.*, **32**, 135–139.
19. Casals-Stenzel, J., Muacevic, G. & Weber, K.-H. (1987). Pharmacological actions of WEB 2086, a new specific antagonist of platelet activating factor. *J. Pharmacol. Exp. Ther.*, **241**, 974–981.
20. Castro-Faria-Neto, H.C., Araújo, C.V., Moreira, S., Bozza, P.T., Thomas, G., Barbosa-Filho, J.M., Cordeiro, R.S.B. & Tibiriçá, E. (1995): Yangambin, a new naturally-occurring platelet-activating factor receptor antagonist : *in vivo* pharmacological studies. *Planta Medica,* **61**, 106–112.
21. Vargaftig, B.B. & Hai, N.D. (1972). Selective inhibition by mepacrine of the release of "rabbit aorta contracting substance" evoked by the administration of bradykinin. *J. Pharm. Pharmacol.*, **24**, 159–161.
22. Blackwell, G.J., Duncombe, W.G., Flower, R.J., Parsons, M.F. & Vane, J.R. (1977). The distribution and metabolism of arachidonic acid in rabbit platelets during aggregation and its modification by drugs. *Br. J. Pharmacol.*, **59**, 353–366.
23. Henson, P., Barnes, P. & Banks-Schlegel, S.P. (1992). Platelet-activating factor: role in pulmonary injury and dysfunction and blood abnormalities. *Am. Rev. Respir. Dis.*, **145**, 726–731.
24. Tramposch, K.M., Chilton, F.H., Stanley, P.L., Franson, R.C, Havens, M.B., Nettleton, D.O., Davern, L.B., Darling, I.M. & Bonney, R.J. (1994). Inhibitor of phospholipase A_2 blocks eicosanoid and platelet-activating factor biosynthesis and has topical anti-inflammatory activity. *J. Pharmacol. Exp. Ther.*, **271**, 852–859.
25. Rao, G.N., Lassegue, B., Alexander, R.W. And Griendlling, K.K. (1994). Angiotensin II stimulates phosphorylation of high-molecular-mass cytosolic phospholipase A_2 in vascular smooth-muscle cells. *Biochem. J.*, **299**, 197–201.
26. Novakova, O., Drnkova, J., Kubista, V. Novak, F. (1994). Regulation of phospholipid degradation and biosynthesis in the heart by isoprenaline: effect of mepacrine. *Physiol. Res.*, **43**,151–156.

PLATELET-ACTIVATING FACTOR AND ANGIOGENESIS

G. Camussi,[1] G. Montrucchio,[2] E. Lupia,[2] M. Arese,[2] and F. Bussolino[2]

[1]Cattedra di Nefrologia, Istituto di Medicina e Sanità Pubblica
II Facoltà di Medicina, Università di Pavia
21100 Varese, Italy
[2]Dipartimento di Fisiopatologia Clinica e Dipartimento di Genetica, Biologia e Chimica Medica
Università di Torino
10126 Torino, Italy

1. INTRODUCTION

Platelet-activating factor (PAF) is a phospholipid autacoid with a spectrum of diverse and potent biological properties relevant for the development of inflammatory reaction and septic shock[1]. Recently, PAF has been also implicated in the embryogenesis and cell differentiation[2,3]. Researchers believe that PAF is a mediator of cell to cell communication which may function both as intercellular and intracellular messenger. Since PAF is produced after appropriate stimulation by many cell types, including most inflammatory cells, it has been implicated in the pathogenesis of a variety of diseases[4]. Moreover, human endothelial cells (EC) were found to produce PAF after stimulation by several inflammatory mediators[5,6]. Many cell types that synthesized PAF, including EC, are also target for PAF[7,8] and express PAF-binding sites on their surface[9]. PAF, indeed, directly stimulates the endothelium, as it enhances the permeability of EC monolayer and induces changes of the cell cytoskeleton, leading to cell retraction and formation of intercellular gaps[8]. Moreover, it was observed that, in sub-confluent EC cultures, the redistribution of cytoskeleton was characteristic of cells undergoing translational movements[8]. These observations prompted us to evaluate whether PAF may stimulate the migration of EC *in vitro*.

2. *IN VITRO* EC MIGRATION

We studied the effect of PAF on EC migration by using Boyden's chambers[10]. PAF, added to the lower well of the Boyden's chamber, induced a dose-dependent migration of

Platelet-Activating Factor and Related Lipid Mediators 2
edited by Nigam *et al.*, Plenum Press, New York, 1996

EC across the 5 μm pore size gelatin-coated polycarbonate filters[10]. The extent of migratory response observed at 10 nM concentration of PAF was similar to that induced by 10 ng/ml bFGF. PAF-induced migration of EC was a receptor-dependent phenomenon, as pretreatment with WEB 2170, a specific PAF receptor antagonist, inhibited the migration and the biologically inactive D-stereoisomer of PAF was ineffective[10]. Moreover, we demonstrated by RT-PCR and Southern blot that human EC expressed the gene for the receptor of PAF[10]. Chemotaxis of EC was also reported for several other mediators such as bFGF, hepatocyte growth factor (HGF), interleukin 8 (IL-8), which were shown to be angiogenic *in vitro* and *in vivo* [11]. This prompted us to study the angiogenic effect of PAF in vivo. In fact, the directional migration of EC within the extracellular matrix is a prerequisite for the neoangiogenesis.

3. *IN VIVO* ANGIOGENIC EFFECT OF PAF

Several experimental evidences indicate that PAF may in vivo play a role in neoangiogenesis[10,12–14]. We recently found that PAF may act as mediator of angiogenesis induced by TNF: PAF, indeed, was actively synthesized *in vivo* during the angiogenic process and PAF-receptor antagonists significantly reduced the angiogenic response to this cytokine[12]. By using a model where Matrigel, injected subcutaneously into mice, is used as vehicle for the delivery of mediators, we observed that PAF induced *in vivo* the formation of canalized vessels[10]. PAF-induced angiogenesis was already evident after 16–24 hours, significantly earlier than that reported for other angiogenic factors such as bFGF, IL-8 or TNF[11]. The minimal concentration of PAF able to induce angiogenesis was 10 nM, an amount compatible with levels of bioactive PAF produced *in vivo*. Thus, PAF synthesized within inflamed tissues may contribute to the angiogenesis observed in chronic inflammation. We have suggested that EC are the primary target for the angiogenic effect of PAF as the majority of cells infiltrating Matrigel were EC and inflammatory cells were absent[10]. PAF was found to induce an in vivo angiogenic response either in an heparin-independent or an heparin-dependent manner, depending on the dose used[10]. At pharmacological concentrations (1–5 μM) PAF induced a heparin-independent angiogenesis. Similar results were obtained by Andrade et al.[13,14], which showed a direct angiogenic effect of repeated administrations of 2 μg PAF in an experimental model of granulomatous reaction induced in mice by the subcutaneous implantation of sponges. In these experimental conditions, however, a role of endogenous heparin can not be excluded, since we found that concentrations as low as 6.4 U/ml were effective[10]. These concentrations are compatible with that reached in inflamed tissues as result of heparin released by mastocytes or EC[15]. Using Matrigel as a substrate, we also observed an heparin-dependent angiogenic response at physiological concentrations of PAF (10–50 nM). Based on these observations, it was suggested that the angiogenic effect of PAF may depend on the secondary involvement of heparin-binding endothelial cell growth factors with angiogenic properties. It was indeed shown that heparin, by forming complex with bFGF and with IL-8, may protect these angiogenic cytokines from inactivation and may favour their delivery to EC[11]. In addition, we recently observed that PAF produced in vitro by Kaposi's sarcoma cells induces and sustains in vivo angiogenesis[16]. It has been suggested that an imbalance in the network of soluble mediators may play a pivotal role in the pathogenesis of Kaposi's sarcoma (KS). We observed that KS cells in culture produced and in part released PAF[16]. The synthesis of PAF was enhanced either by thrombin or cytokines (IL-1, TNF). Moreover, KS cells possessed specific high affinity binding sites for PAF and expresed mRNA for PAF receptor. PAF at nanomolar concentrations stimulates chemotaxis and chemokinesis not only of EC, but also of KS and vascular muscle cells. Using Matrigel as substrate, it was found that the conditionated medium of KS cells stimulated in vivo neoangiogenesis, which was inhibited by the PAF receptor antagonist WEB 2170. The angiogenic effect of PAF was amplified by other factors, such as basic and acid FGF, placental growth factor, vascular endothelial growth factor, hepatocyte growth factor, KC and macrophage inflammatory protein-2. PAF receptor antagonist WEB 2170 abolished the expression of the transcripts of these factors within Matrigel containing the conditionated medium of KS. These results suggest that PAF may cooperate with other cytokines in the development of KS, a highly vascularized tumor composed by spindle-shaped cells (KS cells), vascular smooth muscle cells, endothelial cells and fibroblasts.

In conclusion, PAF may directly provide the signal for migration of EC, which is a prerequisite for the formation of new vessels. However, PAF itself is unable to directly stimulate the proliferation of EC. Therefore, the angiogenesis induced in vivo by PAF need the cooperation of other angiogenic factors able to stimulate the proliferation of EC. It is conceivable that PAF not only may induces the recruitment of EC within Matrigel, but it also stimulates the production by infiltrating EC of heparin-binding growth factors which can act in autocrine manner providing the signal for cell proliferation and final formation of canalized vessels.

4. ACKNOWLEDGMENTS

This work was supported by the Associazione Italiana per la Ricerca sul Cancro, Targeted Project "Angiogenesis" to G.C. and F.B.

5. REFERENCES

1. Snyder, F. 1990. Platelet-activating factor and related acetylated lipids as potent biological acting cellular mediators. Am. J. Physiol. 259: 697.
2. Leidenheimer, N.J., M.D. Browning, and R.A. Harris. 1991. A physiological role for PAF in the stimulation of mammalian embryonic development. TIPS 12: 82.
3. Bussolino, F., G. Pescarmona, G. Camussi, and F. Gremo. 1988. Acetylcholine and dopamine promote the production of platelet activating factor in immature cells of chick embryonic retina. J. Neurochemistry 51: 1755.
4. Bratton, D., and P.M. Henson. Cellular origin of PAF. 1989.In: Platelet-activating factor and human diseases. P.J. Barnes, C.P. Page, and P.M. Henson, eds. Blackwell Scientific Publications, London, UK, p. 23.
5. Camussi, G., M. Aglietta, F. Malavasi, C. Tetta, W. Piacibello, F. Sanavio, and F. Bussolino. 1983. The release of platelet-activating factor from human endothelial cells in culture. J. Immunol. 131: 2397.
6. Prescott, S.M., G.A. Zimmerman, and T.M. McIntyre. 1984. Human endothelial cells in culture produce platelet-activating factor (1-alkyl-2-acetyl-sn-glycero-3-phosphocholine) when stimulated with thrombin. Proc. Natl. Acad. Sci. USA 81: 3534.
7. Lerner, R., P. Lindstrom, and J. Palmblad. 1988. Platelet activating factor and leukotriene B_4 induce hyperpolarization of human endothelial cells but depolarization of neutrophils. Biochem. Biophys. Res. Comm. 153: 805.
8. Bussolino, F., G. Camussi, M. Aglietta, M. Braquet, P. Bosia, G. Pescarmona, F. Sanavio, M. D'Urso, and M. Marchisio. 1987. Human endothelial cells are target for platelet-activating factor. I. Platelet-activating factor induces changes in cytoskeleton structure and promotes albumin diffusion across endothelial monolayer. J. Immunol. 139: 2439.
9. Korth. R., M. Hirafuji, C. Lalau Keraly, D. Delautier, J. Bidault, and J. Benveniste. 1989. Interaction of the antagonist WEB 2086 and its hetrazepine analogue with human platelets and endothelial cells. Biochem. Pharmacol. 44: 223.

10. Camussi G, Montrucchio G, Lupia E, De Martino A, Perona L, Arese M, Vercellone A, Toniolo A, Bussolino F. Platelet-activating fcator directly stimulates in vitro migration of endothelial cells and promotes in vivo angiogenesis by an heparin-dependent mechanism. J Immunol 154: 6492–6501, 1995.
11. Folkman, J., and Y. Shing. 1992. Angiogenesis. J. Biol. Chem. 267: 10931.
12. Montrucchio, G., E. Lupia, E. Battaglia, G. Passerini, F. Bussolino, G. Emanuelli, and G. Camussi. 1994. Tumor necrosis factor α-induced angiogenesis depends on in situ platelet-activating factor biosynthesis. J. Exp. Med. 180: 377.
13. Andrade, S.P., L.B.G.B. Vieira, Y.S. Bakhle, and P.J. Piper. 1994. Effect of platelet activating (PAF) on the formation of blood vessels in subcutaneous implants in mice. J. Lipid. Mediat. 9:117
14. Andrade, S.P., L.B.G.B. Vieira, Y.S. Bakhle, and P.J. Piper. 1992. Effects of platelet activating factor (PAF) and other vasoconstrictors on a model of angiogenesis in the mouse. Int. J. Exp. Path. 73: 503.
15. Azizkhan, R.G., J. Clifford Azizkhan, B.R. Zetter, and J. Folkman. 1980. Mast cell heparin stimulates migration of capillary endothelial cells in vitro. J. Exp. Med. 152: 931.
16. Bussolino F, Arese M, Montrucchio G, Barra L, Primo L, Benelli R, Sanavio F, Aglietta M, Ghigo D, Rola-Pleszczynski M, Bosia A, Albini A, Camussi G. Platelet activating factor produced in vitro by Kaposi-s sarcoma cells induces and sustains in vivo angiogenesis. J Clin Invest 96: 940–952, 1995.

38

ROLE OF SYNERGISTIC ACTION OF PAF AND KININ IN BACTERIAL ENDOTOXIN-INDUCED HYPOTENSION IN RATS

Sachiko Oh-Ishi, Hideki Ishida, and Akinori Ueno

Department of Pharmacology
School of Pharmaceutical Sciences
Kitasato University
Shirokane, Minato-ku, Tokyo 108, Japan

1. SUMMARY

Involvement of PAF and kinin in the endotoxin (LPS)-induced hypotension was examined in the rat strains, Brown Norway (B/N);Kitasato (Normal) and B/N-Kitasato rats, intravenous injection of 10 mg/kg LPS caused a hypotension composed on two phases (15 min and 70 min after LPS injection). However in the kininogen-deficient B/N-Katholiek rats, LPS induced only the second phase. Pretreatment with bradykinin (BK) antagonist, Hoe 140 suppressed LPS-induced hypotension of normal rats to the level of Katholiek rats. TCV 309, a PAF-antagonist, suppressed the LPS-hypotension almost completely in the both strains. The reason why the PAF-antagonist showed almost complete inhibition, could be explained by a synergism. Because concomitant injection of a small amount of BK with PAF caused potentiation of the effect of the PAF action. These results suggest that formation of endogenous BK contributes mainly to the 1st phase of LPS-hypotension, and PAF contributes to the both phases, by either direct and synergistic actions.

2. MATERIALS AND METHODS

Blood pressure was monitored from the catheter inserted into right femoral artery under pentobarbital sodium anesthesia. Systolic or diastolic pressure on the chart was measured and calculated. Bradykinin (Peptide Inst., Minoh), C16-PAF (Funkoshi Pharmaceutical Co.), endotoxin (*E. Coli.* 0.55:B5, LPS, Difco Lab., Detroit), indomethacin (Sigma Chemical. Co., St. Louis, Mo) and dexamethasone (Sigma Chemical Co.) were purchased. Bradykinin antagonist Hoe 140[1] and PAF antagonist TCV 309[2] were kind gifts from Hoechst AG, Frankfurt and Takeda Pharmaceutical Ind., Osaka, respectively. Brown Norway (B/N)-Kitasato, normal rat and B/N-Katholiek, kininogen-deficient rat strains were bred and kept in the animal laboratory of Kitasato University, School of Pharmaceutical Sciences.[3]

3. RESULT AND DISCUSSION

First of all hypotensive responses to bradykinin and PAF were examined in both B/N strain upto 180 min after the rat was anesthetized. There is no significant difference between the dose-response curves of bradykinin and PAF between B/N-Kitasato and B/N-Katholiek strains and rats, and activities of both agents in both strains were observed constant upto 180 min. To elucidate involvement of kinin in endotoxin induced hypotension, the hypotensive response of two strains were compared. As shown in Fig. 1A, in the normal B/N-Kitasato rats injection of LPS at the dose of 10 mg\kg caused hypotension composed of two phases; 1st one at 15 min and 2nd around 70 min. However in the kininogen-deficient B/N-Katholiek rats LPS injection expressed only weaker 2nd phase as shown in Fig. 1A. This result indicates that kinin may be involved in the hypotension of mainly 1st phase. Pretreatment of rats with bradykinin B2-receptor antagonist, Hoe 140 suppressed LPS-induced hypotension of normal rats to the level of kininogen-deficient rats (Fig. 2 a and B). PAF antagonist TCV 309 suppressed almost completely the both phases of the normal rats and the 2nd phase of kininogen-deficient rats. (Fig. 2C and D) These result indicates again kinin may be involved in mainly 1st phase and PAF could be involved in the both phases.

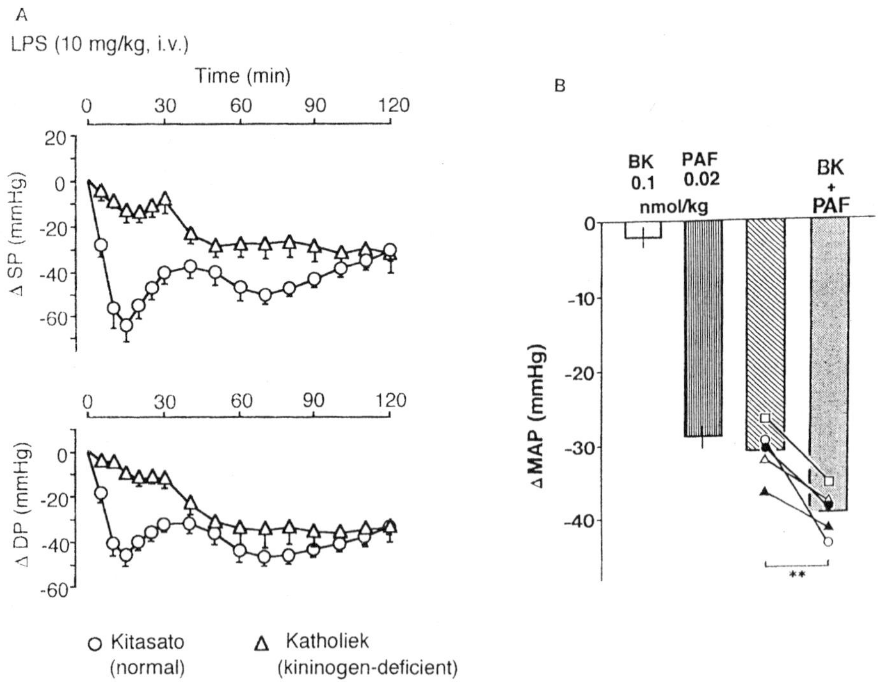

Figure 1. Hypotensive responses induced by LPS, 10 mg/kg, i.v., in B/N-Kitasato (circles) and B/N-Katholiek (kininogen-deficient) rats (triangles) (A). Interaction of bradykinin (BK) and platelet-activating factor (PAF) in the hypotensive response in rates (B).

Then to clarify the reason why PAF-antagonist shows such a complete inhibition, even though there is involvement of bradykinin, we then examined synergism of PAF and bradykinin. As shown in Fig. 1B, a small dose of bradykinin which did not show significant effect was combined with PAF which caused hypotension of approximately 30 mmHg. The concomitant injection of these doses expressed hypotension about 40 mmHg. The third column and symbols in the figure express arithmetic addition of individual effect and their mean. As these data indicate that the combined effect is significantly greater than the addition of these effects, it can be synergistic effect. Therefore these results sug-

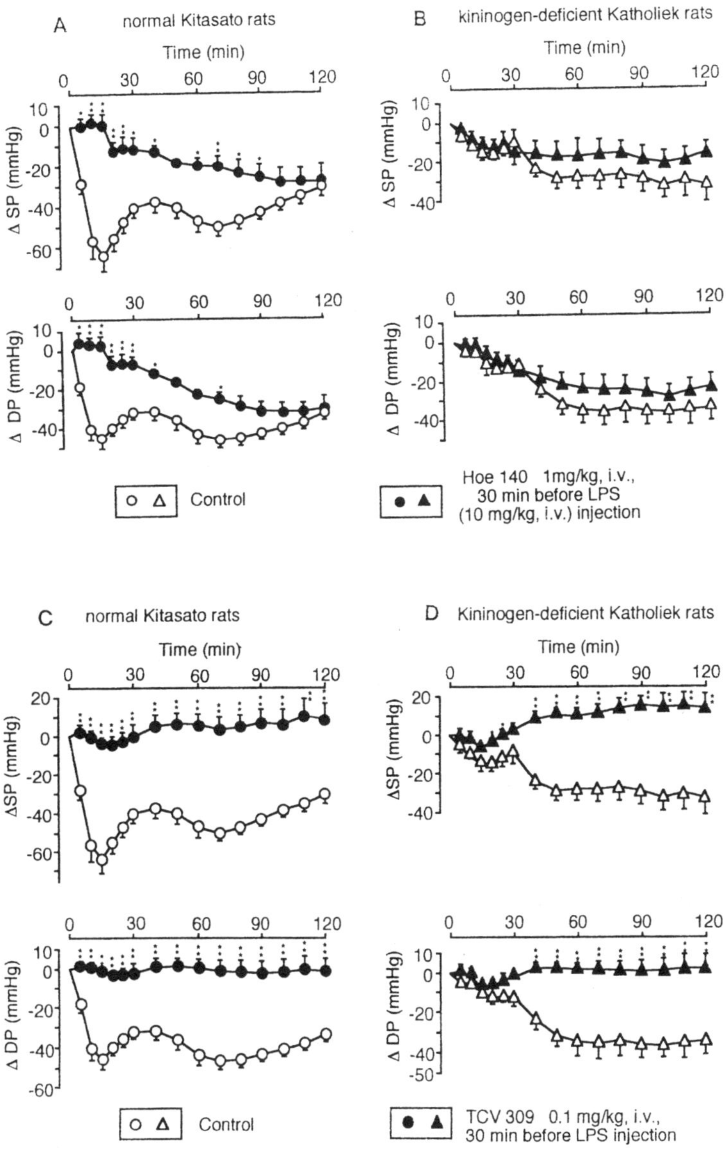

Figure 2. Hypotensive responses induced by LPS, 10 mg/kg, i.v., in B/N-Kitasato (A and C, circles) and B/N-Katholiek rats (B and D, triangles).

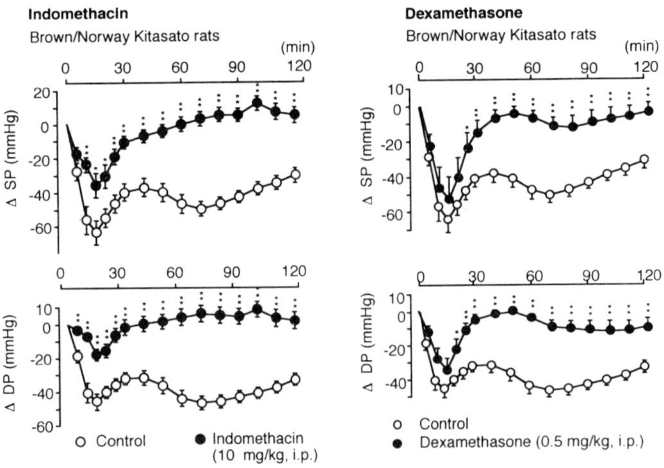

Figure 3. Effect of indomethacine (10 mg/kg, i.p.) and dexamethasone (0.5 mg/kg, i.p.) on the LPS-induced hypotension in B/N-Kitasato rats.

gest that the PAF-antagonist may exsert potent inhibition on the LPS-induced hypotension by inhibiting synergistic action of PAF in addition to the inhibition of the original action of PAF itself.

Effects of indomethacin and dexamethasone on the LPS-induced hypotension were also examined. As shown in Fig. 3, indomethacin, 10 mg/kg attenuated the 2nd phase almost completely and 1st phase partially. Whereas dexamethasone suppressed only the 2nd phase. These results suggest that cyclooxygenase products could be involved in the both phases and especially inducible products may be involved in the 2nd phase.

Involvement of kinin[4,5] and PAF[6] in endotoxin-induced hypotension in rats have been described by many authors. Our result using kininogen-deficient rats clearly demonstrated confirmation of the previous data. Involvement of cyclooxygenase products[7] and nitric oxide[8] were also reported. On the basis of many individual data on each mediator, our present result suggests that PAF may cause hypotension synergistic effect with kinin, arachidonic acid metabolites or nitric oxide in addition to its own original activity.

REFERENCES

1. Hock, F. J., Wirth, K., Albus, U., Linz, W., Gerhards, H. J., Wiemer, G., Henke, S. T., Breiphol, G., Konig, W., Knolle, L. and Scholkens, B. A.: Br. J. Pharmacol., 102, 769–773 (1991).
2. Terashita, Z., Kawamura, M., Takatani, M., Tsushima, S.: Imura, Y., and Nishikawa, K.: J. Pharmacol. Exp. Therap., 260, 748–755 (1991).
3. Oh-Ishi, S., Satoh, K., Hayashi, I., Yamazaki, K. and Nakano, T.; Thrombosis Res., 28, 143–147 (1982).
4. Weipert, J., Hoffmann, H., Siebeck, M. and Whalley. E. T.: Br. J. Pharmacol., 94, 282–284 (1988).
5. Wilson, D. D., Garaviella, L., Kuhn, W., Togo, J., Burch, R. M. and Sterenka, L. R.: Circ. Shock 27, 93–101 (1989).
6. Terashita, Z., Tsushima, S., Yoshioka, Y, Nomura, H., Inada, Y. and Nishikawa, K.: Life Sci., 32, 1975–1982 (1982).
7. Parratt, J. R. and Sturgess, R. M.: Br. J. Pharmacol., 53, 485–488 (1975).
8. Thiemermann, C. and Vane, J. R.: Br. J. Pharmacol., 182, 591–595 (1990).

EPIMER-SPECIFIC ACTIONS OF HEPOXILINS A_3 AND B_3 ON PAF- AND BRADYKININ-EVOKED VASCULAR PERMEABILITY IN THE RAT SKIN *IN VIVO*

Mei Mei Wang, Peter M. Demin, and Cecil R. Pace-Asciak

Research Institute
Hospital for Sick Children
555 University Avenue, Toronto, Canada M5G 1X8 and
Department of Pharmacology, Faculty of Medicine
University of Toronto
Toronto, Canada M5S 1A8

The two epimers of hepoxilins A_3 (8R and 8S) and B_3 (10R and 10S) were tested for effects on vascular permeability in the rat skin assay, a model for acute inflammation. Results showed epimer-specific effects, the R isomer being active, the S isomer being essentially inactive. While the hepoxilins on their own, at a dose of 20 - 40 ng/injection intradermally were essentially inactive, they potentiated the actions of bradykinin (both HxA_3 and B_3 were active) and PAF (only 10R was active) coinjected with the hepoxilins. These results suggest that hepoxilins may mediate (amplify) the actions of inflammatory mediators.

We previously showed that HxA_3 potentiated the inflammatory actions of bradykinin (BK) in the rat skin and that this action of Hx was epimer-specific, the 8R being the active species (1, 2). In this paper, we compared the actions of the two epimers of HxA_3 and its structural analog, HxB_3, on the vascular permeability evoked by both BK as well as PAF (platelet activating factor). Interestingly, both epimer-specificity as well as analog-specificity were observed. While HxA_3 and B_3 were both active in enhancing the BK-evoked permeability, only B_3 enhanced the vascular permeability evoked by PAF. Reviews on the biosynthesis (3) and pharmacology (4) of the hepoxilins have recently appeared.

MATERIALS AND METHODS

Materials: Hepoxilins were chemically prepared in the methyl ester form in our laboratory as previously reported (5, 6). The compounds were stored in benzene solution at -80°C. Before the experiment appropriate amounts were removed, the solvent was

Platelet-Activating Factor and Related Lipid Mediators 2
edited by Nigam *et al.*, Plenum Press, New York, 1996

evaporated, and the residue was dissolved in ethanol at a concentration of 20 ng/μl. Evans Blue, BK and PAF were purchased from Sigma.

Animal preparation: Male Wistar rats (170 - 220 g body weight, Charles River Ltd. St. Constant, Quebec) were anesthetised with Inactin (100 mg/kg, i.p. BYK-Gulden, Constanz, Germany) and their dorsal skin shaved. The trachea and left jugular vein were cannulated for respiration and administration of Evans blue respectively. Body temperature was maintained at 37°C by placing the animals on a thermostatically controlled heating box.

Plasma protein extravasation: Changes in the capillary permeability were based on the leakage of plasma protein-bound dye into the extravascular compartment of the skin. Measurements were made according to Udaka et al (7).with minor modifications. Thirty five mg/kg Evans blue was administered i.v. 5 minutes prior to injection of the test substance. Test substances (40 ng of Hx in 2 μl ethanol, 5 ng PAF and 50 ng BK in Krebs' buffer) alone or in combination were injected at six different areas of the dorsal skin/animal in 100 μl volume of Krebs' solution. A control injection of the Krebs' vehicle containing the amount of ethanol used for the samples was also injected. As ethanol concentrations in higher amounts than used in this study (2 μl), increased plasma leakage, it was important to maintain the amount of ethanol to a minimum. Each concentration of test drug was repeated at least six different times in different rats, each time the location was altered in random fashion. The composition of the modified Krebs' buffer contained in mM: 118 NaCl, 11 glucose, 25 $NaHCO_3$, 4.62 KCl, 1.2 KH_2PO_4, 1.2 $MgSO_4$, 2.5 $CaCl_2$. After 30 minutes, the animals were decapitated and their dorsal skin removed. Circular patches of skin containing the injection site were punched out, weighed and incubated in glass vials containing 3 ml of formamide (BDH, Toronto) at 70°C. An additional untreated patch of skin was collected and used as a measure of background. After 18 h, the vials were centrifuged (5 minutes at 2200 rpm), the supernatant was pipetted into cuvettes and the color read on a Du-65 spectrometer (Beckman) at 620 nm. Protein extravasation was estimated from a comparison of the absorbance of the samples with a standard line of different concentrations of Evans blue. Amount of plasma protein leakage was initially expressed as μg Evans blue/g tissue, then as % response to BK or PAF alone.

Statistics: Student's t-test for unpaired samples were used to compare the color intensity of the test samples with the values for BK or PAF alone. Results were considered significant at the 0.05 level.

RESULTS AND DISCUSSION

Hepoxilins A_3 and B_3 at a dose of 20 - 40 ng injected intradermally in 100 μl Krebs' solution gave essentially no response over that observed with Krebs' alone (data not shown). Preliminary tests with different amounts of PAF and BK showed that amounts causing threshold responses were 5 ng and 50 ng respectively. We therefore employed these doses of PAF and BK coinjected with 40 ng Hx to evaluate whether Hx potentiated the response to these inflammatory mediators. Figure 1 shows the effects of the two epimers of HxA_3 and HxB_3 on the responses elicited by BK (Fig. 1A) and PAF (Fig. 1B). As shown, the R isomer of both hepoxilins potentiated the vascular leakage elicited by BK, although only the R isomer of HxB_3 potentiated the response to PAF. The R-stereospecificity of the hepoxilins suggests that the 8- and 10-hydroxyl group of the structures are important in eliciting the biological action.

We have previously reported that hepoxilins are formed enzymatically from 12(S)-HPETE by an enzyme identified in the pineal gland (8). This enzyme selectively converts

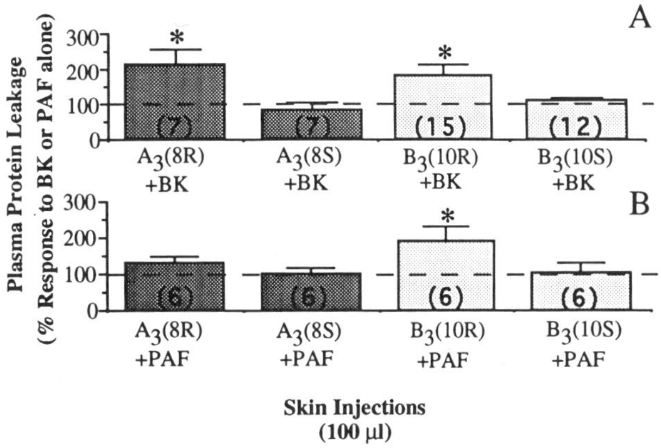

Figure 1. Potentiation by two epimers (R/S) of hepoxilin A_3 and B_3 (40 ng) of the vascular permeability evoked by **A)** bradykinin (BK - 50 ng) or **B)** platelet activating factor (PAF - 5 ng) coinjected with the hepoxilins intradermally (100 μl total volume). The bars show the mean ± SEM % increase in plasma protein exudation as measured by leakage of Evans Blue dye at the site of injection 30 minutes after injection (n=6–15 separate experiments shown in brackets). Note that HxB_3 has effects on both BK and PAF, while HxA_3 only potentiates BK effects. Also note the stereospecificity of action, only the R isomer being active. *p<0.05 vs BK/PAF alone (100%).

the S-epimer of the 12-HPETE substrate, but not the R-epimer to form selectively the S,S-trans epoxide of HxA_3. However, the hydroxyl group at positions 8 in HxA_3 and 10 in HxB_3 formed was racemic. While both Hx epimers possess biological activity in some systems (4) including mobilisation of intracellular calcium in neutrophils, skin as shown in this study appears to be epimer selective. We are currently investigating whether skin selectively forms the biologically active R-epimer.

ACKNOWLEDGMENTS

This study was supported by funds to CRP-A from the MRC of Canada (MT-4181).

REFERENCES

1. Laneuville, O., Corey, E. J. and Pace-Asciak, C. R. (1991) *Prostaglandins, Leukotrienes, Lipoxins & PAF, XIth, Washington International Spring Symposium, Ed. Martyn Bailey, Plenum Press, New York, 1991, pp. 335–338.* Abstract # 155.
2. Laneuville, O., Corey, E. J., Couture, R. and Pace-Asciak, C. R. (1991) *Eicosanoids* 4, 95–97.
3. Pace-Asciak, C. R., Reynaud, D. and Demin, P. M. (1995) *Lipids* 30, 107–114.
4. Pace-Asciak, C. R. (1994) *Biochim. Biophys. Acta* 1215, 1–8.
5. Demin, P. M., Vasiljeva, L. L., Belosludtsev, Y. Y., Myagkova, G. I. and Pivnitsky, K. K. (1990) *Bioorg. Khim.* 16, 571–572.
6. Demin, P. M., Vasilyeva, L. L., Lapitskaya, M. A., Belosludtsev, Y. Y., Myagkova, G. I. and Pivnitskii, K. K. (1990) *Biorg. Khim.* 16, 127–128.
7. Udaka, K., Takeuchi, Y. and Movat, H. Z. (1970) *Proc. Soc. Exper. Biol. Med.* 133, 1384–1386.
8. Reynaud, D., Demin, P. and Pace-Asciak, C. R. (1994) *J. Biol. Chem.* 269, 23976–23980.

40

THE ROLE OF PLATELET-ACTIVATING FACTOR IN THE BIOCOMPATIBILITY OF HEMODIALYSIS MEMBRANES

C. Tetta,[1] N. Haeffner-Cavaillon,[2] C. Navino,[3] S. David,[4] C. Franceschi,[5] F. Mariano,[6] and G. Camussi[6]

[1]Clinical and Laboratory Research Department
Bellco S.p.A., Mirandola (MO), Italy
[2]INSERM, U430, Hopital Broussais, Paris
[3]Civil Hospital, Novara, Italy
[4]Chair of Nephrology
University of Parma, Parma, Italy
[5]Department of Biomedical Sciences
Section of General Pathology, University of Modena, Modena, Italy
[6]Institute of Medicine and Public Health
2nd Faculty of Medicine, University of Pavia at Varese, Varese, Italy

1. INTRODUCTION

Extracorporeal treatments for acute or chronic replacement of organ function still represent a challenge for today's technology. Activation of fluid phase (complement, coagulation, fibrinolysis) and cellular systems (leukocytes, platelets) is known to occur in hemodialysis (1). Among other biochemical indications of leukocyte activation as a conseguence of blood-surface interactions such as oxygen radicals, release of granulocyte proteinases, monocyte prostaglandin release and cytokine generation, platelet-activating factor (PAF) has drawn interest in various biocompatibility studies (reviewed in 1–5). PAF is a phospholipid mediator of inflammation with different biologic properties relevant for the development of inflammation and septic shock (7–11). PAF may act at concentrations of 10^{-12}M and requires an ether linkage at the sn-1 position of the glycerol backbone, a short acyl chain, usually an acetyl residue, at the sn-2 position, and the polar head group of choline or ethanolamine at the sn-3 position (7–8). However, it has been shown recently that PAF belongs to a family of structurally related phospholipid molecules of biologic origin that shares many physiologic activities (12). PAF is considered a mediator of cell-to-cell communication that may function both as an intracellular and intracellular messenger. PAF is produced after immunologic or nor immunologic challenge by a variety of cells such as monocytes/macrophages, polymorphonuclear neutrophils, ba-

Platelet-Activating Factor and Related Lipid Mediators 2
edited by Nigam *et al.*, Plenum Press, New York, 1996

sophils and platelets, that may participate in the development of inflammatory reaction (13). In addition, human endothelial cells were found to produce PAF after stimulation by several inflammatory mediators including thrombin, angiotensin II, vasopressin, leukotriene C4 and D4, histamine, bradykinin, elastase, catheprin G, and plastic (14–18). PAF acts in an autocrine and paracrine way through a specific receptor for which a cDNA has been cloned (19,20). The receptor belongs to the family of "serpentine" receptors containing sever α-helical domains that weave in and out plasma membrane and it interacts with a G protein, which activates a phosphatidylinositol-specific phospholipase C (19,20). PAF receptors exist in all cells that are known targets for PAF. Recently, also human cultured endothelial cells have been shown to express PAF receptors (21). PAF promotes the permeability of the EC monolayer leading to cell retraction and formation of intracellular gaps (22). PAF induces in vitro migration of endothelial cells and promotes in vivo angiogenesis by a heparin-dependent mechanism (21). The aim of the present review is to briefly summarize early evidence for a role of PAF in bioincompatibility events and to highlight recent advances and their possible potential practical application in testing polymer biocompatibility.

2. THE ROLE OF PAF IN BIOCOMPATIBILITY

Historically, leukocyte activation during hemodialysis was inferred from the acute, transient but significant leucopenia occurring with regenerated cellulosic (Cuprophan^{R}) (reviewed in 1–5). Apart from complement-derived activated products (C3a desArg, C5a desArg, terminal complement complex), PAF was shown to be produced by purified human neutrophils incubated with sliced Cuprophan$^{(R)}$ membranes in complete absence of serum, as source of complement. Other Authors later confirmed the occurence of the direct cell-membrane interactions showing that Cuprophan$^{(R)}$ activates arachidonate metabolism and interleukin-1 production.

The interaction of blood with the polymer surface occurs at the earliest stages of the hemodialytic treatment since the reactive sites are masked by protein adsorption. PAF generation occurs in fact very early also in ex vivo experiments ranging from less than 2–3 min to 5–7 min with polyanilonitrile (AN 69) and Cuprophan$^{(R)}$ membranes, respectively (22). The role of PAF has been studied as to its ability to promote adherence of human monocytes to haemodialysis membranes. In haemodialysis, a marked monocytopenia is observed during the overall fall in circulating leukocytes with Cuprophan$^{(R)}$ membranes as an indication that monocytes are involved in the blood-membrane interactions. Several clinical and experimental lines of evidence such as the enhanced plasma values of monocyte-derived cytokines (23–29), the increased spontaneous or stimulated production of cytokines in the supernatants from monocytes obtained after extracorporeal treatment (30–33), and the increased mRNA transcription or protein translation of cytokines in the supernantants of whole blood recirculated in an in vitro closed-loop (33) circuit or perfused in single pass (22). We have recently studied the mechanisms involved in the triggering of monocyte adherence to Cuprophan$^{(R)}$ and other cellulosic membranes as well as synthetic membranes. We have shown that spontaneous adherence at 60 min is significantly higher for regenerated cellulose than AN-69 membranes (34). Tumor necrosis factor-α (TNF-α), interleukin-1 (IL-1) and -6 (IL-6) as well as PAF, but not IL-4, significantly enhanced adherence at 60 min to AN-69. In contrast, adherence was not further inducible in the presence of regenerated cellulose. Both spontaneous and cytokine/bacterial lipopolysaccharide-stimulated adherence were significantly reduced by

SDZ-63072, a specific PAF receptor antagonist. These data have suggested that PAF may act as an adherence mediator. Furthermore, adherence may be significantly enhanced also in the case of synthetic membranes. The biocompatibility of these membranes may be blunted in the presence of bacterial lipopolysaccharide or proinflammatory cytokines. Recently, new cellulosic membranes with different substitution groups have been manufactured and introduced such as diacetate cellulose, benzyl cellulose apart from DEAE-cellulose. Fig. 1 summarizes recent evidence that the type of the substitution group as well as the degree of substitution may indeed modify the spontaneous adherence of mononuclear cells. Another mechanism responsible for monocyte adherence operates via complement-activated, membrane-bound C3 derived products (iC3b, C3c, C3d). Complement activation is known to commonly occur during hemodialysis mainly via the alternate pathway. Although complement-dependent bioincompatibility has been largely related to the generation of C3a and C5a anaphylatoxins, we have recently shown that membrane-bound iC3b and related molecules may induce the production of PAF (35).

PAF production was induced by F(ab)2 fragments of monoclonal antibodies for different epitopes of CR1 and CR3, suggesting that recognition and binding of these epitopes could trigger signal trasduction for the initiation of PAF biosynthesis. The latter was an early (2 min) event that occurred in the presence of both anti-CR1 and - CR3 monoclonal antibodies. Stimulation with anti-CR1 but not anti-CR3 F (ab)2 showed that a late (15 min) peak of PAF production occurred. No PAF was observed at 30 min. These results provided evidence for a possible role of CD11-CD18 integrin complex in the upregulation of monocyte adhesion molecules. The increased adhesiveness of monocytes to Cuprophan(R) could be due to either PAF-induced upregulation of β_2 integrins in monocytes or to as direct effect of PAF on cell adhesion. Flow cytometry analysis of CD11a, CD11b, CD18, and CD35 expression on human monocytes and polymorphonuclear neutrophils demonstrated that these molecules are indeed upregulated by PAF but only in polymorphonuclear neutrophils. In fact there was no increase in the percentage of positive monocytes or in the fluorescence intensity of all the studied molecules on these cells. These data suggest the possibility of a complex co-operation between polymorphonuclear neutrophils and monocytes in the increased adhesiveness of leukocytes to haemodialysis membranes via the release of PAF from both cell types stimulated by their β_2-integrin receptors. The effect of WEB 2170, a PAF receptor antagonist could not be described to downregulation of CD11a/CD18, CD11b/CD18, or CD35 but rather to its ability to an-

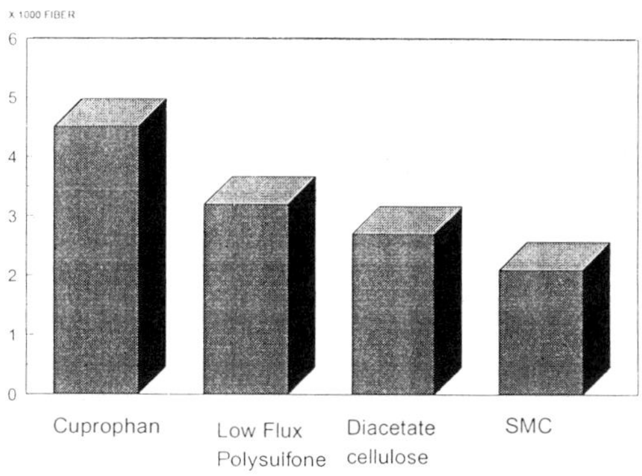

Figure 1. Human mononuclear cell adherence to hemodialysis membranes.

tagonize PAF or PAF related biological activity. We suggest that the production and release of PAF may be an additional mechanism in the increased adhesiveness. Indeed in our study monocyte adherence to fresh plasma-treated membranes was inhibited by preincubation with PAF receptor antagonist. Our data are in agreement with Lo and co-workers (36), who suggested that an additional mechanism may be at work to control CD11b/CD18-dependent adhesion. CD11b/CD18, the most abundant member of CD11/CD18 complex on leukocytes, functions as a receptor for surface-bound iC3b (37). The ability of this receptor to bind the erythrocytes coated with C3bi is transiently stimulated by phorbol esters (38). Phorbol esters are potent activators of PAF biosynthesis (39). Phorbol esters cause no increase in CD11a/CD18 expression but a threefold increase in CD11b/CD18 expression, and a 10-fold increase in adhesion of leukocytes to endothelial cells. Furthermore, adherence of monocytes to complement-activating haemodialysis membranes was inhibited by blocking the interaction between monocyte β_2-integrin and fixed C3-derived activated products (35). Elstad and co-workers showed that CD11b/CD18 integrin acts in concert with β-glucan receptor to induce the synthesis of PAF by monocytes stimulated with complement-opsonised zymosan particles (40). Our data are in agreement with these Authors. However, they reported that ligation of complement receptors of monocytes exposed to an opsonized erythrocytes or erythrocytes exposed to an opsonised erythrocytes or erythrocytes of the bearing IgG, isolated C3b, isolated iC3b, or a combination of IgG with either C3b or iC3b induced less than 5% of the response to the opsonized zymosan after 25 min incubation. In our studies, no detectable released or cell-associated PAF could be observed after 30 min incubation.

The studies concerning the effect of monoclonal antibodies suggest that ligation of CR1 and/or CR3 receptors on complement fractions fixed on Cuprophan$^{(R)}$ membranes may directly transduce the signal that mediate not only degranulation but also neosynthesis of lipid mediators such as PAF.

These studies provided information that PAF production is an early event following ligation of monocyte complement receptors using specific monoclonal antibodies and that this mechanism is operational in monocytes adhering to complement-activating haemodialysis membranes. Previous studies using the same monoclonal antibodies on human normal monocytes showed that ligation of complement receptors could trigger after 24-h incubation the production of IL-1, which is a potent stimulus for PAF production from adherent monocytes. Further studies may clarify the role of the late production of PAF on the functional properties of stimulated monocytes.

REFERENCES

1. Churchill DN. Efficiency and biocompatibility of membranes. In Dialysis Membranes: Structure and Predictions, V. Bonomini and Y. Berland, Editors, Contributions To Nephrology 113, 1994, 60–71.
2. Tetta C., David S., Biancone L., Canino F., Cambi V., Camussi G. Role of platelet activating factor in hemodialysis. Kidney Int 1993; 43 (suppl. 39): S154-S157.
3. Chenoweth DE. Anaphylatoxin formation in extracorporeal circuits. Complement 1986; 3: 162–165.
4. Aljama P. Biocompabilidad. Nefrologia 1990; X (3): 18–22.
5. Kazatchkine MD, Carreno MP. Activation of the complement system at the interface between blood and artificial surfaces. Biomaterials 1988; 9: 30–35.
6. Snyder F. Platelet-activating factor and related acetylated lipids as potent biological acting cellular mediators. Am J Physiol 1990; 259: 697.
7. Pinckard R.N., McManus L., Hanahan D.J. Chemistry and biology of acetyl glyceryl ether phosphorylcholine (platelegt-activating factor). Adv Infflammation Rev 1982; 4: 147.

8. Venable M.E., Zimmerman G.A., McIntyre T.M., Prescott S.M. Platelet-activating factor: a phospholipid autcoid with diverse actions. J Lipid Res 1993; 34: 691.
9. Camussi G., Tetta C., Baglioni C. The role of platelet-activating factor in inflammation. Clin Immun Immunopathol 1990; 57: 331.
10. Yue T.L., Rabinovici R., Feuerstein G. Platelet-activating factor (PAF), a putative mediator in inflammatory tissue injury. Adv Exp Med Biol 1991; 314: 223.
11. McManus L.M., Woodard D.S., Deavers S.I., Pinckard R.N. PAF Molecular heterogeneity: pathobiological implications. Lab Invest 1993; 69: 639.
12. Bratton D., Henson P.M. Cellular origin of PAF. In: Platelet-Activating Factor and human Diseases. P.J. Barnes, C.P. Page, P.M. Henson, eds. Blackwell Scientific Publications, London, 1989, p.23.
13. Camussi G., Aglietta M., Malavasi F., Tetta C., Piacibello W., Sanavio F., Bussolino F. The release of platelet-activating factor from human endothelial cells in culture. J Immunol 1983; 131: 2397.
14. Prescott S.M., Zimmerman G.A. McIntyre T.M. Human endothelial cells in culture produce platewlet-activating factor (1-alkyl-2-acetyl-sn-glycero-3-phosphocholine) when stimulated with thrombin. Proc Natl Acad Sci USA 1984; 81: 3534.
15. McIntyre T.M., Zimmerman G.A., Satoh K., Prescott S.M. Cultured endothelial cells synthesize both platelet-activating factor and prostacyclin in response to histamine, bradykinin, and adenosine triphosphate. J Clin Invest 1985; 76: 271.
16. Camussi G., Tetta C., Bussolino F., Baglioni C. Synthesis and release of platelet-activating factor is ihnibited by plasma α_1-proteinase inhibitor or α_1-antichymotrypsin and is stimulated by proteinases. J Exp Med 1988; 1681: 293.
17. Montrucchio G., Bergerone S., Bussolino F. et al. Streptokinase induces intravascular release of platelet-activating factor in patients with acute myocardial infarction and stimulates its synthesis by cultured human endothelial cells. Circulation 1993; 88: 1476.
18. Nakamura M., Honda Z.I., Izumi T. et al. Molecular cloning and expression of platelet-activating factor receptor from human leukocytes. J Biol Chem 1991; 266: 20400.
19. Ye R.D., Prossitz E.R., Zuo A., Cochrane C.G. Characterization of a human cDNA that encodes a functional receptor for platelet-activating factor. Biochem Biophys Res Commun 1991; 180: 105.
20. Camussi G., Montrucchio G., Lupia E. et al. Platelet-activating factor directly stimulates in vitro migration of endothelial cells and promotes in vivo angiogenesis by a heparin-dependent mechanism. J Immunol 1995; 154: 6492–6501.
21. Camussi G., Turello E., Bussolino F., Baglioni C. Tumor necrosis factor alters cytoskeletal organization and barrier function of endothelial cells. Int Arch Allergy Appl Immunol 1991; 96: 84.
22. Mahiout A., Courtney J.M. Effect of dialyser membranes on extracellular and intracellular granulocyte and monocyte activation in ex vivo pyrogen-free conditions. Biomaterials 1994; 15 (12): 969.
23. Haeffner-Cavaillon N., Cavaillon JM., Ciancioni C., Bacle F., Delous S., Kazatchine MD. In vitro induction of interleukin-1 during hemodialysis. Kidney Int 1989; 35: 121–1218.
24. Tetta C., Segoloni G., Turello E. The production of cytokines in hemodialysis. Blood Purif 1990; 8: 337–346.
25. Bingel M., Lonnemann G., Koch KM., Dinarello CA., Shaldon S. Plasma interleukin-1 activity during hemodialysis: the influences of dialysis membranes. Nephron 1988; 50: 273–282.
26. Herbelin A., Nguyen AT., Zingraff J., Urena P., Descamps-Latscha B. Influence of uremia and hemodialysis on circulating interleukin-1 and tumor necrosis factor-α. Kidney Int 1990; 37: 116–125.
27. Laude M., Haeffner-Cavaillon N., Pusineri C. Induction of interleukin-1 production during hemodialysis with non-complement activating high-permeability membrane (Abstr). lymphokine Res 1988; 7: 335.
28. Herbelin A., Urena P., Nguyen AT., Zingraff J., Descamps-Latscha B. Elevated circulating levels of interleukin-6 in patients with chronic renal failure. Kidney Int 1991; 39: 954–960.
29. Luger A., Kovanik J., Stummvoll HK., Urbanska A., luger T. Blood-membrane interaction in hemodialysis leads to increased cytokine production. Kidney Int 1987; 32: 84–88.
30. Blumenstein M., Schmidt B., Ward RA., Ziegler-Heitbrock HWL., Gurland HJ. Altered interleukin-1 production in patients undergoing hemodialysis. Nephron 1988; 50: 277–281.
31. Chollet-Martin S., Stamatakis G., Bailly S., Mery JP., Gougerot-pacidalo MA. Inducation of tumor necrosis factor-α during hemodialysis: influence of the membrane type. Clin Exp Immunol 1991; 83: 329–332.
32. Pertosa G., Marfella C., Tarantino EA. et al. Involvement of peripheral blood monocytes in haemodialysis: in-vivo induction of tumor necrosis factor alpha, interleukin-6, and β_2-microglobulin. Nephrol Dial Transplant 1991; 2: 18–23.
33. Schindler R., Lonnemann G., Shaldon S., Koch KM, Dinarello CA. Transcription, not synthesis of interleukin-1 and tumor necrosis factor by complement. Kidney Int 1990; 37: 85–93.

34. David S. Tetta C., Camussi G. et al. Adherence of human monocytes to haemodialysis membranes. Nephrol Dial Transplant 1993; 8: 1223–1227.
35. Tetta C., Tropea F., Camussi G. et al. Adherence of human monocytes to haemodialysis membranes: LFA 1 (CD11a/CD18) CR1 (CD35) and CR3 (CD11b/CD18) triggering promotes the biosynthesis of platelet-activating factor and adherence. Nephrol Dial Transplant 1995; 10.
36. Lo SK., Detmers PA., Levin SM., Wright SD. Transient adhesion of neutrophils to endothelium. J Exp Med 1989; 169: 1779–1793.
37. Wright SD., Rao PE., Van Vorrhis WC. et al. Identification of the C3bi receptor of human monocytes and macrophages by using monoclonal antibodies. Proc Natl Acad Sci Usa 1983; 80: 5699.
38. Wright SD., Meyer BC. Phorbal esters cause sequential activation and deactivation of complement receptors on polymorphonuclear leukocytes. J Immunol 1986; 136: 1759.
39. Camussi G. potential role of platelet-activating factor in renal pathophysiology. Kidney Int 1986; 29: 469–475.
40. Elstad MR., Parker CJ., Cowley FS. et al. CD11b/CD18 integrinb and a b-glucan receptors are in concert to induce the synthesis of platelet-activating factor by monocytes. J Immunol 1994; 152: 220–230.

DRUG-INDUCED ALTERATION OF ENDOTHELIAL PERMEABILITY IN THE RAT AORTA

Potential Consequences on the Vessel Wall

Gérard E. Plante, Stéphanie Lehoux, and Pierre Sirois

Departments of Medicine and Pharmacology
University of Sherbrooke
Sherbrooke, Québec, Canada, J1H 5N4

ABSTRACT

We recently reported that some diuretics affect capillary permeability in the normotensive rat. In the present study, we explore the effect of selected antihypertensive drugs administered orally during 10 days, on Evans blue (EB) extravasation within the wall of the thoracic (TA) and abdominal aorta (AA) obtained from spontaneously hyertensive rats (SHR). Description of the EB method has been previously reported. Daily doses (mg/kg) of captopril (CAP: 3.0), perindopril (PER: 0.3), nifedipine (NIF: 1.0), clentiazem (CLE: 0.1), hydralazine (HYD: 0.5), furosemide (FUR: 0.5), cicletanine (CIC: 2.0), hydrochlorothiazide (HCZ: 0.5), and indapamide (IND: 0.04) resulted in comparable blood pressure reduction. Percent changes in EB tissue concentration (measured in ug/g dry tissue) was increased by 24% in both the TA and AA in the untreated SHR. CAP reduced by half EB leakage in the TA, while PER decreased EB extravasation 16% below baseline values. Both angiotensin converting enzyme inhibitors failed to normalize EB leakage in the AA. The calcium channel blockers also normalized EB extravasation in the two segments of the aorta, except that CLE was without effect in the AA. HYD normalized EB leakage in the TA, but not in the AA. All diuretics tested reduced EB extravasation by 48 to 58% below baseline values in the TA, whereas CIC only normalized EB leakage. None of the diuretics affected EB extravasation in the AA of the SHR. In conclusion: 1- the two segments of the aorta were similarly affected in the SHR; 2- despite comparable effect on blood pressure, treatment of the SHR was associated with different responses in the TA and AA; 3- within a given class of drugs, different effects are observed on EB.

Platelet-Activating Factor and Related Lipid Mediators 2
edited by Nigam *et al.*, Plenum Press, New York, 1996

INTRODUCTION

Remodeling of the vascular wall occurs in a variety of morbid conditions, such as aging, arterial hypertension, diabetes mellitus, chronic renal failure. We suggested that damage to other target organs which occurs in these diseases, may apply to the vessel wall, and is most likely related to disruption in the interstitial traffic of vital substrates towards, and/or removal of waste products from the major cellular mass of the blood vessel, the smooth muscle cell [1]. Indeed, the interstitial space of the vascular wall occupies, as it does in all other organs, a strategic position between the capillary network and the intracellular space. In the case of the blood vessel wall, the interstitium represents approximately 50% of the total wet weight [2]. Permeability to fluid, small solutes, and macromolecules across the luminal endothelium, as well as in the adventicial microcirculation of large blood vessels, plays a determinant role in the size, composition, and physico-chemical properties of the vascular interstitial space, or extracellular matrix [3]. It is well known that the latter properties affect the hydraulic conductivity of the interstitial space, which in turn determines the traffic of fluid and solutes from the capillary to the cellular mass. Remodeling of the vascular wall includes trophic changes leading to necrosis of the smooth muscle cell, interstitial swelling, alteration in the interstitial structural proteins [4], as well as subendothelial deposition of abnormal material, such as cholesterol and other lipids [5]. Recent studies revealed that enhanced permeability to albumin-bound Evans blue is associated with significant alteration in the function of vital organs, such as the heart, kidney, gastro-intestinal tract, and the blood vessel wall *per se* [1]. We recently reported that some diuretics, largely utilized in the treatment of arterial hypertension, affect capillary permeability in the normal rat [6]. In the present study, we explore the effect of antihypertensive drugs, belonging to different common classes used in clinical medicine, on Evans blue (EB) extravasation in two distinct vulnerable segments of the aorta (thoracic and abdominal) obtained from spontaneously hypertensive rats (SHR).

METHODS

Eleven groups of rats (each of 6 to 8 animals) weighing between 175 and 200 grams, lodged in individual cages and fed a standard rat-chow diet, were studied over a period of 14 days. After a four-day adaptation period, Wistar Kyoto (WKY) normotensive rats, as well as the SHR received a daily gavage of normal saline (1 ml) during ten consecutive days, whereas each of the seven remaining group was administered (all doses in mg/kg) the following antihypertensive regimes (dissolved in 1 ml of normal saline), also by daily gavage: two angiotensin converting enzyme inhibitors, captopril (CAP: 3.0) and perindopril (PER: 0.3), two calcium channel blockers, nifedipine (NIF: 1.0) and clentiazem (CLE: 0.1), a reference vasodilator, hydralazine (HYD: 0.5), finally four diuretic agents, furosemide (FUR: 0.5), cicletanine (CIC: 2.0), hydrochlorothiazide (HTZ: 0.5), and indapamide (IND: 0.04). Doses were chosen to fit approximately the average therapeutic regimen recommended for human hypertensive subjects. Blood pressure was measured by standard tail cuff plethysmography on each day, before gavage procedures.

Capillary permeability was evaluated by measuring EB-bound albumin interstitial extravasation. In brief, the dye (20 mg/kg) was injected in the caudal vein of unanesthetized animal, exactly ten min before sacrifice. After exsanguination, segments of the thoracic and abdominal aorta are rapidly obtained and weighed immediately. Tissue manipulation, EB extraction by formamide, spectrophotometric analysis, calculations and statisticical methods were described previously [6]. The EB method for evaluation of capillary permeability has been validated in our own [6] and other laboratories [7,8].

RESULTS

Absolute values for EB extravasation in the two selected segments of the rat aorta are shown in Table 1. EB values for the thoracic and abdominal aorta were comparable in normotensive animals. In the SHR however, EB extravasation increased from 76 to 94, and from 86 to 107 ug/g dry tissue, in the thoracic and abdominal aorta, respectively (P<0.01). Angiotensin converting enzyme inhibition with captopril and perindopril reduced EB leakage in the thoracic segment of the aorta. Captopril, significantly reduced EB extravasation from 94 to 85ug/g dry tissue, but the latter value was still above baseline EB obtained in the normotensive rat. Perindopril was more potent in reducing EB value from 94 to 64ug/g dry tissue, the latter value being significantly lower than control (P<0.01). In the abdominal aorta, both drugs failed to affect significantly EB leakage. The two calcium channel blockers tested normalized thoracic aorta endothelial permeability. Clentiazem increased EB extravasation from 86 to 108ug/g dry tissue in the abdominal aorta, a value significantly above control (P<0.01) but not different from EB leakage observed in the SHR. Hydralazine restored EB extravasation from 94 to 80ug/g dry tissue in the thoracic aorta, but failed to influence the abnormal EB value of 107ug/g dry tissue obtained in the abdominal aorta. Interestingly, all four diuretic compounds tested reduced EB extravasation in the thoracic aorta of SHR: all values obtained were either below baseline data observed in the normotensive rat, 58, 48 and 48ug/g dry tissue, for furosemide, hydrochlorothiazide, and indapamide, respectively (P<0.01), or not significantly different from baseline, 70ug/g dry tissue, obtained with cicletanine. In contrast, none of the diuretics tested had any influence on EB leakage associated with arterial hypertension in the abdominal aorta: EB values of 100, 114, 101, and 101ug/g dry tissue, obtained with furosemide, cicletanine, hydrochlorothiazide and indapamide, respectively, are all above baseline normotensive values, but not statistically different from data obtained in the SHR. Drug regimens that were associated with complete correction of EB extravasation in

Table 1. Effect of selected antihypertensive drugs on Evans blue extravasation in segments of the thoracic and abdominal aorta obtained from spontaneously hypertensive rats (SHR). Control values for Evans blue in normotensive rats (Wistar-Kyoto) are given for comparison. Data are given in ug/g dry tissue (+ 1SEM). Refer to text for statistical significance

	Thoracic aorta	Abdominal aorta
Wistar-Kyoto	76+3	86+4
Untreated SHR	94+6	107+4
Captopril	85+3	94+4
Perindopril	64+4	94+3
Clentiazem	83+5	108+5
Nifedipine	71+3	94+2
Hydralazine	80+3	102+3
Furosemide	58+4	100+4
Cicletanine	70+3	114+3
Hydrochlorothiazide	48+3	101+6
Indapamide	48+2	101+3

the thoracic and abdominal aorta, those treatments where EB leakage was reduced below baseline values, finally drugs that were neutral in that respect, are summarized for clarity in Figure 1.

Blood pressure averaged 98 mmHg in the Wistar-Kyoto control rats, and 128 mmHg in the 8-week SHR, a significant elevation (P<0.01). The 10-day daily gavage with all tested treatments, resulted in a comparable reduction in blood pressure. In addition, values obtained after the treatment period were not significantly different from those measured in the Wistar Kyoto rats (data not shown).

DISCUSSION

The results obtained in this *in vivo* study reveals, for the first time, that a number of drugs currently utilized in the treatment of human arterial hypertension, influence endothelial cell permeability in two segments of the aorta, a major conductance artery. Neglected for a while, the role of large conductance arteries in the pathophysiology of hypertension has recently been emphasized [9], particularly with respect to major upstream and downstream morbid consequences which may result from alteration in the compliance of large arteries. Reduction in the elasticity of aortic tissue which develops during uncontrolled hypertension, as well as during "normal" senescence [10], is associated firstly, with enhanced stress on the left ventricle, which may contribute to left ventricular hypertrophy, secondly, with changes in the pattern of blood flow delivery (increased pulse pressure) to microcirculation networks, which may contribute to target organ damage [1]. It is likely that enhanced endothelial permeability to fluid, small solutes and macromolecule in the interstitial space of large arteries, is associated with increased size and compositional changes of this strategic fluid compartment, which occupies half of that vessel wall wet weight.

The experimental approach used in the present study did not allow to identify the site from which EB extravasation occurred in the SHR, and from which side of the blood vessel came the normalization of that permeability defect. Was the luminal endothelial barrier, and/or the adventicial capillary network of the aorta, responsible for the observed phenomena? Previous studies [11,12] and preliminary morphological data from this laboratory [13] indicate that, under normal conditions, more blood (approximately two thirds) enters the vessel wall from the adventicial side, than from the luminal endothelial barrier (approximately one third). This would suggest that, if the same ratio of blood flow distribution ob-

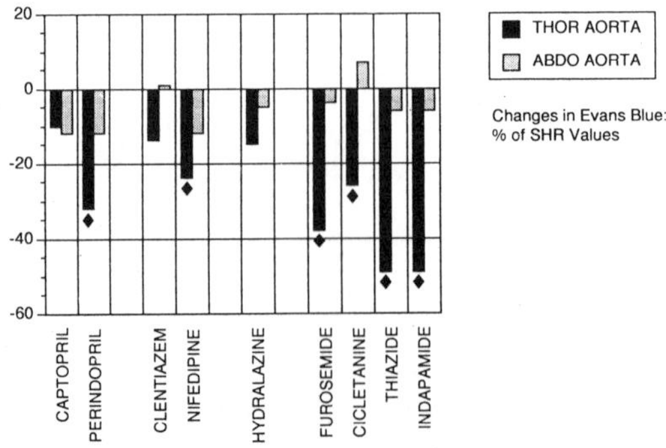

Figure 1. Endothelial permeability in the SHR: comparative effects of treatment.

tains in pathological conditions, more EB extravasation would develop from permeability changes in the adventicial microcirculation. Mechanisms involved in the regulation of the *vasa vasorum* microcirculation are not entirely understood at this point in time Much less understood are the effects of vaso-active xenobiotics on *vasa vasorum* microcirculation. However, the blood flow distribution pattern between the thoracic and abdominal aortic wall has been relatively well established [11,12]. Indeed, the adventicial absolute blood flow is approximately twice as much in the thoracic than in the abdominal aorta. Even if this vascular pattern does not entirely explain the observed features in EB extravasation, it may suggest different control mechanisms, either of endothelial permeability itself, or of pre- and post-capillary resistances in the adventicial microcirculation.

CONCLUSIONS

This study reveals for the first time, three features of pathophysiological interest. Firstly, the two segments of the aorta show similar increased EB extravasation in the untreated SHR. Secondly, despite comparable effect on blood pressure, treatment of the SHR is asociated with different effects on EB vascular wall leakage in the thoracic and abdominal aorta. Thirdly, within a given class of antihypertensive drugs, different effects are observed on EB. On the basis of interstitial abnormalities which may develop due to enhanced EB leakage, it is likely that antihypertensive agents which normalize and/or reverse the enhanced endothelial abnormality observed in the SHR, should be considered as protective of the vascular wall.

REFERENCES

1. Plante, G.E., Chakir, M., Lehoux, S., Lortie, M. Disorders of body fluid balance: a new look into the mechanisms of disease. Can. J. Cardiol. *in press*, 1995.
2. wigg, h., decarlo, m., sibley, l., renkin, E.M. Interstitial exclusion of albumin in rat tissues measured by a continuous infusion method. Am. J. Physiol. 263:H1222–1233, 1992.
3. Comper, W.D., Laurent, T.C. Physiological function of connective tissue polysaccharides. Physiol. Rev. 58:255–315, 1978.
4. Levy, B., Michel, J.B., Salzmann, I.L., Azizi, M., Poitevin, P., Safar, M.E. Camilleri, J.P. Effects of chronic inhibition of converting enzyme on mechanical and structural properties of arteries in rat renovascular hypertension. Circ. Res. 63:227–239, 1988.
5. Ross, R., The pathogenesis of atherosclerosis. An update. N. Engl. J. Med. 314:488–500, 1986.
6. Lehoux, S., Sirois, M.G., Sirois, P., Plante, G.E. Acute and chronic diuretic treatment selectively affects vascular permeability in the unanesthetized normal rat. J. Pharmacol. Exp. Therap. 269:1094–1099, 1994.
7. Lehoux, S., Plante, G.E., Sirois, M.G., Sirois, P., D'orleans-Juste, P. Phosphoramidon blocks big-endothelin-1 but not endothelin-1 enhancement of vascular permeability in the rat. Br. J. Pharmacol. 107:996–1000, 1992.
8. Patterson, C.E., Rhoades, R.A., Garcia, J.G.N. Evans blue as a marker of albumin clearance in cultured endothelial monolayer and isolated lung. J. Appl. Physiol. 72:865–873, 1992.
9. Asmar, R.G., Pannier, B., Santoni, J.P., Laurent, S., London, G.M., Levy, B., Safar, M.E. Reversion of cardiac hypertrophy and reduced arterial compliance after converting enzyme inhibition in essential hypertension. Circulation 4:941–950, 1988.
10. Folkow, B., Svanborg, A. Physiology of cardiovascular aging. Physiol. Rev. 73:725–764, 1993.
11. Heistad, D.D., Marcus, M.L., Law, E.G., Armstrong, M.L., Ehrhardt, J.C., Abboud, F.M. Regulation of blood flow to the aortic media in dogs. J. Clin. Invest. 62:133-, 1978.
12. Williams, J.K., Armstrong, M.L., Heistad, D.D. Blood flow through new microvessels: factors that affect regrowth of vasa vasorum. Am. J. Physiol. 254:H126, 1988.
13. Lehoux, S., Larouche, A., Cadieux, A., Plante, G.E. Perméabilité endothéliale de l'aorte du rat spontanément hypertendu: effets de divers antihypertenseurs. Arch. Mal. Coeur *in press*, 1995.

PAF PROMOTES THE DEVELOPMENT OF THE PREIMPLANTATION EMBRYO

C. O'Neill

Human Reproduction Unit
Royal North Shore Hospital of Sydney
St. Leonards, NSW, 2065, Australia

1. INTRODUCTION

The preimplantation embryo of mammalian species represents the period from fertilisation until the implantation of the embryo into the uterus at the blastocyst stage. In species such as experimental rodents and humans, implantation occurs 4–6 days after fertilisation at which time the embryo has 60–100 cells. During this time the embryo remains free-living within the reproductive tract and there is an exponential increase in cell numbers. Experiments *in vitro* have shown that growth and development of the embryo during this period is relatively autonomous requiring simple defined medium. There is no apparent absolute requirement for exogenous growth factors. Such autonomy suggests that the stimulus for growth during this period may be endogenous to the embryo.

Several candidate autocrine growth factors have been proposed, for example insulin-like growth factor (IGF)-I, IGF-II and TGF-α (review [1]), although none have yet been unequivocally shown to be essential for embryo development. In particular, gene knock-out experiments for several growth factors or their receptors have not been shown to prevent this stage of embryo development. Studies in this laboratory in recent years have examined the role of PAF in early embryo development with recent evidence suggesting that embryo-derived PAF may act as an autocrine embryotrophic factor.

2. BIOSYNTHESIS OF PAF BY THE PREIMPLANTATION EMBRYO

There are now numerous reports of the release of PAF by the preimplantation embryo from a number of mammalian species. These reports have used a variety of different assay procedures and include several independent confirmations (review [2]). There have been several reports failing to detect the release of PAF, but these may be due to the vagaries of the PAF extraction and assay procedures together with the nature of PAF's binding to its carrier molecule[3]. PAF release *in vitro* requires extracellular albumin and approximately 41% of PAF produced by the 2-cell embryo remains associated with the embryo[3].

Recent studies show that PAF is the product of biosynthesis by the embryo rather then release of pre-stored PAF. Incubation of 2-cell mouse embryos with ^{14}C-hexadecanol, 1-*0*-^{3}H-alkyl-lysoPAF, 1-*o*-^{3}H-alkyl-acetyl-glycerol and methyl-^{3}H choline chloride resulted in the incorporation of these radiolabelled substrates into radiolabelled PAF released by the embryo[4]. Radioactive carbon was also incorporated into PAF from the simple energy substrates pyruvate, lactate and glucose. The specific activity of incorporation was greatest for pyruvate and 2405-times less for glucose, reflecting the energy substrate preference for this stage embryo[4]. The incorporation of ^{14}C-hexadecanol into PAF infers the presence of alkyldihydroxyacetone-phosphate synthase in the embryo. It was also observed that both 1-*o*-^{3}H-alkyl-acetyl-glycerol and methyl-^{3}H-choline resulted in the production of both radiolabelled lysoPAF and PAF. The accumulation of lysoPAF under these circumstances suggested that in the presence of excess substrate, the enzyme lysoPAF:acetyltransferase may limit PAF synthesis.

Examination of enzyme activity within embryos supported this observation[5]. Two enzymes were assessed being dithiothreitol-insensitive cytidinediphospho-choline:1-*O*-alkyl-2-acetyl-*sn*-glycerol cholinephosphotransferase (cholinephosphotransferase, EC 2.7.8.16) of the '*de novo*' pathway and acetyl-coenzyme A: 1-*O*-alkyl-2-lyso-*sn*-glycero-3-phosphocholine acetyltransferase (acetyltransferase, EC 2.3.1.67) of the 'remodelling' pathway. Activity of both enzymes was present in embryos. Cholinephosphotransferase required Mg^{2+} and was inhibited by Ca^{2+}, while acetyltransferase required the presence of NaF (a phosphatase inhibitor). The activity of cholinephosphotransferase was not different between unfertilised oocytes and zygotes, and did not change significantly with advancing developmental stage (Table 1). The specific activity of PAF:acetyltransferase was significantly lower ($p< 0.03$) in unfertilised oocytes than in zygotes of corresponding age. During the preimplantation period, activity was higher at the 2-cell stage than for morulae while activity in blastocysts was intermediate (Table 1). For both enzymes no product was formed in the absence of exogenous substrate suggesting that under the assay conditions negligible endogenous substrate was available. The activation of lysoPAF:acetyltransferase following fertilisation indicates that this enzyme may be of regulatory significance in the initiation of PAF biosynthesis by zygotes, while its phosphatase sensitivity suggests that phosphorylation may be a mechanism for its activation in the zygote.

Table 1. The specific activity of lysoPAF:acetyltransferase and DTT insensitive cholinephosphotransferase in mouse oocytes and preimplantation embryos. Adapted from Wells and O'Neill[5]

	Acetyltransferase		Cholinephosphotransferase	
Embryo stage	N	Specific activity (fmol PAF/embryo/min)	N	Specific activity (fmol PAF/embryo/min)
Oocyte	4	0.078 ± 0.044	3	3.298 ± 1.080
Zygote	6	0.358 ± 0.097	3	2.307 ± 0.400
2-cell	6	0.591 ± 0.120	4	5.005 ± 1.000
Morulae	7	0.11 ± 0.052	4	3.810 ± 1.720
Blastocyst	3	0.210 ± 0.050	3	1.485 ± 0.495

3. METABOLISM OF EMBRYO-DERIVED PAF IN THE REPRODUCTIVE TRACT

The essentially ubiquitous nature of the high capacity enzyme PAF:acetylhydrolase means that PAF is generally considered to be a mediator with a short half-life *in vivo*. This obviously has the potential to limit the concentration of PAF achieved *in situ* and thus limit PAF's site of action to the immediate vicinity of the site of production. The capacity for PAF produced by embryos to act as a mediator will be dependent upon its half-life and hence effective local concentration.

We have recently examined the activity of PAF:acetylhydrolase in the uterus of mice during early pregnancy and throughout the luteal phase[6]. PAF:acetylhydrolase activity was found to be present in uterine lumenal fluid and in endometrial tissue throughout the reproductive cycle of mice. Its activity was not affected by cations (Mg^{++} and Ca^{++}) and was not inhibited by bromophenacyl bromide, a phospholipase A_2 inhibitor. A significant reduction in uterine PAF:acetylhydrolase activity occurred in the uterus during the preimplantation phase of pregnancy (Day 1–4) while activity rapidly increased after the time of implantation (Table 2). This change may have been under steroidal hormone regulation since in ovariectomised animals treatment with 2 mg progesterone/day for 3 days significantly suppressed PAF:acetylhydrolase activity in uterine fluids while one days treatment with 0.1–10 μg estradiol increased activity. The relatively low levels of PAF:acetylhydrolase activity in the uterus during the preimplantation phase corresponded with the time when PAF produced by the uterus and the embryo may be highest. In the rabbit[7] it was shown that the PAF concentration in endometrial tissue increased throughout the preimplantation period reaching a peak on day 4. In pregnant animals the uterine concentration declined dramatically following implantation. Thus the amount of PAF present in the uterus was high when PAF:acetylhydrolase activity in mice was relatively low and PAF decreased when PAF:acetylhydrolase was at its maximum.

It is also likely that metabolism of embryo-derived PAF is reduced by its association with a carrier molecule. We have found that mouse embryo-derived PAF is associated with albumin *in vitro* and limited proteolytic analysis demonstrated that it was almost en-

Table 2. The specific activity of PAF:acetylhydrolase activity in uterine lumenal fluid and endometrial tissue form mice throughout the reproductive cycle. Adapted from O'Neill[6]

	Endometrial tissue		Lumenal fluid	
Day	N	Specific activity (pmol/μg/min)	N	Specific activity (pmol/μg/min)
−2	13	0.967 ± 0.45	11	12.98 ± 13.36
−1	11	0.917 ± 0.53	10	2.65 ± 2.01
0	10	0.767 ± 0.43	8	2.35 ± 1.33
1*	4	0.390 ± 0.18	5	5.32 ± 2.38
2*	7	0.520 ± 0.24	8	1.22 ± 0.89
3*	4	0.460 ± 0.08	6	0.95 ± 1.44
4*	5	0.440 ± 0.20	4	0.77 ± 0.19
5	4	2.550 ± 0.40	4	3.42 ± 1.14
6	7	4.390 ± 1.58	5	24.05 ± 4.30
7	4	2.970 ± 0.80	6	35.79 ± 3.73
8	4	1.650 ± 0.82	6	10.38 ± 5.72
9	Not performed		8	6.89 ± 3.88

*Represents the preimplantation phase in mice.

tirely bound to domain II of the albumin molecule[3]. Furthermore, this binding bestowed protection on PAF from serum PAF:acetylhydrolase *in vitro*. Interestingly, synthetic PAF bound by albumin in solution did not have access to domain II and was not protected from PAF:acetylhydrolase activity.

Should such 'cryptic' binding of PAF to domain II of albumin also occur *in vivo*, the resulting resistance to hydrolysis by PAF:acetylhydrolase together with the markedly reduced PAF:acetylhydrolase activity present within the reproductive tract during the preimplantation phase of pregnancy may mean that embryo-derived PAF has an extended half-life and thus may accumulate to significant local concentrations within the reproductive tract.

4. ACTIONS OF PAF ON THE PREIMPLANTATION EMBRYO

Examination of the effects of PAF on embryos relies on *in vitro* culture procedures. Supplementation (0.1–1.0 μg PAF/ml) of culture medium with exogenous PAF (but not lysoPAF) significantly increased glucose metabolism in both mouse[8] and human embryos[9], and for mouse embryos this effect was attenuated by the competitive receptor-antagonist, SRI 63–441[8]. Such supplementation with PAF also resulted in a significant enhancement of embryo viability as assessed by their pregnancy potential following embryo transfer, which in mice resulted in a 24–34% increase in implantation and fetal development rates and in a human IVF program resulted in a 69% increase in pregnancy rate[9] and a 101% increase in the live birth rate following embryo transfer[10].

It is significant in considering the mechanism for such beneficial effects of PAF that there are long standing observations that culture *in vitro* adversely affects embryo development[11]. Cell doubling times of mouse embryos from the 2-cell to the blastocyst stage *in vitro* have been shown to be significantly longer for embryos developing *in vitro* compared with those developing *in vivo*. It was also shown that there was a linear reduction of PAF release in culture medium when 2-cell mouse embryos were cultured *in vitro* for up to 72h[12]. Embryos freshly collected from the reproductive tract at developmental stages equivalent to 48h (morulae) and 72h (blastocysts) culture released higher levels of PAF than the corresponding embryos produced by culture *in vitro*[12]. Thus, the reduced developmental potential of embryos was correlated with reduced release of embryo-derived PAF and this could be at least partially overcome by supplementing media with exogenous synthetic-PAF. Since culture reduced cell doubling times we also studied the effect of PAF on the embryo cell-cycle *in vitro*.

The mitotic index was used to assess the rate of cell-cycle progression for various stage embryos in the presence or absence of PAF[13]. Mitotic index was measured by cultur-

Table 3. The effects of PAF (1.86 μM) and WEB 2086 (33 μM) on the mitotic index of 8-cell mouse embryos following culture *in vitro*. Adapted from Roberts et al.[13]

Treatments		Mitotic index	
PAF	WEB	Median	(25th percentile–75 percentile)
–	–	0.438	(0.125–0.75)
+	–	0.75	(0.375–1.0)
+	+	0.375	(0.125–0.75)
–	+	0.0	(0.0–0.125)

ing embryos in the presence of colchicine, a mitotic spindle inhibitor. Thus, cells were trapped in metaphase. Fixation and nuclear staining of embryos at various times after culture and visualisation of mitotic figures allowed the mitotic index to be measured. The mitotic index was calculated as the number of cells per embryo in metaphase divided by the total number of cells in the embryo. The rate of progression to metaphase of 8-cell embryos was significantly faster for embryos cultured in PAF than in control media. However, by 16 h there was no difference between the two treatments showing that although PAF increased the rate of cell-cycle progression it did not effect the total number of cells eventually dividing. The action of PAF on the rate of cells reaching metaphase was blocked by the PAF-receptor antagonists WEB 2086 (33μM). Furthermore, in the absence of exogenous PAF, WEB 2086 caused a depression of the mitotic index below control levels, possibly indicating the inhibition of the action of endogenous embryo-derived PAF. This technique also showed that at the blastocyst-stage, embryos cultured in the presence of PAF (0.1–10μg/ml) had a significantly higher mitotic index than those cultured without PAF. There was a negative correlation between the mitotic index and the number of cells within the blastocysts. However, this negative correlation was greater for blastocysts cultured in media supplemented with PAF compared with control media. Thus, PAF stimulated mitosis to a greater degree in small embryos (those with fewer cells and hence retarded for their developmental age) than for blastocysts with a high cell number.

These results show that a role for PAF in promoting embryo development and viability was by its action to stimulate cell-cycle progression at various stages during preimplantation development. The effect was apparently receptor mediated since PAF's action was inhibited by a putative receptor antagonist WEB 2086. These results, however, do not indicate whether PAF acted specifically as a mitogen or growth factor affecting specific control points in the cell cycle or was secondary to general metabolic effects on the embryo, such as the enhanced oxidative metabolism of carbohydrates which has been demonstrated.

5. CONCLUSION

There is now sufficient evidence to conclude that PAF is synthesised by the mammalian preimplantation embryo and that a significant amount of this is released by the embryo. The reduced activity of PAF:acetylhydrolase in the reproductive tract during this time and the apparent protection of released PAF from PAF:acetylhydrolase by virtue of its binding to domain II of albumin suggests that sufficient PAF may accumulate in the reproductive tract to allow it to act as a mediator. PAF has been shown to cause a variety of modifications to the reproductive tract and maternal physiology including: localised platelet activation[14], leading to transient systemic thrombocytopenia in early pregnancy[15]; reduced vascularity of the oviduct, apparently due to vasoconstriction during the preimplantation phase[14]; induction of the expression of an immunosuppressed state known as 'early pregnancy factor'[16]; and modulation of prostaglandin release by the reproductive tract[17, 18]. Yet studies with PAF-receptor antagonists suggest that an important target for PAF is the embryo itself rather then the preparation of the reproductive tract for pregnancy[19]. Consistent with such data are the observations that PAF enhanced the metabolic rate of embryos at various stages of preimplantation embryo development *in vitro* and that it also increased the rate of cell-cycle progression by embryos. Such events were correlated with an increase in the viability and pregnancy potential of embryos following their transfer back to the uterus. An autocrine embryotrophic role for PAF can thus be assigned.

The extent to which embryo-derived PAF is essential for embryo development can not be defined however at this time. Since the embryo produces PAF and is also its target for action, it is experimentally difficult to define the developmental fate of embryos in the complete absence of PAF. To show that PAF is essential for dèvelopment this would be required. Specific and potent inhibitors PAF synthesis are required to produce embryos in which PAF has been 'knocked-out'. Blocking PAF's action with receptor antagonists has provided conflicting data, probably because these agents appear to act as partial agonists rather then true antagonists in this model[20]. Thus, a complete definition of PAF's role in preimplantation embryo development awaits better experimental tools for manipulating PAF's synthesis and action on cells.

REFERENCES

1. Adamson ED, 1993. Activities of growth factors in preimplantation embryos. J Cell Biochem 53:280–287
2. Ryan JP, O'Neill C, Ammit AJ, Roberts CG, 1992. Metabolic and developmental responses of preimplantaion embryos to platelet activating factor. Reprod Fertil Dev 4:387–398
3. Ammit AJ, O'Neill C, 1995. PAF released by preimplantation embryos binds to albumin. In: Nigam S, 3. Kunkel G, Prescott SM, Vargaftig BB (eds.), Platelet-activating factor and related lipid mediators in health and disease. New York: Plenum Publishing,
4. Wells XE, O'Neill C, 1992. Biosynthesis of platelet-activating factor by the mouse two-embryo. J Reprod Fertil 96:61–71
5. Wells XE, O'Neill C, 1994. Detection and preliminary characterization of two enzymes involved in biosynthesis of platelet-activating factor in mouse oocytes,zygotes and preimplantation embryos: dithiothreitol-insensitive cytidinediphospho-choline:1-*o*-alkyl-2-acetyl-*sn*-glycerol cholinephosphotransferase and acetyl-coenzyme A:1-*o*-alkyl-2-lyso-*sn*-glycero-3-phosphocholine acetyltransferase. J Reprod Fertil 101:385–391
6. O'Neill C, 1995. Activity of platelet-activating factor acetylhydrolase in the mouse uterus during the estrous cycle, throughout the preimplantation phase of pregnancy, and throughout the luteal phase of pseudopregnancy. Biol Reprod 52:965–971
7. Angle MJ, Jones MA, McManus LM, Pinckard RN, Harper MJK, 1988. Platelet-activating factor in the rabbit uterus during early pregnancy. J Reprod Fertil 83:711–722
8. Ryan JP, O'Neill C, Wales RG, 1990. Oxidative metabolism of energy substrates by preimplantation mouse embryos in the presence of platelet activating factor. J Reprod Fertil 89:301–307
9. O'Neill C, Ryan JP, Collier M, Saunders DM, Ammit AJ, Pike IL, 1989. Supplementation of IVF culture media with platelet activating factor (PAF) increased the pregnancy rate following embryo transfer. Lancet ii:769–772
10. O'Neill C, Ryan JP, Collier M, Saunders DM, Ammit AJ, Pike IL, 1992. Outcome of pregnancies resulting from a trial of supplementing human IVF culture media with platelet-activating factor. Reprod Fertil Develop 4:109–112
11. Bowman P, McLaren A, 1970. Viability and growth of mouse embryos after *in vitro* culture and fusion. J Embryol exp Morph 23:693–704
12. Ryan JP, Spinks NR, O'Neill C, Ammit AJ, Wales RG, 1989. Platelet activating factor (PAF) production by mouse embryos in-vitro and its effects on embryonic metabolism. J Cell Biochem 40:387–395
13. Roberts C, O'Neill C, Wright L, 1993. Platelet activating factor (PAF) enhances mitosis in preimplantation mouse embryos. Reprod Fertil Dev 5:271–279
14. Stein BA, O'Neill C, 1994. Morphometric evidence of changes in the vasculature of the uterine tube of mice induced by the 2-cell embryo on the second day of pregnancy. J Anat 185:397–403
15. O'Neill C, 1985. Thrombocytopenia is a initial maternal response to fertilisation in mice. J Reprod Fert 73:559–566
16. Orozco C, Perkins T, Clarke FM, 1986. Platelet activating factor induces the expression of early pregnancy factor activity in female mice. J Reprod Fertil 78:549–555
17. Smith SK, Kelly RW, 1988. Effect of platelet-activating factor on the release of PGF-2a by separated cells of human endometrium. J Reprod Fertil 82:271–276

18. Battye KM, O'Neill C, Evans G, 1992. Evidence that platelet activating factor suppresses uterine oxytocin-induced 13,14-dihydro-15-keto-prostaglandin F2a release and phosphatidylinositol hydrolysis in the ewe. Biol Reprod 47:213–219
19. Spinks NR, Ryan JP, O'Neill C, 1990. Antagonists of embryo-derived platelet activating factor act by inhibiting the ability of the mouse embryo to implant. J Reprod Fertil 88:241–248
20. O'Neill C, 1995. Platelet-activating factor-antagonists reduce implantation in mice at low doses only. Reprod Fertil Develop 7:51–57

PAF RELEASED BY PREIMPLANTATION EMBRYOS BINDS TO ALBUMIN

A. J. Ammit and C. O'Neill

Human Reproduction Unit
Royal North Shore Hospital of Sydney
St. Leonards, NSW 2065, Australia

1. BACKGROUND

The preimplantation embryo from a variety of mammalian species, including humans[1, 2], mice[3, 4], rabbit[5] and sheep[6], have been reported to release PAF *in vitro*. It has been demonstrated that embryo-derived PAF may act as an autocrine stimulant of embryonic metabolism and growth[7, 8] and the release of PAF by human embryos created by *in vitro* fertilization appears to be correlated with embryonic viability and pregnancy potential[1, 9, 10]. PAF has a range of effects on maternal physiology which include: thrombocytopenia and splenic contraction[11]; modulation of prostaglandin F_{2a} secretory activity of the uterus *in vitro*[12] and *in vivo*[13]; changes in the vasculature of the uterine tube of mice[14]; and the induction of early pregnancy factor expression[15].

Despite this evidence indicating that PAF may play a number of roles during early pregnancy, controversy exists regarding the production of PAF by embryos. Several reports have failed to detect PAF in embryo-conditioned media[16–18]. Amiel *et al.*[16] tested human embryo conditioned media in the splenectomized mouse bioassay[19] and an *in vitro* platelet aggregation bioassay, following PAF extraction and HPLC. PAF activity was detected by the splenectomized mouse bioassay, but not in the platelet aggregation bioassay. Similar results were found by Adamson *et al.*[18]. Smal *et al.*[17] failed to detect PAF following extraction of murine embryo culture medium and quantification using the NEN Du Pont RIA.

Possible explanations for the negative results include: (i) the platelet agonist released from embryos is not PAF; (ii) PAF inhibitors were present; and (iii) PAF carrier molecule/s exist.

1.1 Characterization of Embryo-Derived PAF

It was proposed that the bioactive fraction of embryo-conditioned media was not PAF[18], but a substance capable of inducing platelet aggregation *in vivo*. O'Neill[3] reported that embryo-derived PAF possessed similar chemical, biochemical and physiological properties to PAF. The identity of embryo-derived PAF has been confirmed by pharmacological characterization with a specific PAF receptor antagonist, SRI 63–441[20]. Further, selectivity for PAF-induced platelet aggregation was ensured by inhibition of the other major pathways of platelet aggregation in the bioassay (the arachidonic acid- and adenosine diphosphate-dependant pathways[21]).

In 1990, Smal *et al.*[17] suggested that the disparity between their results and those previously published using platelet aggregation bioassays[19, 20, 22] were due to a different molecular nature of embryo-derived PAF, and that the molecular species of embryo-derived PAF was not immunoreactive in the RIA. Since 1990, PAF production by embryos has been measured by PAF RIA in two independent studies[23, 24]. Furthermore, the molecular species composition of embryo-derived PAF released from murine embryos has been characterized by high performance liquid chromatography[25]. Murine embryos release a heterogenous mixture of C16:0 and C18:0 PAF. Although the RIA anti-PAF antibody had different affinity for C16:0 and C18:0 PAF (EC_{50} were 1.5ng PAF/mL and 6ng PAF/mL, respectively[25]), both molecular species were immunoreactive. The structure of murine embryo-derived PAF has also been confirmed by electron impact mass spectrometry[4].

1.2 PAF Inhibitors

It is also possible that endogenous PAF inhibitors could be present which interfere with PAF measurement techniques. Lipid inhibitors of PAF have been found in human saliva[26], rat uterus[27] and rat liver[28]. While the inhibitors from rat uterus[27] and human saliva[26] can be separated from PAF by normal-phase HPLC, the rat liver source of PAF inhibitor co-migrated with PAF on TLC using conventional solvent systems (chloroform:methanol:water; 65:35:7, v/v/v)[28]. However, it was possible to detect synthetic PAF exogenously added to murine embryo-conditioned media by RIA[17] and bioassay[21] suggesting that the presence of inhibitors was not affecting detection in that system.

1.3 Evidence to Suggest the Existence of a PAF Carrier Molecule

Human embryo-derived PAF has been shown to be produced by embryos created by IVF[1, 2]. IVF embryos are cultured for 24hr in media containing 10% (v/v) autologous serum. Serum contains the enzyme PAF:acetylhydrolase[29] (E.C. 3.1.1.47) (PAF:AH), which removes the acetate group from the C2 position of PAF to leave inactive lysoPAF[30]. PAF:AH is present in plasma[29], cells[31] and tissues[31]. In plasma, the half-life ($t_{1/2}$) of exogenously added synthetic PAF was shown to be 7.8 min[32], however, PAF released from human embryos created by IVF could be detected after 24hr culture in the presence of 10% (v/v) serum[1, 2]. To exclude the possibility that the IVF procedure and superovulation regimen may result in reduced serum levels of PAF:AH, PAF:AH specific activity was measured in serum obtained from the female partner of couples undergoing IVF (at the Human Reproduction Unit, Royal North Shore Hospital of Sydney, NSW, Australia). The average PAF:AH specific activity was 7.624 +/- 0.648pmol acetate released/min/mg protein (mean +/- s.e.m.) (n = 18). This was not significantly different (P = 0.11, t-test) from the average PAF:AH specific activity for a group of a normal volunteers, i.e. 7.808 +/- 0.473pmol acetate released/min/mg protein (mean +/- s.e.m.) (n = 20). Further, the detection of PAF released by embryos after 24hr exposure to serum did not appear to be due to the presence of an inhibitor of PAF:AH, since the $t_{1/2}$ of exogenously added synthetic PAF in IVF culture media was 13min[33]. These observations were not unique to IVF, since mouse embryo-derived PAF was detected after 1hr exposure to 100% (v/v) normal mouse plasma[34]. These results suggest the existence of a carrier molecule for embryo-derived PAF. Association of embryo-derived PAF with this molecule may cause resistance to hydrolysis by PAF:AH.

1.3.1 PAF Carrier Molecules. A number of compounds have been shown to bind PAF, and may therefore act as candidates for embryo-derived PAF carrier molecules. Most studies have examined the binding of PAF to protein constituents in human plasma or serum.

Albumin is present in human plasma at a concentration of ~ 0.5mM[35]. The binding of PAF to albumin has been indirectly demonstrated in a number of studies. The presence of albumin was required in order to detect the platelet-stimulating activity of PAF released from rabbit basophils[36]. Albumin has been shown to solubilize PAF in aqueous media, since non-specific binding of PAF to glass[37] can be reversed with increasing albumin concentration[38]. PAF exogenously added to plasma comigrates with the albumin fraction[39]. Albumin was necessary for PAF synthesis and release from neutrophils[39].

More directly, the number of binding sites for PAF on albumin have recently been quantified[35]. PAF binds to albumin on at least four binding sites per albumin molecule of protein, with relatively high affinity (average equilibrium dissociation constant of 1×10^{-7}M)[35].

PAF-binding factor/PAF-releasing factor[40, 41] also acts as a carrier molecule for PAF. This factor was discovered during purification of PAF:AH in human serum[40]. It was found that recovery of enzyme activity was 1.6 to 2 times higher than in the original serum, suggesting the possible occurrence of an inhibitor of PAF:AH. After purification of the putative inhibitor, it was found that the effect was due to the binding of PAF to a molecule that protected it from the effect of PAF:AH, rather than inhibition of the activity of PAF:AH. This factor was termed PAF-binding factor[40]. After a further study[41], that suggested that this factor induced the release of PAF from stimulated neutrophils, the factor was referred to as PAF-releasing factor and its molecular weight was estimated to be 240kDa[41]. Although PAF associated with PAF-releasing factor is protected from degradation by serum PAF:AH, the PAF/PAF-binding factor complex was still able to induce aggregation of washed rabbit platelets[41].

PAF-like activity has been found associated with the lipoproteins[42, 43]. It has also been demonstrated that PAF can bind to alpha-1-acid glycoprotein *in vitro*[44]. *In vivo,* alpha-1-acid glycoprotein functions to transport endogenous substances, such as endogenous amines, in serum and extravascular spaces. The physiological and pathological significance of the *in vitro* interaction of PAF with alpha-1-acid glycoprotein remains to be evaluated.

2. ALBUMIN IS A CARRIER MOLECULE FOR EMBRYO-DERIVED PAF

To identify the putative carrier molecule for embryo-derived PAF, murine embryos were cultured in a defined media where albumin was the only exogenous macromolecule present. PAF release from murine embryos increased in a dose-dependent manner with increasing albumin concentration in the culture media. Embryo-derived PAF concentration in the presence of 3mg bovine serum albumin (BSA)/mL was significantly higher ($P < 0.01$, Mann-Whitney) than obtained in the absence of BSA. The median concentrations were 4.5ng PAF/mL (Q1 - Q3, 1.3 - 7.9ng PAF/mL) and 0.2ng PAF/mL (Q1 - Q3, 0.0 - 1.5ng PAF/mL), respectively. Size fractionation, gel filtration and affinity chromatography showed that the carrier molecule for embryo-derived PAF was albumin. BSA is a single chain polypeptide with 583 amino acid residues with 9 loop-link-loop structures (stabilized by disulphide bridges) arranged in 3 repeating domains[45, 46]. Limited proteolysis with pepsin and trypsin demonstrated that the major binding site for embryo-derived PAF on albumin was located within a hydrophobic region on domain II. This binding protects PAF from hydrolysis by PAF:AH and also prevents its detection in immunoassays and platelet activation bioassays. Partial reduction with dithiothrietol suggested that binding of PAF by albumin may be due to conformational changes involving disulphide bonds. These changes may make the binding site of BSA accessible to embryo-derived PAF *in vivo*. In contrast, aqueous suspensions of synthetic PAF do not readily interact with this albumin binding site *in vitro*.

3. CONCLUSIONS

The preferential localization of PAF released from embryos to the hydrophobic core of albumin may explain reports which failed to detect PAF in embryo-conditioned media[16–18]. Embryo-released PAF may be present in a 'cryptic' form, which may make its detection difficult in some assay protocols, and may have implications for its actions and half-life *in vivo*. Further, the absence of apparent conformational changes to albumin in the absence of the embryo may explain the dissimilar biochemical and pharmacological responses of embryo-derived PAF and synthetic PAF, and may suggest a novel mechanism for the binding of PAF released from cells by albumin.

REFERENCES

1. O'Neill, C., and Saunders, D.M. (1984) Lancet, ii, 1034–1035
2. Punjabi, U., Vereecken, A., Delbeke, L., Angle, M., Gielis, M., Gerris, J., Johnston, J., and Buytaert, P. (1990) J. IVF & ET. 7, 321–326
3. O'Neill, C. (1985) J. Reprod. Fert. 75, 375–380
4. Kodama, H., Muto, H., and Maki, M. (1989) Acta Obstet. Gynae. Jpn. 41, 899–906
5. Minhas, B.S., Zhu, Y-P., Kim, H-N., Burwinkel, T.H., Ripps, B.A., and Buster, J.E. (1993) J. Assist. Reprod. Genet. 10, 366–370
6. Battye, K.M., Ammit, A.J., O'Neill, C., and Evans, G. (1991) J. Reprod. Fert. 93, 507–514
7. Ryan, J.P., O'Neill, C., and Wales, R.G. (1990) J. Reprod. Fert. 89, 301–307
8. Roberts, C., O'Neill, C., and Wright, L. (1993) Reprod. Fertil. Dev. 5, 271–279
9. O'Neill, C., Gidley-Baird, A.A., Pike, I.L., and Saunders, D.M. (1987) Fertil. Steril. 47, 969–975
10. Ryan, J.P., Spinks, N.R., O'Neill, C., andWales, R.G. (1990) J. Reprod. Fert. 89, 309–315
11. O'Neill, C. (1985) J. Reprod. Fert. 73, 559–566
12. Smith, S.K., and Kelly, R.W. (1988) J. Reprod. Fert. 82, 271–276
13. Battye, K.M., O'Neill, C., and Evans, G. (1992) Biol. Reprod. 47, 213–219
14. Stein, B.A., and O'Neill, C. (1994) J. Anat. 185, 397–403
15. Orozco, C., Perkins, T., and Clarke, F.M. (1986) J. Reprod. Fert. 78, 549–555
16. Amiel, M-L., Testart, J., and Benveniste, J. (1991) Fertil. Steril. 56, 62–65
17. Smal, M.A., Dziadek, M., Cooney, S.J., Attard, M., and Baldo, B.A. (1990) J. Reprod. Fert. 90, 419–425
18. Adamson, L.M., Hanf, V., Mittman, S.G., and Tinneberg, H-R. (1992) Eur. J. Obstet. Gynae. Reprod. Biol. 46, 19–24
19. O'Neill, C. (1985) J. Reprod. Fert. 73, 567–577
20. Collier, M., O'Neill, C., Ammit, A.J., and Saunders, D.M. (1990) Hum. Reprod. 5, 323–328
21. Ammit, A.J., and O'Neill, C. (1991) Hum. Reprod. 6, 872–878

22. Collier, M., Ammit, A.J., O'Neill, C., and Saunders, D.M. (1988) Hum. Reprod. 3, 993–998
23. Battye, K.M., Ammit, A.J., O'Neill, C., and Evans, G. (1991) J. Reprod. Fert. 93, 507–514
24. Nakatsuka, M., Yoshida, N., and Kudo, T. (1992) Hum. Reprod. 7, 1435–1439
25. Ammit, A.J., Wells, X.E., and O'Neill, C. (1992) Hum. Reprod. 7, 865–870
26. Smal, M.A., and Baldo, B.A. (1991) Lipids, 26, 1144–1147
27. Nakayama, R., Yasuda, K., and Saito, K. (1987) J. Biol. Chem. 262, 13174–13179
28. Miwa, M., Hill, C., Kumar, R., Sugatani, J., Olson, M.S., and Hanahan, D.J. (1987) J. Biol. Chem. 262, 527–530
29. Farr, R.S., Cox, C.P., Wardlow, M.L., and Jorgensen, R. (1980) Clin. Immunol. Immunopathol. 15, 318–330
30. Blank, M.L., Lee, T-C., Fitzgerald, V., and Snyder, F. (1981) J. Biol. Chem. 256, 175–178
31. Stafforini, D.M., Prescott, S.M., Zimmerman, G.A., and McIntyre, T.M. (1991) Lipids, 126, 979–985
32. Stafforini, D.M., Carter, M.E., Zimerman, G.A., McIntyre, T.M., and Prescott, S.M. (1989) Proc. Natl. Acad. Sci. 86, 2393–2397
33. Pike, I.L., Ammit, A.J., and O'Neill, C. (1992) Reprod. Fertil. Dev. 4, 399–410
34. Adamson, L.M., Podsiadly, B., Smart, Y.C., Stanger, J.D., and Roberts, T.K. (1991) Mol. Reprod. Dev. 30, 207–213
35. Clay, K.L., Johnson, C., and Henson, P. (1990) Biochem. Biophys. Acta, 1046, 309–314
36. Benveniste, J., Henson, P.M., and Cochrane, C.G. (1972) J. Exp. Med. 136, 1356–1377
37. Blank, M.L., Lee, T-C., Fitzgerald, V., and Snyder, F. (1981) J. Biol. Chem. 256, 175–178
38. Ludwig, J.C. McManus, L.M., and Pinckard, R.N. (1986) Advances in Inflammation Research, 11, ed. Otterness, I. pp. 111–125. New York: Raven Press
39. Ludwig, J.C., Hoppens, C., McManus, L.M., Mott, G.E., and Pinckard, R.N. (1985) Arch. Biochem. Biophys. 241, 337–347
40. Matsumoto, M., and Miwa, M. (1985) Advances in Prostaglandin, Thromboxane, and Leukotriene Research, 15, ed. Hayashi, O., and Yamamoto, S. pp. 705–706. New York: Raven Press
41. Miwa, M., Sugatani, J., Ikemura, T., Okamoto, Y., Ino, M., Saito, K., Suzuki, Y., and Matsumoto, M. (1992) J. Immunol. 148, 872–880
42. Benveniste, J., Nunez, D., Duriez, P., Korth, R., Bidault, J., and Fruchart, J-C. (1988) FEBS Letters, 26, 371–376
43. Korth, R., Zimmerman, K., and Richter, W.O. (1994) Chem. Phys. Lipids, 70, 109–119.
44. McNamara, P.J., Brouwer, K.R., and Gillespie, M.N. (1986) Biochem. Pharmacol. 35, 621–624
45. Brown, J.R. (1976) Fed. Proc. 35, 2141–2144
46. Carter, D.C., He, X-M., Munson, S.H., Twigg, P.D., Gernert, K.M., Broom, M.B., and Miller, T.Y. (1989) Science, 244, 1195–1198

44

ROLE OF PLATELET-ACTIVATING FACTOR IN PARTURITION

Hisashi Narahara,* Kiwamu Toyoshima,† and John M. Johnston

Departments of Biochemistry and
Obstetrics-Gynecology and
The Cecil H. and Ida Green Center for Reproductive Biology Sciences
The University of Texas
Southwestern Medical Center at Dallas
5323 Harry Hines Boulevard, Dallas, Texas 75235-9051

Platelet activating factor (PAF) has been implicated in a number of processes in reproductive biology including the synthesis and release of PAF by human spermatozoa,[1] enhancement of sperm motility,[2] ovulation[4,5] and penetration into the oocyte. PAF has also been detected in the uterine tissue of a number of species[6,7] and PAF receptors have been identified in endometrium,[8] and myometrium.[9] The relationship of PAF to implantation has received special attention. O'Neill and colleagues first reported[10] that PAF was secreted by the embryo (for review see[11]) and that the implantation rate was increased in proportion to the PAF secreted into the media. A positive correlation was found also between the pregnancy outcome and PAF secretion by the embryo.[11,12]

PAF is also involved in parturition. It was reported that the eicosanoid precursor, arachidonic acid, was increased in amniotic fluid during early labor.[13] Although approximately 50% of the arachidonic acid in amnion tissue was associated with the ethanolamine plasmalogens, the arachidonic acid was decreased during early labor only in the diacylphosphatidylethanolamine and phosphatidylinositol.[14] A model system was proposed to explain its selective release from these two phospholipids in which Ca^{2+} occupied a central role.[15] PAF facilitates the increase of Ca^{2+} in a number of systems.[16] The PAF concentration in the amniotic fluid was therefore examined. PAF was increased in the amniotic fluid obtained from women in active labor.[13] PAF was also shown to stimulate PGE_2 formation in amnion tissue.[17]

* Present address: Department of Ob/Gyn, Oita Medical University, 1-Hazama-cho, Oita-gun, Oita 879-55, Japan.
† Present address: Department of Ob-Gyn, Keio University, Tokyo, Japan.

1. ORIGIN OF PAF IN AMNIOTIC FLUID: THE ROLE OF THE FETAL LUNG

At term, amniotic fluid is enriched with surfactant, a lipoprotein complex synthesized by the type II pneumonocytes of the fetal lung and transported by fetal breathing to the amniotic fluid in the form of lamellar bodies. When the distribution of PAF between the lamellar bodies and the supernatant fractions was determined in amniotic fluid, approximately half of the PAF was associated with the lamellar body-enriched fraction.[13] The activities of several of the enzymes involved in PAF biosynthesis were assayed in lung tissue throughout gestation. Several of the critical enzymes involved in PAF biosynthesis were shown to increase several-fold late in gestation.[18,19] During fetal lung development, the type II pneumonocytes undergo a remarkable alteration, from cells rich in glycogen to cells that synthesize large quantities of dipalmitoylphosphatidylcholine (DPPC). It has been suggested[20] that lung glycogen may serve as the major source of the palmitate of DPPC in the type II pneumonocyte of the fetus.

The role of PAF in glycogen mobilization in fetal lung tissue was, therefore, assessed. PAF was injected into 24-day-old fetal rabbits in utero and it was found that the glycogen concentrations were reduced in association with the increase of DPPC in fetal lung.[18] PAF also caused a dose-dependent increase of surfactant secretion by isolated type II cells.[21]

2. PAF METABOLISM IN AMNION AND WISH CELLS

The communication between the fetus and mother may occur via the amniotic fluid. PAF was present in the amniotic fluid obtained from women in labor. We had also noted that the ethanolamine plasmalogen increased in amnion tissue during gestation.[22] Employing WISH cells (a cell line derived from amnion), it was found that PAF could be converted to ethanolamine phospholipids, especially the plasmalogen species.[23] In Figure 1 is the reaction sequence that has been established for this conversion of PAF to ethanolamine plasmalogens in these cells.

In the synthesis of PAF via the remodeling pathway from alkyl-acyl-glycerophosphocholine, the first step was thought to be the conversion of alkyl-acyl-GPC to lysoPAF. This reaction was originally thought to be catalyzed by a phospholipase A_2. Recently, a second pathway for the synthesis of lysoPAF has been demonstrated in the membrane fractions of platelets,[24] alveolar macrophages,[25,26] HL-60 cells,[27] and neutrophils,[28] in which the acyl group of alkyl-acyl-GPC is transferred to a lyso acceptor in a CoA-independent transacylation. In Figure 2 is illustrated an abbreviated form of the transacylase cycle. The plasmalogen of 2-lyso-phosphatidylethanolamine was one of the more active acyl acceptors. *sn*-2-Arachidonoyl-containing phosphatidylcholines were the most active donors in the transacylation reaction.[26]

This mechanism outlined in Figure 2 would explain the enrichment of the plasmalogen species of PE with arachidonic acid, since the transacylase is rather specific for polyunsaturated fatty acids.[29] Moreover, the relative specific release of arachidonic acid as mediated by the phospholipase A_2, is known to be present in amnion.[30] By this reaction, the necessary substrate for the synthesis of eicosanoids would be produced.[20] The transacylase activity was assayed, in a human amnion derived cell line, in the WISH cells. The results are illustrated in Figure 3. No differences in enzyme specificity was found for

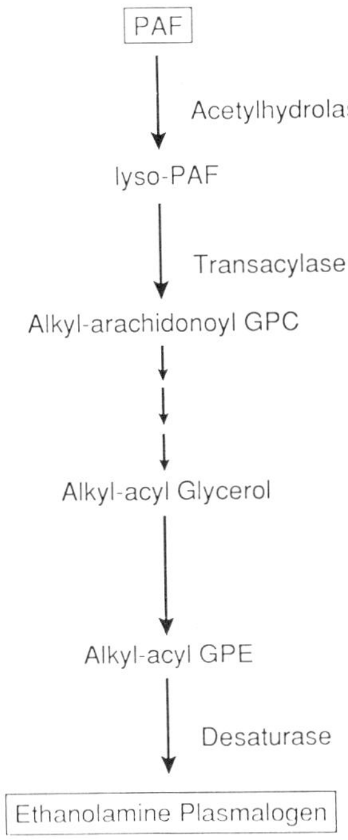

Figure 1. Proposed pathway for the conversion of PAF into ethanolamine plasmalogens.

the plasmalogen vs. diacyl species if both contained arachidonate in the *sn*-2 position.[30] The presence of lysoPAF:acetyl CoA acetyltransferase activity in amnion had been previously reported.[31] Thus, this tissue contains transacylase as well as all the necessary enzymes for the simultaneous formation of eicosanoids and PAF as illustrated in Figure 2.

The concentration of PAF may be altered by either its rate of biosynthesis and degradation or both. The enzyme that degrades and biologically inactivates PAF is PAF-acetylhydrolase (PAF-AH). The plasma activity of PAF-AH in the rabbit was assayed throughout pregnancy.[38] It was found that the activity decreased from a pre-insemination

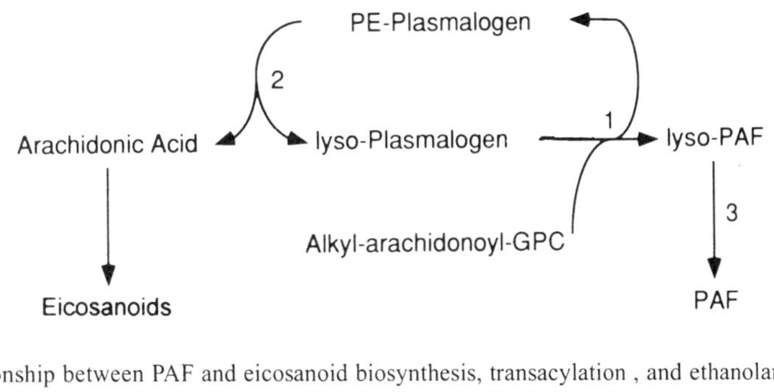

Figure 2. Relationship between PAF and eicosanoid biosynthesis, transacylation , and ethanolamine plasmalogen formation . Lyso-plasmalogen formed by the hydrolytic cleavage of ethanolamine plasmalogen. as catalyzed by a phospholipase A_2 (reaction 2) can be reacylated specifically with arachidonic acid by a coenzyme A-independent transacylase (reaction 1); the donor of arachidonic acid for this reaction is alkyl-arachidonoyl-GPC. The latter reaction will result in the production of lysoPAF that can be converted to PAF by lysoPAF:acetyl CoA acetyltransferase (reaction 3).

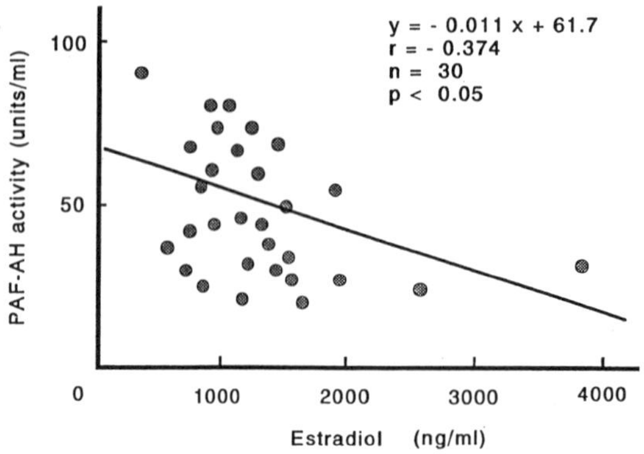

Figure 3.

value of about 130 nmol/min/mL of plasma to approximately 10 nmol/min/mL of plasma on or about the 29th day of pregnancy. The activity returned to control values following delivery. A decrease in the activity of the plasma PAF-AH during pregnancy have been observed in mice,[33] rats,[34] and humans.[35] It has been suggested that the decidua is a site of PAF inactivation, due to its abundant blood supply.[36] The prevention of PAF from reaching the myometrium would assure a quiescent uterus throughout most of gestation. The marked decrease in plasma PAF-AH activity that occurs late in gestation may result in a higher steady-state concentration of PAF, resulting in heightened myometrial contractility and, therefore, contribute to the initiation of parturition.

3. ROLE OF MACROPHAGES IN PAF METABOLISM

Decidual tissue contains an abundant macrophage population. The observation that differentiation of human peripheral monocytes into macrophages was accompanied by a increased secretion of PAF-AH[37] led us to examine macrophages as a source of PAF-AH. Therefore, a change in PAF-AH secreted by these cells may regulate the PAF present in the uterine cavity. The decidual macrophages secrete the plasma-type isozyme of PAF-AH.[38]

The inactivation of PAF in the decidua prior to the time of parturition may not only be due to the plasma PAF-AH, but also due to the PAF-AH secreted by decidual macrophages.

We have also described a number of factors including various endotoxins and cytokines that may alter PAF-AH secretion by decidual macrophages.[39] It was demonstrated that LPS inhibited the secretion of PAF-AH by decidual macrophages. LPS also stimulates the monocyte/macrophage system to induce the release of cytokines, such as tumor necrosis factor (TNF-α) and interleukin-1 (IL-1).. IL-1α, IL-1β, TNF-α, and interferon-γ (IFN-γ) also inhibited the release of PAF-AH by decidual macrophages. It has been reported that IL-1 receptor antagonist is normally present in human amniotic fluid,[40] and that INF-γ enhances production of TNF-α and IL-1; therefore, the positive feedback loop among TNF-α, IL1, and INF-γ may lead to an augmentation of the LPS-induced inflammatory response resulting in an elevated PAF level concentration in certain disease states, such as chorioamnionitis.

These findings provide further support for the suggestion that PAF not only participates in the events of normal labor, but may also be involved also in the pathogenesis of preterm labor or preterm rupture of membranes caused by endotoxins and the activation of the cytokine network.

4. REGULATION OF THE ENYZMES INVOLVED IN PAF METABOLISM

In view of the changes in PAF-AH activity described previously, and of the pronounced changes in maternal and fetal hormones that occur late in pregnancy, the effect of several steroid hormones on the enzyme activity was studied. It was found that 17α-ethynylestradiol administration to adult rats decreased the enzyme activity to one-fifth of the normal value.[41] Estriol, 17β-estradiol, and estrone also decreased the activity, but to a lesser extent. A prolonged elevation of the activity was observed in response to the injection of medroxyprogesterone. We have suggested that the progesterone-to-estrogen ratio that occurs late in gestation, may contribute to the events leading to the initiation of parturition by increasing the concentration of PAF. PAF-AH also may play a role in certain complications of pregnancy, such as premature delivery and pregnancy-induced hypertension.[35]

As an outgrowth of our studies on the hormonal regulation of PAF-AH, it was also observed that the administration of dexamethasone caused a three-fold increase in plasma PAF-AH activity. This effect of dexamethasone may account, in part, for the anti-inflammatory properties of glucocorticoids.

REFERENCES

1. E. Baldi, C. Falsetti, C. Krausz, G. Gervasi, V. Carloni, R. Casano, and G. Forti, Stimulation of platelet-activating factor synthesis by progesterone and A23187 in human spermatozoa, *Biochem. J.* 292:209 (1993).
2. D.D. Ricker, J.L. Robertson, B.S. Minhas, M.G. Dodson, and R. Kumar, The effects of platelet-activating factor on the motility of human spermatozoa, *Fertil. Steril.* 52:655 (1989).
3. A.O. Abisogun, P. Braquet, and A. Tsafriri, The involvement of platelet-activating factor in ovulation, *Science* 243:381 (1989).
4. X.-M. Li, N. Sagawa, Y. Ihara, A. Okagaki, M. Hasegawa, K. Inamori, H. Itoh, T. Mori, and C. Ban, The involvement of platelet-activating factor in thrombocytopenia and follicular rupture during gonadotropin-induced superovulation in immature rats, *Endocrinology* 129:3132 (1991).
5. B.S. Minhas, Platelet-activating factor treatment of human spermatozoa enhances fertilization potential, *Am. J. Obstet. Gynecol.* 168:1314 (1993).
6. K. Yasuda, K. Satouchi, and K. Saito, Platelet-activating factor in normal rat uterus, *Biochem. Biophys. Res. Commun.* 138:1231 (1986).
7. A.A. Alecozay, B.G. Casslen, R.M. Riehl, F.D. DeLeon, M.J.K. Harper, M. Silva, T.A. Nouchi, and D.J. Hanahan, Platelet-activating factor in human luteal phase endometrium, *Biol. Reprod.* 41:578 (1989).
8. G.B. Kudolo, and M.J.K. Harper, Estimation of platelet-activating factor receptors in the endometrium of the pregnant rabbit: Regulation of ligand availability and catabolism by bovine serum albumin, *Biol. Reprod.* 43:368 (1990).
9. Y.-p. Zhu, R.A. Word, and J.M. Johnston, The presence of PAF binding sites in human myometrium and its role in uterine contraction, *Am. J. Obstet. Gynecol.* 166:1222 (1992).
10. C. O'Neill, Partial characterization of embryo-derived platelet-activating factor in mice, *J. Reprod. Fertil.* 75:375 (1985)
11. C. O'Neill, PAF and the establishment of pregnancy, in: *Platelet-Activating Factor and Human Disease*, P.J. Barnes, C.P. Page, P.M. Henson, eds, Blackwell Scientific Publishing, Oxfore, p 282 (1989).

12. U. Punjabi, A. Vereecken, L. Delbeke, M. Angle, M. Gielis, J. Gerris, J. Johnston, and Ph Buytaert, Embryo-derived platelet-activating factor, a marker of embryo quality and viability following ovarian stimulation for *in vitro* fertilization. *J. In Vitro Fertil. Embryo Transfer* 7:321 (1990).
13. M.M. Billah, and J.M. Johnston, Identification of phospholipid platelet-activating factor (1-*O*-alkyl-2-acetyl-*sn*-glycero-3-phosphocholine, PAF) in human amniotic fluid and urine, *Biochem. Biophys. Res. Commun.* 113:51 (1983).
14. J.R. Okita, P.C. MacDonald, and J.M. Johnston, Mobilization of arachidonic acid from specific glycerophospholipids of human fetal membranes during early labor, *J. Biol. Chem.* 257:14,029 (1982).
15. J.E. Bleasdale, and J.M. Johnston, Prostaglandins and human parturition: Regulation of arachidonic acid mobilization, in *Reviews in Perinatal Medicine*, vol. 1, Alar R. Liss, New York, p 789 (1982).
16. F. Snyder, Chemical and biochemical aspects of platelet-activating factor: A novel class of acetylated ether-linked choline-phospholipids, *Med. Res. Rev.* 5:107 (1985).
17. C. Morris, H. Khan, M.H.F. Sullivan, and M.G. Elder, Effects of platelet-activating factor on prostaglandin E_2 production by intact fetal membranes, *Am. J. Obstet. Gynecol.* 166:1228 (1992).
18. D.R. Hoffman, C.T. Truong, and J.M. Johnston, Metabolism and function of platelet-activating factor in fetal rabbit lung development, *Biochim. Biophys. Acta* 879:88 (1986).
19. D.R. Hoffman, C.T. Truong, and J.M. Johnston, The role of platelet-activating factor in human fetal lung metabolism, *Am. J. Obstet. Gynecol.* 155:70 (1986).
20. K. Toyoshima, H. Narahara, M. Furukawa, R.A. Frenkel, and J.M. Johnston, Platelet-activating factor. Role in fetal lung development and relationship to normal and premature labor, *Clinics in Perinatology* 22:263 (1995).
21. H. Eguchi, R.A. Frenkel, and J.M. Johnston, Binding and metabolism of platelet-activating factor (PAF) by isolated rat type Ii pneumonocytes, *Arch Biochem. Biophys.* 308:426 (1994).
22. J.M. Johnston, and N. Maki, PAF and fetal development, in *PAF in Health and Disease*, P.J. Barnes, C.P. Page, P.C. Henson, eds, Blackwell Scientific Publishers, Oxford, p 297 (1989).
23. R.A. Frenkel, and J.M. Johnston, Metabolic conversion of platelet-activating factor into ethanolamine plasmalogen in an amnion-derived cell line, *J. Biol. Chem.* 267:19186 (1992).
24. R.M. Kramer, and D. Deykin, Arachidonoyl transacylase in human platelets. Coenzyme A-independent transfer of arachidonate from phosphatidylcholine to lysoplasmenylethanolamine, *J. Biol. Chem.* 258:13806 (1983).
25. M. Robinson, M.L. Blank, and F. Snyder, Acylation of lysophospholipids by rabbit alveolar macrophages, *J. Biol. Chem.* 260:7889 (1985).
26. T. Sugiura, Y. Masuzawa, Y. Nakagawa, and K. Waku, Transacylation of lyso-platelet-activating factor and other lysophospholipids by macrophage microsomes, *J. Biol. Chem.* 262:1199 (1987).
27. Y. Uemura, T.-c. Lee, and F. Snyder, A coenzyme A-independent transacylase is linked to the formation of platelet-activating factor (PAF) by generating the lyso-PAF intermediate in the remodeling pathway, *J. Biol. Chem.* 266:8268 (1991).
28. M.L. Nieto, M.E. Venable, S.A. Bauldry, D.G. Greene, M. Kennedy, D.A. Bass, and R.L. Wykle, Evidence that hydrolysis of ethanolamine plasmalogens triggers synthesis of platelet-activating factor via a transacylation reaction, *J. Biol. Chem.* 266:18699 (1991).
29. F. Snyder, The role of transacylases in the metabolism of arachidonate and platelet-activating factor, *Prog. Lipid Res.* 31:65 (1992).
30. K. Toyoshima, H. Narahara, R.A. Frenkel, and J.M. Johnston, Coenzyme A-independent transacylation in amnion-derived (WISH) cells, *Arch. Biochem. Biophys.* 314:224 (1994).
31. M.M. Billah, G.C. Di Renzo, C. Ban, C.T. Truong, D.R. Hoffman, M.M. Anceschi, J.E. Bleasdale, and J.M. Johnston, Platelet-activating factor metabolism in human amnion and the responses of this tissue to extracellular platelet-activating factor, *Prostaglandins* 30:841 (1985).
32. N. Maki, D.R. Hoffman, and J.M. Johnston, PAF (platelet-activating factor) acetylhydrolase activities in maternal, fetal, and newborn rabbit plasma during pregnancy and lactation, *Proc. Natl. Acad. Sci. USA* 85:728 (1988).
33. A.A. Saleh, M.W. Church, and J.M. Johnston, The effect of alcohol on platelet-activating factor acetylhydrolase-activity in pregnant and nonpregnant mice, *Alcohol Clin, Exp. Res.* 18:1009 (1994).
34. K. Yasuda, H. Eguchi, H. Narahara, and J.M. Johnston, Platelet-activating factor: Its regulation in parturition, in *Eicosanoids and Other Bioactive Lipids in Cancer, Inflammation and Radiation Injury*, S. Nigam, K.V. Honn, L.J. Marnett, *et al.*, eds, Kluwer Academic, Norwell MA, p 727 (1993).
35. N. Maki, R.R. Magness, S. Miyaura, N.F. Gant, and J.M. Johnston, Platelet-activating factor acetylhydrolase activity in normotensive and hypertensive pregnancies, *Am. J. Obstet. Gynecol.* 168:50 (1993).

36. J.M. Johnston, R.A. Frenkel, and H. Narahara, Platelet-activating factor metabolism in intrauterine tissue, in *Molecular Aspects of Placental and Fetal Membrane Autacoids*, G.E. Rice, S.P. Brennecke, eds, CRC Press, Boca Raton, p 97 (1993).
37. M.R. Elstad, D.M. Stafforini, T.M. McIntyre, S.M. Prescott, and G.A. Zimmerman, Platelet-activating factor acetylhydrolase increases during macrophage differentiation: A novel mechanism that regulates accumulation of platelet-activating factor, *J. Biol. Chem.* 264:8467 (1989).
38. H. Narahara, Y. Nishioka, and J.M. Johnston, The role of platelet-activating factor in parturition: Secretion of platelet-activating factor acetylhydrolase by human decidual macrophages, *J. Clin. Endocrinol. Metab.* 77:1268 (1993).
39. H. Narahara, and J.M. Johnston, Effects of endotoxins and cytokines on the secretion of platelet-activating factor-acetylhydrolase by human decidual macrophages. *Am. J. Obstet. Gynecol.* 169:531 (1993).
40. R. Romero, W. Sepulveda, M. Mazor, F. Brandt, D.B. Cotton, C.A. Dinarello, and M.D. Mitchell, The natural interleukin-1 receptor antagonist in term and preterm parturition, *Am. J. Obstet. Gynecol.* 167:863 (1992).
41. S. Miyaura, N. Maki, W Byrd, and J.M. Johnston, The hormonal regulation of platelet-activating factor-acetylhydrolase activity in plasma, *Lipids* 26:1915 (1991).

45

THE ROLE OF PLATELET-ACTIVATING FACTOR AND ITS RECEPTOR IN ENDOMETRIAL RECEPTIVITY

Asif Ahmed and Sharon Dearn

Reproductive Physiopathology Group
Department of Obstetrics and Gynaecology
The Medical School, University of Birmingham
Edgbaston, Birmingham, B15 2TT, United Kingdom

1. PLATELET-ACTIVATING FACTOR AND IMPLANTATION

1.1 Background

Successful mammalian pregnancy requires the preparation of a receptive endometrium into which the embryo attaches and invades. Disorders of implantation cause infertility and may be the cause of life-threatening complications in pregnancy[1]. Impaired implantation is a major cause of peri-implantation embryonic loss[2]; about 30 % of all pregnancies end before they are clinically detected[3]. Abnormalities might arise in the preparation of the endometrium resulting in failed adhesion or abnormal development of pre or post implanted embryos.

Implantation is characterised by an inflammatory-type response with expansion of extracellular fluid volume, increased vascular permeability and vasodilatation[4,5]. Morphological and animal studies show that the concentration of prostaglandins (PG) E_2, $PGF_{2\alpha}$ and 6-keto-$PGF_{1\alpha}$ are high at implantation sites[6]. PGE_2 induces changes in local vascular permeability and stromal oedema which accompany implantation[7]. Indomethacin, an inhibitor of PG synthesis, prevents implantation in the mouse[8], rabbit[9] and rat[10].

1.2 Embryo-Derived PAF

There is reasonable evidence to suggest that platelet-activating factor (PAF), an inflammatory mediator with potent vasodilatory and vascular permeability properties, may be the first physiological signal produced by the embryo for the maternal recognition of pregnancy. Mice and human pre implantation embryos synthesise PAF[11,12]. Culture fluid from mice embryos induces platelet activation and thrombocytopenia, biochemical and pharmacological characterisation revealing the active agent capable of inducing systemic platelet depletion to be homologous to synthetic PAF[13], and in both species transient thrombocytopenia occurs in the mother in early pregnancy. Addition of PAF to culture medium increases the metabolic rate, cleavage rate and implantation potential of mouse embryos cultured *in vitro* [14,15]. Studies in women confirm these findings[16,17]. Murine and rat implantation is abolished by PAF antagonists[18,19], a finding which could not however be reproduced in the rabbit[20], and successful implantation of human embryos broadly correlates with their ability to synthesise PAF in culture[12]. Until recently the embryo was the main focus of attention with respect to PAF, however the finding that PAF stimulates phospholipase C (PLC) activity only in the secretory endometrium has led to the proposal that the principal role of PAF lies in the preparation of a receptive endometrium for implantation[21].

1.3 PAF Production in the Endometrium

PAF is not only produced by the embryo during early pregnancy, but also by the stroma of human secretory endometrium where its levels are hormonally regulated and progesterone and PGE_2 augment these levels[22,23]. Uterine PAF production is highest just prior to implantation and drops in areas of the uterus adjacent to the embryo at the time of implantation[24]. PAF causes a dose-dependant increase in the synthesis of PGE_2 by enriched glandular, but not stromal, fractions removed in the secretory phase of the menstrual cycle[25]. In light of the requirement for local elevation of endometrial prostaglandin synthesis at the site of implantation it is likely that the cellular mechanisms whereby PAF enhances prostaglandin biosynthesis may play a pivotal role in local cellular events at the time of implantation[26].

2. INTRACELLULAR REGULATION BY PAF

2.1 PAF Receptors in the Endometrium

PAF and other calcium (Ca^{2+}) mobilising agonists transmit their intracellular messages by binding to specific receptors on the cell surface. The ligand-bound receptors activate effector systems such as phospholipases including phospholipase A_2, phospholipase C, and phospholipase D via a receptor coupled G-protein to generate second messengers. Autoradiographic binding studies in the rabbit have shown binding of PAF to endometrial epithelium and to the embryonic disc of Day 6 blastocysts[27], however the presence of a specific PAF receptor and its precise location in specific cells of the human endometrium has not previously been determined. Molecular cloning of the PAF receptor from a human leukocyte cDNA library has identified a seven transmembrane spanning G protein-linked receptor[28]. By using RT-PCR, mRNA encoding the PAF receptor has been detected in both proliferative and secretory endometrium and in a transformed cell line derived from human endometrial epithelium, HEC-1B[29] (Figure 1), a cell line which has been used to generate substantial data concerning the role of PAF in endometrium. The expression of PAF receptor mRNA in the epithelial cell line suggests that PAF receptors may be located on the epithelial cells in normal human endometrium. This has recently been confirmed by *in situ* hybridisation (Figure 2) which showed that PAF receptor mRNA is localised predominantly in the glandular epithelium of secretory endometrium (unpublished data) thereby providing justification for the choice of the HEC-1B cell line for studying PAF receptor expression and function in human endometrium. Weak signal was observed in the proliferative endometrium suggesting that PAF receptor expression may be under ovarian steroid regulation.

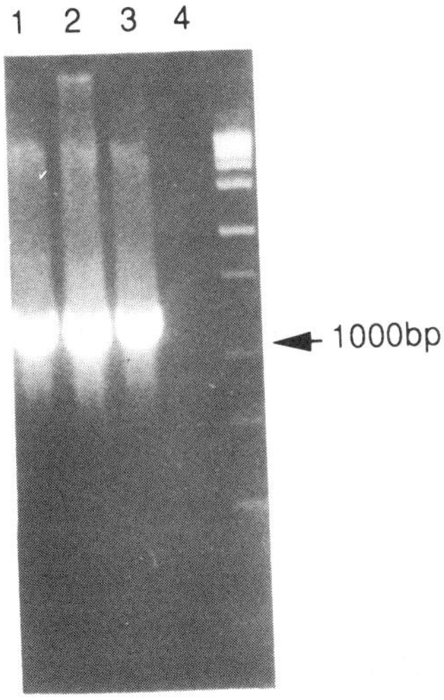

Figure 1. Reverse transcription polymerase chain reaction (RT-PCR) for PAF receptor. Samples of total RNA extraceted from HEC-1A and HEC-1B cell lines, from lymphocytes from blood and from endometrium throughout the menstrual cycle were subjected to nested RT-PCR using PAF receptor specific primers as previously described[29]. Control reaction were with no input cDNA. A single band of 1150 bp can be seen. Sequencing confirmed that this encodes the PAF receptor.

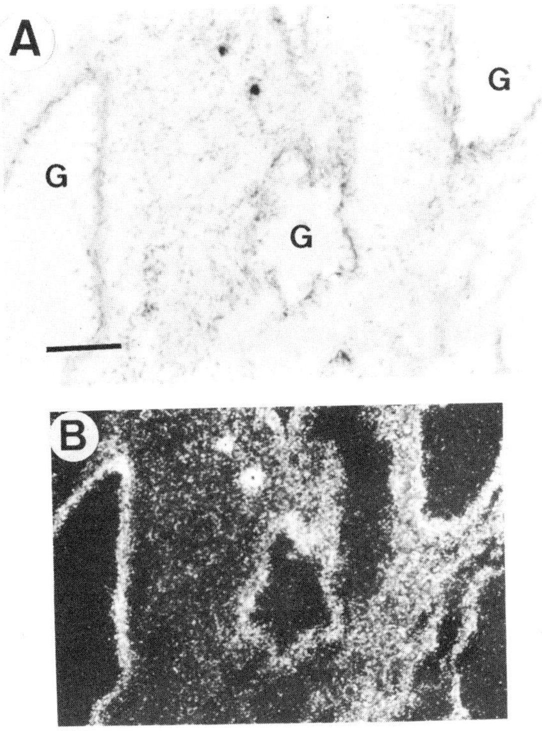

Figure 2. Localisation of PAF receptor in secretory endometrium. Bright-field (A) and dark-field (B) photomicrographs of 10 μm cryostat sections of endometrium after *in situ* hybridisation with ^{35}S-labelled antisense PAF receptor cRNA probe. The hybridisation signal is shown by white silver grains in darkfield view. The PAF receptor gene was expressed mainly by the glandular epithelium. Scale bar= 100 μm.

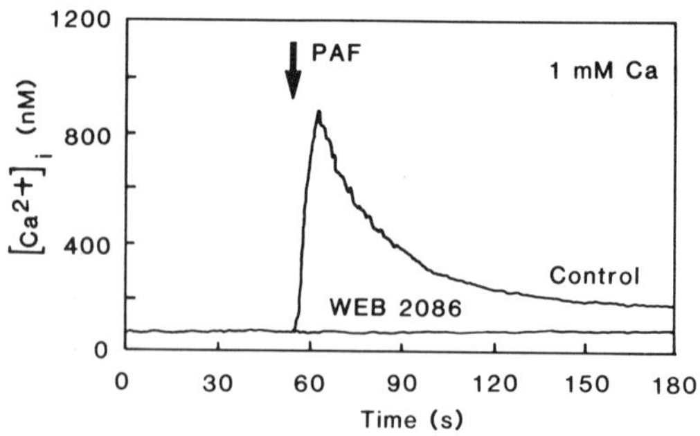

Figure 3. Effect of the PAF antagonist on PAF-evoked rise in intracellular Ca^{2+} concentration in HEC-1B cells. Fura-2-loaded cells were stimulated with 10 nM PAF in the presence of 1 mM external Ca^{2+} after preincubation with 10 μM WEB 2086 for 2 min (WEB 2086) or the vehicle alone (control).

2.2 Stimulation of Uterine PLC and PtdIns 4,5 P_2 Hydrolysis

The hydrolysis of phosphatidylinositol 4,5 bisphosphate (PtdIns 4,5 P_2) by PtdIns 4,5 P_2-specific phospholipase C yields two second messengers, 1,2-diacylglycerol (DAG), the main physiological activator of phosphatidylserine-dependant protein kinase C, and inositol 1,4,5-trisphosphate (Ins(1,4,5)P_3). Ins(1,4,5)P_3 induces an elevation of intracellular calcium, $[Ca^{2+}]_i$, required for Ca^{2+}-dependent processes[30]. Expression of functional PAF receptors in HEC-1B cells was demonstrated by the ability of PAF to stimulate PLC activity and Ca^{2+} mobilisation since both processes were inhibited by WEB 2086 (Figure 3). In HEC-1B cells, the formation $[^3H]$Ins(1,4,5)P_3 in response to PAF appears to be mediated by a pertussis toxin-insensitive G-protein as is the case for PAF-evoked $[Ca^{2+}]_i$ or inositol phosphate synthesis in U-937 cells[31]. In many tissues, including human[32] and guinea-pig[33] endometrium, phospholipase A_2 activity is Ca^{2+}-dependent and therefore influenced by the release of Ca^{2+} from intracellular stores. Ca^{2+}-stimulation of phospholipase A_2 activity after exposure to PAF represents an additional potential source of arachidonic acid for prostaglandin synthesis. Thus PAF-induced prostaglandin synthesis may arise either by the release of arachidonic acid from DAG or by increased phospholipase A_2 activity.

2.3 Stimulation of PLD and PtdCho Hydrolysis

PAF-stimulated PtdIns 4,5 P_2 hydrolysis is a transient process in human endometrial cells (Figure 4). Sustained DAG formation from phospholipase D (PLD)-catalysed phosphatidylcholine (PtdCho) hydrolysis may cause prolonged activation of PKC which regulates events such as proliferation and secretion[34]. PAF has been shown to stimulate the breakdown of PtdCho by multiple phospholipase pathways in endometrial explants during the secretory phase of the menstrual cycle[35]. The generation of $[^3H]$choline in response to stimulation with PAF in $[^3H]$choline-labelled endometrial explants preceded $[^3H]$glycerophosphocholine ($[^3H]$GPC) and $[^3H]$phosphocholine formation suggesting PLD-catalysed PtdCho breakdown. Stimulation of PLD-catalysed PtdCho hydrolysis was confirmed by PAF-evoked $[^3H]$Phosphatidylbutanol ($[^3H]$PBut) formation via the transphosphatidylation reaction, a reaction unique to the activation of PLD. $[^3H]$PBut formation in response to PAF was rapid, dose-related and attenuated by the specific PAF receptor antagonist indicating a receptor mediated mechanism[26]. Although the rise in intracellular $[^3H]$choline and $[^3H]$PBut formation were similar, suggesting that the initial route of hydrolysis is PLD-catalysed PtdCho breakdown, PAF-stimulated $[^3H]$PBut accumulation was found to be kinetically downstream of inositol phosphate formation in human endometrium[36], a result in

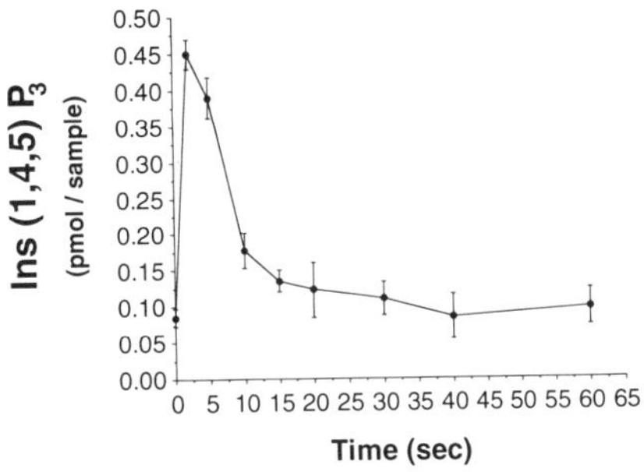

Figure 4. Kinetics of PAF-stimulated Ins(1,4,5)P_3 formation in endometrial glands. Cells were stimulated with 100 nM PAF for defined periods and Ins(1,4,5)P_3 produced was assayed by competitive displacement of [^{3}H]Ins(1,4,5)P_3 binding to bovine adrenal-cortex microsomes, quantified by using an Ins(1,4,5)P_3 standard curve.

agreement with that obtained in other systems suggesting that in the endometrium PLC is activated before PLD. The slower rise observed in intracellular [^{3}H]phosphocholine in response to PAF in [^{3}H]choline-labelled endometrial explants may be due to the activation of a phosphatidylcholine specific-PLC, but could also be due to phosphorylation of liberated [^{3}H]choline by choline kinase[37], while the delayed increase in intracellular [^{3}H]GPC may be attributed to sequential activation of phospholipase A_2 and lysophospholipase, the enzymes responsible for the sequential deacylation of PtdCho[38]. Activation of phospholipase A_2 activity after exposure to PAF represents a potential source of arachidonic acid for prostaglandin synthesis from PtdCho breakdown. Furthermore, lysophosphatidylcholine formed by the activation of phospholipase A_2 may ensure local cell fusion and invasion of the endometrium at the site of implantation[39].

As mentioned earlier, and in contrast to PAF receptor distribution, the responsiveness of the endometrium to PAF appears to be confined to the secretory phase of the menstrual cycle, providing further support for the hypothesis that PAF-mediated responses in the endometrium are under ovarian steroid regulation. Furthermore, PAF production in endometrium has been shown to be regulated by ovarian steroids[22]. Pretreatment of HEC-1B cells with 17β–oestradiol enhances PAF-evoked PLD activity (Figure 5) while leaving PAF-evoked PLC activity, as measured by [^{3}H]Ins1,4,5)P_3 accumulation, unaffected (Fig-

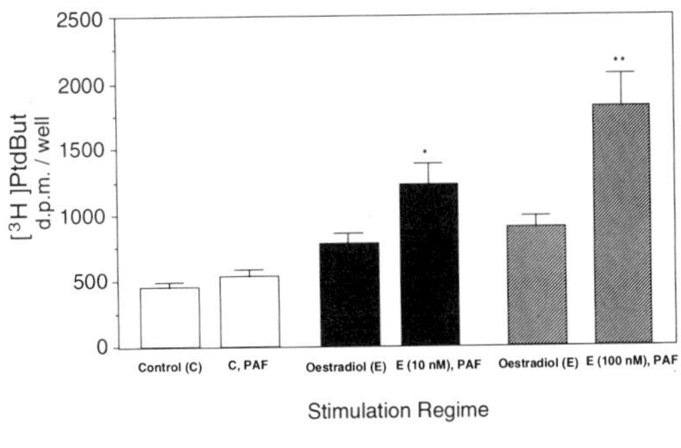

Figure 5. Effect of 17β-oestradiol (E) on PAF-stimulated phospholipase D activity in HEC-1B cells. Untreated cells (open column) and cells pre-treated with 10 nM 17β-oestradiol (solid column) and 100 nM 17β-oestradiol (hatched column) were pre-labelled with [^{3}H]myristic acid for 48 h at 37°C. Cells were stimulated with 100 nM PAF for 20 min in the presence of 30 mM butan-1-ol. Each point represents the mean ± s.e.m. of triplicate determination of three separate experiments. Statistical analysis was performed using Student paired t-test. *$p<0.05$ and **$p<0.02$ when compared with 17β-oestradiol-pretreated cells under unstimulated condition.

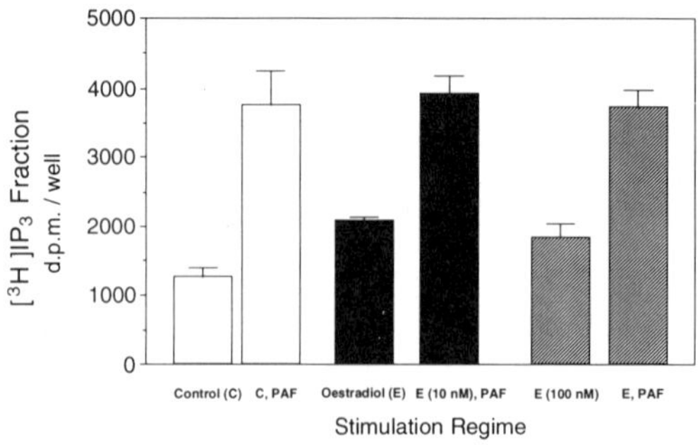

Figure 6. Effect of 17β-oestradiol on PAF stimulated [^{3}H]inositol trisphosphate accumulation in HEC-1B cells. Untreated cells (open column) and cells pre-treated with 10 nM 17β-oestradiol (solid column) and 100 nM 17β-oestradiol (hatched column) were pre-labelled with [^{3}H]inositol for 48 h at 37 °C. Cells were stimulated with 100 nM PAF for 20 min in the presence of 10 mM LiCl. Each point represents the mean ± s.e.m. of triplicate determination of three separate experiments.

ure 6). This response may by due to increased PKC activity since 17β–oestradiol also enhances phorbol ester (TPA)-stimulated PLD activity. That the effect of TPA is mediated by increased C-kinase activity is supported by the lack of effect of β-phorbol. This effect, however, is abolished at high concentrations of 17β-oestradiol (Figure 7) suggesting a different mechanism of activation.

There are a number of possible mechanisms for the induction of PAF-evoked phospholipase D in ovarian steroid-treated cells. As the steroid hormones have no effect on PAF-evoked phospholipase C activation, it is possible that they may upregulate G-protein coupling to phospholipase D but not phospholipase C. Another possibility is that the steroids induce the synthesis of factors which in turn modulate phospholipase D activity. Recently, it was observed that 17β-oestradiol increased the mRNA of a specific isoform of the α-isoenzyme of phospholipase C in the rat brain and uterus[40] suggesting that the modi-

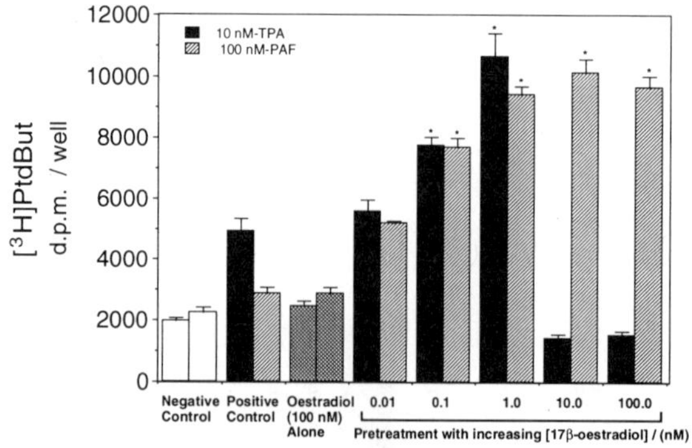

Figure 7. Effect of 17β-oestradiol on PAF- and TPA-stimulated phospholipase D activity in HEC-1B cells. Cells were pre-labelled with [^{3}H]myristic acid and the cells were stimulated as indicated in the presence of 30 mM butan-1-ol for 15 min with vehicle (open column), 10 nM TPA (solid column) or 100 nM PAF (hatched column) in the absence and following pretreatment with increasing concentrations of 17β-oestradiol for 30 min at 37°C. Each point represents the mean ± s.e.m. of triplicate determination of two separate experiments. Statistical analysis was performed using Student paired t-test. *$p<0.05$ when compared with TPA- and PAF-evoked [^{3}H]phosphatidylbutanol values of untreated cells.

fication of the PtdIns(4,5)P_2 signal transduction pathway by steroid hormones can occur at the pre-translational level. Steroid hormone action can also induce the synthesis of peptide factors which in turn modulate PtdIns(4,5)P_2 metabolism[41]. Such mechanisms may also prevail for phospholipase D. As phospholipase D activation appears to be down-stream to the activation of protein kinase C[36,42], the most likely explanation may be that 17β-oestradiol upregulates protein kinase C activity thereby inducing phospholipase D activity in these cells. Such a mechanism would be consistent with increased PKC activity[43] and translocation of PKC from cytoplasm to membrane[44] evoked by 17β-oestradiol in isolated plasma membranes.

2.4 PAF-Stimulated Protein Tyrosine Phosphorylation

PAF has also been shown to stimulate the activation of the tyrosine kinase $pp60^{c\text{-}src}$ in rabbit platelets causing its rapid translocation from the cytosol to the membranes[45]. In HEC-1B cells PAF stimulates the tyrosine phosphorylation of two major proteins of approximately 80KDa and 42–44KDa (Figure 8) Although these proteins remain to be identified, studies have suggested that pp44 is isoform of mitogen-activated protein (MAP) kinase[46]. This enzyme family has been shown to play a pivotal role in the regulation of cell division in a number of cell types[47,48]. PAF also stimulates the tyrosine phosphorylation of pp80. In other cells this protein has been shown to be strongly activated by classical G-protein-linked agonists and phorbol myristate acetate (PMA)[49]. It is unclear if the activation of pp80 is controlled by a PKC-dependent kinase upstream of pp80 or if pp80 itself contains sites which are regulated by phosphorylation upon serine residues[50]. Taken together these results suggests that functional PAF receptors may play a role in the regulation of growth and division in endometrial cell by coupling to a tyrosine kinase pathway.

In addition to MAP kinase, a number of tyrosine kinase substrates such as focal adhesion kinase and cytoskeletal proteins may also be regulated by phosphorylation of tyrosine residues[50]. As PAF stimulates tyrosine phosphorylation of a number of proteins, it is reasonable to speculate that PAF may also regulate endometrial cell motility and adhesion by activation of this pathway. PAF antagonists have been shown to inhibit trophoblast outgrowth *in vitro* in a dose-dependent manner[51] suggesting that PAF may be involved in the

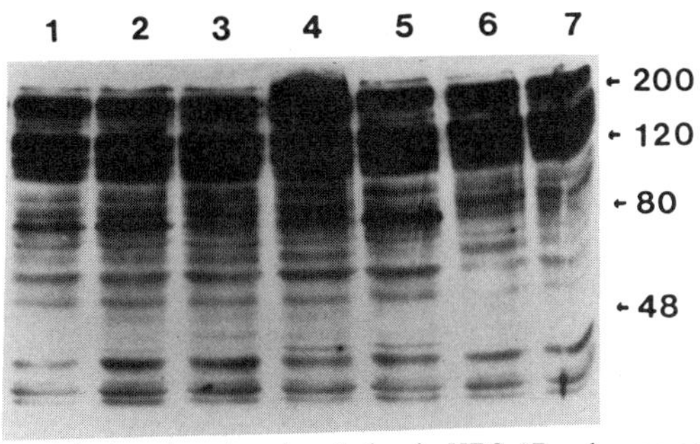

Figure 8. Agonist-evoked protein tyrosine phosphorylation in HEC-1B, a human endometrial epithelial cell line. Cells were stimulated in a time-dependent fashion with 100 nM PAF for 5 min and 15 min (Lane 1 and 2), with 20 ng/ml epidermal growth factor (EGF) for 5 min and 10 min (Lane 3 and 4), with 100 nM PMA for 5 min (Lane 5) and with DMSO 0.05% (Lane 6 and 7). Cell extracts were assayed for phosphotyrosine contents as outlined in the Materials and Methods. This experiment is a representation of three separate experiments. Reprobing the anti-phosphotyrosine blot with anti-MAP kinase antibody confirmed the identify of the band as the 42–44 KDa form of this kinase (data not shown).

activation of blastocyst attachment and the resulting differentiation of the trophectoderm into invasive trophoblast. PAF receptor mRNA has recently been localised by *in situ* hybridisation and was predominantly localised in the syncytiotrophoblast and cytotrophoblast bilayer of first trimester placenta further suggesting a role for PAF in early pregnancy and possibly in trophoblast growth and differentiation (unpublished data).

2.5 PAF and the NO Signalling System

Recent studies have identified a further intracellular signalling system which may mediate the actions of PAF in a variety of cell types. PAF has previously been shown to dilate some vascular beds and constrict others[52,53], however the mechanisms by which it produces these effects have, until now, been elusive. It has since been shown that PAF-induced dilatation of the mesenteric arterial bed is attenuated by detergent treatment[53], suggesting the requirement of an intact endothelium and a role for endothelium-derived relaxing factor/nitric oxide (NO) in PAF-induced vasodilatation. Similarly, PAF-induced vasodilatation in the rabbit afferent arteriole has been shown to be highly dependant upon intact endothelium-derived NO synthesis[54]. Consistent with this view, PAF has been reported to stimulate the release of NO in cultured endothelial cells derived from the bovine renal glomerulus[55]. The actions of NO with respect to PAF however extend beyond its vasodilator potential. Recent findings suggest an important role for NO in the regulation of albumin extravasation. Inhibition of endogenous NO synthesis potentiates PAF-elicited albumin extravasation in selected beds of the conscious rat[56]. Furthermore, it has been reported that NO may be a normal inhibitor of cell proliferation *in vivo* [57,58] since low levels of NO have been shown to inhibit the growth of mesangial cells[57] and vascular smooth muscle[58].

Recent work reported the presence of an L-arginine-NO-cyclic guanosine monophosphate system in rat uterus[59], suggesting that locally synthesised NO may play a role in

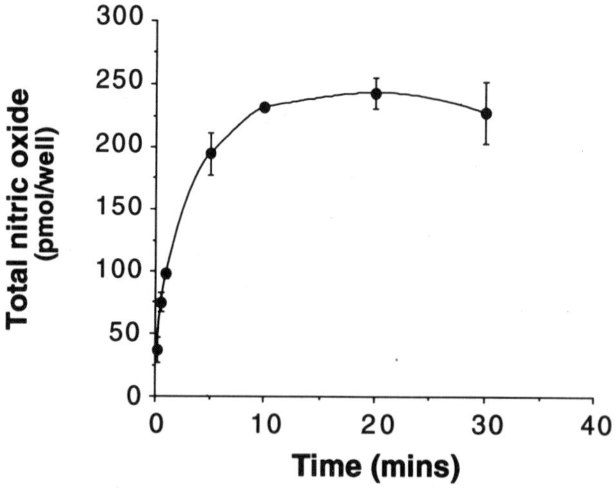

Figure 9. Time course of platelet-activating factor (PAF)-mediated NO release. HEC-1B cells were plated at a density of 300 000 cells ml^{-1} in 24 well plates and grown to 80% confluency in Dulbecco's modified Eagle's medium (DMEM) containing 10% fetal calf serum. Following serum starvation, confluent monolayers were stimulated with 1 nM PAF for the times indicated. Reactions were terminated by aspiration of the conditioned media, which was subsequently stored at -70 °C pending NO analysis using a Sievers NOA 270B chemiluminescent analyser. The results are expressed as means±s.e.m. of a typical experiment (triplicate determinations per experiment; 4 similar experiments).

the control of the uterine vascular bed. Animal studies support such a role for NO, attenuation of oestrogen-induced uterine vasodilatation by inhibition of NO synthesis is observed in the sheep[60], while PGE_2-induced vasodilatation in the uterine artery is mediated by the release of NO[61] in the dog. Nitric oxide synthase (NOS) mRNA and protein have recently been identified in the human uterus[62], and although the production of NO in human endometrium has not yet been addressed, PAF has been shown to stimulate the release of NO from human endometrial epithelial cells (unpublished work). NO generation in response to PAF was rapid and consistent with activation of a constitutive isoform of NOS, reaching a peak within 10 minutes of stimulation (Figure 9). PAF-mediated NO release was attenuated both by the specific PAF receptor antagonist, WEB 2170, indicating a receptor-mediated mechanism, and by N^G-monomethyl-L-arginine (L-NMMA), an L-arginine analogue, indicating NOS activation (Figure 10).

In light of these observations, and in view of the vasodilator potential of NO, it is likely that uterine-derived NO augments the vasodilatory effect of PAF, thereby supporting our hypothesis that the fundamental role of PAF in the uterus lies in the preparation of a receptive endometrium for implantation. Nitric oxide synthase is a highly regulated enzyme, the primary amino acid sequence contains consensus sequence binding sites for nicotinamide adenine dinucleotide phosphate (NADPH), flavin adenine dinucleotide (FAD), flavin mononucleotide (FMN) and calmodulin, as well as sites for protein phosphorylation by cAMP-dependant protein kinase A, PKC and calcium/calmodulin protein kinase II[63,64]. These recognition sites not only provide multiple means for regulation of NO levels, but also a means for 'cross communication' between the phosphoinositide and NO signalling systems, thereby co-ordinating the effects of PAF in the endometrium.

3. PAF IN PREGNANCY

Considerable information points to an essential role for PAF in the implantation process. A critical role for this autocoid in the initiation and maintenance of parturition has also

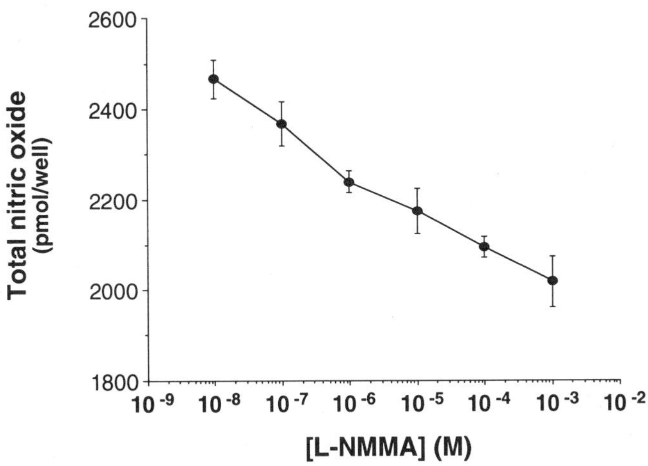

Figure 10. Effect of inhibition of NO synthesis by L-NMMA on PAF-mediated NO release. HEC-1B cells were stimulated for 30 min with 1 nM PAF in the presence of increasing concentrations of L-NMMA. Basal NO release was 2150 pmol/well and PAF-mediated NO release 2536 pmol/well.

been demonstrated. The first suggestion that PAF may be involved in parturition was based upon the observation that PAF was present in the amniotic fluid of women at term and in active labour[65] though not in the fluid obtained from women at term who were not in labour, observations which were later confirmed[66,67]. A significant proportion of this PAF was shown to be associated with surfactant-enriched lamellar bodies, implicating the fetal lung as the source. Furthermore, PAF concentrations in amniotic fluid from women destined for preterm delivery as a consequence of premature rupture of membranes and preterm labour, were found to be several fold higher[68]. PAF has been shown to stimulate PGE_2 production in amnion tissue and cells[69,70] and it has also been documented that PAF can stimulate myometrial contraction in rats[66], guinea pigs[71] and humans[72]. Moreover, a PAF receptor has been identified and characterised in the plasma membranes of human myometrium[73] and shown to have characteristics similar to those reported for other tissues[74]. Prolongation of parturition has been accomplished by administration of the PAF receptor antagonist, L659,989 to pregnant rats[75], providing further support for a role of PAF in parturition, however failure to block parturition completely in this manner, suggests the involvement of multiple mechanisms.

PAF metabolism in both fetal and maternal compartments is likely to be of importance throughout gestation. An elevated PAF biosynthetic capacity and content is observed in several fetal tissues during the later stages of gestation[76,77], while the specific activity of PAF acetylhydrolase, the major enzyme involved in the inactivation of PAF, is decreased to 10% of the non-pregnant level in maternal rabbit plasma during late pregnancy[78], returning to pre-pregnancy levels within 24 hours of delivery. A study in women confirms these findings[79]. Such observations have led to the suggestion that the combination of increased PAF synthesis in fetal compartments, in conjunction with the decreased degredative capacity of the maternal compartment, may promote PAF delivery, as well as that of other autocoids such as the prostaglandins, to the myometrium where they can then act to stimulate myometrial contraction[80]. PAF acetylhydrolase secretion has been demonstrated by human decidual macrophages and the presence of autocrine or paracrine regulation of PAF concentration in decidual tissue suggested[81]. Furthermore, it has been suggested that PAF is metabolised to ethanolamine plasmalogens, a rich source of arachidonic acid, in amnion[82]. At term in labour, released arachidonate, as a result of PLA_2 action, may then be used in the synthesis of eicosanoids, while the resulting lysoplasmalogens may then stimulate PAF synthesis via the remodelling pathway[83,84].

Although these studies implicate PAF in the initiation and maintenance of parturition, its precise role in these processes remains controversial, particularly in light of recent findings in which, contrary to earlier studies, PAF failed to stimulate the contraction of either labouring or non-labouring human pregnant myometrium (unpublished work), questioning a role for this autocoid in the initiation of parturition in humans. Further studies are required to elucidate its precise physiological role in pregnancy.

REFERENCES

1. Reginald PW, Beard RW, Chapple J, Forbes PB, Liddel HS, Mowbray JF and Underwood JL (1987) Outcome of pregnancies progressing beyond 28 weeks gestation in women with a history of recurrent miscarriage British Journal of Obstetrics and Gynaecology 94 643–648.
2. Yovich JL and Matson PL (1988) Early pregnancy wastage after gamete manipulation British Journal Obstetrics and Gynaecology 95 1120–27.
3. Wilcox AJ, Weinberg CR, O'Connor JF, Baird DD, Schlatterer JP, Canfield RE, Armstrong EG and Nisula BC (1988) Incidence of early pregnancy loss New England Journal of Medicine 31 189–194.

4. Hertig AT (1964) Gestational hyperplasia of endometrium; a morphologic correlation between ova, endometrium, and corpora lutea during early pregnancy Laboratory Investigation 13 1153–1191.
5. McRae AC and Heap RB (1988) Uterine vascular permeability, blood flow and extracellular fluid space during implantation in rats Journal of Reproduction and Fertility 82 617–625.
6. Kennedy TG and Zamecnik J (1978) The concentration of 6-keto-prostaglandin F1 alpha is markedly elevated at the site of blastocyst implantation in the rat Prostaglandins 61 599–605.
7. Kennedy TG (1983) Embryonic signals and the initiation of blastocyst implantation Australian Journal of Biological Science 36 531–543.
8. Kinoshita K, Satoh K, Ishihara O, Tsutsumi O, Nakayama M, Kashimura F and Mizuno M (1985) Involvement of prostaglandins in implantation in the pregnant mouse Advances in Prostaglandin, Thromboxane and Leukotriene Research 15 705–607.
9. Snabes MC and Harper MJK (1984) Site of action of indomethacin on implantation in the rabbit Journal of Reproduction and Fertility 71 559–65:
10. Acker G, Hecquet F, Etienne A, Braquet P and Mencia-Huerta JM (1988) Role of platelet-activating factor (PAF) in the ovoimplantation in the rat: effect of the specific PAF-acether antagonist, BN 52021 Prostaglandins 35 233–241.
11. O'Neill C (1985) Partial characterisation of the embryo-derived platelet-activating factor in mice Journal of Reproduction and Fertility 75 375–380.
12. O'Neill C, Gidley-Baird AA, Pike IL and Saunders DM (1987) Use of a bioassay for embryo-derived platelet-activating factor as a means of assessing quality and pregnancy potential of human embryos Fertility and Sterility 47 969–975.
13. Collier M, O'Neill C, Ammit AJ and Saunders DM (1988) Biochemical and pharmacological characterisation of human embryo-derived platelet-activating factor Human Reproduction 3 993–998.
14. Ryan JP, O'Neill C and Wales RG (1990a) Oxidative metabolism of energy substrates by peri-implantation mouse embryos in the presence of platelet-activating factor Journal of Reproduction and Fertility 89 301–307.
15. Ryan JP, Spinks NR, O'Neill C & Wales RG (1990b) Implantation potential and fetal viability of mouse embryos cultured in media supplemented with platelet-activating factor Journal of Reproduction and Fertility 89 309–315.
16. O'Neill C, Ryan JP, Collier M, Saunders DM, Ammit AJ and Pike IL (1989) Supplementation of in-vitro fertilisation culture media with platelet-activating factor Lancet ii 769–772.
17. O'Neill C, Ryan JP, Collier M, Saunders DM, Ammit AJ and Pike IL (1992) Outcome of pregnancies resulting from a trial of supplementing human IVF culture media with platelet-activating factor Reproduction, Fertility and Development 4 109–112.
18. Spinks NR and O'Neill CO (1987) Embryo-derived PAF activity is essential for the establishment of pregnancy in the mouse Lancet i 106–107.
19. Acker G, Hecquet F, Etienne A, Braquet P and Mercia-Huerata JM (1988) Role of platelet-activating factor (PAF) in ovoimplantation in the rat: effect of the specific PAF-acether antagonist, BN 52021 Prostaglandins 35 233–241.
20. Norris CJ, Peairs WA, Kudolo GB, Newton ER and Harper MJK (1994) Platelet-activating factor antagonists and implantation in rabbits Journal of Reproduction and Fertility 100 395–401.
21. Ahmed A and Smith SK (1992) Platelet-activating factor stimulates phospholipase C activity in human endometrium Journal of Cellular Physiology 152 207–214.
22. Alecozay AA, Harper MJK, Schenken RS and Hanahan DJ (1991) Paracrine interactions between platelet-activating factor and prostaglandins in hormonally-treated human luteal phase endometrium in vitro Journal of Reproduction and Fertility 91 301–312.
23. Harper MJK, Kudolo GB, Alecozay AA and Jones MA (1989) Platelet-activating factor (PAF) and blastocyst-endometrial interactions Prog Clin Biol Res 294 305–315.
24. Johnston JM, Noriei M, Angle MJ and Hoffman DR (1990) Regulation of the arachidonic acid cascade and PAF metabolism in reproductive tissues. In : Eicosanoids in Reproduction M.D Mitchell ed CRC Press Boston 5–37.
25. Smith SK and Kelly RW (1988) Effect of platelet-activating factor on the release of $PGF_{2\alpha}$ and PGE_2 by separated cells of human endometrium. Journal of Reproduction and Fertility 82 271–276.
26. Ahmed A and Smith SK (1992a) The endometrium: prostaglandins and intracellular signalling at implantation Bailliere's Clinical Obstetrics and Gynaecology 6 731–754.
27. Kudolo GB, Kasamo M and Harper MJK (1991) Autoradiographic localization of platelet-activating factor (PAF) binding sites in the rabbit endometrium during the peri-implantation period Cell Tissue Research 265 231–241.

28. Nakamura M, Honda Z, Izumi T, Sakanaka C, Mutoh H, Minami M, Bito H, Seyama Y, Matsumoto T, Noma M and Shimiza T (1991) Molecular cloning and expression of platelet-activating factor receptor from human leukocytes Journal of Biological Chemistry 266 20400–20405.
29. Ahmed A, Sage, SO, Plevin R, Shoaibi MA, Sharkey AM and Smith SK (1994) Functional platelet-activating factor receptors linked to inositol lipid hydrolysis, calcium mobilisation and tyrosine kinase activity in the human endometrial HEC-1B cell line Journal of Reproduction and Fertility 101 459–466.
30. Berridge MJ and Irvine RF (1989) Inositol phosphates and cell signalling Nature 341 197–205.
31. Barzaghi G, Sarau HM and Mong S (1989) Platelet-activating factor-induced phosphoinositide metabolism in differentiated U-937 cells in culture Journal of Pharmacology and Experimental Therapeutics 248 559–566.
32. Bonney RC (1985) Measurement of phospholipase A2 activity in human endometrium during the menstrual cycle Journal of Endocrinology 107 183–189.
33. Downing I and Poyser NL (1983) Estimation of phospholipase A_2 activity in guinea-pig endometrium on day 7 and 16 of the estrous cycle Prostaglandins Leukotrienes in Medicine 12 107–117.
34. Exton JH (1990) Signalling through phosphatidylcholine breakdown Journal of Biological Chemistry 265 1–4.
35. Ahmed A, Ferriani RA, Plevin R and Smith SK (1992) Platelet-activating factor mediated phosphatidylcholine hydrolysis by phospholipase D in human endometrium Biology of Reproduction 47 59–65.
36. Ahmed A, Ferriani RA and Smith SK (1991) Stimulation of phosphatidylinositol 4,5-bisphosphate and phosphatidylcholine by platelet-activating factor (PAF) in human endometrium Journal of Reproduction and Fertility Abstract Series 8: abstract 105.
37. Cook SJ and Wakelam MJO (1989) Analysis of the water-soluble products of phosphatidylcholine breakdown by ion-exchange chromatography Biochemical Journal 263 581–587.
38. Billah MM and Anthes JC (1990) The regulation and cellular functions of phosphatidylcholine hydrolysis Biochemical Journal 269 281–291.
39. Morin C, Langlais J and Lambert RD (1992) Possible implication of lysophosphatidylcholine in cell fusion accompanying implantation in the rabbit Satellite Symposium of the Ninth International Congress of Endocrinology (Endocrinology of Embryo-Endometrial Interactions Bordeaux France Abstract 19.
40. Mobbs CV, Kaplitt M, Kow LM and Pfaff DW. (1991) Mol. Cell. Endocrinol. 80 C187-C191.
41. Lippman M, Bolan G and Huff K (1976) Cancer Research 36 4595–4601.
42. Ahmed A, Plevin R, Shoaibi MA, Fountain SA, Ferriani RA and Smith SK (1994) Basic FGF activates phospholipase D in endothelial cells in the absence of inositol lipid hydrolysis. American Journal of Physiology 266 C206-C212.
43. Morazova TM, Mitina RL, RauVA and Sidorkina OM (1989) Biokhimiya 54 610–618.
44. Sidorkina OM, MorazovaTM and Rau VA (1988) Biokhimiya 53 406–411.
45. Dhar A and Shukla SD (1991) Involvement of pp60c-src in platelet-activating factor-stimulated platelets. Evidence for translocation from cytosol to membrane Journal of Biological Chemistry 266:18797–801.
46. Ray LB and Sturgill TW (1987) Rapid stimulation by insulin of a serine/threonine kinase in 3T3-L1 adipocytes that phosphorylates microtubule-associated protein in vitro Proceedings of the National Academy of Science USA 84 1502–1506.
47. Pulvener BJ, Kymakis JM, Avruch J, Nikolakaki E, and Woodgeti JR (1991) Phosphorylation of c-jun by MAP kinase Nature 353 670–674.
48. Wood KW, Sarnecki C, Roberts TM and Blenis J (1992) Ras mediates nerve growth factor receptor modulation of three signal-transducing protein kinases: MAP kinase, Raf-1 and RSK Cell 68 1041–1050.
49. Zachary I, Gil J, Lehmann W, Sinnett-Smith J and Rosengurt F (1990) Bombesin, vasopressin and endothelin rapidly stimulate tyrosine phosphorylation in intact Swiss 3T3 cells Proceedings of the National Academy of Science 88 4577–4581.
50. Saville MK, Paterson A, Malarkey K and Plevin R (1994) Regulation of endothelin-1- and lysophosphatidic acid tyrosine phosphorylation of focal adhesion kinase ($pp125^{FAK)}$ in Rat-1 fibroblasts Biochemical Journal 301 (pt2) 407–14.
51. Spinks NR and O'Neill C (1988) Antagonists of embryo-derived platelet-activating factor prevent implantation of mouse embryos Journal of Reproduction and Fertility 84 89–98.
52. Bracquet P, Touqui L, Shen Y and Vargaftig BB (1987) Perspectives in platelet-activating research Pharmacological Revue 39 97–145.
53. Kamata K, Mori T, Shigenobu K and Kasuya Y (1989) Endothelium-dependant vasodilator effects of platelet-activating factor on rat resistance vessels British Journal of Pharmacology 98 1360–1364.
54. Juncos LA, Ren Y, Arima S and Ito S (1993) Vasodilator and Constrictor Actions of Platelet-activating Factor in the Isolated Microperfused Afferent Arteriole of the Rabbit Kidney. Role of Endothelium-derived

Relaxing Factor/Nitric Oxide and Cyclooxygenase Products Journal of Clinical Investigation 91 1374–1379.

55. Morsden PA, Tommy A, Brock TA and Ballerman BJ (1990) Glomerular endothelial cells respond to calcium mobilising agonists with release of EDRF American Journal of Physiology 258 F1295-F1303.
56. Filep JG, Foldes-Filep E and Sirois P (1993) Nitric oxide modulates vascular permeability in the rat coronary circulation British Journal of Pharmacology 108 323–326.
57. Garg UC and Hassid A (1989) Inhibition of rat mesangial cell mitogenesis by nitric oxide-generating vasodilators American Journal of Physiology 257 (Renal Fluid Electrolyte Physiology 26) F60-F66.
58. Garg UC and Hassid A (1989) Nitric oxide-generating vasodilators and 8-bromo-cyclic guanosine monophosphate inhibit mitogenesis and proliferation of cultured rat vascular smooth muscle cells Journal of Clinical Investigation 83 1774–1777.
59. Yallampalli C, Izumi H, Byam-Smith M and Garfield RE (1994) An L-arginine-nitric oxide-cyclic guanosine monophosphate system exists in the uterus and inhibits contractility during pregnancy American Journal of Obstetrics and Gynaecology 170 175–185.
60. Van Buren GA, Yang DA and Clark KE (1992) Estrogen-induced uterine vasodilatation is antagonised by L-arginine methyl ester, an inhibitor of nitric oxide synthesis American Journal of Obstetrics and Gynaecology 167 828–833.
61. Kimura T, Yoshida Y and Toda N (1992) Mechanisms of relaxation induced by prostaglandins in isolated canine arteries American Journal of Obstetrics and Gynaecology 167 1409–1416.
62. Telfer JF, Lyall F, Norman JE and Cameron IT (1995) Identification of nitric oxide synthase in human uterus Human Reproduction 10 (1) 19–23.
63. Giovanelli J, Campos KL and Kaufman S (1991) Proceedings of the National Academy of Science USA 88 7091–7095.
64. Bredt DS, Ferris CD and Snyder SH (1992) Nitric Oxide Synthase Regulatory Sites: Phosphorylation by cyclic AMP-dependant Protein Kinase, Protein Kinase C, and Calcium/Calmodulin Protein Kinase; Identification of Flavin and Calmodulin Binding Sites Journal of Biological Science 267 (16) 10976–10981.
65. Billah MM and Johnston JM (1983) Identification of phospholipid platelet-activating factor (1-O-alkyl-2-acetyl-*sn*-glycero-3-phosphocholine) in human amniotic fluid and urine Biochemical Biophysical Research Communications 113 51–8.
66. Nishihira J, Ishibashi T, Imai Y and Muramatsu T (1984) Mass spectrophotometric evidence for the presence of platelet-activating factor (1-O-alkyl-2-acetyl-*sn*-glycero-3-phosphocholine) in human amniotic fluid during labour Lipids 19 907–10.
67. Bernal AL, Newman GE, Phizackerley PJR, Laird E, Poss C and Barlow DH (1992) Platelet-activating factor levels in human follicular and amniotic fluids European Journal of Obstetrics and Gynaecology and Reproductive Biology 46 39–44.
68. Hoffman DR, Romero R and Johnston JM (1990) Detection of platelet-activating factor in amniotic fluid of complicated pregnancies American Journal of Obstetrics and Gynaecology 162 (2) 525–528.
69. Billah MM, Di Renzo GC, Ban C, Truong CT, Hoffman DR, Anceschi MM, Bleasdale JE and Johnston JM (1985) Platelet-activating factor metabolism in human amnion and the responses of this tissue to extracellular platelet-activating factor Prostaglandins 30 841–850.
70. Bleasdale JE, Bala GA, Thakur NR and McGuire JC (1989) Regulation of prostaglandin production by intrauterine tissue In: Cosmi E, Di Renzo GC (eds) Proceedings of the European Congress of Perinatal Medicine 828–834.
71. Montrucchio G, Alloatti G, Tetta C, Roffinello C, Emanuelli G and Camussi G (1986) In vitro contractile effect of platelet-activating factor on guinea pig myometrium Prostaglandins 32 539–54.
72. Tetta G, Montrucchio G, Alloatti G, et al (1986) Platelet-activating factor contracts human myometrium in vitro Proceedings of the Society for Experimental Biology and Medicine 183 376–81.
73. Zhu Y, Word RA and Johnston JM (1992) The presence of platelet-activating factor binding sites in human myometrium and their role in uterine contraction American Journal of Obstetrics and Gynaecology 166 (4) 1222–1227.
74. Hwang S, Lee C, Cheah MN and Shen TY (1983) Specific receptor sites for 1-O-alkyl-2-acetyl-*sn*-glycero-3-phosphocholine (platelet-activating factor) on rabbit platelet and guinea pig smooth muscle membranes Biochemistry 22 4756–63.
75. Zhu Y, Hoffman DR, Hwang SB, Miyaura S and Johnston JM (1991) Prolongation of parturition in the pregnant rat following treatment with platelet-activating factor (PAF) receptor antagonists Biology of Reproduction 44 39–42.
76. Hoffman DR, Truong CT and Johnston JM (1986) Metabolism and function of platelet-activating factor in rabbit fetal lung development Biochim Biophys 980 (1) 88–96.

77. Hoffman DR, Truong CT and Johnston JM (1986) The role of platelet-activating factor in human fetal lung maturation and the initiation of parturition American Journal of Obstetrics and Gynaecology 155 (1) 70–75.
78. Maki N, Hoffman DR and Johnston JM (1988) PAF (platelet-activating factor) acetylhydrolase activities in maternal, fetal and newborn rabbit plasma during pregnancy and lactation Proc Natl Acad Sci USA 85 728–732.
79. Johnston JM and Maki N (1980) PAF and fetal development In: Barnes PJ, Page CP (eds) PAF in Health and Disease Oxford UK; Blackwell Scientific Publications Ltd 297–322.
80. Johnston MJ and Miyaura S (1990) Platelet-activating factor: the alpha and omega of reproductive biology In: PAF antagonists: now development for clinical application The Woodlands Texas: Portfolio 139–60.
81. Narahara H, Nishioka Y and Johnston JM (1993) Secretion of Platelet-Activating Factor Acetylhydrolase by Human Decidual Macrophages Journal of Clinical Endocrinology and Metabolism 77 (5) 1258–1262.
82. Frenkel RA and Johnston JM (1992) Metabolic conversion of platelet-activating factor into ethanolamine plasmalogen in an amnion-derived cell line Journal of Biological Chemistry 267 19186–91.
83. Uemura Y, Lee T and Snyder F (1991) A co-enzyme A-independent transacylase is linked to the formation of platelet-activating factor (PAF) by generating the lyso-PAF intermediate in the remodelling pathway Journal of Biological Chemistry 266 8268–72.
84. Nieto ML, Venable ME, Bauldry SA et al (1991) Evidence that hydrolysis of ethanolamine plasmalogens triggers synthesis of platelet-activating factor via a transacylation reaction Journal of Biological Chemistry 266 18699–706.

46

NON-SPECIFIC PAF BINDING TO EMBRYONAL F9 CELLS AND PROSTACYCLIN SYNTHESIS IN HUMAN UMBILICAL VEIN ENDOTHELIAL CELLS

Ruth-Maria Korth*

Forschung in der Allgemeinmedizin FIDA
Palestrinastr. 7A, D-80639 München, Germany

1. INTRODUCTION

The platelet activating factor (Paf) is chemically 1-O-alkyl-2-acetyl-*sn*-glycero-3-phosphocholine with saturated[1] or non-saturated alkyl groups in position 1 of the molecule[2–4].

Besides various inflammatory activities, the synthesis of Paf in human embryos plays a role during in vitro fertilization[5,6]. Paf analogues with antagonist potency such as SRI 63–441 might reduce the implantation rate in mice. However, embryos are not desensitized to a second Paf exposure arguing against putative Paf receptors in embryonal cells[5].

Paf analogues such as CV 3988 inhibit the specific Paf acetylhydrolase and the synthesis of prostacyclin in endothelial cells[7,8]. Paf is metabolized in cells to arachidonate-phosphocholines a common precursor of Paf and $PGF_{2\text{-alpha}}$ and the concerned enzyme system is under hormonal control, particularly from estradiol[9]. Intracellular Paf receptors express collagenase mRNA in epithelial cells which are inhibited by the hydrophobic (intracellular) Paf receptor antagonist BN-50730[10] and collagenase mRNA increased in embryos at the time of implantation in the uterine wall explaining a role of Paf in successful pregnancy.

Hydrophilic hetrazepines such as WEB 2086 inhibit Paf receptors only on the outer plasmamembranes of intact human blood cells and they do not interfere with the specific Paf degrading acetylhydrolase activity[11–15]. Very high affinity Paf receptors on endothelial cells mediate cellular calcium flux and WEB 2086 is inhibitory[11,14,15.] Paf metabolism has been paralleled with prostacyclin synthesis[14] whereas Paf analogues not metabolized by the acetylhydrolase has been used as an argument in favour of a Paf receptor-dependent prostacyclin release[16].

* Address for correspondence: Ruth-Maria Korth M.D., Forschung in der Allgemeinmedizin FIDA, Palestrinastr. 7A, D-80639 München, Germany. Tel. 89/ 17 56 50, Fax 89/ 17 79 09.

The present study investigates the specific Paf receptor antagonist WEB 2086 on labeled Paf binding to intact mouse embryonal carcinoma F9 cells and prostacyclin synthesis of human umbilical vein endothelial cells because WEB 2086 did not interfere with Paf metabolism[11].

2. METHODS

Mouse embryonal carcinoma F9 cells were obtained from ATCC (USA). F9 cells were cultured as described with media changes every 2 days[17]. In short, flasks were coated with gelatin (Sigma Chemicals, Australia) and routine culture was performed in Dulbecco's medium (4500 mg/l D-glucose and glutamine, NaHCO3 3.7 g/l, penicillin and streptomycin 50 µg/ml, 10% heat-inactivated foetal calf serum). When cultures reached confluence, cells were harvested by exposure to phosphate-buffered saline containing EDTA (5 mM) for 30 min at 0°C. Embryonal F9 cells were washed in Tyrode's Hepes buffer without Ca^{2+} (pH 6.4) and resuspended in the same buffer brought to pH 7.4 containing Ca^{2+} (1.3 mM) as described[18]. F9 cells (10^7 cells per ml) were added to the same buffer containing [^{3}H]Paf (3.5 nM) in the presence of indicated concentrations of fatty acid-free bovine serum albumin (BSA, 0.025–0.25 %, w/v, Sigma Chemicals).

Embryonal F9 cells were cultured in the presence of retinoic acid (RA, 0.01 µM per 10^5 cells per Petri dish for 48 h) as decribed[17]. These cells were added to the same buffer containing various concentrations of [^{3}H]Paf (0.44, 0.88, 1.8, 3.6, 5.4, 7.1 nM) with or without WEB 2086 (2 µM) in the presence of 0.05 nM BSA.

Following incubation for 30 min at 20°C without stirring, the cells were separated from the suspension by dilution (1:1) with the same buffer and centrifugation (11000 x g, 1min). The free radioactivity and the cell-bound radioactivity were counted using standard conditions in scintillation fluid (BCS). Values were measured in triplicates and calculated as means.

Human umbilical vein endothelial cells were prepared as described before and the presence of very high affinity Paf receptors was shown before in parallel experiments[14]. After washing three times as described endothelial cells ($3x10^4$/ml) were incubated for 30 min with Paf (5 nM) with indicated concentrations of WEB 2086 (10, 100, 1000 nM) or vehicle. At the end of the incubation period, the supernatants were removed and stored at -20°C until tested for 6-keto-$PGF_{1\text{-alpha}}$ using antibodies developed by Dr. F. Dray (Pasteur Institute) as described[8,14,16].

3. RESULTS

Prostacyclin release: The amount of the stable prostacyclin metabolite 6-keto-$PGF_{1\text{-alpha}}$ in the supernatants of human umbilical vein endothelial cells increased from 60 ± 17 pg/ml to 133 ± 10 pg/ml in the presence of 5 nM Paf. The metabolite 6-keto-$PGF_{1\text{-alpha}}$ increased to 134 ± 5 pg/ml and 145 ± 11 pg/ml in the presence of 10 and 100 nM WEB 2086 and decreased to 116 ± 12 pg/ml in the presence of 1000 nM WEB 2086. The high dose WEB 2086 inhibition was statistically not significant.

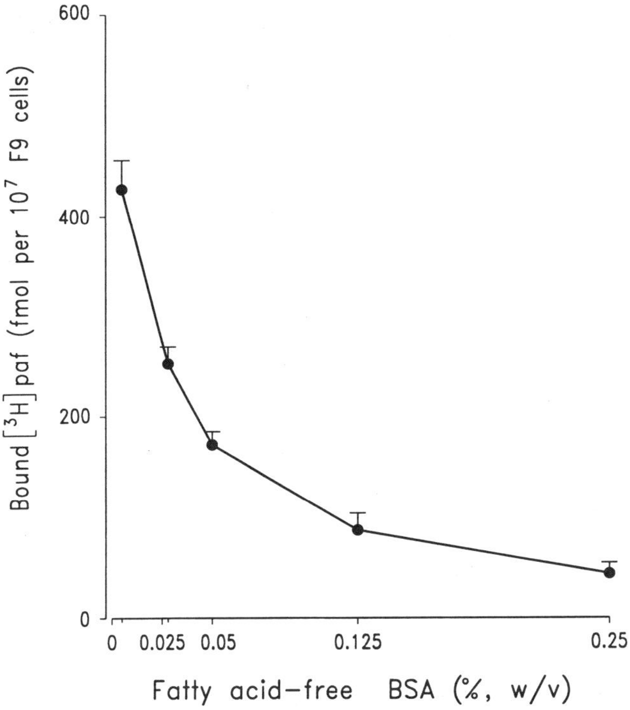

Figure 1. Indicated concentrations of fatty acid-free serum albumin inhibited the total [^{3}H]Paf binding (o) to embryonal F9 cells in a concentration-dependent manner. WEB 2086 (2 μM) was not inhibitory (not shown). Values are means ± 1 S.D. from three experiments.

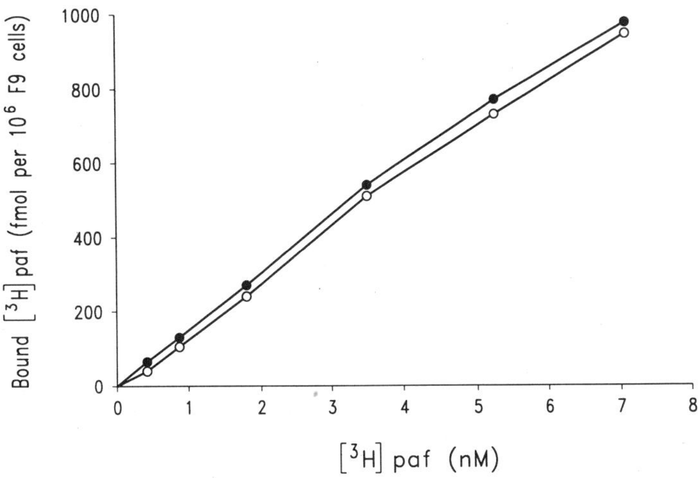

Figure 2. Embryonal F9 cells cultured in the presence of retinoic acid bound [^{3}H]Paf in a concentration-dependent manner (0.05 % BSA, o). WEB 2086 (2 μM, o) did not inhibit the [^{3}H]Paf binding in a statistically relevant manner. The values are taken from one experiment and are representative of three. The total [^{3}H]Paf binding to RA-treated F9 cells increased as compared with non-treated F9 cells shown in Figure 1.

4. DISCUSSION

Our data suggest only non-specific [^{3}H]Paf binding to embryonal F9 cells. Embryonal F9 cells bound [^{3}H]Paf with low affinity so that serum albumin could exhibit its non-specific inhibitory effect and the specific Paf receptor antagonist WEB 2086 failed to inhibit [^{3}H]Paf binding to embryonal F9 cells.

Human umbilical vein endothelial cells synthesize prostacyclin in response to Paf as described before[14] with inhibitory effects of the Paf analogue CV 3988[8]. The specific Paf receptor antagonist WEB 2086 failed here to inhibit prostacyclin synthesis probably because it did not interfere with Paf metabolism[11]. We are presently investigating prostacyclin synthesis and non-specific Paf binding using monocyte U 937 cells or immortalized human umbilical vein endothelial cells because these cells express high or very high affinity Paf receptors only after long time incubation with growth factors such as lipoproteins or insulin[12,14].

5. SUMMARY

Our data suggest non-specific Paf binding to intact embryonal F9 cells. Non-specific Paf binding might have some physiological relevance when Paf metabolites interfere with prostacyclin synthesis, for example, in umbilical vein endothelial cells.

ACKNOWLEDGMENTS

The experiments with F9 cells were performed in the Royal North Shore Hospital in Sydney (Australia) and we thank Chris O'Neill for the discussions. The 6-keto-PGF$_{1\text{-alpha}}$ assay was performed in the laboratories of INSERM in France. We thank Jacques Benveniste from INSERM U 200 and Francoise Russo-Marie from INSERM U 332 for the discussions. We gratefully acknowledge the expert technical asstistance of Brigitte Canton (INSERM U 332). This work is dedicted to Elisabeth von Rosenstiel in Munich.

REFERENCES

1. Benveniste, J. Paf-acether (platelet-activating factor). In: *Advances in Prostaglandin, Thromboxane, and Leukotriene Research* (Eds. Samuelsson B., Wong P.Y.-K. and Sun F.F.) pp. 355–358, Raven Press, New York, 1989.
2. Korth, R., Hillmar, I., Muramatsu, T. and Zöllner, N. 1-O-alkyl-2-acetyl-sn-glycero-3-phosphocholines: Influence on aggregation and [^{3}H]serotonin release of human thrombocytes. *Chem. Phys. Lipids* 33: 47–53, 1983.
3. Korth, R., Mangold, H.K., Muramatsu, T. and Zöllner, N. Unsaturated 1-O-alkyl-2-acetyl-sn-glycero-3-phosphocholines (unsaturated platelet-activating factor): Aggregation of human platelets after incubation with indomethacin, creatinephosphate/creatinephosphokinase, xylocain and hirudine, and serotonin release after incubation with indomethacin. *Chem. Phys. Lipids* 36: 209–214, 1985.
4. Korth, R., Riess, H., Brehm, G. and Hiller, E. Unsaturated platelet-activating factor: Influence on aggregation, serotonin release and thromboxane synthesis of human thrombocytes.*Thrombos. Res.* 41: 699–706, 1986.
5. Ryan, J.P., O'Neill, C. and Wales, R.G. Oxidative metabolism of energy substrates by preimplantation mouse embryos in the presence of platelet-activating factor *J. Reprod. Fert.* 89: 301–307, 1990.

6. O' Neill, C., Ryan, J.P., Collier, M., Saunders, D.M., Ammit, A.J., and Pike, I.L. Supplementation of in-vitro fertilization culture medium with platelet activating factor. *The Lancet* ii: 769–772, 1989.
7. Miwa, M., Miyake, T., Yamanaka, T., Sugatani, J., Suzuki, Y., Sakata, S., Araki, Y. and Matsumoto, M. Characterization of serum platelet-activating factor (PAF) acetylhydrolase: correlation between deficiency of serum PAF acetylhydrolase and respiratory symptoms in asthmatic children. *J. Clin. Invest.* 82, 1983–1991, 1988.
8. Korth, R., Hirafuji, M., Bidault, J., Canton, B., Russo-Marie, M. and Benveniste, J. Comparison between Paf-acether receptors on intact washed human platelets and human endothelial cells in culture. In: *Vascular Endothelium. Receptors and transduction mechanisms* (Eds. Catravas, J.D., Gillis, C.N. and Ryan, U.S.), pp 89–98, Plenum press New York and London, 1989.
9. Nakayama, R., Yasuda, K., Okumura, T. and Saito, K. Effect of 17ß-estradiol on PAF and prostaglandin levels in oophorectomized rat uterus. *Biochim. Biophys. Acta* 1085: 235–240, 1991.
10. Bazan, H.E.P., Tao, Y. and Bazan, N.G. Platelet-activating factor induces collagenase expression in corneal epithelial cells. *Proc. natl. Acad. Sci.* 90: 8688–8682, 1993.
11. Korth, R., Hirafuji, M., Lalau Keraly, C., Delautier, D., Bidault, J. and Benveniste, J. Interaction of the Paf antagonist WEB 2086 and its hetrazepine analogues with human platelets and endothelial cells. *Brit. J. Pharmacol.* 98: 653–661, 1989.
12. Korth, R. and Middeke, M. Long time incubation of monocytic U 937 cells with LDL increases specific Paf-acether binding and the cellular acetylhydrolase activity. *Chem. and Phys. Lipids* 59: 207–213, 1991.
13. Korth, R., Zimmermann, K. and Richter, W.O. Lipoprotein-associated paf (LA-paf) was found in washed human platelets and monocyte/macrophage like U 937 cells. *Chem. and Phys. Lipids* 72: 109–119, 1994.
14. Korth, R.M., Hirafuji, M., Benveniste, J., and Russo-Marie, F. Human umbilical vein endothelial cells: specific binding of platelet-activating factor and cytosolic calcium flux. *Biochem. Pharmacology* 49: 1793–1799, 1995.
15. Korth R.-M. Verwendung von paf-Acether-Antagonisten zur Herstellung eines Arzneimittels und Verfahren zu deren Wirksamkeitsbestimmung. *Europäische Patentschrift* 88117022.9, 1988.
16. D'Humière, S.D., Russo-Marie, F. and Vargaftig, B.B. Paf-acether induced synthesis of prostacyclin by human endothelial cells. *Eur. J. Pharmacol.* 131: 13–19, 1986.
17. American Type Culture Collection, 6th edition, p 167, 1988.
18. Korth R. and Benveniste J. BN 52021 displaces ^{3}H-Paf-acether from, and inhibits its binding to intact human platelets. *Eur. J. Pharmacol.* 142: 331- 341, 1987.

47

PLATELET-ACTIVATING FACTOR (PAF): SIGNALLING AND ADHESION IN CELL-CELL INTERACTIONS*

G. A. Zimmerman,[1,4] M. R. Elstad,[1,3,4] D. E. Lorant,[1,6]
T. M. McIntyre,[1,4,5] S. M. Prescott,[2,3,4] Matthew K. Topham,[1,4]
A. S. Weyrich,[1,4] and Ralph E. Whatley[1,4]

[1]Nora Eccles Harrison Cardiovascular Research and Training Institute
[2]Program in Human Molecular Biology and Genetics
Departments of [3]Biochemistry, [4]Medicine, [5]Pathology, and [6]Pediatrics
University of Utah
Salt Lake City, Utah
[7]Veterans Affairs Medical Center
Salt Lake City, Utah

1. PAF IN CELL-CELL INTERACTIONS

PAF is a signalling molecule[1] that mediates interactions between cells[2]. Cell-cell interactions are critical in both physiologic and pathologic processes[3,4]. Many of the best-characterized effects of PAF involve cellular signalling in the inflammatory and vascular systems, but it also accomplishes signalling functions in the reproductive system, the brain, and in many other organs[2]. It has the remarkable property of being able to act as an endocrine, paracrine, autocrine, or juxtacrine signal depending on the cell of origin and the specific conditions under which it interacts with target cells[5]. The amphipathic properties of PAF and its ability to interact with both lipids and proteins contribute to this versatility.

One of the most compelling pieces of evidence that PAF plays a role in physiologic cell-cell interactions is that there is a precise system of checks and balances that governs its biologic activities[2,6]. It seems unlikely that this intricate control system would have evolved if PAF does not have specific physiologic roles. Major features of the regulatory

* Supported by the Nora Eccles Treadwell Foundation, individual grants from the National Institutes of Health and the Department of Veterans Affairs Medical Research Fund, an NIH Training Grant in Cellular and Physiologic Responses in Lung Disease, an NIH Special Center of Research in ARDS, and an NIH Center of Excellence in Molecular Hematology.

Platelet-Activating Factor and Related Lipid Mediators 2
edited by Nigam *et al.*, Plenum Press, New York, 1996

system include: 1) tightly regulated synthetic pathways; 2) cell-specific expression and control of the receptor for PAF; 3) a group of degradative enzymes—the PAF acetylhydrolases—that are expressed in a developmentally-regulated cell- and tissue-specific manner. Together, these components constitute a powerful system for control of the signalling actions of PAF. A defect in one of the regulatory components, together with a stimulus for its production, results in tissue injury in animal models of pathologic inflammation[2]. These studies and preliminary clinical observations suggest that breakdown of the tight constraints on the biologic actions of PAF may contribute to certain human diseases[7].

There is also an additional control mechanism. In many cases, PAF appears to act in a spatially-regulated fashion as a juxtacrine signal when cells of the inflammatory system adhere to one another[4,5]. The signalling roles of PAF are also closely aligned with intercellular adhesion in a variety of other ways. The next sections will briefly review some of the relationships between PAF, intercellular signalling, and cell-cell adhesion.

2. JUXTACRINE SIGNALLING BY PAF: ACTIONS AT THE INTERFACES BETWEEN CELLS

In early studies of the production of PAF by human endothelial cells in primary culture, we found that newly-synthesized PAF remains associated with the cell monolayer[8,9]. PAF that was metabolically-labeled with tritated acetate was not released, even in the presence of physiologic concentrations of human serum albumin added to serve as an "acceptor" (i.e. to allow it to transfer from the plasma membranes of the cells to albumin in solution). The accumulation of PAF in endothelial cells stimulated by thrombin or histamine was paralleled by adhesion and activation of isolated human polymorphonuclear leukocytes (PMNs; neutrophils)[10]. The signals for these neutrophil responses were delivered by the endothelial cell—that is, they were "endothelial cell-dependent"—and the signalling molecule remained associated with the endothelial monolayer. The kinetics of the endothelial cell-dependent PMN adhesion and of PAF accumulation suggested that the two processes are related. A variety of experimental strategies then documented that PAF that is retained by the stimulated endothelial cells binds to its receptor on the neutrophils, resulting in activation of the leukocytes[10,11]. These experiments indicated that PAF acts in a "juxtacrine" fashion at the interface between the endothelial cell and the leukocyte[12]. Juxtacrine signalling molecules are those that bind to signal transducing receptors on target cells and induce functional or phenotypic changes in them while remaining associated with the cell of origin (the signalling cell)[4,5] (also see below).

A consequence of PMN activation by PAF at the surfaces of stimulated endothelial cells is "inside-out signalling" of β_2 integrins[1,3], making them competent to bind to ligands on the endothelial surface (Fig. 1)[12]. This is one component of the adhesive interaction between the two cells. There is also a second critical molecular interaction in this adhesive process. PAF acts in concert with a glycoprotein, P-selectin (formerly called GMP-140)[13], that is coexpressed with it on the plasma membranes of stimulated endothelial cells[12]. P-selectin tethers the PMN to the endothelial cell[1,13]. Together, P-selectin on the endothelial cell and the β_2 integrins on the activated PMN act to mediate tight adhesion of the leukocytes[12]. *In vivo*, such tight adhesion is required to allow leukocytes to stop rolling along the intimal surface of an inflamed vessel, become firmly adherent, and transmigrate[1,14]. P-selectin and PAF on the surfaces of stimulated endothelial cells, together with their counterstructures on the neutrophil—the ligand for P-selectin, PSGL-1[15,16], and the PAF receptor[2]—form a juxtacrine system (Figure 2). Tethering of PMNs by P-selectin and sig-

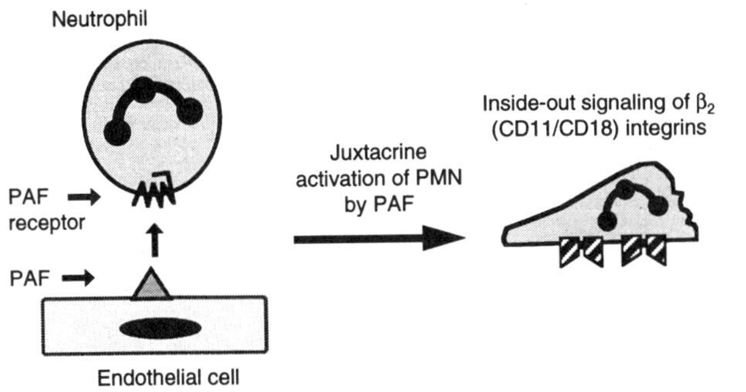

Figure 1. PAF is a juxtacrine signalling molecule that is expressed by stimulated human endothelial cells and induces inside-out signalling of β_2 integrins on neutrophils. (Modified from reference 2 with permission.)

nalling by PAF defines a paradigm by which critical events in inflammation can be accomplished in a spatially-specific fashion[1,2,5].

Other activation responses besides inside-out signalling of β_2 integrins, each of which is important in acute inflammation[2], are induced in the adherent PMNs (Figure 2). In each instance, PAF provides the critical signal. In contrast, P-selectin does not directly signal when presented to the PMNs in purified form or in model membranes, and does not induce the activation responses shown in Figure 2[17]. However, tethering of the PMN by P-selectin is required for PAF to signal efficiently and, in some cases, appears to facilitate responses to signals delivered by PAF[12,17] (also see below). Activation of PMNs by PAF also modifies the distribution and function of PSGL-1 on their surfaces (Figure 2), a regulatory feature that may allow "release" of the P-selectin component of the adhesive bond so that movement and transmigration can occur[18].

In addition to their physiologic significance, the experiments summarized above provide powerful evidence that PAF activates PMNs in a localized fashion at the endothelial surface, rather than acting free in solution. Otherwise, tethering by P-selectin would

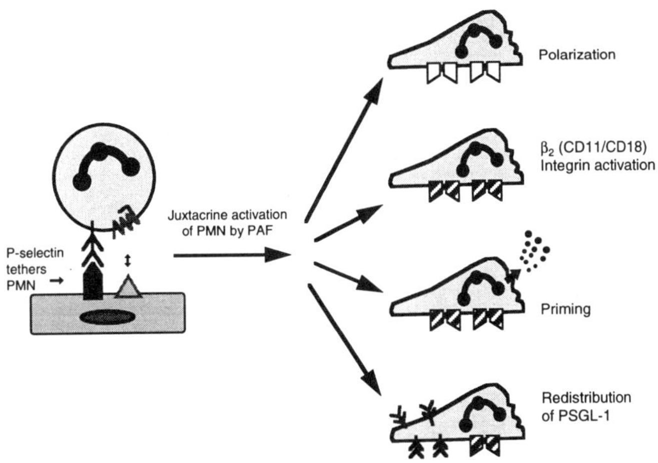

Figure 2. PAF and P-selectin are components of a juxtacrine signalling system that mediate adhesion of neutrophils and their local activation. See text for details. In addition to the activation responses shown, an increase in intracellular Ca^{2+} is induced in the adherent neutrophils. (Modified from reference 17 with permission.)

not be required[12,17]. Other laboratories have also reported evidence that PAF can act as a juxtacrine signal[19–24]. It is unknown if PAF is located in the outer phospholipid leaflet of the plasma membrane of the signalling cell, or instead is noncovalently linked to a "presenting protein", in these localized signalling interactions[11].

When PAF acts at the interfaces between adherent cells (Fig. 1) it does not physically tether the cells together so far as is known. This is different from the original example of juxtacrine signalling by Anklesaria, Massagué, and co-workers, where a membrane-anchored polypeptide, pro-TGF_{α}, mediated both adhesion and signalling[25,26]. The difference illustrates the variable ways in which adhesion is related to this mechanism of signalling. Juxtacrine signalling molecules can induce the action of an adhesion molecule (Figure 1), act in concert with a separate adhesion factor (Figure 2), or directly mediate adhesion—as in the case of proTGFα. PAF acts in two of these modes: it both induces the action of β_2 integrins (Figure 1), and acts in concert with P-selectin (Figure 2). In signalling interactions between endothelial cells and leukocytes, a juxtacrine *system* such as that made up of P-selectin and PAF (Figure 2) may have biologic advantages over a single juxtacrine signalling molecule and its receptor (Figure 3)[5]. For example, tethering by P-selectin provides specificity by "selecting" target cells that have PSGL-1 (or other ligands that specifically recognize P-selectin) and also bear the serpentine receptor for PAF (for example, PMNs, monocytes). In contrast, cells that have only the PAF receptor, such as platelets, are not signalled (our unpublished observations). This may be one mechanism by which targeting and activation of PMNs and monocytes is accomplished without inducing platelet aggregation and thrombosis in regulated acute inflammatory responses.

Localized actions of PAF at interfaces between cells may have clinical importance. High local concentrations of PAF at cellular surfaces may prevent, or reduce, blockade of the PAF receptor on target cells by competitive antagonists[2,7]. This would be expected to be particularly pronounced for antagonists with affinities for the receptor that are lower than that of PAF itself, and might be influenced by whether the antagonist is water- or lipid-soluble. A second point of clinical relevance is that disruption of juxtacrine signalling, resulting from release of PAF into solution in the blood, may be a mechanism of disease[4,5]. Under these conditions indiscriminate activation of leukocytes and platelets in flowing blood, rather than activation in a spatially-regulated fashion, may lead to their aggregation and sequestration resulting in occlusion and injury of vessels[4,5]. PAF may be released by endothelial cells injured by certain bacterial toxins that cause leukocyte sequestration *in vivo* (M. Bunting, SM Prescott, et al., manuscript in preparation). Also, oxidatively-modified "PAF-like" phospholipids are released from endothelial cells injured by oxidants[27,28].

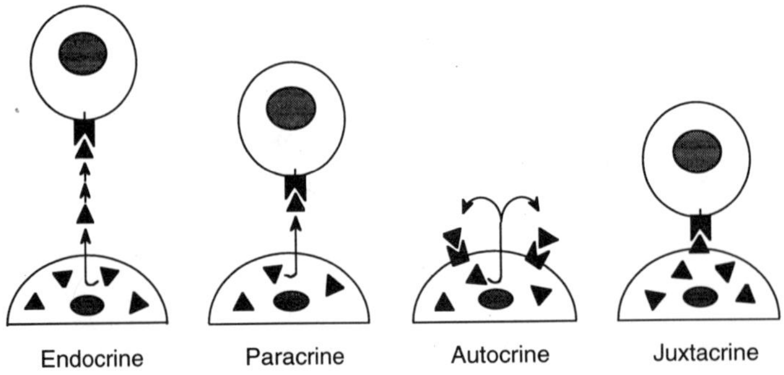

Figure 3. Mechanisms of cellular signalling. (Modified from reference 5 with permission.)

3. ADHESION MODIFIES SIGNALS DELIVERED THROUGH THE PAF RECEPTOR

Human monocytes bear receptors for PAF[2]; they also constitutively express ligands for P-selectin, including the high affinity ligand, PSGL-1[15,16,29,30] (Weyrich et al, manuscript in preparation). We have examined signalling in these cells using secretion of monokines (chemokines and cytokines synthesized by monocytes) as the index response. Certain monokines, including MCP-1, TNF_{α}, and IL-8, are coded by immediate-early genes that are quiescent in circulating monocytes. PAF alone is a weak or ineffective agonist for synthesis and secretion of these monokines by freshly-isolated monocytes in suspension[29]. However, adhesion of monocytes to P-selectin makes them responsive to PAF, and synthesis and secretion of the factors are induced[29]. Thus, tethering (adhesion) of the monocyte has modified its ability to respond to an extracellular signal—in this case, one delivered through the PAF receptor. This *adhesion-dependent signal modification* is specific for P-selectin and is not indiscriminantly conferred by adhesion to control surfaces (Figure 4 and Weyrich et al, unpublished experiments). The mechanism, in part, involves intracellular integration of the signal delivered through the PAF receptor on the adherent monocytes leading to facilitated nuclear translocation of a critical transcription factor, NF-κB[29]. Kravchenko, Ye, and co-workers have also reported activation of the NF-κB system by the PAF receptor[31]. For monokine secretion (Figure 4), engagement of PSGL-1 by P-selectin is required for adhesion-dependent signal modification, and there may be multiple points of regulation (Weyrich et al., manuscript in preparation). Integration and amplification of signalling responses may also occur in PMNs when they are engaged by P-selectin under certain conditions[12,17].

4. THE SYNTHESIS OF PAF IS INFLUENCED BY ADHESION

Opsonized particles and microorganisms are recognized by adhesion molecules on the surfaces of monocytes and macrophages, and can activate these leukocytes in an in-

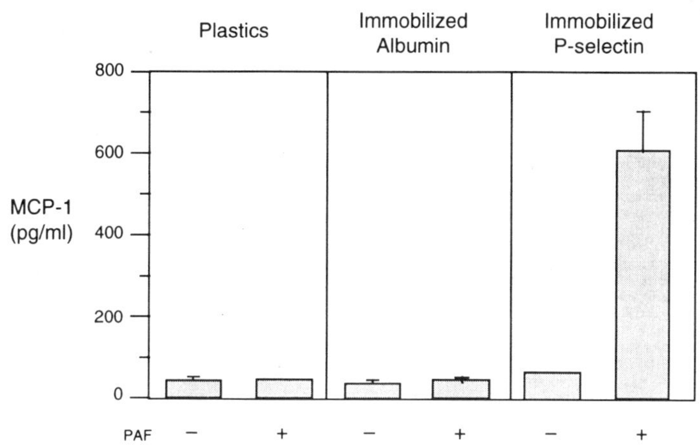

Figure 4. Adhesion of human monocytes to purified immobilized P-selectin, but not to uncoated plastic or immobilized albumin, results in responsiveness to PAF and MCP-1 secretion. See text and reference 23 for details (modified from reference 23 with permission).

flammatory milieu. Zymosan, a fungal wall component containing β-glucan, induces synthesis of PAF by monocytes[32,33]; the synthesis and secretion of PAF are dramatically enhanced by opsonization of the zymosan particles[33]. Binding of iC3b on the opsonized particle by the β_2 integrin, α_M/β_2, on the surface of the monocyte is required for maximal synthesis[33]. However, the critical signal that induces PAF synthesis appears to be delivered through a β-glucan receptor on the leukocyte (Figure 5)[33]. Thus, both adhesive and signalling events are required, and engagement of an adhesion molecule —α_M/β_2 integrin—modifies the signalling.

Adhesion of monocytes to purified, immobilized P-selectin enhances PAF synthesis induced by opsonized β-glucan[30]. Adhesion to P-selectin does not directly induce quantitative upregulation or enhanced recognition of iC3b by α_M/β_2 integrin on the monocytes. However, binding of opsonized particles via α_M/β_2 integrin is clearly required for a maximal generation of PAF[30]. Thus, adhesion of monocytes to P-selectin and engagement of α_M/β_2 integrin may coordinately modify kinases or other intracellular components that are sites of signal integration[4], leading to enhanced synthesis of PAF when the β-glucan receptor is activated in this model.

5. SUMMARY

Signalling by PAF is closely linked to adhesive interactions between cells of the inflammatory and vascular systems. It acts as a juxtacrine signal that alters the activity of β_2 integrins on myeloid leukocytes (Figure 1), and works in concert with P-selectin at the surfaces of endothelial cells (Figure 2 and text). Observations in models of flow[23] and *in vivo*[33] support the original experiments using cultured endothelium under static conditions that indicated that PAF acts at this vascular interface. P-selectin modifies and integrates signals delivered through the PAF receptor on monocytes (Figure 4). Adhesion via P-selectin and engagement of β_2 integrins modify signals leading to PAF synthesis (text and Figure 5). The intimate relationship between adhesive events and signalling by PAF may be a critical determinant in its roles in physiologic and pathologic responses.

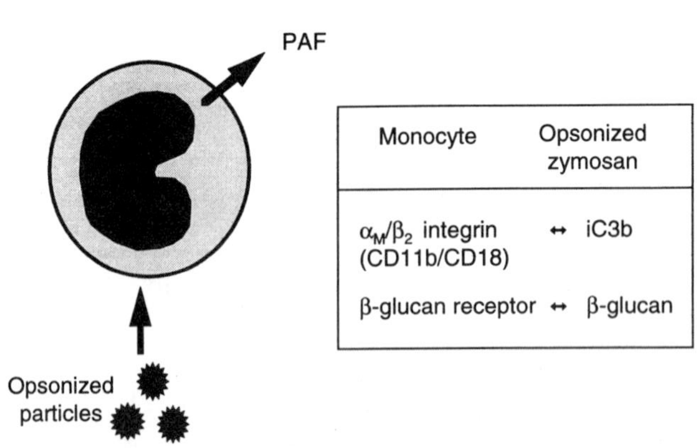

Figure 5. α_M/β_2-integrin and a β-glucan receptor on monocytes mediate maximal PAF synthesis when the cells bind opsonized zymosan particles. See text and reference 27 for details.

REFERENCES

1. Zimmerman GA, Prescott SM and McIntyre TM. Endothelial cell interactions with granulocytes: Tethering and signaling molecules. *Immunology Today* 1992; 13(3):93–100.
2. Zimmerman GA, McIntyre TM, Prescott SM: Platelet-activating factor: A fluid-phase and cell-associated mediator of inflammation. In *Inflammation: Basic Principles and Clinical Correlates, Second Edition.* JI Gallin, IM Goldstein, R Snyderman, editors, Raven Press, New York, pp 149–176, 1992
3. Hynes RO and Lander AD. Contact and adhesive specificities in the associations, migrations and targeting of cells and axons. *Cell* 1992; 68:303–322.
4. Zimmerman GA, McIntyre TM and Prescott SM. Cell-to-cell communication. In *The Lung: Scientific Foundations. Second Edition* RG Crystal, JB West, ER Weibel, PJ Barnes, eds. Raven Press, in press 1995.
5. Zimmerman GA, Lorant DE, McIntyre TM and Prescott SM. Juxtacrine intercellular signaling: Another way to do it. *Am J Resp Cell Mol Biol* 1993; 9:573–577.
6. Prescott SM, Zimmerman GA and McIntyre TM. Platelet-activating factor (PAF). *J Biol Chem* 1990; 265:17381–17384.
7. Imaizumi T, Stafforini DM, Yamada Y, McIntyre TM, Prescott SM and Zimmerman GA. Platelet-activating factor: A mediator for clinicians. *J Internal Med* 1995; 238:5–20.
8. Prescott SM, Zimmerman GA and McIntyre TM: Human endothelial cells in culture produce platelet-activating factor (1-alkyl-2-acetyl-*sn*-glycero-3-phosphocholine) when stimulated with thrombin. *Proc Natl Acad Sci USA* 1984; 81:3534–3538.
9. McIntyre TM, Zimmerman GA and Prescott SM. Leukotrienes C_4 and D_4 stimulate human endothelial cells to synthesize platelet-activating factor and bind neutrophils. *Proc Natl Acad Sci USA* 1986; 83:2204–2208.
10. Zimmerman GA, McIntyre TM and Prescott SM. Thrombin stimulates the adherence of neutrophils to human endothelial cells *in vitro*. *J Clin Invest* 1985; 76:2235–2246.
11. Zimmerman GA, McIntyre TM, Mehra M and Prescott SM. Endothelial cell-associated platelet-activating factor: a novel mechanism for signaling intercellular adhesion. *J Cell Biol* 1990; 110:529–540.
12. Lorant DE, Patel KD, McIntyre TM, McEver RP, Prescott SM and Zimmerman GA. Coexpression of GMP-140 and PAF by endothelium stimulated by histamine or thrombin: A juxtacrine system for adhesion and activation of neutrophils. *J Cell Biol* 1991; 115:223–234.
13. Geng J-G, Bevilacqua MP, Moore KL, McIntyre TM, Prescott SM, Kim JM, Bliss GA, Zimmerman GA and McEver RP. Rapid neutrophil adhesion to activated endothelium mediated by GMP-140. *Nature* 1990; 343:757–760.
14. McEver RP. Leukocyte-endothelial cell interactions. *Curr Opin Cell Biol* 1992; 4:840–849.
15. Sako D, Chang X-J, Barone KM, Vachino G, White HM, Shaw G, Veldman GM, Bean KM, Ahern TJ, Furie B, Cumming DA and Larsen GR. Expression cloning of a functional glycoprotein ligand for P-selectin. *Cell* 1993; 75:1179–1186.
16. Moore KL, Patel KD, Bruehl RE, Fugang L, Johnson DA, Lichenstein HS, Cummings RD, Bainton DF and McEver RP. P-selectin glycoprotein ligand-1 mediates rolling of human neutrophils on P-selectin. *J Cell Biol* 1995; 128:661–671.
17. Lorant DE, Topham MK, Whatley RE, McEver RP, McIntyre TM, Prescott SM and Zimmerman GA. Inflammatory roles of P-selectin. *J Clin Invest* 1993; 92:559–570.
18. Lorant DE, McEver RP, McIntyre TM, Moore KL, Prescott SM and Zimmerman GA. Activation of polymorphonuclear leukocytes reduces their adhesion to P-selectin and causes redistribution of ligands for P-selectin on their surfaces. *J Clin Invest* 1995; 96:171–182.
19. Vercellotti GM, Wickham NWR, Gustafson KS, Yin HQ, Herbert M and Jacob HS. Thrombin-treated endothelium primes neutrophil functions: inhibition by platelet-activating factor receptor antagonists. *J Leukoc Biol* 1989; 45:483–490.
20. Ninio E, Leyravaud S, Bidault J, Jurgens P and Benveniste J. Cell adhesion by membrane-bound paf-acether. *International Immunology* 1991; 3:1157–1163.
21. Kuijpers TW, Hakkert BC, Hart MHL and Roos D. Neutrophil migration across monolayers of cytokine-prestimulated endothelial cells: a role for platelet-activating factor and IL-8. *J Cell Biol* 1992; 117:565–572.
22. Zhou W, Javors MA and Olson MS. Platelet-activating factor as an intercellular signal in neutrophil-dependent platelet activation. *J Immunol* 1992; 149:1763–1769.
23. Macconi D, Foppolo M, Paris S, Noris M, Aiello S, Remuzzi G and Remuzzi A. PAF mediates neutrophil adhesion to thrombin or TNF-stimulated endothelial cells under shear stress. *Am J Physiol* 1995; 269(*Cell Physiol* 38):C42-C47.

24. von Asmuth EJU and Buurman WA. Endothelial cell associated platelet-activating factor (PAF), a costimulatory intermediate in TNF-α-induced H_2O_2 release by adherent neutrophil leukocytes. *J Immunol* 1995; 154:1383–1390.
25. Anklesaria P, Texidó J, Laiho M, Pierce JH, Greenberger JS and Massagué J. Cell-cell adhesion mediated by binding of membrane-anchored transforming growth factor α to epidermal growth factor receptors promotes cell proliferation. *Proc Natl Acad Sci* 1990; 87:3289–3293.
26. Massagué J. Transforming growth factor-α. A model for membrane-anchored growth factors. *J Biol Chem* 1990; 265(35):21393–21396.
27. Patel KD, Zimmerman GA, Prescott SM and McIntyre TM. Novel leukocyte agonists are released by endothelial cells exposed to peroxide. *J Biol Chem* 1992; 267:15168–15175.
28. McIntyre TM, Patel KD, Smiley PL, Stafforini DM, Prescott SM and Zimmerman GA. Oxidized phospholipids with PAF-like bioactivity. *J Lipid Mediators Cell Signalling* 1994; 10:37–39.
29. Weyrich AS, McIntyre TM, McEver RP, Prescott SM and Zimmerman GA. Monocyte tethering by P-selectin regulates monocyte chemotactic protein-1 and tumor necrosis factor-α secretion: Signal integration and NF-κB translocation. *J Clin Invest* 1995; 95:2297–2303.
30. Elstad MR, LaPine TR, Cowley FS, McEver RP, McIntyre TM, Prescott SM and Zimmerman GA. P-selectin regulates platelet-activating factor synthesis and phagocytosis by monocytes. *J Immunol* 1995; 155:2109–2122.
31. Kravchenko VV, Pan Z, Han J, Herbert J-M, Ulevitch RJ and Ye RD. Platelet-activating factor induces NF-κB activation through a G protein-coupled pathway. *J Biol Chem* 1995; 270:14928–14934.
32. Elstad MR, Prescott SM, McIntyre TM and Zimmerman GA. Synthesis and release of platelet-activating factor by stimulated human mononuclear phagocytes. *J Immunol* 1988; 140:1618–1624.
33. Elstad MR, Parker CJ, Cowley FS, Wilcox LA, McIntyre TM, Prescott SM and Zimmerman GA. CD11b/CD18 integrin and a β-glucan receptor act in concert to induce the synthesis of platelet-activation factor by monocytes. *J Immunol* 1994; 152:220–230.
34. Coughlan AF, Hau H, Dunlop LC, Berndt MC, Hancock WW. P-selectin and platelet-activating factor mediate initial endotoxin-induced neutropenia. *J Exp Med* 1994; 179:329–334.

48

IDENTIFICATION AND MOLECULAR CHARACTERIZATION OF THE CalB DOMAIN OF THE CYTOSOLIC PHOSPHOLIPASE A_2 ($cPLA_2$) IN HUMAN NEUTROPHILS

Hong Zhang, Birgit Wendel, Veronika van Wyk, and Santosh Nigam*

Eicosanoid Research Division
Department of Gynecology
University Medical Hospital Benjamin Franklin
Free University Berlin
D-12200 Berlin, Germany

INTRODUCTION

The receptor-mediated activation of human neutrophils by ligands such as chemotactic peptide N-formyl-methionine-leucine-phenylalanine (fMLP) is coupled with the activation of phospholipases C and A_2 (1,2). Phospholipase A_2 (PLA_2) hydrolyzes arachidonic acid (AA) and lysophospholipid from the *sn*-2 position of 1-alkyl-2-arachidonoyl-*sn*-3-glycerophosphocholine of the membrane (3). Currently, it is accepted that the cytosolic PLA_2 ($cPLA_2$) and not the secretory PLA_2 ($sPLA_2$) is primarily responsible for the release of AA (4) in most of the cell systems. The reasons for this assumption are based on the observations such as low requirements of Ca^{++} concentrations (micromolar for $cPLA_2$ vs milimolar for $sPLA_2$), fatty acid preference of $cPLA_2$ at the *sn*-2 position (5), resistance of $cPLA_2$ to disulfide-reducing agents (4) and increased concentrations of cytosolic Ca^{++} (5,6). In addition, it is stimulated by phosphorylation through protein kinase C (PKC) , but inhibited by staurosporine, an inhibitor of PKC (7). Conversely, previous studies from our laboratory have shown that the inhibition of PKC by staurosporine potentiated the activation of PLC and PLA_2 in fMLP-challenged human PMN (8–10).

In order to find whether the $cPLA_2$ in human neutrophils is a different isoform than described in the literature, we investigated the molecular structure of the $cPLA_2$ in human neutrophils and compared with the known $cPLA_2$ of U937 cell line. The present results demonstrate that the $cPLA_2$ in human neutrophils is identical to that one found in U937 cell line.

* Address correspondence to Dr. S. Nigam.

Platelet-Activating Factor and Related Lipid Mediators 2
edited by Nigam *et al.*, Plenum Press, New York, 1996

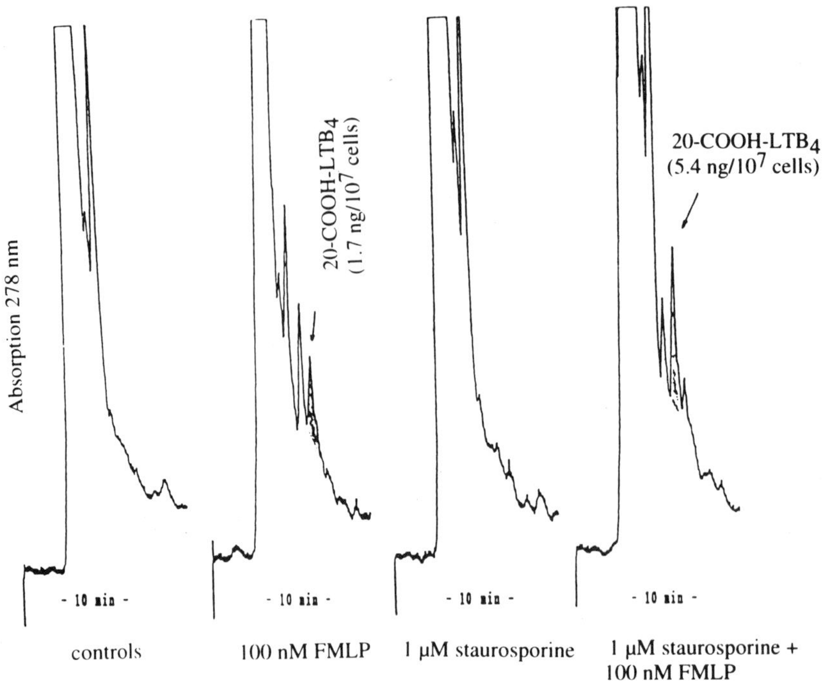

Figure 1. Influence of Staurosporin on the formation of 20-COOH-LTB_4 human neutrophils following stimulation with FMLP.

RESULTS AND DISCUSSION

Fig. 1 shows the potentiating effect of 1 μM staurosporine (3-fold) on the formation of 20-carboxy-LTB_4, a metabolite of leukotriene B_4 (LTB_4) in fMLP-activated human neutrophils. This observation is, however, in contradiction to the currently reported mechanism for the release of AA and PAF in different cell systems, in which the inhibition of PKC has been shown to decrease the AA release (7).Thus, it is evident that the $cPLA_2$ described in the literature may not explain alone the mechanism for eicosanoid and PAF formation in human neutrophils.

It was therefore of interest to elucidate the structure of $cPLA_2$ in human neutrophils and compare it with the known one. We synthesized four different primers A_1, A_2, B_1 and B_2 from the known $cPLA_2$-sequence of human monocytic cell line U937 (Fig. 2) and expected three different fragments by using different constellation of the primers. The polymerase chain reaction (PCR) of the cDNA from human neutrophils enriched three different fragments with specific combination of the primers. The combination of primers A_1 and A_2 gave a small fragment of approx. 0.4 kb (Fig.3, right panel). The $cPLA_2$ gene from U937 cell line served as control (Fig.3, left panel). This fragment contained the CalB-domain, which was then cloned and sequenced.

After translation of the $cPLA_2$-sequence from human neutrophils in the amino acid sequence, the first 44 amino acids were used for an alignment with the known $cPLA_2$-sequences of U937 cell line and other Ca^{2+}-dependent enzymes. The alignment shown in

```
A1 ------------->
ATG CTT GAT ACT CCA GAT CCC TAT GTG GAA CTT TTT ATC TCT ACA ACC CCT GAC AGC AGG AAG AGA ACA AGA CAT TTC AAT AAT
 M   L   D   T   P   D   P   Y   V   E   L   F   I   S   T   T   P   D   S   R   K   R   T   R   H   F   N   N

GAC ATA AAC CCT GTG TGG AAT GAG ACC TTT GAA TTT ATT TTG GAT CCT AAT CAG GAA AAT GTT TTG GAG ATT ACG TTA ATG GAT
 D   I   N   P   V   W   N   E   T   F   E   F   I   L   D   P   N   Q   E   N   V   L   E   I   T   L   M   D

GCC AAT TAT GTC ATG GAT GAA ACT CTA GGG ACA GCA ACA TTT ACT GTA TCT TCT ATG AAG GTG GGA GAA AAG AAA GAA GTT CCT
 A   N   Y   V   M   D   E   T   L   G   T   A   T   F   T   V   S   S   M   K   V   G   E   K   K   E   V   P
                                            B1 ------------------->
TTT ATT TTC AAC CAA GTC ACT GAA ATG GTT CTA GAA ATG TCT CTT GAA GTT TGC TCA TGC CCA GAC CTA CGA TTT AGT ATG GCT
 F   I   F   N   Q   V   T   E   M   V   L   E   M   S   L   E   V   C   S   C   P   D   L   R   F   S   M   A
                         <----------------------- A2
CTG TGT GAT CAG GAG AAG ACT TTC AGA CAA CAG AGA AAA GAA CAC ATA AGG GAG AGC ATG AAG AAA CTC TTG GGT CCA AAG AA
T
 L   C   D   Q   E   K   T   F   R   Q   Q   R   K   E   H   I   R   E   S   M   K   K   L   L   G   P   K   N

AGT GAA GGA TTG CAT TCT GCA CGT GAT GTG CCT GTG GTA GCC ATA TTG GGT TCA GGT GGG GGT TTC CGA GCC ATG GTG GGA TTC
 S   E   G   L   H   S   A   R   D   V   P   V   V   A   I   L   G   S   G   G   G   F   R   A   M   V   G   F

TCT GGT GTG ATG AAG GCA TTA TAC GAA TCA GGA ATT CTG GAT TGT GCT ACC TAC GTT GCT GGT CTT TCT GGC TCC ACC TGG TAT
 S   G   V   M   K   A   L   Y   E   S   G   I   L   D   C   A   T   Y   V   A   G   L   S   G   S   T   W   Y

ATG TCA ACC TTG TAT TCT CAC CCT GAT TTT CCA GAG AAA GGG CCA GAG GAG ATT AAT GAA GAA CTA ATG AAA AAT GTT AGC CAC
 M   S   T   L   Y   S   H   P   D   F   P   E   K   G   P   E   E   I   N   E   E   L   M   K   N   V   S   H

AAT CCC CTT TTA CTT CTC ACA CCA CAG AAA GTT AAA AGA TAT GTT GAG TCT TTA TGG AAG AAG AAA AGC TCT GGA CAA CCT GTC
 N   P   L   L   L   L   T   P   Q   K   V   K   R   Y   V   E   S   L   W   K   K   K   S   S   G   Q   P   V

ACC TTT ACT GAT ATC TTT GGG ATG TTA ATA GGA GAA ACA CTA ATT CAT AAT AGA ATG AAT ACT ACT CTG AGC AGT TTG AAG GAA
 T   F   T   D   I   F   G   M   L   I   G   E   T   L   I   H   N   R   M   N   T   T   L   S   S   L   K   E

AAA GTT AAT ACT GCA CAA TGC CCT TTA CCT CTT TTC ACC TGT CTT CAT GTC AAA CCT GAC GTT TCA GAG CTG ATG TTT GCA GAT
 K   V   N   T   A   Q   C   P   L   P   L   F   T   C   L   H   V   K   P   D   V   S   E   L   M   F   A   D

TGG GTT GAA TTT AGT CCA TAC GAA ATT GGC ATG GCT AAA TAT GGT ACT TTT ATG GCT CCC GAC TTA TTT GGA AGC AAA TTT TTT
 W   V   E   F   S   P   Y   E   I   G   M   A   K   Y   G   T   F   M   A   P   D   L   F   G   S   K   F   F

ATG GGA ACA GTC GTT AAG AAG TAT GAA GAA AAC CCC TTG CAT TTC TTA ATG GGT GTC TGG GGC AGT GCC TTT TCC ATA TTG TTC
 M   G   T   V   V   K   K   Y   E   E   N   P   L   H   F   L   M   G   V   W   G   S   A   F   S   I   L   F

AAC AGA GTT TTG GGC GTT TCT GGT TCA CAA AGC AGA GGC TCC ACA ATG GAG GAA GAA TTA GAA AAT ATT ACC ACA AAG CAT ATT
 N   R   V   L   G   V   S   G   S   Q   S   R   G   S   T   M   E   E   E   L   E   N   I   T   T   K   H   I

GTG AGT AAT GAT AGC TCG GAC AGT GAT GAT GAA TCA CAC GAA CCC AAA GGC ACT GAA AAT GAA GAT GCT GGA AGT GAC TAT CAA
 V   S   N   D   S   S   D   S   D   D   E   S   H   E   P   K   G   T   E   N   E   D   A   G   S   D   Y   Q

AGT GAT AAT CAA GCA AGT TGG ATT CAT CGT ATG ATA ATG GCC TTG GTG AGT GAT TCA GCT TTA TTC AAT ACC AGA GAA GGA CGT
 S   D   N   Q   A   S   W   I   H   R   M   I   M   A   L   V   S   D   S   A   L   F   N   T   R   E   G   R

GCT GGG AAG GTA CAC AAC TTC ATG CTG GGC TTG AAT CTC AAT ACA TCT TAT CCA CTG TCT CCT TTG AGT GAC TTT GCC ACA CAG
 A   G   K   V   H   N   F   M   L   G   L   N   L   N   T   S   Y   P   L   S   P   L   S   D   F   A   T   Q

GAC TCC TTT GAT GAT GAT GAA CTG GAT GCA GCT GTA GCA GAT CCT GAT GAA TTT GAG CGA ATA TAT GAG CCT CTG GAT GTC AAA
 D   S   F   D   D   D   E   L   D   A   A   V   A   D   P   D   E   F   E   R   I   Y   E   P   L   D   V   K

AGT AAA AAG ATT CAT GTA GTG GAC AGT GGG CTC ACA TTT AAC CTG CCG TAT CCC TTG ATA CTG AGA CCT CAG AGA GGG GTT GAT
 S   K   K   I   H   V   V   D   S   G   L   T   F   N   L   P   Y   P   L   I   L   R   P   Q   R   G   V   D

CTC ATA ATC TCC TTT GAC TTT TCT GCA AGG CCA AGT GAC TCT AGT CCT CCG TTC AAG GAA CTT CTA CTT GCA GAA AAG TGG GCT
 L   I   I   S   F   D   F   S   A   R   P   S   D   S   S   P   P   F   K   E   L   L   L   A   E   K   W   A

AAA ATG AAC AAG CTC CCC TTT CCA AAG ATT GAT CCT TAT GTG TTT GAT CGG GAA GGG CTG AAG GAG TGC TAT GTC TTT AAA CCC
 K   M   N   K   L   P   F   P   K   I   D   P   Y   V   F   D   R   E   G   L   K   E   C   Y   V   F   K   P

AAG AAT CCT GAT ATG GAG AAA GAT TGC CCA ACC ATC ATC CAC TTT GTT CTG GCC AAC ATC AAC TTC AGA AAG TAC AAG GCT CCA
 K   N   P   D   M   E   K   D   C   P   T   I   I   H   F   V   L   A   N   I   N   F   R   K   Y   K   A   P

GGT GTT CCA AGG GAA ACT GAG GAA GAG AAA GAA ATC GCT GAC TTT GAT ATT TTT GAT GAC CCA GAA TCA CCA TTT TCA ACC TTC
 G   V   P   R   E   T   E   E   E   K   E   I   A   D   F   D   I   F   D   D   P   E   S   P   F   S   T   F

AAT TTT CAA TAT CCA AAT CAA GCA TTC AAA AGA CTA CAT GAT CTT ATG CAC TTC AAT ACT CTG AAC AAC ATT GAT GTG ATA AAA
 N   F   Q   Y   P   N   Q   A   F   K   R   L   H   D   L   M   H   F   N   T   L   N   N   I   D   V   I   K

GAA GCC ATG GTT GAA AGC ATT GAA TAT AGA AGA CAG AAT CCA TCT CGT TGC TCT GTT TCC CTT AGT AAT GTT GAG GCA AGA AGA
 E   A   M   V   E   S   I   E   Y   R   R   Q   N   P   S   R   C   S   V   S   L   S   N   V   E   A   R   R

TTT TTC AAC AAG GAG TTT CTA AGT AAA CCC AAA GCA TAG
 F   F   N   K   E   F   L   S   K   P   K   A   *
                                <------------- B2
```

Figure 2. Nucleotide sequence and deduced amino acid sequence of human $cPLA_2$.

Fig. 4 illustrates the homology between $cPLA_2$s from human neutrophils and U937 cell line, i.e. they possess 100% identical amino acids. Thus, we conclude that the $cPLA_2$ from human neutrophils is identical to that reported from U937 cell line.

ACKNOWLEDGMENT

Authors wish to thank Dr. James D. Clark, Genentech, Boston for providing $cPLA_2$ plasmid and antibodies. The technical assistance of Vessna Mirkovic and Janet Scharein is thankfully acknowledged.

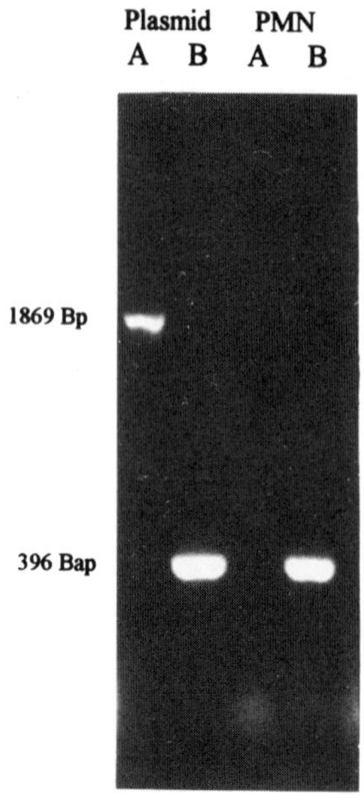

Figure 3. Amplification of $cPLA_2$ gene coding the CalB domain of human netrophils.

```
cPLA2(PMN)     M---L-DTPDPYVELFIS--TTPDSRKRTRHFN-NDINPVWNET-FEFIL-DP      1-  44
cPLA2(U937)    M---L-DTPDPYVELFIS--TTPDSRKRTRHFN-NDINPVWNET-FEFIL-DP     38-  81
PKC-γ          MDP-N-GLSDPYVKLKLIPDPRNLTKQKTRTVK-ATLNPVWNET-FVFNI-KP    186- 233
p65a           M---G-GTSDPYVKVFLL--PEKKKKFETKVHR-KTLNPVFNEQ-FTFKV--P    173- 215
p65b           MDV-G-GLSDPYVKIHLMQNGKRLKKKKTTIKK-HTLNPYYNES-FSFEV--P    302- 348
GAP            LPV-K-HFTNPYCNIYLN----SVQVAKTHAR--EGQNPVWSEE-FVFDD-LP    603- 645
PLC-γ1         LPNRGRGIVCPFVEIEVAGAEYDSIKQKTEFVVDNGLNPVWPAKPFHFQISNP   1098-1150
PLC-γ1         MRD-E-AF-DPFDKSSL------------------------------------   1067-1080
```

Figure 4. Amino acid alignment of homologous CalB domain.

This study was supported by the Deutsche Forschungsgemeinschaft, Bonn, Germany (Ni 242/9–1).

REFERENCES

1. Abdel-Latif, AA (1986) Pharmacol. Rev., 38, 227–272.
2. Berridge, MJ (1987) Ann. Rev. Biochem., 56, 159–193.
3. Hanahan, DJ (1986) Ann. Rev. Biochem., 55, 483–509.
4. Lin, LL et al. (1993) Cell, 72, 269–278.
5. Clark, JD et al. (1991) Cell, 65, 1043–1051.
6. Channon, JY & Leslie, CC (1990) J. Biol. Chem., 265, 5409–5413.
7. Lin, LL et al. (1992) P.N.A.S. USA, 89, 6147–6151.
8. Nigam et al. (1992) Biochim. Biophys. Acta, 1135, 301–308.
9. Müller, S & Nigam, S (1992) Eur. J. Pharmacol., 218, 251–258.
10. Nigam, S. et al. (1995) Adv. Prostagl. Thr. Leukotr. Res., 23, 471–475.
11. Clark, JD et al. (1991) Cell, 65, 1043
12. Sharp, JD et al. (1991) J. Biol. Chem., 266, 14850–14853

CHARACTERIZATION OF PLASMALOGEN-SELECTIVE PHOSPHOLIPASE A_2 FROM BOVINE BRAIN

Hsiu-Chiung Yang,[1] Akhlaq A. Farooqui,[1] and Lloyd A. Horrocks[1,2]

[1]Department of Medical Biochemistry
Ohio State University
1645 Neil Ave., Room 479, Columbus, Ohio 43210–1218
[2]Neurovation Inc.
1275 Kinnear Road, Columbus, Ohio 43212–1155

1. ABSTRACT

Plasmalogens are hydrolyzed by a plasmalogen-selective phospholipase A_2. This enzyme, purified from bovine brain, does not require Ca^{2+} and is localized in cytosol. It has a molecular mass of 39 kDa and is strongly inhibited by glycosaminoglycans, gangliosides, and sialoglycoproteins. These molecules may be involved in the regulation of its enzymic activity. Plasmalogen-selective phospholipase A_2 plays an important role in the release of free fatty acids and platelet-activating factor during trauma.

2. INTRODUCTION

Plasmalogens are a special type of glycerophospholipid with a vinyl ether linkage at the *sn*-1 position of the glycerol back bone. They are found in heart, brain, muscle cells, immune cells, neutrophils, and macrophages, with ethanolamine plasmalogens ten-fold higher than choline plasmalogens except in muscle (Horrocks and Sharma, 1982). The role of plasmalogens in mammalian metabolism is not fully understood. Because of the *sn-1* vinyl ether linkage, plasmalogens are more susceptible to oxidative stress than other phospholipids and cells with a high plasmalogen content are more resistant to photosensitized killing. Plasmalogens may act as a singlet oxygen quencher to protect Chinese hamster ovary cells from oxidative stress. Plasmalogens have also been reported to act as chain-breaking antioxidant and protect LDL from oxidative stress (Engelmann et al., 1994; Horrocks and Sharma, 1982; Paltauf, 1994). However, recent studies suggest that plasmalogen epoxides, peroxidized plasmalogens, which are converted into α-hydroxyaldehydes, are extremely toxic to cells. The α-hydroxyaldehydes increase at a higher rate in

Platelet-Activating Factor and Related Lipid Mediators 2
edited by Nigam *et al.*, Plenum Press, New York, 1996

LDL compared to VLDL and HDL. This may explain the deleterious effects of peroxidized LDL in atherosclerosis. Thus, peroxidation of plasmalogens seems to be more harmful than protective for the human body (Felde and Spiteller, 1995).

Plasmalogens act as a reservoir for polyunsaturated fatty acids and may play important roles physiologically and pathologically. Thus in addition to serving as building blocks of cellular membranes, plasmalogens provide precursors for the generation of eicosanoids and platelet-activating factor (PAF). Low levels of these second messengers have neurotrophic effects, but at high concentration they are cytotoxic and may be involved in allergic response, inflammation, and neural trauma (Farooqui et al., 1995). Plasmalogens may also be involved in the transport of ions across cell membrane and membrane fusion (Paltauf, 1994; Farooqui et al., 1995). This article discusses the enzymic degradation of plasmalogens by phospholipases and the properties and regulation of these enzymes.

3. DISCUSSION

3.1 Degradation of Plasmalogens by Phospholipases

Plasmalogens are hydrolyzed by plasmalogen-selective phospholipase A_2 (PLA_2), plasmalogenase, and lysoplasmalogenase. Plasmalogen-selective PLA_2 cleaves the *sn-2* ester bond with the generation of free fatty acid and lysoplasmalogen. The *sn-1* alk-1′-enyl ether bond of plasmalogens and lysoplasmalogens is hydrolyzed by plasmalogenase and lysoplasmalogenase, respectively. Choline and ethanolamine lysoplasmalogen-specific lysoplasmalogenases have been partially purified from rat liver and heart microsomes (Jurkowitz-Alexander et al., 1989). The role of these enzymes in heart and brain tissue is not known.

The degradation of plasmalogens by phospholipases may be a receptor-mediated process. It has been proposed that the stimulation of receptors on the neuronal cell surface by agonists results in the stimulation of the Ca^{2+}-independent plasmalogen-selective PLA_2. The activation of this PLA_2 causes the hydrolysis of plasmalogens with the liberation of free fatty acids, mainly arachidonic acid, and lysoplasmalogen. Arachidonic acid is metabolized to eicosanoids and lysoplasmalogen can be either reacylated or hydrolyzed by lysoplasmalogenase. These events may be the first wave of second messenger generation during receptor stimulation and signal transduction. The lysoplasmalogen may change the membrane fluidity and permeability and allow Ca^{2+} entry. Ca^{2+} ions may act as second messengers initially but a high level of Ca^{2+} can produce cytotoxic effects (Farooqui et al., 1995).

3.2 Properties of Plasmalogen-Selective Phospholipase A_2

To date, two Ca^{2+}-independent plasmalogen-selective PLA_2 have been purified, one from heart, and the other from brain. The properties of these two enzymes are shown in Table 1. The myocardial Ca^{2+}-independent plasmalogen-selective PLA_2 was purified from canine, human, and rabbit heart. This enzyme has a molecular mass of 40 kDa and is tightly associated with a polypeptide that is apparently identical to 85 kDa phosphofructokinase. The association between the myocardial PLA_2 and phosphofructokinase suggests that the myocardial PLA_2 may be regulated by glycolysis. The myocardial PLA_2 is activated by ATP and other nucleotides and inhibited by DTNB and Triton X-100. This en-

Table 1. Physicochemical and kinetic properties of Ca^{2+}-independent plasmalogen-selective PLA_2 from human myocardium and bovine brain

Property	Human myocardial PLA_2	Bovine brain PLA_2
Localization	Cytosol	Cytosol
Molecular Mass	40 kDa	39 kDa
Ca^{2+} Effect	No effect	No effect
ATP	Stimulates	Inhibits
DTNB	Inhibits	Inhibits
Triton X-100	Inhibits	Stimulates
pH optimum	7.0	7.4
Km value (μM)	4.0	40.0
Vmax (nmol/min/mg)	194,000	65.0
Substrate	Choline plasmalogen	Ethanolamine plasmalogen

zyme is strongly inhibited by the suicide inhibitor, bromoenol lactone, which is the specific inhibitor for this enzyme (IC_{50} ~30 nM) (Hazen and Gross, 1992).

The other plasmalogen-selective PLA_2 has been purified from the cytosolic fraction of bovine brain. It has a molecular mass of 39 kDa. In contrast to the myocardial enzyme, the brain PLA_2 is not affected by ATP and other nucleotides in the micromolar range, but is markedly inhibited by these nucleotides at 2 mM or above, the normal intracellular concentration of ATP. Non-ionic detergents, Triton X-100 and Tween-20, stimulate the enzymic activity. Other detergents, such as sodium deoxycholate, taurocholate, and octylglucoside inhibit this enzyme. Unlike the myocardial enzyme, the suicide inhibitor, bromoenol lactone, has no effect on brain PLA_2. Like heart PLA_2, the bovine brain enzyme is inhibited by the SH-group blocking agent, DTNB. Besides DTNB, other SH-group blocking agents, iodoacetate and N-ethylmaleimide (NEM), inhibit brain PLA_2 as well, but with a lesser degree of inhibition. The 39 kDa bovine brain cytosolic PLA_2 is inhibited by various polyvalent anions (citrate > sulfate > phosphate), and metal ions (Ag^+, Hg^{2+} and Fe^{3+}). Mg^{2+} does not affect the enzymic activity. The inhibition of the brain PLA_2 by polyvalent anions and metal ions may be involved in the regulation of this enzyme.

Glycosaminoglycans (GAGs) and sialoglycoconjugates are highly negatively charged molecules present on the cell surface. They play an important role in the cell adhesion, recognition and receptor functions. Our previous studies (Yang et al., 1994) have shown that bovine brain cytosolic PLA_2 is markedly inhibited by these molecules. The inhibition pattern of GAGs on the brain PLA_2 was heparan sulfate > hyaluronic acid > chondroitin sulfate > heparin. The 39 kDa PLA_2 is also inhibited by N-acetylneuraminic acid (NANA), gangliosides (GM3 > GM1), and sialoglycoproteins (mucin > Cowper's gland mucin > fetuin), with the pattern of NANA > gangliosides > sialoglycoproteins. Other glycoconjugates, such as colominic acid (poly 2,8-N-acetylneuraminic acid), cerebroside, ceramide, and sulfatide have no effect on PLA_2 activity. The simple sugars, glucose and sucrose, have no effect on enzymic activity.

3.3 Plasmalogen-Selective Phospholipase A_2 and Platelet-Activating Factor (PAF)

The synthesis of PAF and release of arachidonate is closely linked. Thus the stimulation of neutrophils activates a phospholipase A_2 which releases arachidonate from 1-O-alk-1′-enyl-2-acyl-Gro*P*Etn with the formation of alkenyl-lyso-Gro*P*Etn. The latter serves as an acceptor for the transfer of arachidonate from 1-O-alkyl-2-arachidonyl-Gro*P*Cho. This reaction releases lyso-PAF which is acetylated to PAF. Thus PLA_2 acting on ethanolamine plasmalogen may play a role in the direct hydrolysis of 1-O-alkyl-2-arachidonoyl-Gro*P*Cho by Ca^{2+}-independent phospholipase A_2 in PAF synthesis (Nieto et al., 1988; Uemura et al., 1991). This mechanism does not eliminate a role of direct hydrolysis of 1-O-alkyl-2-arachidonyl-Gro*P*Cho by Ca^{2+}-dependent PLA_2 (Kramer et al., 1988).

3.4 Regulation of Plasmalogen-Selective Phospholipase A_2

The release of arachidonic acid from neuronal membrane may be a receptor-mediated process. In a rat thoracic aortic smooth muscle cell line (A-10), cytosolic PLA_2 is linked to the vasopressin receptor. The release of arachidonic acid upon the stimulation of vasopressin receptors can be blocked by a Ca^{2+}-independent plasmalogen-selective PLA_2 inhibitor, bromoenol lactone. In neuronal cells, the release of arachidonic acid is coupled to the stimulation of NMDA and muscarinic receptors and is linked to Ca^{2+}-independent PLA_2 (Farooqui and Horrocks, 1991). The receptor for brain plasmalogen-selective PLA_2 is not known. However, in general activities of phospholipases are regulated by the elevation of intracellular Ca^{2+}, interactions with stimulatory and inhibitory proteins, and covalent modification of these enzymes.

Bovine brain plasmalogen-selective Ca^{2+}-independent PLA_2 is inhibited by GAGs and sialoglycoconjugates. The interaction of PLA_2 with these molecules may be involved in anchoring during translocation of this enzyme to the inner leaflet of the lipid bilayer and this may be involved in regulation. Ligand-induced alterations in protein-protein interactions may also be involved in the modulation of myocardial plasmalogen-selective PLA_2. Neuropeptides with stimulatory and inhibitory properties may also be involved in the regulation of brain PLA_2. The regulation of PLA_2 is very complex. It may vary from one cell type to another. It is very likely to have superimposed mechanisms. This is very important for the understanding of signal transduction and the dynamics of neuronal cell regulation.

ACKNOWLEDGMENT

Supported by research grants NS-10165 and NS-29441 from the National Institutes of Health, U. S. Public Health Service.

4. REFERENCES

Engelmann, B., Bräutigam, C., and Thiery, J. 1994, Plasmalogen phospholipids as potential protectors against lipid peroxidation of low density lipoproteins, *Biochem. Biophys. Res. Commun.* 204:1235.

Farooqui, A.A., Yang, H.-C., and Horrocks, L.A. 1995, Plasmalogens, phospholipases A2, and signal transduction, *Brain Res. Rev.* (in press).

Farooqui, A.A. and Horrocks, L.A. 1991, Excitatory amino acid receptors, neural membrane phospholipid metabolism and neurological disorders, *Brain Res. Rev.* 16:171.

Felde, R. and Spiteller, G. 1995, Plasmalogen oxidation in human serum lipoproteins, *Chem. Phys. Lipids* 76:259.

Hazen, S.L. and Gross, R.W. 1992, Identification and characterization of human myocardial phospholipase A2 from transplant recipients suffering from end-stage ischemic heart disease, *Circ. Res.* 70:486.

Horrocks, L.A. and Sharma, M. 1982, Plasmalogens and O-alkyl glycerophospholipids, in: "Phospholipids, New Comprehensive Biochemistry, Vol. 4," 51–93, J.N. Hawthorne, G.B. Ansell, eds., Elsevier Biomedical Press, Amsterdam.

Jurkowitz-Alexander, M., Ebata, H., Mills, J.S., Murphy, E.J., and Horrocks, L.A. 1989, Solubilization, purification, and characterization of lysoplasmalogen alkenylhydrolase (lysoplasmalogenase) from rat liver microsomes, *Biochim. Biophys. Acta* 1002:203.

Kramer, R.M., Jakubowski, J.A., and Deykin, D. 1988, Hydrolysis of 1-alkyl-2-arachidonoyl-sn-glycero-3-phosphocholine, a common precursor of platelet-activating factor and eicosanoids, by human platelet phospholipase A2, *Biochim. Biophys. Acta* 959:269.

Nieto, M.L., Velasco, S., and Crespo, M.S. 1988, Biosynthesis of platelet-activating factor in human polymorphonuclear leukocytes, *J. Biol. Chem.* 263:2217.

Paltauf, F. 1994, Ether lipids in biomembranes, *Chem. Phys. Lipids* 74:101.

Uemura, Y., Lee, T., and Snyder, F. 1991, A coenzyme A-independent transacylase is linked to the formation of platelet-activating factor (PAF) by generating the lyso-PAF intermediate in the remodeling pathway, *J. Biol. Chem.* 266:8268.

Yang, H.-C., Farooqui, A.A., and Horrocks, L.A. 1994, Effects of glycosaminoglycans and glycosphingolipids on cytosolic phospholipases A2 from bovine brain, *Biochem. J.* 299:91.

50

STRUCTURE AND FUNCTION OF PHOSPHOLIPASE A_2 RECEPTOR

Kohji Hanasaki and Hitoshi Arita

Shionogi Research Laboratories
Shionogi & Co., Ltd.
5-12-4, Sagisu, Fukushima-ku, Osaka 553, Japan

1. INTRODUCTION

Phospholipase A_2 (EC 3.1.1.4: PLA2) is the group name of enzymes that cleave an acyl ester bond at the *sn*-2 position of glycerophospholipid (1). Mammalian PLA2s are classified into two types, secretory 14 kDa PLA2 and cytosolic higher molecular weight of PLA2 (2). Secretory PLA2 are further classified into two subgroups, group I and group II, on the basis of their characteristics in primary structure (3). Group II PLA2 (PLA2-II) is abundant at inflammatory loci and its expression is modulated by various inflammatory cytokines, and thus this type of PLA2 is thought to play some roles in the pathogenesis of inflammation (4,5). On the other hand, group I PLA2 (PLA2-I) is mainly secreted from the pancreas and the major physiological function of PLA2-I has long been thought to be digestion of phospholipids in nutrients (6). However, recent studies have revealed the presence of this type of PLA2 in several non-digestive organs such as lung and spleen (7). What is the biological role of PLA2-I in non-digestive organs? This question led us to initiate the series of studies on the PLA2 receptor.

2. FUNCTION OF PLA2 RECEPTOR

2.1 Characterization of PLA2 Receptor

In several cell types, we provided the evidence of the presence of specific receptor for PLA2-I in a binding assay system using ^{125}I-labeled PLA2-I as a ligand. The receptor was found to recognize a mammalian mature type of PLA2-I with a dissociation constant of about 1 nM (8). Mature form of PLA2-I was known to be generated from its inactive precursor proenzyme (proPLA2-I) via proteolytic enzymes, such as trypsin in the intestine (9) or plasmin in the tissues (10). The PLA2 receptor did not recognize proPLA2-I as well as PLA2-II and commercial available snake and bee venom PLA2. The receptor is expressed in various cell

Platelet-Activating Factor and Related Lipid Mediators 2
edited by Nigam *et al.*, Plenum Press, New York, 1996

types prepared from rat, mouse, bovine, porcine and human, but not expressed in single cells such as erythrocytes, platelets, mast cells and macrophages (11).

2.2 Biological Responses via PLA2 Receptor

Rat renal glomerular mesangial cells possessed the PLA2 receptor with a Kd value of 1 nM and 4,000 sites per cell (12). Since mesangial cells are well known as an abundant source for eicosanoids production, we examined the effect of PLA2-I on the prostaglandin E_2 (PGE2) production. PLA2-I was found to elicit PGE2 production with a lag time of about 10 hr after its addition in cultured rat mesangial cells. This enhanced effect was specific to PLA2-I, and its inactive precursor proPLA2-I and PLA2-II could not induce PGE2 generation. These data were correlated well with the binding specificity of PLA2 receptor. During this activation process, PLA2-I did not significantly induce or activate both cytosolic PLA2 and cyclooxygenase. Northern blotting analysis and immunoblotting analysis revealed that PLA2-I enhanced the PLA2-II mRNA levels and the secretion of PLA2-II protein in a similar time course as PGE_2 production. Thus, PLA2-I elicits PLA2-II induction and secretion via the PLA2 receptor in rat mesangial cells, which might involve in the PGE_2 production. Glucocorticoid derivative, dexamethasone, was found to suppress these PLA2-I-induced increases in PGE_2 production as well as PLA2-II gene expression. In order to study a possible involvement of phospholipid-hydrolyzing activity into the receptor-mediated activation process, several mutants of porcine PLA2-I was prepared (13,14). We focused on Gly^{30}, which is located in the Ca^{2+} binding loop and directly involves in the catalytic activity, as well as Leu^{31}, that also forms a part of interfacial binding surface and involves in the substrate binding (15). Site-directed mutagenesis of Leu^{31} to Arg^{31} or Ser^{31} resulted in the decrease in both receptor binding and enzymatic activities. In contrast, when Gly^{30} was replaced by a Ser^{30}, this mutant almost completely lost its enzymatic activity, but retained full receptor binding activity (14). In rat mesangial cell system, this enzymatically inactive Gly^{30} mutant induced PGE_2 synthesis with the same potency as the wild-type PLA2-I. Since Gly^{30} is located in the Ca^{2+} binding loop, the mutation probably reduces the Ca^{2+} binding affinity that is important for enzymatic activity but not for binding activity. Thus, these data clearly demonstrate that PLA2-I-induced biological responses via the receptor do not depend on its catalytic activity.

Table 1 summarizes the PLA2-I-induced biological responses via the PLA2 receptor so far identified. PLA2-I can induce various cellular responses such as proliferation (11,

Table 1. Physiological responses against PLA2-I stimulation via the PLA2 receptor

Responses	Cells and tissues
Proliferation	Swiss 3T3 cells
	Rat vascular smooth muscle cells
	Rat chondrocytes
	Rat synovial cells
Chemokinetic cell migration	A7r5 cells (Rat embryonic aortic cell line)
Eicosanoid formation	Guinea pig parenchyma
	Porcine cerebral arteries
	Mouse osteoblastic cells
	Rat renal mesangial cells
Progesterone release	Rat corpora lutea

16, 17), chemokinetic cell migration (18), eicosanoid formation (12, 19–21) and steroid hormone release (22). These effects are elicited by mammalian mature type of PLA2-I, and not by proPLA2-I, PLA2-II and venom PLA2, thus demonstrating the receptor-mediated responses. The differences in the PLA2-elicited responses among several cell types might depend on the discrepancies in the signal transduction machinery under the PLA2 receptor-mediated activation process, which remains obscure at present.

3. STRUCTURE OF PLA2 RECEPTOR

The PLA2 receptor was purified from the membranes of bovine corpus lutea using a PLA2-I affinity column chromatography (23). Based on partial amino acid sequences from the purified receptor, its cDNA was cloned and sequenced (24). The deduced primary structure contains all of the partial amino acid sequences determined by protein chemical method. The bovine PLA2 receptor composed of 1,463 amino acid residues with canonical motifs of glycosylation, internalization as well as signal and transmembrane sequences. In separate experiments, we found that N-glycosylation of the receptor protein is essential for the ligand recognition (23,25) and also found that PLA2-I was actually internalized and sequentially degraded in the cultured cells after receptor binding (11).

By computer homology search, the structure of the PLA2 receptor was found to be homologous to that of mammalian mannose receptor that is one of the calcium dependent animal lectins with tandem eight CRD domains (26). The domain organization of the PLA2 receptor is then schematically illustrated according to that of mannose receptor (Figure. 1). Although the sequence identities between CRD of mannose receptor and CRD-like domains of PLA2 receptor are not so high, the correspondence can be verified by the fact that CRD-like domains conserve many invariant residues observed in CRD of C-type lectins. In order to confirm the assignment of transmembrane and cytoplasmic domains, genetically engineered mutants were transiently expressed in COS-7 cells. The mutant lacking the predicted cytoplasmic portion was retained on the cell surface, whereas the mutant lacking both the predicted cytoplasmic and transmembrane portions was secreted into the culture medium, both of which had the same binding affinity as the membrane-bound form. In addition, the minimum essential binding sites were found to be the CRD-like domains 3 to 5 in the PLA2 receptor (27). We also calculated the sequence identity between our bovine PLA2 receptor and rabbit M-type PLA2 receptor which has been reported by Dr. Lazdunski's group (28). Overall identity was 81% and the homology was also observed throughout the entire molecule, thus these receptors are supposed to be the products of the same gene.

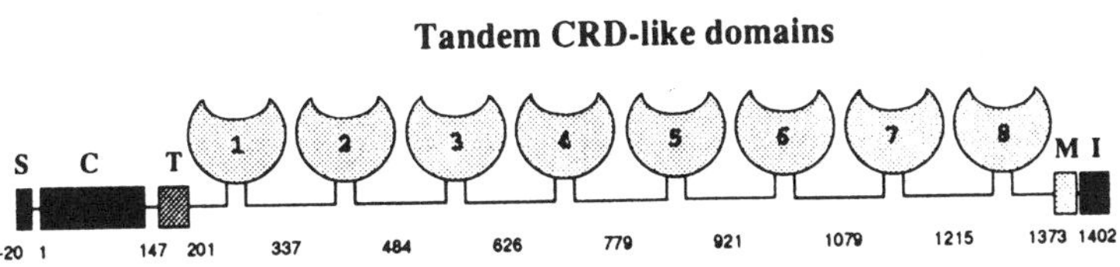

Figure 1. Schematic representation of the domain organization of the PLA2 receptor. The amino acid residue numbers corresponding to the NH_2-terminal residue in each domain are indicated below the schematic drawing. S, signal sequence; C, Cys-rich domain; T, fibronectin type II repeat-like domain; 1–8, CRD-like domains 1–8; M, membrane-spanning region; I, intracellular region.

Drickamer recently categorized CRD-containing proteins into seven subgroups based either on the overall domain organization or on the degree of the similarities in CRD sequences (29). According to this classification, the macrophage mannose receptor belongs to group VI because of being type I receptor as well as having multiple CRD domains. The sequence comparison between the mannose and PLA2 receptors unambiguously indicates that the PLA2 receptor is the second member in this unique C-type lectin subgroup VI. However, the critical five amino acid residues directly involved in Ca^{2+}-dependent sugar binding located in carboxyl terminal parts of CRD domains were found to be totally replaced in the PLA2 receptor, indicating that the PLA2 receptor does not possess the Ca^{2+}-dependent sugar binding activity. In rabbit M-type PLA2 receptor, Lazdunski *et al.* (28) have reported the inhibitory action of BSA-conjugated mannose or galactose against snake venom PLA2 binding. However, these suppressive effects of sugar conjugates were not observed in our PLA2-I receptor system, and labeled mannosylated BSA did not bind to the purified PLA2 receptor. In addition, PLA2-I binding to the receptor is a Ca^{2+}-independent reaction. Taken together, these findings demonstrate the functional diversity of the PLA2 receptor from other C-type lectins.

We then cloned and sequenced mouse and human PLA2 receptors. The overall identity between the amino acid sequences of the mouse and bovine receptors is 70%. Northern blot analysis revealed the presence of mouse PLA2 receptor mRNA in various mouse tissues, especially abundant in liver and kidney (27). Multiple transcripts were observed, which are probably generated through alternative processing after transcription of a single PLA2 receptor gene. We are now making mice that lack the PLA2 receptors by means of a gene-targeting method, and also preparing polyclonal antibody against the mouse PLA2 receptor using a recombinant soluble receptor as an antigen. These approaches will allow us to understand the physiological function of PLA2-I through the binding to the PLA2 receptor.

REFERENCES

1. Arita, H., Nakano, T. and Hanasaki, K. Prog. Lipid Res. **28**, 273–301 (1989).
2. Dennis, E. A. J. Biol. Chem. **269**, 13057–13060 (1994).
3. Kudo, I., Murakami, M., Hara, S. and Inoue, K. Biochim. Biophys. Acta **117,** 217-231 (1993).
4. Nakano, T., Ohara, O., Teraoka, H. and Arita, H. J. Biol. Chem. **265**, 12745–12748 (1990).
5. Oka, S. and Arita, H. J. Biol. Chem. **266**, 9956–9960 (1991).
6. Slotboom, A. J., Verheij, H. M. and De Haas, G. H. New Comprehensive Biochemistry, vol 4, pp. 359–434. Elsevier Science, Amsterdam, 1982.
7. Sakata, T., Nakamura, E., Tsuruta, Y., Tamaki, M., Teraoka, H., Tojo, H., Ono, T. and Okamoto, M. Biochim. Biophys. Acta **1007**, 124–126 (1989).
8. Ohara, O., Ishizaki, J. and Arita, H. Prog. Lipid Res. **34**, 117–138 (1995).
9. Ohara, O., Tamaki, M., Nakamura, E., Tsuruta, Y., Fujii, Y., Shin, M., Teraoka, H. and Okamoto, M. J. Biochem. **99**, 733–739 (1986).
10. Nakano, T., Fujita, H., Kikuchi, N. and Arita, H. Biophys. Biochem. Res. Commun. **198**, 10–15 (1994).
11. Hanasaki, K. and Arita, H. J. Biol. Chem. **267**, 6414–6420 (1992).
12. Kishino, J., Ohara, O., Nomura, K. and Arita, H. J. Biol. Chem. **269,** 5092–5098 (1994).
13. Ishizaki, J., Kishino, J., Teraoka, H., Ohara, O. and Arita, H. FEBS Lett. **324**, 349-352 (1993).
14. Kishino, J., Kawamoto, K., Ishizaki, J., Verheij, H. M., Ohara, O. and Arita, H. J. Biochem. **117**, 420–424 (1995).
15. Dijkstra, B. W., Drenth, J. and Kalk, K. H. Nature **289**, 604–606 (1981).
16. Arita, H., Hanasaki, K., Nakano, T., Oka, S., Teraoka, H. and Matsumoto, K. J. Biol. Chem. **266,** 19139–19141 (1991).
17. Kishino, J., Tohkin, M. and Arita, H. Biochem. Biophys. Res. Commun. **186**, 1025-1031 (1992).
18. Kanemasa, T., Hanasaki, K. and Arita, H. Biochim. Biophys. Acta **1125**, 210–214 (1992).

19. Kanemasa, T., Kishino, J. and Arita, H. FEBS Lett. **303**, 217–220 (1992).
20. Nakajima, M., Hanasaki, K. and Arita, H. FEBS Lett. **309,** 261–264 (1992).
21. Tohkin, M., Kishino, J., Ishizaki, J. and Arita, H. J. Biol. Chem. **268**, 2865–2871 (1993).
22. Nomura, K., Fujita, H. and Arita, H. Endocrinology **135**, 603–609 (1994).
23. Hanasaki, K. and Arita, H. Biochim. Biophys. Acta **1127**, 233–241 (1992).
24. Ishizaki, J., Hanasaki, K., Higashino, K., Kishino, J., Kikuchi, N., Ohara, O. and Arita, H. J. Biol. Chem. **269,** 5897–5904 (1994).
25. Fujita, H., Kawamoto, K., Hanasaki, K. and Arita, H. Biochem. Biophys. Res. Commun. **209**, 293–299 (1995).
26. Taylor, M. E., Conary, J. T., Lennartz, M. R., Stahl, P. D. and Drickamer, K. J. Biol. Chem. **265**, 12156–12162 (1990).
27. Higashino, K., Ishizaki, J., Kishino, J., Ohara, O. and Arita, H. Eur. J. Biochem. **225**, 375–382 (1994).
28. Lambeau, G., Ancian, P., Barhanin, J. and Lazdunski, M. J. Biol. Chem. **269,** 1575-1578 (1994).
29. Drickamer, K. Curr. Opin. Struct. Biol. **3**, 393–400 (1993).

51

PAF-INDUCED MAPK ACTIVATION IS INHIBITED BY WORTMANNIN IN NEUTROPHILS AND MACROPHAGES

Ingvar Ferby, Iwao Waga, Kazuhiko Kume, Chie Sakanaka, and Takao Shimizu

Department of Biochemistry
Faculty of Medicine
The University of Tokyo
Hongo 7-3-1, Bunkyo-ku, Tokyo 113, Japan

1. INTRODUCTION

The signal transduction mediating PAF-induced MAPK activation in CHO-cells stably expressing cloned PAF-receptor, has been extensively studied in our group (1). Pertussis toxin (PTX) blocked about 50% of the PAF-induced MAPK activation, without altering phosphatidylinositol (PI) hydrolysis and calcium responses. It was therefore speculated that the PAF-receptor mediates MAPK activation through one PTX-insensitive pathway that yields PI hydrolysis and possibly activation of conventional protein kinase C (cPKC), and another PTX-sensitive pathway that is likely to be independent of calcium and cPKC but mainly remains obscure.

Wortmannin, a potent inhibitor of PI 3-K (2–4) has been reported to block many functional responses in neutrophils (5, 6). Furthermore, a product of PI 3-K has been shown to stimulate an atypical PKC (PKCζ) in a calcium-insensitive manner *in vitro* (7). These observations raised the possibility that PI 3-K might be involved in a calcium-independent activation of MAPK by PAF.

We here provide evidence that the PAF receptor mediates activation of MAPK, through two distinct pathways, one calcium-dependent and the other calcium-independent, in guinea-pig neutrophils and P388D1 macrophage-like cells. We further demonstrate that a PI 3-K inhibitor, wortmannin, inhibits MAPK activation, on a target directed to the calcium-independent pathway.

Platelet-Activating Factor and Related Lipid Mediators 2
edited by Nigam *et al.*, Plenum Press, New York, 1996

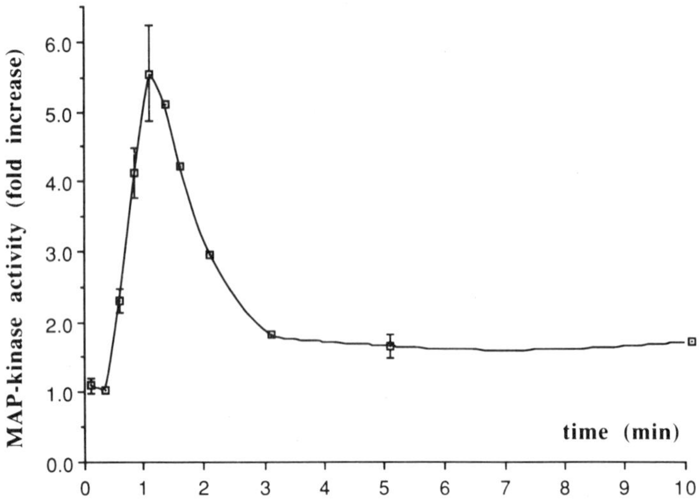

Figure 1. Activation of MAPK by PAF in guinea-pig neutrophils.

Guinea-pig neutrophils (2 x 10^6 cells/ ml) were challenged with 100 nM PAF, and MAPK activity measured in the cell lysates at the indicated times. (Open circles and vertical bars denote the mean and SD of three experiments). Basal radioactivity was 1.3 x 10^4 cpm.

2. ACTIVATION OF MAPK BY PAF

Guinea-pig peritoneal neutrophils were prepared and examined for the activation of MAPK, measured as *in vitro* phosphorylation of a highly specific peptide substrate, following addition of 100 nM PAF. PAF evoked a rapid and highly transient activation of MAPK. The activity peaked 1 min after stimulation, with a 5–6 fold increase, and returned to near basal level already after 3 min (Fig. 1). Like a majority of early responses to PAF, the activity reached a maximal level at a dose of 10–100 nM (data not shown).

Furthermore, PAF evoked a 2–3 fold activation of MAPK in the murine macrophage like cell-line P388D1, in a similar rapid manner as observed for neutrophils (8).

3. WORTMANNIN INHIBITS MAPK ACTIVATION

Figure 2 shows the effect of various doses of wortmannin on PAF-induced MAPK activation in guinea-pig neutrophils. Wortmannin partially (50–60%) inhibited the MAPK activation with a half maximal inhibition observed with 200–300 nM. Wortmannin alone did not alter basal MAPK activity, nor did it when included in the cell extracts after stimulation have any effect on stimulated MAPK activity (not shown).

In the following we examined the effect of wortmannin on phorbol myristate acetate (PMA)-induced activation of MAPK. The results are presented in figure 3. While 1 μM wortmannin partially inhibited PAF-induced MAPK activity, it did not inhibit but rather

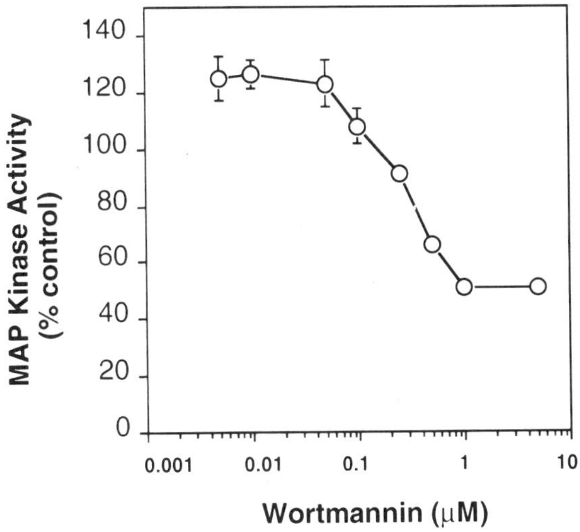

Figure 2. Inhibition of MAPK activation by wortmannin. Neutrophils were incubated with wortmannin for 10 min at 37°C, and stimulated with PAF for 1 min. The cell lysates were then subjected to MAPK measurements. (Open circles and vertical bars denote the mean and SD of three experiments).

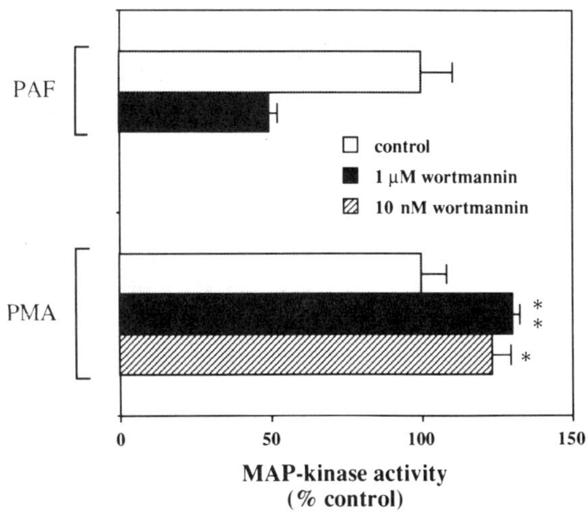

Figure 3. Effects of wortmannin on PAF- or PMA-induced activation of MAPK. MAPK activity of neutrophils stimulated with either 100 nM PAF for 1 min, or 100 nM PMA for 3 min. (The columns and horizontal bars indicate the mean and SD of three experiments). (* <0.05, ** <0.01, Student's t-test). Basal activity (0 % control) correspond to 3.4×10^4 cpm, and 100 % activation with PAF and PMA was 1.3×10^5 and 2.0×10^5 cpm respectively.

slightly enhance the PMA-induced activation. Since PMA activates cPKC and novel PKC these results imply that wortmannin does not affect any signaling components between these PKC isoforms and MAPK.

4. ROLE OF CALCIUM IN THE ACTIVATION OF MAPK

Apparently, a part of the PAF-induced activation of MAPK is mediated through a pathway insensitive to wortmannin. Thus, we next adressed the possibitity that this wortmannin-insensitive pathway might instead depend on the transient increase in intracellular calcium. Loading our neutrophils with the intracellular calcium chelator, BAPTA/AM completely abolished the PAF-induced response (Fig. 4A). However, cells preloaded with BAPTA/AM still responded to PAF with a 40–50% activation of MAPK compared with non-treated cells. When combining BAPTA/AM loading with wortmannin treatment a nearly complete inhibition of the MAPK response occured (Fig. 4B and C), indicating that wortmannin and BAPTA target on two separate pathways both mediating the PAF-induced response. Similar results were obtained for P388D1 cells (8). Treatment with wortmannin did not affect PAF-induced calcium response in guinea-pig neutrophils (Fig. 4A).

5. IN SUMMARY

In the present study we examined the mechanism by which PAF activates MAPK in native cells such as guinea-pig neutrophils and P388D1 macrophage-like cells. We found that PAF activates MAPK through two distinct pathways. One calcium-dependent pathway that likely involves cPKC, and another calcium-independent but wortmannin-sensitive pathway. Using molecular biological methods we are presently examining whether hetrodimeric (p85/p110) type PI 3-kinase is the actual target of wortmannin involved in PAF mediated activation of MAPK.

Figure 4. Effect of wortmannin and BAPTA/AM on Ca^{2+} rises and MAPK activation: **A.** Ca^{2+} transient in cells treated with wortmannin or BAPTA/AM: Guinea pig neutrophils (2 x 10^6 cells/ml) were preincubated with or without 20 μM BAPTA/AM at 25°C for 30 min and/or 1 mM wortmannin at 37°C for 10 min. Fura-2 monitored intracellular Ca^{2+}-increases after addition of 100 nM PAF are shown. *Trace a*, control cells; *trace b*, wortmannin-treated cells; and *trace c*, BAPTA/AM-loaded cells. PAF was added at the time points indicated by the arrows. **B.** Effect of BAPTA/AM and/or wortmannin on MAPK activation: Lanes: 1) Non-stimulated neutrophils; 2) Cells stimulated with 100 nM PAF for 1 min; PAF-stimulated cells pretreated with: 3) wortmannin ; or 4) BAPTA/AM; or 5) both. Preincubational conditions are the same as in Fig. 4 A. The MAPK activity induced by PAF in the absence of any compound was defined as 100% activity. The columns and vertical bars shows the mean and SD of three experiments. (Basal and 100 % activity correspond to $2.6x10^4$ and $1.1x10^5$ cpm respectively). **C.** Effect of BAPTA/AM and/or wortmannin on MAPK gel-shift: Cell lysates (same as Fig. 4 B) were subjected to western blotting with polyclonal anti-MAPK antibody.

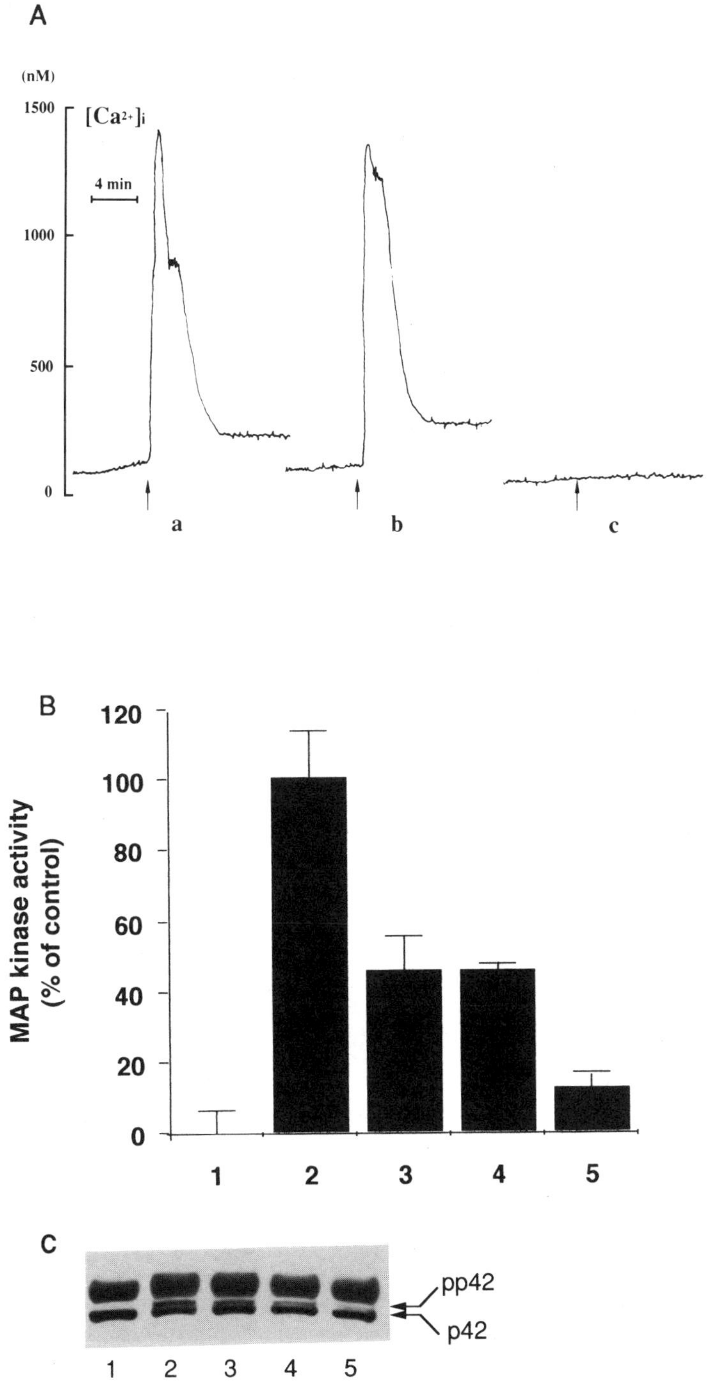
A
(nM)
1500
[Ca2+]i
4 min
1000
500
0
a
b
c
B
MAP kinase activity
(% of control)
120
100
80
60
40
20
0
1
2
3
4
5
C
pp42
p42
1
2
3
4
5

REFERENCES

1. Honda, Z., Takano, T., Gotoh, Y., Nishida, E., Ito, K., and Shimizu, T. (1994) *J. Biol. Chem.* **269**, 2307–2315
2. Arcaro, A., and Wymann, M. P. (1993) *Biochem. J.* **296**, 297–301
3. Okada, T., Sakuma, L., Fukui, Y., Hazeki, O., and Ui, M. (1994) *J. Biol. Chem.* **269**, 3563–3567
4. Woscholski, R., Kodaki, T., McKinnon, M., Waterfield, M. D., and Parker, P. J. (1994) *FEBS Lett.* **342**, 109–114
5. Dewald, B., Thelen, M., and Baggiolini, M. (1988) *J. Biol. Chem.* **263**, 16179–16184
6. Reinhold, S. L., Prescott, S. M., Zimmerman, G. A., and McIntyre, T. M. (1990) *FASEB J.* **4**, 208–214
7. Nakanishi, H., Brewer, K. A., and Exton, J. H. (1993) *J. Biol. Chem.* **268**, 13–16
8. Ferby, I., Waga, I., Kume, K. and Shimizu, T. *Manuscript submitted for publication elsewere*

52

ACTIVATION OF 85 kDa PLA_2 BY EICOSANOIDS IN HUMAN NEUTROPHILS AND EOSINOPHILS

Robert L. Wykle, Jonny Wijkander, Andrew B. Nixon, Larry W. Daniel, and Joseph T. O'Flaherty

Department of Biochemistry
Bowman Gray School of Medicine
Medical Center Blvd.
Winston-Salem, North Carolina 27157–1016

1. INTERRELATIONSHIPS OF PAF SYNTHESIS AND ARACHIDONATE METABOLISM

The initial step in the remodeling pathway of PAF synthesis is catalyzed by phospholipase A_2 (PLA_2).[1] It has generally been accepted that PLA_2 acts directly on 1–*0*-alkyl-2-arachidonoyl-*sn*-glycero-3-phosphocholine (alkyl-2-AA-GPC) to form lyso-PAF (1-alkyl-2-lyso-GPC) which is then acetylated to yield PAF. More recent studies suggest that PAF synthesis may also be initiated indirectly through the action of CoA-independent transacylase following the hydrolysis of 1–*0*-alk-1′-enyl-2-AA-*sn*-glycero-3-phosphoethanolamine (alkenyl-2-AA-GPE) by PLA_2.[2,3] In the putative indirect pathway, an accumulation of alkenyl-2-lyso-GPE triggers the transfer of AA from alkyl-2-AA-GPC to the alkenyl-2-lyso-GPE thus forming lyso-PAF. It seems logical that both the direct and indirect pathways may contribute to PAF synthesis. However, the relative physiological importance of the two pathways has not been established.

Both the direct and indirect routes of synthesis require the action of PLA_2. A number of studies have demonstrated the close interrelationship of PAF and arachidonic acid (AA) metabolism.[4,5] Our studies have focused on human neutrophils, where both PAF and lipoxygenase products are synthesized in parallel upon stimulation of the cells. Further, evidence indicates that the AA-linked molecular species of 1-alkyl-2-acyl-GPC is the obligate precursor of PAF. In other studies 5-hydroxyeicosatetraenoic acid (5-HETE) and leukotriene B_4 (LTB_4) were shown to enhance PAF synthesis in response to other stimuli.[6,7] Evidence from the earlier studies suggested that the lipoxygenase products increased the production of lyso-PAF through the action of PLA_2.

Platelet-Activating Factor and Related Lipid Mediators 2
edited by Nigam *et al.*, Plenum Press, New York, 1996

2. CoA-INDEPENDENT TRANSACYLASE STUDIES

In ionophore A23187-stimulated neutrophils both alkenyl-2-lyso-GPE and lyso-PAF are observed to accumulate. If the CoA-independent transacylase pathway were solely responsible for PAF synthesis, one would expect to see the hydrolysis of the ethanolamine-linked class (PE) before hydrolysis of the choline-linked phosphoglycerides (PC) and appearance of lyso-PAF. We have developed a labeling scheme for labeling the PE class using 1–*0*-[^{3}H]alkyl-2-lyso-GPE in a manner similar to that used earlier to label cells with 1–*0*-alkyl-2-AA-GPC.[8] The plasmalogen, alkenyl-2-acyl-GPE, makes up two thirds of the PE class in human neutrophils and contains greater than 80% of the AA found in PE, the largest pool of AA in the cells.[9] We compared labeling of the cells with [^{3}H]alkenyl-2-lyso-GPE and [^{3}H]alkyl-2-lyso-GPE and found that both are taken up by the cells and specifically acylated with AA. After labeling with these precursors the cells thus contain either [^{3}H]alkenyl-2-AA-GPE or [^{3}H]alkyl-2-AA-GPE incorporated into their membranes. Upon stimulation, both labeled subclasses behaved in the same manner and were deacylated to yield the corresponding ^{3}H-lysophospholipids, which accounted for 15 to 30% of total labeled species. The labeled alkyl-linked precursor is easier to prepare in large quantities and is more stable than the alkenyl-linked subclass upon storage. Labeling with [^{3}]H-alkyl-2-lyso-GPE thus provides a good model for studying PE hydrolysis in the stimulated neutrophils. We carried out studies using the cells labeled with 1-[^{3}H]alkyl-2-AA-GPC and 1-[^{3}H]alkyl-2-AA-GPE. Upon stimulation, [^{3}H]alkyl-2-lyso-GPE and [^{3}H]lyso-PAF both appeared very rapidly and we were unable to conclude whether the transacylase appeared to be responsible for the lyso-PAF formation. Since both products appear to be formed simultaneously, it may be that both the direct and indirect pathways are operative. Our findings have not ruled out either pathway.

3. PRESENCE AND PROPERTIES OF 85 kDa CYTOSOLIC PLA_2 IN NEUTROPHILS

The 85 kDa PLA_2 is selective for AA and is activated by a number of stimuli.[10] Although the AA-specific PLA_2 activity first described by Leslie *et al.*[11] in neutrophils was almost certainly due to the 85 kDa enzyme, until recently, the presence of the 85 kDa PLA_2 in human neutrophils was controversial. Although a number of our earlier studies failed to demonstrate the enzyme in neutrophils, we now routinely observe the activity, possibly because a number of protease and phosphatase inhibitors are now included in the homogenization buffer. The neutrophil enzyme has the same properties as described for the 85 kDa enzyme from platelets and other cells. Thus, when using phosphatidylcholine vesicles as a substrate, it is activated by submicromolar concentrations of Ca^{2+}, is selective for AA species, is not inhibited by dithiothreitol, and elutes like the 85 kDa PLA_2 upon gel chromatography. A band migrating with authentic 85 kDa PLA_2 and binding antibody to the 85 kDa protein was observed upon electrophoresis and Western blot analysis. A shift in its electrophoretic mobility was observed upon stimulation of the neutrophils with the chemotactic peptide fMLP. The shift in mobility suggested phosphorylation as described by others[12] and was accompanied by a 2-fold increase in activity. A number of studies demonstrating the presence of the 85 kDa PLA_2 in human neutrophils have recently been reported.[13]

4. ACTIVATION OF THE 85 kDa PLA_2 BY AA METABOLITES

We found that addition of low concentrations of free AA (<1 μM) to human neutrophils resulted in a 2-fold increase in the PLA_2 activity as measured employing AA-containing PC vesicles. We at first speculated that the AA might be acting directly through protein kinase C to activate the PLA_2. However, we found that 5-lipoxygenase inhibitors blocked the ability of the AA to activate the PLA_2.[13]

In addition, linolenic acid did not activate the PLA_2. The lipoxygenase products LTB_4 and 5-HETE, as well as the recently described[14] 5-oxo-ETE [5-oxo-6,8,11, 14-(E,Z,Z,Z)-eicosatetraenoic acid] were tested and were all found to activate the 85 kDa PLA_2. Maximal activation (2-fold increase) was obtained at 1 nM LTB_4, whereas 0.5 to 1 μM concentrations of 5-oxo-ETE and 5-HETE were required with 5-HETE being the least active. The lipoxygenase inhibitors did not block the activity of the LTB_4, 5-HETE or 5-oxo-ETE. A time course study revealed that LTB_4 almost fully activated the PLA_2 within 1 min while activation by AA required 2 min, in agreement with other indications that conversion of AA to an active eicosanoid was required for it to activate the PLA_2.

We carried out studies using a LTB_4 receptor antagonist LY 255283 and found that it blocked activation of the PLA_2 by AA and by LTB_4 but not by 5-HETE. This finding suggests that the AA acts via formation of LTB_4 and its interaction with its plasma membrane receptor.[13]

We examined the release of AA mass by GC/MS in neutrophils stimulated by LTB_4, 5-HETE or 5-oxo-ETE alone or in combination with fMLP. The lipoxygenase products alone led to little release of endogenous AA, but when added in combination with fMLP resulted in a 2- to 4-fold increase in release over that observed with fMLP alone. In the priming response 5-oxo-ETE was most active followed by 5-HETE and LTB_4.[13] The primed release of endogenous AA in the intact cells appears to require phosphorylaiton and Ca^{2+} influx (fMLP).

These studies suggested that an initial release of AA can lead to further activation of the 85 kDa PLA_2 by conversion to LTB_4 and its interaction with the LTB_4 receptor. This combination of actions could act as a positive feedback mechanism. Alternatively, release of AA by secreted low molecular weight PLA_2 might activate the 85 kDa PLA_2 as observed upon addition of free AA. In order to examine this possibility, pancreatic PLA_2 was added to the cells and shown to activate the 85 kDa PLA_2. The activation by the pancreatic PLA_2 was blocked by the LTB_4 receptor antagonist.

5. PHOSPHORYLATION OF PLA_2 AND MAP KINASE IN RESPONSE TO LIPOXYGENASE PRODUCTS AS INDICATED BY ELECTROPHORETIC MOBILITY SHIFTS

Mobility shifts indicating phosphorylation of the 85 kDa PLA_2 paralleled the increases in activity observed with AA, LTB_4, 5-HETE and 5-oxo-ETE. The mobility shift in response to LTB_4, but not 5-HETE, was blocked by the LTB_4 receptor antagonist. The findings suggest that separate receptors exist for LTB_4 and 5-HETE. A mobility shift of the 85 kDa PLA_2 was also observed in response to the pancreatic phospholipase A_2.

Three MAP kinase bands were observed by Western blot analysis in neutrophils, P40, P42 and P44. The MAP kinase bands also shifted in response to LTB_4, 5-HETE and 5-oxo-ETE. The MAP kinase appeared to be phosphorylated more rapidly than the 85 kDa

PLA_2 but the PLA_2 remained phosphorylated for a longer period. The activation and phosphorylation of the 85 kDa PLA_2 is suggested by these findings to be linked through activation of the MAP kinase cascade.

6. INFLUENCE OF 5-oxo-ETE IN HUMAN EOSINOPHILS

Eosinophils were impressively sensitive to activation by 5-oxo-ETE. Only two bands of MAP kinase were observed in eosinophils by Western blot analysis. A gel shift of MAP kinase was observed in response to as low as 10 pM 5-oxo-ETE. Western blot analysis also indicated the presence of the 85 kDa PLA_2 in eosinophils. The PLA_2 was phosphorylated in response to 5-oxo-ETE as suggested by gel shifts. It thus appears the same pathway for activation of the 85 kDa PLA_2 of neutrophils in response to 5-lipoxygenase products is also operative in eosinophils, but differs markedly in its sensitivity to 5-oxo-ETE.

7. RESPONSE OF THE 85 kDa PLA_2 OF NEUTROPHILS TO PRIMING BY ALKYLACYL- AND DIACYL-DIGLYCERIDES

Neutrophils release both diacylglycerol and 1-alkyl-2-acylglycerol upon stimulation by ionophore or fMLP. The alkylacylglycerol is derived from the action of phospholipase D and a phosphohydrolase. We have found different priming responses to the two subclasses of diglycerides. In studies carried out by Bauldry and coworkers,[15,16] diacylglycerol primed neutrophils to release 2- to 3-fold more AA in response to fMLP than was observed in unprimed cells. Alkylacylglycerol similarly primed the cells to release more AA in response to fMLP. However, priming by the diacylglycerol resulted in significant production of 5-HETE and LTB_4, whereas priming by alkylacylglycerol did not yield 5-lipoxygenase products. These results might possibly be explained if the two diglycerides prime for AA release from different pools, which might result through activation of different phospholipase A_2's. Alternatively, the diacylglycerol might elicit activation of the 5-lipoxygenase, whereas the alkylacylglycerol does not. We have not ruled out the latter possibility but have found different affects on activation of the 85 kDa PLA_2.

Treatment of neutrophils with diacylglycerol resulted in increased 85 kDa PLA_2 activity accompanied by its phosphorylation as evidenced by the gel shift assay. No increase in activity or phosphorylation was observed in response to alkylacylglycerol. These findings suggest that the two diglycerides may be priming for increased AA release by different enzymes and may provide a tool for examining the role of the different PLA_2's and conversion of different pools of AA to lipoxygenase products.

8. CONCLUSIONS

The findings demonstrate that human neutrophils contain the 85 kDa PLA_2 which behaves in the same manner as the enzyme from other cells. The 85 kDa PLA_2 was shown to be activated by exposure of the cells to exogenous AA, which in turn, required conversion to 5-lipoxygenase products to activate the 85 kDa PLA_2. In neutrophils, LTB_4 was the most potent product while 5-oxo-ETE and 5-HETE were less potent. The phosphorylation of the 85 kDa PLA_2 as shown by gel shifts paralleled its activation. The eicosanoids tested also resulted in phosphorylation of MAP kinase as suggested by gel shifts. Exogenous pancreatic PLA_2 activated the 85 kDa PLA_2 presumably through AA release. The findings suggest a positive feedback mechanism through which newly released AA is converted to LTB_4 and other eicosanoids that can activate the intercellular enzyme through surface receptors. Further, secreted PLA_2 could cross-talk with the 85 kDa PLA_2 by this mechanism. Human eosinophils also contained the 85 kDa PLA_2 and MAP kinase components which were potently activated by 5-oxo-ETE. The studies suggest that the 85 kDa PLA_2 may be an enzyme responsible for PAF synthesis since it is activated in conjunction with activation of PAF synthesis by the 5-lipoxygenase products.

REFERENCES

1. Prescott, S. M., Zimmerman, G. A., and McIntyre, T. M.: Platelet-activating factor. J. Biol. Chem. *265*:17381–17384 (1990).
2. Nieto, M. L., Venable, M. E., Bauldry, S. A., Greene, D. G., Kennedy, M., Bass, D. A., and Wykle, R. L.: Evidence that hydrolysis of ethanolamine plasmalogens triggers synthesis of platelet activating factor via a transacylation reaction. J. Biol. Chem. *266*:18699–18706 (1991).
3. Uemura, Y., Lee, T.-C., and Snyder, F.: A coenzyme A-independent transacylase is linked to the formation of platelet-activating factor (PAF) by generating the lyso-PAF intermediate in the remodeling pathway. J. Biol. Chem. *266*:8268–8272 (1991).
4. Ramesha, C. S., and Pickett, W. C.: Platelet-activating factor and leukotriene biosynthesis is inhibited in polymorphonuclear leukocytes depleted of arachidonic acid. J. Biol. Chem. *261*:7592–7595 (1986).
5. Suga, K., Kawasaki, T., Blank, M. L., and Snyder, F.: An arachidonoyl (polyenoic)-specific phospholipase A_2 activity regulates the synthesis of platelet-activating factor in granulocytic HL-60 cells. J. Biol. Chem. *265*:12363–12371 (1990).
6. Billah, M. M., Bryant, R. W., and Siegel, M. I.: Lipoxygenase products of arachidonic acid modulate biosynthesis of platelet-activating factor (1-*0*-alkyl-2-acetyl-*sn*-glycerol-3-phosphocholine) by human neutrophils via phospholipase A_2. J. Biol. Chem. *260*:6899–6906 (1985).
7. Tessner, T. G., O'Flaherty, J. T., and Wykle, R. L.: Stimulation of platelet-activating factor synthesis by a nonmetabolizable bioactive analog of platelet-activating factor and influence of arachidonic acid metabolites. J. Biol. Chem. *264*:4794–4799 (1989).
8. Chilton, F. H., O'Flaherty, J. T., Ellis, J. M., Swendsen, C. L., and Wykle, R. L.: Selective acylation of lyso platelet activating factor by arachidonate in human neutrophils. J. Biol. Chem. *258*:7268–7271 (1983).
9. Mueller, H. W., O'Flaherty, J. T., Greene, D. G., Samuel, M. P., Wykle, R. L.: 1–*0*-alkyl-linked glycerophospholipids of human neutrophils: Distribution of arachidonate and other acyl residues in the ether-linked and diacyl species. J. Lipid Res. *25*:383–388 (1984).
10. Ramesha, C. S. and Ives, D. L.: Detection of arachidonoyl-selective phospholipase A_2 in human neutrophil cytosol. Biochim. Biophys. Acta. *1168*:37–44 (1993).
11. Alonso, F., Henson, P. M., and Leslie, C. C.: A cytosolic phospholipase in human neutrophils that hydrolyzes arachidonoyl-containing phosphatidylcholine. Biochim. Biophys. Acta. *878*:273–280 (1986).
12. Lin, L.-L., Lin, A. Y., and Knopf, J. L.: Cytosolic phospholipase A_2 is coupled to hormonally regulated release of arachidonic acid. *Proc. Natl. Acad. Sci. 89*:6147–6151 (1992).
13. Wijkander, J., O'Flaherty, J. T., and Wykle, R. L.: 5-Lipoxygenase products modulate the activity of the 85 kDa phospholipase A_2 in human neutrophils. J. Biol. Chem., in press.
14. Powell, W. S., Gravel, S., MacLeod, R. J., Mills, E., and Hashefi, M.: Stimulation of human neutrophils by 5-oxo-6,8,11,14-eicosatetraenoic acid by a mechanism independent of the leukotriene B_4 receptor. J. Biol. Chem. *268*:9280–9286 (1993).
15. Bauldry, S. A., Wykle, R. L., and Bass, D. A.: Phospholipase A_2 activation in human neutrophils. Differential actions of diacylglycerols and alkylacylglycerols in priming cells for stimulation by *N*-formyl-met-leu-phe. J. Biol. Chem. *263*:16787–16795 (1988).
16. Bauldry, S. A., Wykle, R. L., and Bass, D. A.: Differential actions of diacyl- and alkylacylglycerols in priming phospholipase A_2, 5-lipoxygenase and acetyltransferase activation in human neutrophils. Biochim. Biophys. Acta *1084*:178–184 (1991)

53

PHOSPHATIDYLCHOLINE BIOSYNTHESIS AS A TARGET FOR PHOSPHOLIPID ANALOGUES

Christoph C. Geilen, Thomas Wieder, and Constantin E. Orfanos

Department of Dermatology
University Medical Center Benjamin Franklin
The Free University of Berlin
Hindenburgdamm 30, D-12200 Berlin, Germany

In current anticancer therapies, most cytostatic agents impair cell division by crosslinking DNA (e.g. cis-platin or alkylating agents), disrupting the cytoskeleton (e.g. vinblastine) or rectifying the cytoskeleton (e.g. taxol). In a new approach to cancer chemotherapy, the cell membrane was described as a target for cytostatic agents. It is known that alkyl-lysophospholipids possess antineoplastic properties in vitro and in vivo[1], leading to the development of an other class of antiproliferative phospholipid analogues, the alkylphosphocholines. One of these phospholipid analogues, hexadecylphosphocholine (HePC), has been shown to inhibit cell proliferation and tumour growth [2–4].

We could demonstrate that HePC inhibits the biosynthesis of phosphatidylcholine by disturbing the translocation process of the rate limiting enzyme, CTP: phosphocholine cytidylyltransferase (EC 2.7.7.15), and that this inhibition correlates with the antiproliferative effect of this analogue in different cell lines[5–8]. Therefore, we systematically investigated the effect of different phospholipid analogues on cell proliferation and phosphatidylcholine biosynthesis. The chemical structures of the analogues are shown in figure 1. Using alkylphosphocholines[9], 1-O-octadecyl-2-O-methylglycero-3-phosphocholine (ET-18-OCH_3)[10] and N-acetyl-sphingosine-1-phosphocholine[11], it could be demonstrated that inhibition of phosphatidylcholine biosynthesis was paralleled by inhibition of cell proliferation (Fig. 2 and Fig.3). It is interesting to note that all biologically active phospholipid analogues contained a phosphocholine head group. Investigating hexadecylphosphoethanolamine (HePE) and hexadecylphosphoserine (HePS) neither cell proliferation nor phosphatidylcholine biosynthesis was inhibited (Fig. 2 and Fig. 3).

In further experiments, inhibition of the rate-limiting enzyme of phosphatidylcholine biosynthesis, CTP: phosphocholine cytidylyltransferase was shown to be the mechanism underlying the inhibition of phosphatidylcholine biosynthesis for all three different classes of phospholipid analogues . Incubation of cells with the different analogues reduced the active, membrane-bound form of cytidylyltransferase (Fig. 4). Therefore, we suggest a link between the regulation of phosphatidylcholine biosynthesis by cytidylyltransferase

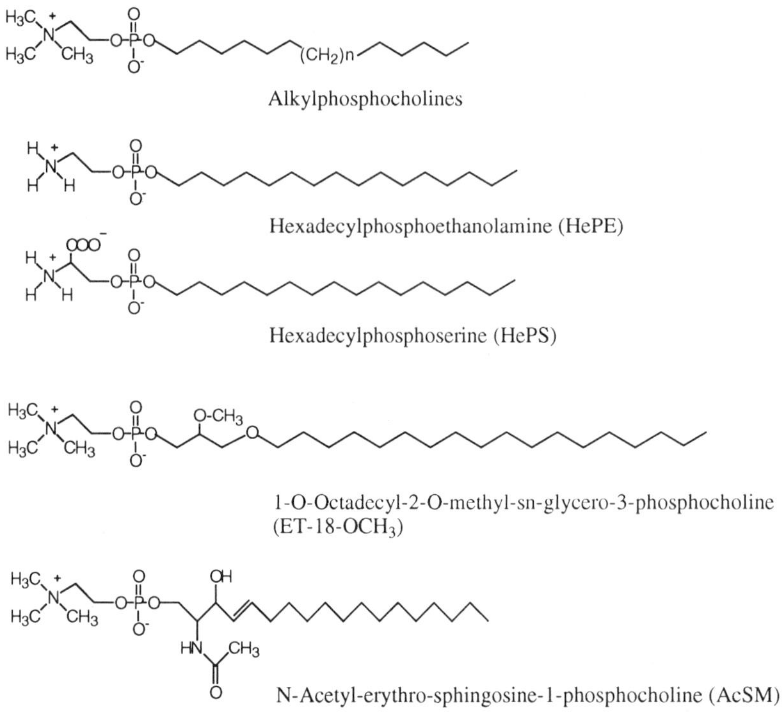

Figure 1. Chemical structures of different phospholipid analogues. Alkylphosphocholines: dodecylphosphocholine (DoPC; n=1), tetradecylphospho-choline (TePC; n=3), hexadecylphosphocholine (HePC; n=5), heptadecylphospho-choline (HepPC; n=6), octadecylphosphocholine (OcPC, n=7) and eikosadecyl-phosphocholine (EikPC; n=9).

and the proliferative properties of cells. Further evidence for this hypothesis is provided by a very recent study which demonstrated that inhibition of phosphatidylcholine biosynthesis results in an arrest of the cell cycle[12].

In conclusion, interfering with the cell membrane metabolism of phospholipids by the use of membrane-active agents, such as phospholipid analogues, may be an useful, novel approach to cancer treatment.

ACKNOWLEDGMENT

This study was supported by a grant of the Deutsche Forschungsgemeinschaft (Ge 641/3–3), research grants from the University Medical Center Benjamin Franklin and the Berliner Krebsgesellschaft. The authors want to thank Dr. D. Arndt and Dr. R. Zeisig (Max-Delbrück Center of Molecular Medicine, Berlin-Buch, Germany), Dr. R.T.C. Huang (Institute for Molecular Biology and Biochemistry, Berlin-Dahlem, Germany) and ASTA Medica AG (Frankfurt a. M., Germany) for providing phospholipid analogues. HaCaT cells were a gift from Dr. N.E. Fusenig (German Cancer Research Center, Heidelberg, Germany).

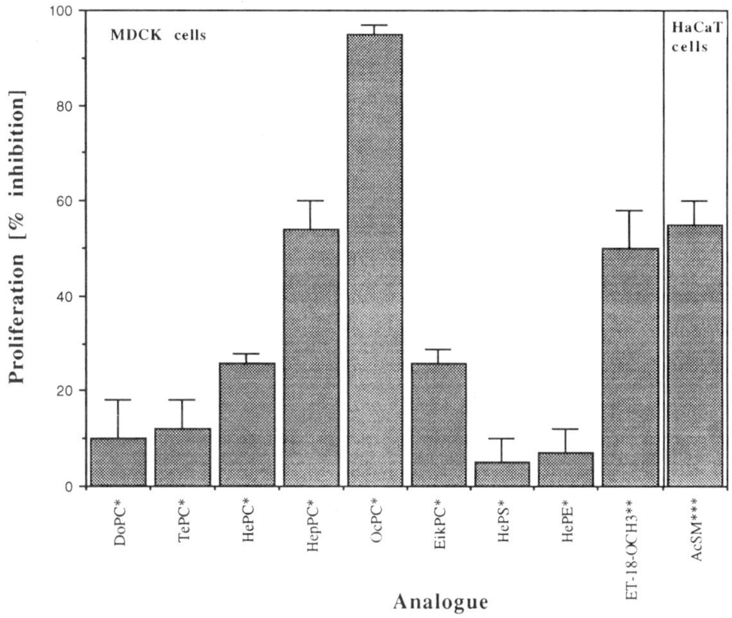

Figure 2. Antiproliferative effect of different phospholipid analogues. Proliferating MDCK cells or HaCaT cells (20 000 cells/cm^2) were incubated with medium containing different phospholipid analogues or no additional supplements as control. After 24 h, the proliferation rate was determined by crystal violet assay as described in detail [9]. Values are given as % inhibition of cell proliferation compared with control values ± s.d. (n=4). The concentrations of the different analogues were *100 μmol/l, **75 μmol/l and ***10 μmol/l.

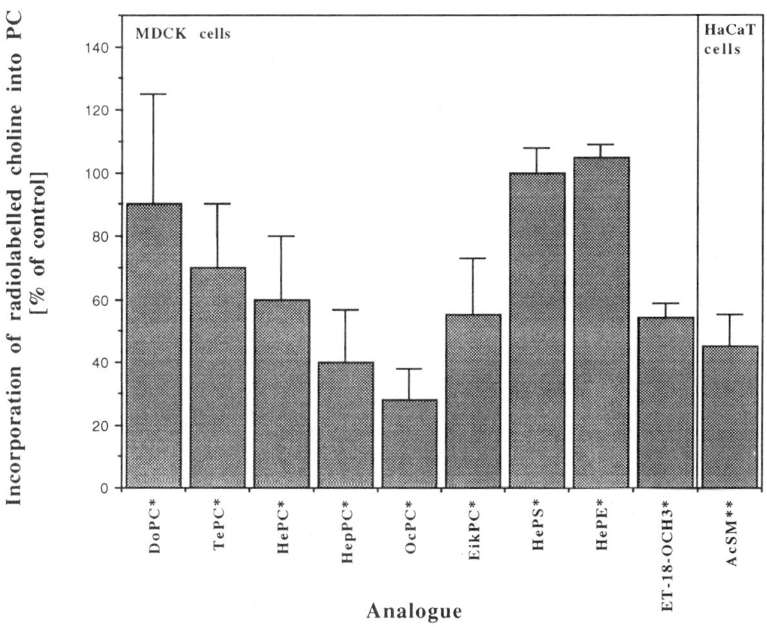

Figure 3. Effect of different phospholipid analogues on choline incorporation into phosphatidylcholine. Confluent MDCK cells or HaCaT cells were incubated for 6 h with pulse-medium containing 2 μCi/ml [methyl-^{3}H] choline chloride and different phospholipid analogues. The cells were harvested, lipids were extracted, and the choline incorporation into phosphatidylcholine was determined by HPTLC and radioscanning as described elsewere[10]. For control values no phospholipid analogue was added. Values of incorporated radioactivity are given as % control ± s.d. (n=3). The concentrations of the different analogues were *50 μmol/l and **20 μmol/l.

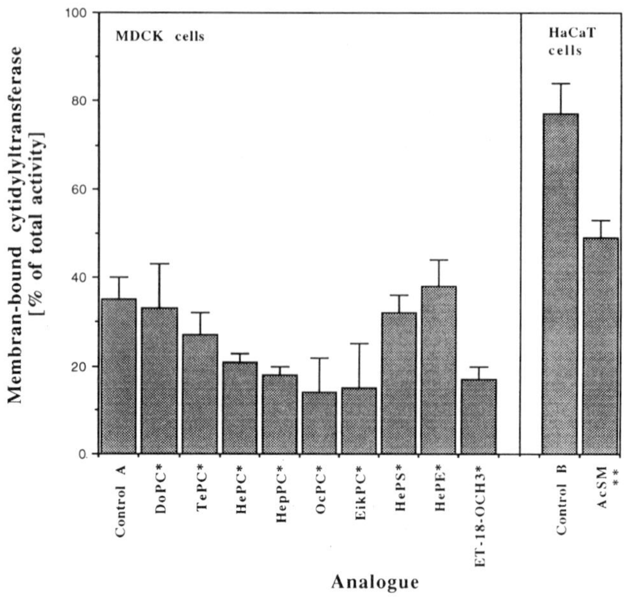

Figure 4. Effect of different phospholipid analogues on the subcellular distribution of the CTP:phosphocholine cytidylyltransferase. Confluent MDCK cells or HaCaT cells were incubated for 6 h with medium containing different phospholipid analogues or no additional supplements as control. Then the subcellular distribution of cytidylyltransferase was determined by digitonin-mediated release and measurement of cytidylyltransferase activity in the cytosolic supernatant and in the remaining cell ghosts as described in detail[10]. The values of membrane-bound enzyme are given in % of total activity of cytidylyltransferase ± s.d. (n=3). The concentrations of the different analogues were *50 μmol/l and **10 μmol/l.

REFERENCES

1. Berdel, W. E., Andreesen, R., and Munder, P.G. (1985) in Phospholipids and cellular Regulation Kuo, J. F. et. al. Vol. 2, pp. 41–73, CRC Press, Boca Raton, FL.
2. Unger, C., Fleer. E.A.M., Kötting, J., Neumüller, W., and Eibl, H. (1992) Prog. Exp. Tumor Res. 34: 25–32.
3. Geilen, C. C., Haase, R., Buchner, K., Wieder, Th., Hucho, F., and Reutter, W. (1991) Eur. J. Cancer 27, 1650–1653.
4. Unger, C., Eibl, H., Breiser, H.W., van Heyden, H.W., Engel, J., Hilgard, P., Sindermann, H., Peukert, M., and Nagel, G.S. (1988) Onkologie 11: 295–296.
5. Haase, R., Wieder, Th., Geilen, C.C. , and Reutter, W. (1991) FEBS Lett. 288, 129–132.
6. Geilen, C.C., Wieder, Th., and Reutter, W. (1992) J. Biol. Chem. 267, 6719–6724.
7. Wieder, Th., Geilen, C. C., and Reutter, W. (1993) Biochem. J. 291, 561–567.
8. Detmar, M., Geilen, C.C., Wieder, Th., Orfanos, C.E., and Reutter, W. (1994) J. Invest. Dermatol. 102: 490–494.
9. Wieder, Th., Haase, A., Geilen, C.C., and Orfanos, C.E. (1995) Lipids 30: 389–393.
10. Geilen, C.C., Haase, A., Wieder, Th., Arndt, D., Zeisig, R., and Reutter, W. (1994) J. Lipid Res. 35: 625–632.
11. Wieder, Th., Perlitz, C., Wieprecht, M., Huang, R.T.C., Geilen, C.C., ans Orfanos, C.E. (1995) Biochem. J., in press.
12. Boggs, K.P., Rock, C.O., and Jackowski, S. (1995) J. Biol. Chem. 270: 7757- 7764.

54

ENDOTHELIN-1 AS ONE OF THE MEDIATORS OF THE INTERACTION BETWEEN ENDOTHELIUM AND PLATELETS IN HUMANS

Ralf Knöfler,[1] Yumiko Takada,[2] Akikazu Takada,[2] and Gerhard Weissbach[1]

[1]Department of Pediatrics
Medical Faculty of the Technical University Dresden, Germany
[2]Department of Physiology
Hamamatsu University School of Medicine, Japan

1. INTRODUCTION

Endothelin-1 (ET-1) is an endothelium-derived peptide with multiple functions affecting the cardiovascular, renal, neuroendocrine, pulmonary and gastrointestinal system[1,2]. Elevated plasma levels of ET-1 were found in diseases associated with the activation of platelets, e.g. myocardial infarction, disseminated intravascular coagulation and diabetes. These findings prompted us to investigate the ET-1 effect on platelet aggregation in human whole blood and in platelet rich plasma (PRP).

2. MATERIALS AND METHODS

2.1 Blood Sampling

Venous blood was collected from antecubital veins of 39 healthy persons (18–38 years). Nine volumes of blood were anticoagulated with one volume of 3.8% trisodium citrate.

2.2 Measurement of Platelet Aggregation

Whole blood aggregation was monitored by the impedance method using a lumi aggregometer type 500 VS (Chronolog Corp., Havertown, USA). The maximal aggregation was determined 6 min after the addition of agonist. According to the producer's direction the excursion of 100 mm (25 Ohms) was expressed as 100% of the maximal response.

PRP was prepared by centrifugation at 200 x g for 10 min, then blood was centrifuged further to yield platelet poor plasma (PPP). PRP was diluted with autologous PPP

Platelet-Activating Factor and Related Lipid Mediators 2
edited by Nigam *et al.*, Plenum Press, New York, 1996

and platelet count was adjusted to 250,000/μl. Platelet aggregation was measured by the optical method with an aggregometer type PAT-2A (Elma Corp., Tokyo, Japan). Aggregation was expressed as maximal increase in light transmission within 3 min after the addition of agonist.

Furthermore it was tried to detect the release of substances from blood cells which are responsible for the enhancement of ADP-induced aggregation in whole blood. For this purpose whole blood from 8 different donors was divided into two portions followed by the preparation of PRP from one part and the ET-1 incubation (0.2μM for 5min) at room temperature of the other portion. Then from both of them PPP was prepared and the platelet count of PRP was adjusted to 250,000/μl using PPP obtained from ET-1 incubated- or non-incubated whole blood. The two different diluted PRP preparations were compared measuring simultaneously platelet aggregation by ADP at 0.25, 0.5, 1.0, 2.0 and 5.0μM.

Preliminary experiments showed that ET-1 at the concentrations used did not induce platelet aggregation neither in whole blood nor in PRP.

2.3 Substances

The aggregation by ADP (for whole blood from Chronolog Corp., for PRP from Daiichi Chemical Company, Kyoto, Japan) was performed with two threshold (5.0 and 7.5μM in whole blood; 0.25 and 0.5μM in PRP) and one suprathreshold concentration (10.0μM in whole blood; 5.0μM in PRP).

The aggregation by collagen (whole blood-Chronolog Corp., PRP-Daiichi Chemical Company) was induced by one low (0.5μg/ml in whole blood, 1.0μg/ml in PRP) and one high suprathreshold concentration (5.0μg/ml in whole blood, 2.0μg/ml in PRP).

Samples were preincubated for 1 and 5 min with ET-1 (Peptide Institute Corp., Osaka, Japan) at 0.1 and 0.2μM.

MCI-9042, a 5-HT_2 receptor antagonist (10μM; Mitsubishi Kasei Corp., Japan), H-7, a protein kinase C inhibitor (100μM; Seikagaku Corp., Japan) and BQ-485Na, an ET_A receptor antagonist (0.34μM; Banyu Pharmaceutical Corp., Japan) were added 1 min before ET-1 was supplied to the sample. The concentrations for these substances were more than 10 times higher than the specific K_i or IC_{50} values.

2.4 Statistics

Statistical analysis was performed by means of WILCOXON rank sum test; $p<0.05$ was considered to be significant.

3. RESULTS

3.1 Aggregation Induced by ADP

3.1.1. Whole Blood Aggregation. The preincubation with ET-1 at 0.1 and 0.2μM for 5min markedly enhanced the whole blood aggregation induced by ADP at 5 and 7.5μM (Table 1). As depicted in Table 1, the enhancement of ADP-induced whole blood aggregation could be demonstrated only at threshold (5.0 and 7.5μM), but not at suprathreshold concentration (10.0μM) of the agonist. This phenomenon was time- (Fig.1) and concentration-dependent (Fig.2). The ET-1 effect could not be abolished by H-7, MCI-9042 or BQ-485Na (Table 2).

Table 1. ADP- and collagen-induced aggregation in whole blood and PRP without (control) and with ET-1 preincubation at 0.1 and 0.2μM for 1 and 5 min. Data are expressed as means ±SEM

Final concentration	Control (%)	+ET-1 0.1μM for 1 min (%)	+ET-1 0.1μM for 5 min (%)	+ET-1 0.2μM for 1 min (%)	+ET-1 0.2μM for 5 min (%)
Whole blood (% maximal aggregation)					
ADP 5.0 μM	**4.7** ± 2.05 (n=10)	6.75 ± 1.89 (n=4)	**22.7** ± 6.5[b] (n=10)	7.21 ± 2.85 (n=4)	**27.67** ± 7.5[a] (n=6)
ADP 7.5 μM	**6.18** ± 1.71 (n=12)	6.25 ± 1.44 (n=4)	**21.58** ± 4.6[b] (n=12)	8.5 ± 2.1 (n=4)	**30.19** ± 4.3[c] (n=21)
collagen 0.5 μg/ml	**50.18** ± 2.9 (n=17)	37.91 ± 4.8 (n=11)	**35.0** ± 5.5[a] (n=11)	**32.5** ± 6.8[a] (n=6)	**35.36** ± 2.9[a] (n=11)
Platelet Rich Plasma (% light transmission)					
collagen 1.0 μg/ml	**49.23** ± 5.4 (n=26)	47.38 ± 1.4 (n=10)	51.25 ± 6.5 (n=14)	35.7 ± 10.6 (n=7)	**36.37** ± 6.9[a] (n=15)
collagen 2.0 μg/ml	**70.43** ± 2.7 (n=29)	60.46 ± 6.3 (n=11)	**56.4** ± 5.6[b] (n=15)	**56.9** ± 5.8[a] (n=8)	**62.43** ± 4.7[a] (n=17)

[a]p<0.05; [b]p>0.01; [c]p<0.001 vs control

3.1.2. Aggregation in PRP. In contrast to whole blood, ET-1 at 0.1 and 0.2μM did not modify the aggregatory reponse of platelets in PRP to ADP at 0.25, 0.5 and 5.0μM. No significant differences in the extent of aggregation were observed between the two different PRP preparations (diluted with PPP which was obtained from ET-1 incubated and non-incubated whole blood). The data are not shown.

3.2 Aggregation Induced by Collagen

3.2.1. Whole Blood Aggregation. Whole blood aggregation in response to collagen at 0.5μg/ml was slightly inhibited after the preincubation with ET-1 at 0.1 and 0.2μM for 1 and 5 min (Table 1), whereas the aggregation by collagen at 5μg/ml was not different between controls and the preincubated samples (data not shown).

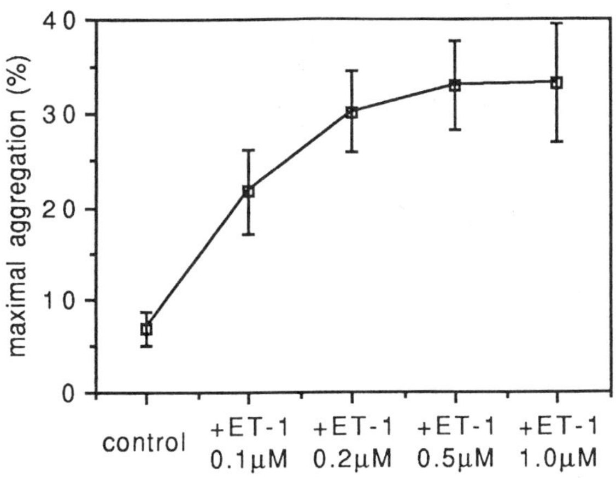

Figure 1. Time dependency of the effect of ET-1 at 0.1μM on whole blood aggregation induced by ADP at 5μM. The extent of enhancement is directly related to the prolongation of the preincubation time.

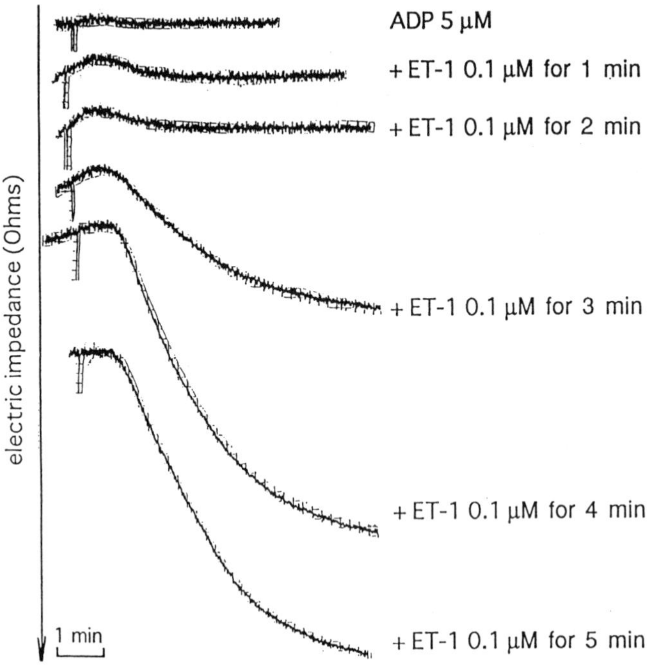

Figure 2. Relationship between ET-1 concentration and extent of enhanced aggregation in whole blood induced by ADP at 7.5μM. Data are expressed as means ±SEM from 4–12 donors.

Table 2. Effect of ET-1 on the agonist-induced aggregation in relation to the additional preincubation with H-7, MCI-9042 and BQ-485Na. Data are expressed as means ±SEM

Final concentration	Control (%)	+ET-1 0.2μM/5min (%)	+H-7 100μM/6min +ET-1 0.2μM/5min (%)	+MCI-9042 10μM/6min +ET-1 0.2μM/5min (%)	+BQ-485 Na 0.34μM/6min +ET-1 0.2μM/5min (%l)
Whole blood (% maximal aggregation)					
ADP 7.5μM	6.18 ± 1.71 (n=27)	30.19 ± 4.3 (n=21)	25.64 ± 5.5 (n=14)	27.83 ± 8.5 (n=6)	36.17 ± 7.94 (n=6)
Collagen 0.5 μg/ml	50.18 ± 2.9 (n=17)	35.36 ± 2.9 (n=11)	36.86 ± 5.9 (n=7)	28.8 ± 6.78 (n=5)	33.67 ± 3.71 (n=3)
Platelet Rich Plasma (% light transmission)					
Collagen 2.0 μg/ml	70.43 ± 2.7 (n=29)	62.43 ± 4.7 (n=17)	42.92 ± 6.7 (n=6)	48.12 ± 7.7 (n=6)	40.94 ± 9.61 (n=4)

[a]p<0.05; [b]p<0.01 vs control

3.2.2. Aggregation in PRP. After preincubation with ET-1 the collagen-induced aggregation in PRP at 1 and 2μg/ml was inhibited. H-7, MCI-9042 and BQ-485Na were not able to abolish the inhibitory ET-1 effect on collagen-induced aggregation neither in whole blood nor in PRP (Table 2).

4. DISCUSSION

In confirmation with the data from the literature[3–6], ET-1 itself was not able to induce platelet aggregation neither in whole blood nor in PRP. However, the present study demonstrates for the first time that ET-1 affects the agonist-induced aggregation in human whole blood. A comparable study was performed by Battistini et al.[3] who could not detect an ET-1 effect on ADP- and collagen-induced whole blood aggregation. In contrast to our methods, they used high suprathreshold concentrations of the agonists inducing 60 to 80% of maximal aggregation. Ostensibly this phenomenon can be observed only at lower (threshold and low suprathreshold) concentrations of the aggregating agents which is supported by data obtained from aggregation in PRP[7,8].

In contrast to whole blood, the ADP-induced aggregation in PRP was not influenced by ET-1 which ist in agreement with some previous studies[3,4,6]. The discrepancy between the results obtained from ADP-induced aggregation in whole blood and in PRP suggests that in whole blood ET-1 interacts with leukocytes and/or erythrocytes. The samples of PRP diluted with PPP obtained from ET-1 incubated and non-incubated whole blood showed no difference in platelet aggregation. Therefore the release of mediators by other blood cells, e.g. platelet activating factor from neutrophils[9], seems not to be the main underlying mechanism. It is more likely that ET-1 interacts directly with leukocytes and/or erythrocytes which are involved as cells in the aggregates formed in whole blood in response to the agonist[10]. It should also be considered that ET-1 may increase the sensitivity of ADP receptors. But this is unlikely, because platelets are more responsive to ADP in PRP compared to whole blood[11] and in PRP no ET-1 effect was observed.

In contrast to ADP, the collagen-induced aggregation was slightly inhibited in whole blood and in PRP after preincubation with ET-1. This suggests a direct ET-1-mediated inhibition of a platelet activation pathway which is involved in the collagen-, but not the ADP-induced platelet activation. Our findings are supported by other authors[8,12] who used thrombin as a comparable strong aggregating agent. It should be hypothesized that in case of the collagen-induced aggregation in whole blood the inhibitory effect overcomes the ET-1-induced stimulation of other blood cells. If ET-1 interacts with receptor(s) on platelets or affects transmembrane signal transduction or intracellular activation pathway(s) remains to be studied. Touyz et al.[12] and Pietraszek et al.[13] suggested that ET-1 activates the protein kinase C (PKC) in platelets which does not cause a measurable platelet aggregation. It would suppress signal transduction pathways leading to an inhibited aggregation after addition of the agonist. Therefore we tried to abolish the ET-1 effect using H-7, a PKC inhibitor. But in contrast to the authors[12,13] we could not succeed. The reason for this discrepancy is not clear. Probably it was due to the short duration of preincubation time or the reported weakness and PKC-non-selectivity of this substance in platelets[14]. Pietraszek et al.[13] suggested that ET-1 may interact with the platelet 5-HT_2 receptor, but according to the results obtained from the experiments with MCI-9042, a selective 5-HT_2 receptor antagonist[15], this could not be confirmed. The ET-1 effect also was not abolished by preincubation with an ET_A receptor antagonist, suggesting that platelets do not possess such receptors.

In conclusion, the results of this study support the hypothesis that ET-1 is able to modulate platelet activation by the interaction with platelets, but with other blood cells too. This might be of importance under the pathophysiological conditions of an elevated plasma level. It should be stressed that the measurement of platelet aggregation in whole blood and in PRP represents *in vitro* methods, but especially the whole blood aggregation may reflect some of the processes in human blood vessels.

REFERENCES

1. Yanagisawa, M., Kurihara, H., Kimura, S., Tomobe, Y., Kobayashi, M., Mitsui, Y., Goto, K.. and Masaki, T. (1988) Nature, 332, Botelho, L.H.P.
2. Rubanyi, G.M. and Botelho, L.H.P. (1991) FASEB J., 5, 2713–2720
3. Battistini, B., Filep, J.G., Herman, F. and Sirois, P. (1990) Thromb.Res., 60, 105–108
4. Patel, A., Fairbanks, L., Gordge, M.P. and Neild, G. (1989) Thromb.Res., 56, 769–770
5. Dockrell, M.E., Haynes, W.G., Williams, B.C. and Webb,D.J. (1993) J.Cardiovasc.Pharmacol., 22 Suppl.8, S204-S206
6. Edlund, A. and Wennmalm, A. (1990) Clin.Physiol., 10, 585–590
7. Matsumoto, Y., Ozaki, Y., Kariya, T. and Kume, S. (1990) Biochem.Pharmacol., 40, 909–911
8. Astarie-Dequeker, C., Iouzalen, L., David-Dufilho, M. and Devynck, M.A. (1992) Br.J.Pharmacol., 106, 966–971
9. Gomez-Garre, D., Guerra, M., Gonzales, E., Lopez-Farre, A., Riesco, A., Caramelo, C., Escamero, J. and Egido, J. (1992) Eur.J.Pharmacol., 224, 167–172
10. Joseph, R., Welch, K.M.A., D'Andrea, G. and Riddle, J.R. (1989) Thromb.Res., 53, 485–491
11. Ingerman-Wojenski, C.M. and Silver, M.J. (1984) Thromb. Haemostas., 51, 154–156
12. Touyz, R.M. and Schiffrin, E.L. (1995) Clin.Sci.Colch., 88, 277–283
13. Pietraszek, M.H., Takada, Y. and Takada, A. (1992) Eur.J. Pharmacol., 219, 289–293
14. Spacey, G.D., Bonser, R.W., Randall, R.W. and Garland, L.G. (1990) Cell.Signal., 2, 329–338
15. Hara, H., Osakabe, M., Kitajima, A. Tamao, Y. and Kikumoto, R. (1991) Thromb.Haemostas., 65, 415–420

EVIDENCE FOR ACTIVATION OF CYCLOOXYGENASE-1/-2 BY ENDOGENOUS NITRIC OXIDE IN ADJUVANT ARTHRITIC LEWIS RATS

Hermann-Josef Thierse,[1,3] Walter-Gunar Friebe,[1] Werner Scheuer,[2] Wolfgang Voelter,[3] and Ulrich Tibes[1]

[1]Boehringer Mannheim GmbH
Department of Preclinical Research, 68298 Mannheim, Germany
[2]Boehringer Mannheim GmbH
Department of Immunology/Oncology, 82377 Penzberg, Germany
[3]University of Tübingen, Department of Biochemistry
Hoppe-Seyler-Str. 4 72076 Tübingen, Germany

1. INTRODUCTION

Increased enzyme activities of cytosolic phospholipase A_2 ($cPLA_2$) and nitric oxide synthase (NOS) have been implicated in the pathophysiology of rheumatoid arthritis (RA)[1,2,3,4] and the alteration of immunological responses[5,6]. As determined on messenger RNA (mRNA) and protein level, in human rheumatoid synoviocytes, both, $cPLA_2$ and growth factor-responsive prostaglandin H synthase-2 (PGHS-2 or cyclooxygenase-2, COX-2) activity were induced by interleukin-1ß, while secretory PLA_2 ($sPLA_2$) and constitutive prostaglandin H synthase-1 (PGHS-1 or cyclooxygenase-1, COX-1) remained unaffected. Reflecting the enhanced activity of this pathway, its induction was associated with high levels of prostaglandin-2 (PGE_2), a mediator of pain and inflammation in the joints of patients with RA[2]. In addition, patients with RA showed high levels of nitrite (NO^-_2, breakdown product of nitric oxide) and 3-nitrotyrosine, indicating elevated NOS activity and nitric oxide ($NO^\bullet$) dependent oxidative damage[3,7]. In order to evaluate a physiological link between $cPLA_2$, COX-1/-2 and NOS pathway in vivo, we investigated the effect of L-nitro-arginine-methyl-ester (L-NAME), an inhibitor of NOS, in adjuvant-induced arthritis in the rat, while dexamethasone served as control drug, This model exhibits several pathological features similar to those occurring in autoimmune reactive RA in humans, characterized by chronic inflammation of the joints. In our studies following parameters have been determined: i) PGE_2 as a product of the PLA_2 and COX-1/-2 pathway, ii) platelet-activating-factor (PAF, PAF-acether) as product of the PLA_2 and acetyl-

Platelet-Activating Factor and Related Lipid Mediators 2
edited by Nigam *et al.*, Plenum Press, New York, 1996

transferase pathway, iii) NO^-_2 as product of the NOS pathway, iv) paw swelling as an indicator of in vivo response.

NO^-_2 was quantified by Griess reaction[10]. PLA2 activity was measured according to Rehfeldt et al.[11].

2. MATERIALS AND METHODS

2.1 Adjuvant Arthritis

Male Lewis rats (200–230g; Charles River) were housed in propylene cages with food and water ad libitum and maintained on a 12-hours light/12-hours dark cycle in a temperature controlled room (21–25°C). Adjuvant arthritis was induced by a single intradermal injection (100 μl) of complete Freund's adjuvant (CFA, 0.1mg inactivated Mycobacterium tuberculosis, Difco Laboratories, Detroit, Michigan, USA, incomplete Freund's adjuvant, Sigma, St. Louis, Missouri, USA) into the right hindpaw at day 1. Starting at day 10, animals daily received either L-NAME (100mg/kg i.p., Calbiochem-Novabiochem, San Diego, California, USA) or dexamethasone (dexa; 1mg/kg p.o., Sigma) till day 22. One placebo treated arthritic group served as treatment control (placebo control) another group of healthy animals, receiving neither CFA-injection nor treatment served as control for arthritis induction (control).

2.2 PGE_2, PAF, NO_2^-, Paw Swelling and Statistics

After Folch lipid extraction PGE_2 was quantified by a commercial ELISA (Boehringer Mannheim GmbH, Mannheim, Germany) and calculated for ng/mg rat paw tissue. Since PAF cannot be extracted directly from biological tissues the phospholipid was enriched by solid-phase extraction. In this study we used prepacked silica cartridges according to D.Janero & C.Burghardt[8,9] (Figure 1). Eluted PAF was detected by antibody and quantified by a scintillation proximity assay (SPA technology) from Amersham™(Cardiff, Wales, UK). To examine the development of arthritis, the volumes of left and right hindpaw were analyzed at days 10 and 22 (μl volume difference) by hydroplethysmometry (TSE systems, Kronberg, Germany).

Data are expressed as mean ± SEM for (n=6) animals. Comparisons were made between the healthy control and arthritic group at day 22. The results were analyzed by Student's unpaired t test for unequal variances to determine the significant difference between means.

3. RESULTS

Compared to healthy controls PGE_2 ($p<0,001$), PAF ($p<0,05$) and NO^-_2 ($p<0,05$) levels were increased in the CFA-injected arthritic right hindpaw (primary lesion) as determined on day 22 after arthritis induction (Figure 2). In the non-injected arthritic left hindpaw (secondary lesion, immunological side) PGE_2 ($p<0,01$) and NO^-_2 ($p<0,05$) levels increased significantly after 22 days, whereas PAF levels did not change (Figure 2). Paw swelling increased in both arthritic hindpaws ($p<0,05$) between day 10 and 22 (Table 1). In relation to placebo-treated arthritic rats L-NAME reduced PGE_2 ($p<0,01$) in both hindpaws (Table 1).In the left hindpaw (secondary lesion, immunological side) L-NAME de-

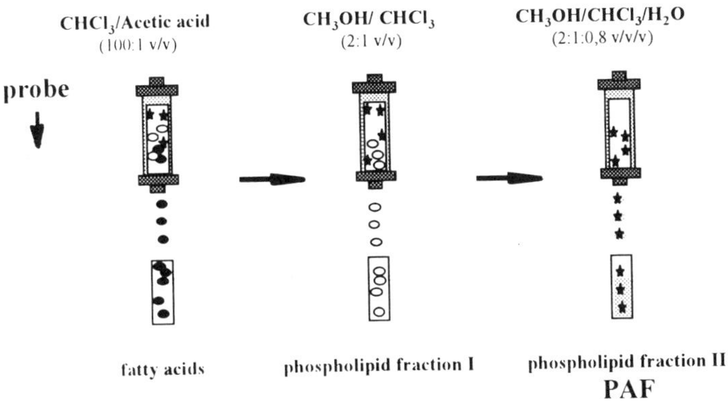

Figure 1. PAF enrichment by solid-phase extraction using silica cartridges.

creased not only PGE_2 ($p<0,01$) but also NO^-_2 ($p<0,05$) and paw swelling ($p<0,05$) (Table 1). In the right hindpaw (primary lesion) L-NAME decreased NO^-_2 as well as paw swelling, but not significantly (Table 1). Dexamethasone decreased PGE_2 synthesis ($p<0,05$) in both right and left hindpaw (Table 1). Like L-NAME dexamethasone reduced NO^-_2 levels ($p<0,01$) on the left side (secondary lesion, immunological side) compared to placebo-treated rats (Table 1).

Paw swelling decreased in right ($p<0,001$) and left hindpaw ($p<0,01$) after dexamethasone treatment (Table 1). In both hindpaws neither L-NAME nor dexamethasone suppressed PAF synthesis (Table 1). Tested in vitro L-NAME did not effect recombinant $cPLA_2$ activity (Figure 3), whereas a standard $cPLA_2$ inhibitor (BM16.5354) dose-dependently suppressed $cPLA_2$ activity[1].

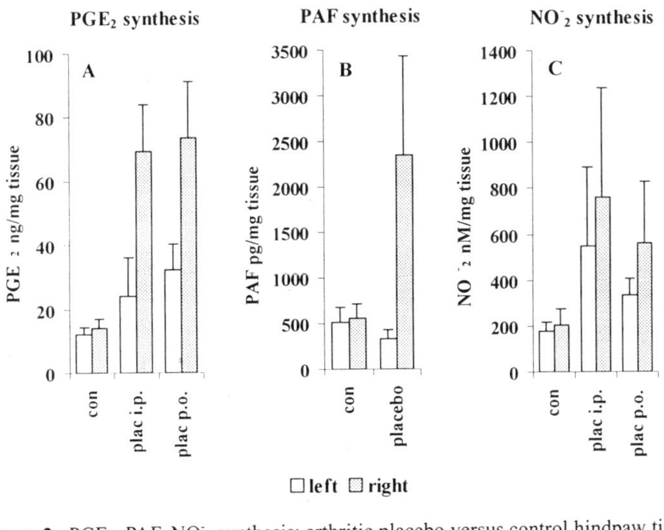

Figure 2. PGE_2, PAF, NO^-_2 synthesis: arthritic placebo versus control hindpaw tissue.

Table 1.

	Healthy control		Placebo control		l-name		Dexa	
	Left	Right	Left	Right	Left	Right	Left	Right
PGE_2	12.1	13.9	24.1	69.2	16.3	27.3	22.8	44.9
ng/mg tissue	(±2)	(±3)	(±12)	(±15)	(±3,4)	(±2,7)	(±7)	(±15)
% inhibition			**	*	**	**	***	***
					33	61	30	39
PAF	515	556	336	2354	506	2619	280	2053
pg/mg tissue	(±172)	(±163)	(±96)	(±1082)	(±411)	(±1352)	(±65)	(±885)
% inhibition			ns	***	ns	ns	ns	ns
					0	0	17	30
NO_2^-	179	204	334	615	226	337	210	744
nM/mg tissue	(±37)	(±73)	(±75)	(±262)	(±45)	(±78)	(±57)	(±189)
% inhibition								
			***	***	***	ns	**	ns
					59	46	37	0
paw	55	65	582	575	115	150	−162	−1030
vol.diff. /μl[1]	(±21)	(±41)	(±429)	(±342)	(±240)	(±367)	(±66)	(±209)
% inhibition			***	***	***	ns	**	*
					80	74	>100	>100

*p<0.001; **p<0.01; ***p<0.05; ns= not significant. [1]vol.diff.= volume difference

4. CONCLUSIONS

Our data demonstrate an inhibition of PGE_2 synthesis by L-NAME in chronic experimental inflammation. In contrary we did not observe suppression of PAF synthesis suggesting a minor role of PAF in chronic inflammation. However, since PAF plays a role in acute inflammation[12] it may contribute to that portion of paw inflammation remaing after NOS inhibition. As expected L-NAME inhibited NO^-_2 production in the arthritic tissue. Inhibition of PGE_2 synthesis may be due to i) reduced activation of COX-1/2 by decreased $NO^{\bullet}$-synthesis, ii) decreased formation of arachidonic acid (AA) by PLA_2, or iii) both. Ac-

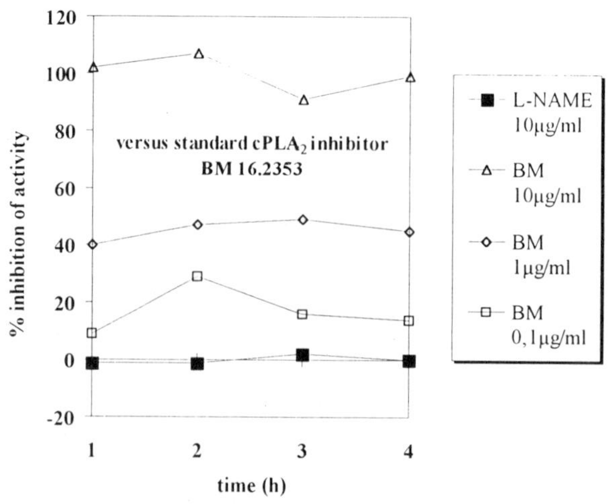

Figure 3. Recombinant $cPLA_2$ is not effected by L-NAME.

tivation of COX by NO• has been shown by Salvemini et al. in vitro and in vivo[13,14]. In addition we recently discovered besides PGE_2 reduction a complete inhibition of AA release in L-NAME treated arthritic rats[15]. Thus, both mechanisms may operate, demonstrating a cross talk between NOS on the one side, COX and PLA_2 on the other side. Based on the results we obtained by dexamethasone treatment (Table 1) and previous data[1] we suggest that iNOS, COX-2, $cPLA_2$, not $sPLA_2$ are the isoforms responsible. Thus, there is evidence of a cross talk between iNOS, COX-2 and $cPLA_2$ in adjuvant arthritis. Besides $cPLA_2$ and/or COX-2 inhibition, iNOS may be an exciting target in the development of new drugs for the therapy of rheumatoid arthritis and other immunologic disorders[16,17,18].

ACKNOWLEDGMENT

We are grateful to A.Litters and E.Winter from the Preclinical Research (BM) for technical assistance. Studies were supported by a grant from Boehringer Mannheim GmbH to H.-J.T.

REFERENCES

1. U. Tibes, A. Vondran, E. Rodewald, W.-G. Friebe, W. Schäfer, and W. Scheuer, Inhibition of allergic and non-allergic inflammation by phospholipase A_2 inhibitors. *Int.Arch.Allergy Immunol.* 107.432–434 (1995).
2. K.I. Hulkower, S.J. Wertheimer,W. Lewin, J.W. Coffey, C.M. Anderson, T. Chen, D.L. DeWitt, R.M. Crowl, W.C. Hope, and D.W. Morgan, Interleukin-1ß induces cytosolic phospholipase A_2 and prostaglandin H synthase in rheumatoid synovial fibroblasts. *Arthritis Rheum.* 37.653–661 (1994).
3. H. Kaurr and B. Halliwell, Evidence for nitric oxide-mediated oxidative damage in chronic inflammation. *FEBS* 350.9–12. (1994).
4. A. Ialenti, S. Moncada, and M. Di Rosa, Modulation of adjuvant arthritis by endogenous nitric oxide. Br.J.Pharmacol. 110.701–706. (1993).
5. X. Wei, I.G. Charles, A. Smith, J. Ure, G.-j. Feng, F.-p. Huang, D. Xu, W. Muller, S. Moncada, and F.Y Liew, Altered immune responses in mice lacking inducible nitric oxide synthase. *Nature* 375.408–411. (1995).
6. J.D. MacMicking, C. Nathan, G. Hom, N. Chartrain, D.S. Fletcher, M. Trumbauer, K. Stevens, Q. Xie, K. Sokol, N. Hutchinson, H. Chen, and J.S. Mudgett, Altered responses to bacterial infection and endotoxic shock in mice lacking inducible nitric oxide synthase. *Cell* 81.641–650. (1995).
7. A.J. Farrell, D.R. Blake, R.M.J. Palmer, and S. Moncada, Increased concentrations of nitrite in synovial fluid and serum samples suggest increased nitric oxide synthesis in rheumatic diseases. *Ann. Rheum. Dis.* 51.1219–1222. (1992).
8. D.R. Janero and C. Burghardt, Solid-phase extraction on silica cartridges as an aid to platelet-activating factor enrichment and analysis. *J* Chromatogr. 526.11–24. (1990).
9. D.J. Hanahan and S.T. Weintraub, Platelet-Activating Factor Isolation, Identification, and Assay. *Methods of Biochem. Anal.* 31.195–219. (1985).
10. M. Lepoivre, H. Boudbid, and J.-F. Petit, Antiproliferative activity of g-Interferon combined with lipopolysaccharide on murine adenocarcinoma: dependence on an L- arginine metabolism with production of nitrite and citrulline. *Cancer Res.* 49.1970- 1976. (1989).
11. W. Rehfeldt, R. Hass, and M. Goppelt-Struebe, Charaterization of phospholipase A_2 in monocytic cell lines. Functional and biochemical aspects of membrane association. Biochem. J. 276.631–636. (1991).
12. L.W. Tjoelker, C. Wilder, C. Eberhardt, D.M. Stafforini, G. Dietsch, B. Schimpf, S. Hooper, H.L. Trong, L.S. Cousens, G.A. Zimmerman, Y.Yamada, T.M. McIntyre, S.M. Prescott, and P.W. Gray, Anti-inflammatory properties of a platelet-activating factor acetylhydrolase. *Nature* 374.549–553. (1995).
13. D. Salvemini, T.P. Misko, J.L. Masferrer, K. Seibert, M.G. Currie, and P. Needleman, Nitric oxide activates cyclooxgenase enzymes. *Proc. Natl. Acad. Sci. USA* 90.7240–7 244. (1993).
14. D. Salvemini, P.T. Manning, B.S. Zweifel, K. Seibert, J. Connor, M.G. Currie, P. Needleman, and J.L. Masferrer, Dual inhibition of nitric oxide and prostaglandin production contributes to the antiinflammatory properties of nitric oxide synthase inhibitors. *J. Clin. Invest.* 96.301 (1995).

15. H.-J. Thierse, W. Scheuer, A. Krusch, G.-W. Friebe, P. Gierschick, and U. Tibes, Major role of cPLA2, secondary of COX and NOS, minor roles of sPLA2 and PLC in experimental inflammation. *Eicosanoids and other bioactive lipids in cancer, inflammation and radiation injury: 4th Intern. Conf. Hongkong.* 42. (1995).
16. J.R. Vane, J.A. Mitchell, I. Appleton, A. Tomlinson, D. Bishop-Bailey, J. Croxtall and D.A. Willoughby, Inducible isoforms of cyclooxygenase and nitric-oxide synthase in inflammation. *Proc. Natl. Acad. Sci. USA* 91.2046–2050 (1994).
17. I. Wicks, G. McColl, and L. Harrison, New perspectives on rheumatoid arthritis. *Immunol. Today* 15.553–556 (1994).
18. J.R. Connor, P.T. Manning S.L. Settle, W.M. Moore G.M. Jerome, R.K. Webber, F. Siong Tjoeng, and M.G. Currie, Suppression of adjuvant-induced arthritis by selective inhibition of inducible nitric oxide synthase. *Eur. J. Pharmacol.* 273.15–24 (1995).

56

PHOSPHATIDYLINOSITOL-3-KINASE PATHWAY IS STIMULATED BY FUNGAL LIPID CONTAINING COCOA BUTTER EQUIVALENTS (CBE) IN FMet-Leu-Phe (FMLP)- AND PHORBOLESTER (PMA)-CHALLENGED HUMAN NEUTROPHILS

M. P. Roux,[1] J. L. F. Kock,[2] and S. Nigam[1*]

[1]Eicosanoid Research Division
Department of Gynecology
University Medical Hospital Benjamin Franklin
D-12200 Berlin, FRG
[2]Department of Microbiology and Biochemistry
University of the Orange Free State
Bloemfontein, South Africa

INTRODUCTION

It is well established that fungi produce fats and oils, although a commercial utilization of these processes with rare exception could not yet be achieved. The reason hereto lied primarily on the high costs of production. Recently, a novel biotechnological process for the production of fungal lipids from *Mucor circinelloides* by using acetic acid (a cheap carbon source) at sub-toxical levels as the starting substrate has been described (1). The products obtained were oil- and butter-like compounds depending upon the dissolved oxygen tension maintained (Figure 1). The analyses of the lipid demonstrated a γ-linolenic acid (GLA)-rich lipid (so-called fungal oil) and a stearic acid-rich lipid (so-called fungal butter) (2). This first report of low cost production of high value lipids thus raised a caveat for more consideration.

Since cocoa butter is extensively used in the chocolate industry (3), Ratledge et al., (4)], it was of primary interest for us to investigate the effect of fungal lipid containing CBE on the cellular functions of blood cells. In the present study, we describe the modula-

* Address correspondence to Dr. S. Nigram.

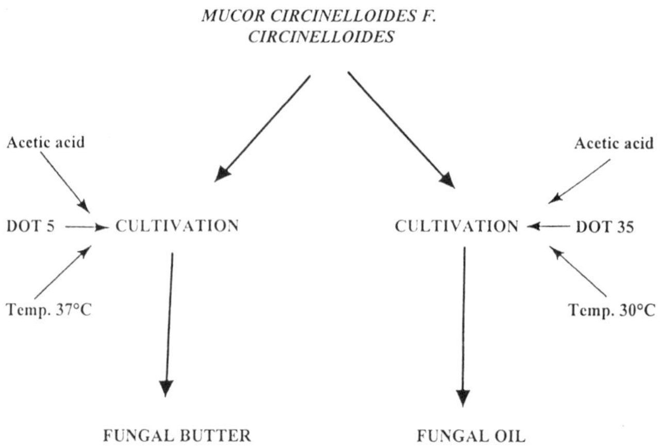

Figure 1. Cultivation parameters for the production of fungal butter and oil by *M. circinelloides f. circinelloides.*

tory role of fungal butter on the receptor-mediated respiratory burst, which is a prerequisite for the host-defence system of the individual, in human neutrophils.

MATERIALS AND METHODS

FMLP, staurosporine, cytochrome C, superoxide dismutase and 1α-lysophosphatidylcholine were purchased from Sigma, FRG. PMA, Wortmannin and LY 294002 were supplied by Biomol, FRG. Ficoll-Hipaque and Dextran 500 were obtained from Pharmacia, FRG.

Stock solutions of FMLP, PMA, Staurosporine, Wortmannin and LY 294002 were prepared in dimethyl sulfoxide (DMSO) and appropriate concentrations were made by dilution with Ca^{2+}-free phosphate-buffered saline (PBS). Final concentrations of DMSO in the incubation sample (< 0.2%) did not affect superoxide anion generation in our experiments.

Preparation of Lipid Emulsion

The fungal butter and 1-α phosphatidylcholine (1:4) were dissolved in chloroform by gentle stirring. The chloroform was evaporated under a stream of N_2 gas, followed by heating (50°C) the samples in a rotary vacume evaporater for at least 3h. Hereafter, PBS was added to the precipitate, which was heated and sonicated till a suspension was obtained. This suspension was centrifuged for 5 min. at 1000 rpm tp precipitate the phosphatidylcholine and the liquid phase was decanted and used during the experiments as an oil emulsion.

Isolation of Human Neutrophils

Blood (40 ml) was obtained from healthy donors and the neutrophils were isolated by Ficoll-paque separation as described (5).

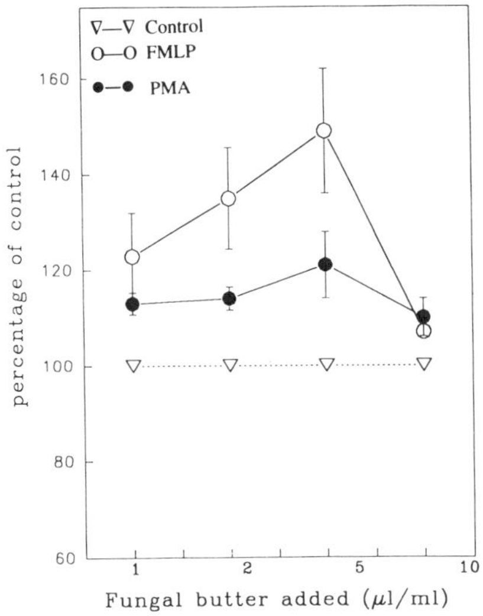

Figure 2. Effect of fungal butter on the superoxide anion generation in human neutrophils challenged by FMLP and PMA.

Superoxide Anion Production

The production of superoxide anion was determined by SOD-inhibitable reduction of cytochrome c, determined spectrophotometrically at 550 nM with a microplate reader (6). Kinetic measurements using 2. 10^6 cells/200μl were performed between 0.5 min. and 5.0 min after addition of 100 nM PMA and fMLP, respectively. Assays were performed in duplicate.

RESULTS AND DISCUSSION

Upon stimulation with 100 nM FMLP and PMA, human neutrophils pretreated with fungal butter showed a significant increase of O_2^- production in a dose-dependent manner (*Figure 2*).

At higher dosage of fungal butter the enhancement of O_2^--production was drastically reduced, because the viability of the cells was affected. Moreover, FMLP-induced generation of O_2^- was more pronounced as compared with the PMA-stimulated O_2^--production. Since O_2^--production is catalyzed by NADPH-oxidase, a protein kinase C (PKC)-dependent enzyme, it was of interest to investigate the effect of staurosporine, an inhibitor of PKC, on the fungal butter-mediated enhancement of O_2^-- generation in human neutrophils challenged with FMLP and PMA. Surprisingly, whereas staurosporine at a concentration of 300 nM completely inhibited O_2^--production by FMLP or PMA in untreated human neutrophils, it could suppress O_2^--production only partially in neutrophils pretreated with 5 μl/ml of fungal butter (*Figure 3*). This suppression measured approx. 35 % and 20 % for FMLP and PMA, respectively. The application of other PKC inhibitors revealed similar effect (data not shown). These data implicated a different mechanism for the generation of O_2^-, because the phosphorylation of the 47-kDa subunit of the NADPH-oxidase system (p47phox), which is responsible for the O_2^--production, requires the Ca^{2+}-activated ß-iso-

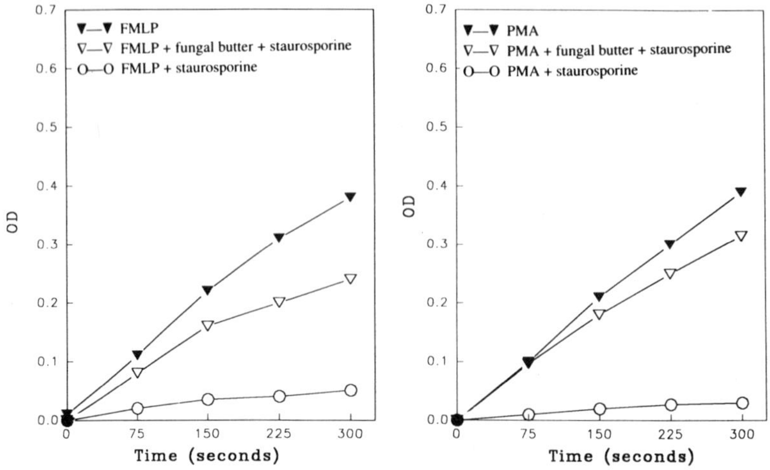

Figure 3. Effect of fungal butter on FMLP- and PMA-induced superoxide anion generation in human neutrophils pretreated with protein kinase C inhibitor staurosporine.

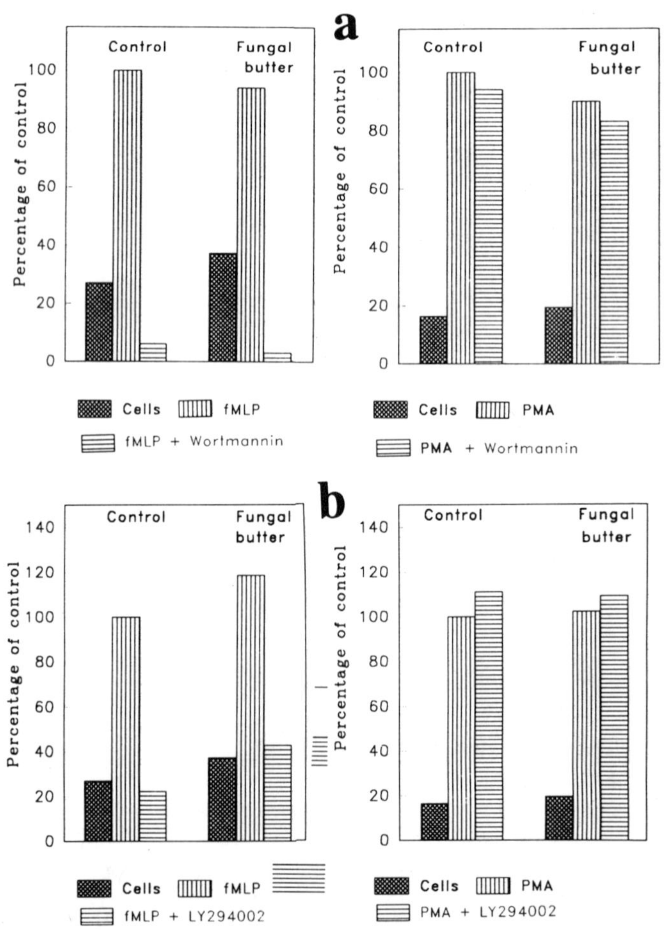

Figure 4. Effect of fungal butter on FMLP- and PMA-induced superoxide anion generation in human neutrophils pretreated with PI-3-kinase inhibitor (a) Wortmannin or (b) LY294002.

form of PKC. However, Ca^{2+}-activated isoforms of PKC are not activated by FMLP (7), and fungal butter did not affect intracellular Ca^{2+} or Ca^{2+}-fluxes across the membrane (data not shown). Interestingly, Wortmannin and LY 294002, inhibitors of PI-3-kinase, abolished O_2^--production by FMLP in neutrophils pretreated with fungal butter (*Figure 4a*). Moreover, the inhibition of PMA-induced O_2^--production by Wortmannin and LY 294002 was only minimal (*Figure 4b*). This is fully in agreement with the recent finding that FMLP-stimulated but not PMA-stimulated neutrophils exhibit a rapid and transient increase in PI(3,4,5)P_3 (8), which is completely blocked by the PI-3-kinase inhibitors. These data strongly suggest that PI-3-kinase activity is potentiated by fungal butter predominantly in FMLP- but to a lesser extent also in PMA-challenged human neutrophils.

ACKNOWLEDGMENTS

This study was supported in part by the Deutsche Forschungsgemeinschaft, Bonn (Ni 242/9–1).

REFERENCES

1. Kock JLF and Botha A, (1993) *S. African J. Sci., 89, 465.*
2. Roux MP et al., (1994) *World J. Microbiol. Biotech., 10, 417–422.*
3. Beavan M et al., (1992) *in: Industrial Applications of Single Cell Oils, Kyle DG and Ratledge C (Eds.), Am. Oil Chemists Soc., p.156–184.*
4. Ratledge C, (1992) *in: Industrial Applications of Single Cell Oils, Kyle DG and Ratledge C (Eds.), Am. Oil Chemists Soc., p.1–15.*
5. Müller S and Nigam S (1992) *Eur. J. Pharmacol., 218, 251–258.*
6. Pick E and Mizel D (1981) *J. Immunol. Methods, 46, 211–226.*
7. Toker et al., (1994) *J. Biol. Chem., 269, 32358–32367.*
8. Traynor-Kaplan AE et al., (1989) *J. Biol. Chem., 264, 15668–15673.*

57

PLATELET-TYPE ARACHIDONATE 12-LIPOXYGENASE

Michihiro Nakamura,[1] Natsuo Ueda,[1] Shozo Yamamoto,[1]
Kazunori Ishimura,[2] Kenjiro Tomo,[3] and Minoru Okuma[3]

[1]Department of Biochemistry
[2]Department of Anatomy
Tokushima University, School of Medicine
Tokushima 770
[3]The First Division, Department of Internal Medicine
Faculty of Medicine, Kyoto University
Kyoto 606-01, Japan

1. INTRODUCTION

In human platelets arachidonic acid is released from membrane phospholipids by the catalysis of phosopholipase A_2, and then oxygenated either by cyclooxygenase or by 12-lipoxygenase. It is well known that the cyclooxygenase pathway leads to the synthesis of pro-aggregatory and vasoconstrictive thromboxane A_2. 12-Lipoxygenase introduces one oxygen molecule regiospecifically and stereoselectively into the carbon 12 of arachidonic acid. The primary product is 12*S*-hydroperoxy-5, 8, 14-*cis*-10-*trans*-eicosatetraenoic acid (12-HPETE), which is reduced to 12*S*-hydroxy acid (12-HETE) in a whole cell preparation. A number of papers have reported biological activities of either 12-HPETE or 12-HETE[1]. However, most of them could not be generalized in terms of the cell type and animal species, and the general biological significance of the enzyme has not been well understood[1,2]. Recently two groups of bioactive arachidonate metabolites were identified as the products of the 12-lipoxygenase pathway. Namely, transcellular synthesis of lipoxins via leukotriene A_4 requires the participation of platelet 12-lipoxygenase[3], and 12-HPETE is enzymatically transformed to hepoxilins[4]. Our group has demonstrated the occurrence of two distinct isoforms of 12-lipoxygenase in mammalian tissues[1,5]. The leukocyte-type 12-lipoxygenase has a broad substrate specificity reacting with linoleic and linolenic acids with 18 carbon atoms as well as arachidonic acid with 20 carbon atoms. In contrast, the platelet-type enzyme is much less active with the C18 fatty acids. The leukocyte-type enzyme is also found in other tissues in addition to leukocytes[1]. The platelet-type enzyme was found earlier in platelets of various animal species, and found recently in epidermis[6–8]. For a better understanding of the physiological role of the platelet-type 12-

Platelet-Activating Factor and Related Lipid Mediators 2
edited by Nigam *et al.*, Plenum Press, New York, 1996

lipoxygenase we planned immunohistochemical studies of the enzyme. However, purification of this enzyme from platelets was difficult, and no polyclonal antibody with a high affinity for the enzyme had not been available until recently. Therefore, we attempted to purify a recombinant enzyme as antigen.

2. PREPARATION OF THE ANTISERUM AGAINST PLATELET 12-LIPOXYGENASE

The 12-lipoxygenase cDNA previously cloned from human erythroleukemia cells[9] was constructed so that the enzyme with a 6-histidine tag at the amino-terminus could be expressed in *E. coli*. This recombinant enzyme was purified by ammonium sulfate fractionation and Ni^{2+}-NTA resin affinity chromatography which could bind specifically to the 6-histidine tag. The purified enzyme showed a specific activity of about 120 μmol/min/mg protein, and a single protein band on SDS-polyacrylamide gel electrophoresis. A polyclonal antibody was raised in a rabbit against this purified enzyme. Upon immunoblotting with the purified recombinant 12-lipoxygenase as a standard, the antiserum gave one major band of about 75 kDa with the lysates of human and mouse platelets. Thus, our antiserum against the human platelet enzyme cross-reacted with the mouse enzyme.

3. IMMUNOHISTOCHEMISTRY OF PLATELET-TYPE 12-LIPOXYGENASE

Based on these preliminary experiments, we performed immunohistochemical studies on the 12-lipoxygenase of mouse bone marrow cells and skin. Each tissue was fixed with 4% paraformaldehyde, and immunostained by the avidin-biotin-peroxidase complex method. Peroxidase reaction was performed with diaminobenzidine and hydrogen peroxide as substrates.

3.1 Bone Marrow Cells

When mouse bone marrow cells were immunostained and observed by light microscopy, huge cells presumed to be megakaryocytes were positively stained, and some smaller cells were also stained. A control experiment using non-immune rabbit serum indicated that the positive staining was not attributed to an endogenous peroxidase. When the smears of human bone marrow cells were also applied to immunohistochemistry, megakaryocytes were clearly stained with this antiserum. For the following immunoelectron microscopic observation, mouse bone marrow cells were treated with osmium tetroxide and embedded in Epon epoxy resin. Platelets were positively stained (Fig. 1a) in comparison with a control using a non-immune rabbit serum (Fig. 1b). It should be noted that 12-lipoxygenase was detected in the cytoplasm of platelets. The plasma membranes and subcellular organelles were hardly immunostained. The subcellular localization of the 12-lipoxygenase in platelets has long been a subject of discrepancy. On the basis of differential centrifugation some investigators reported the association of a considerable 12-lipoxygenase activity with the membrane fraction as well as the cytosol fraction[10], whereas other investigators reported its predominant localization in the cytosol[5, 11–13] or in the membrane[14]. In our present work, 12-lipoxygenase was localized predominantly in the cytoplasm of mouse platelets.

In addition to platelets, eosinophils were positively stained, and the immunoperoxidase reaction products were detected in the cytoplasm. Erythrocytes, lymphocytes and neutrophils were negative. Megakaryocytes were also immunostained among mouse bone marrow cells. These megakaryocytes were characteristic of the demarcation membranes with numerous invaginations of the plasma membranes, and irregularly lobulated nuclei (Fig. 1c). These findings suggested maturation of the megakaryocyte. As for the subcellular localization, 12-lipoxygenase was detected in the cytoplasm, but not in the nucleus, subcellular organelles and membrane structures. A control using a non-immune rabbit serum gave no immunoperoxidase products. Furthermore, there was a large immunostained cell which had a kidney-shaped nucleus and poorly-developed demarcation membranes (Fig. 1d). These features are of the early stage of megakaryocyte maturation, and indicated that 12-lipoxygenase was expressed at the early stage in thrombocytopoiesis. It is possible that 12-lipoxygenase and its metabolites play a certain role in the course of thrombocytopoiesis.

3.2 Skin

Previously our laboratory reported that platelet-type 12-lipoxygenase was expressed not only in platelets but also in epidermis[6]. The 12-lipoxygenase was found in human epidermal cells, and a greater part of the enzyme activity was detected in the microsomal fraction as examined by differential centrifugation in contrast to the platelet enzyme[6, 8]. To

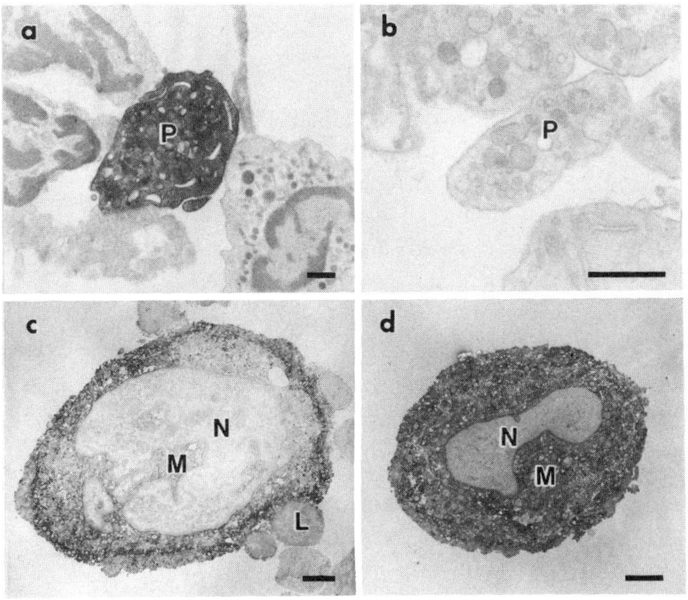

Figure 1. Electron Microscopic Immunocytochemistry of 12-Lipoxygenase in Bone Marrow Cells. Immunoreaction products for 12-lipoxygenase are present in the cytoplasm of mouse platelets (a), and mature (c) and immature (d) megakaryocytes. No reaction products are present in platelets in the control with non-immune rabbit serum (b). P, platelets; M, megakaryocytes; L, lymphocyte; N, nucleus. Original magnifications : a x 7600; b x 21000; c x 2900; d x 3400. Bars : a, b = 1 μm; c, d = 3 μm.

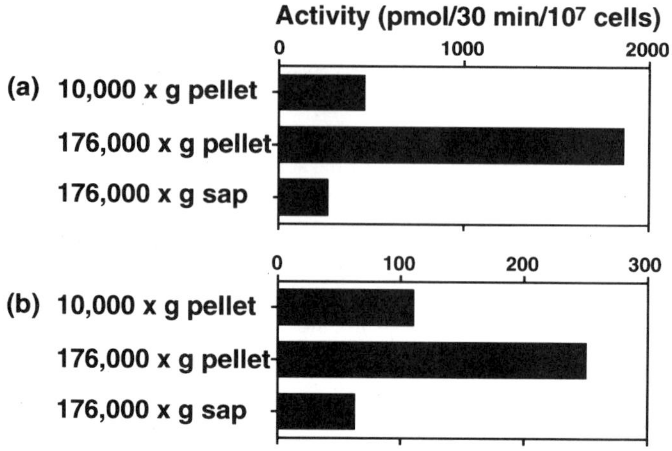

Figure 2. Subcellular Localization of 12-Lipoxygenase in Mouse Keratinocytes. Transfected (a) and non-transfected (b) keratinocytes were homogenized and subjected to differential centrifugations followed by 12-lipoxygenase activity assay.

investigate the intracellular localization of 12-liopxygenase, we performed immunohistochemistry with mouse skin. The 12-lipoxygenase was detected in keratinocytes of germinal layer. However, the staining was not sufficient for the observation by electron microscopy. Therefore, we attempted to over-express 12-lipoxygenase in the keratinocytes. The cDNA encoding human platelet 12-lipoxygenase[9] was inserted into an expression vector, pCMV-5. A primary culture of mouse keratinocytes was transfected with this pCMV/HP12LO vector by the lipofection method. Intracellular localization of the over-expressed 12-lipoxygenase was examined by differential centrifugation, followed by enzyme activity assay and immunohistochemistry. The 12-lipoxygenase activity in non-transfected cells was lost during the primary culture for 30 hours. In contrast, the enzyme activity increased in the transfected cells. When the cells were homogenized and subjected to differential centrifugations, most of the 12-lipoxygenase activity was detected in the particulate fraction shown as 10,000 x g pellet and 176,000 x g pellet (Fig. 2a). This subcellular localization of the enzyme activity was similar to that of pre-transfected cells (Fig. 2b).

The keratinocytes over-expressing 12-lipoxygenase were applied to immunohistochemistry. Light microscopic observation revealed positively stained cells with a larger number of negative cells. In non-transfected culture, no positive cells were observed. Electron microscopy detected the recombinant 12-lipoxygenase mainly in the cytoplasm, but not in the nucleus, subcellular organelles and membrane structures (Fig. 3). Non-transfected cells gave no immunoperoxidase products. Thus, the morphological observation showed the localization of recombinant 12-lipoxygenase in the cytoplasm of keratinocytes as in the case of platelets. The discrepancy between the immunohistochemistry and the differential centrifugation remains unsolved. It is possible that the enzyme was localized originally in the cytosol but was bound to the particulate fraction during the procedures of homogenization and centrifugation.

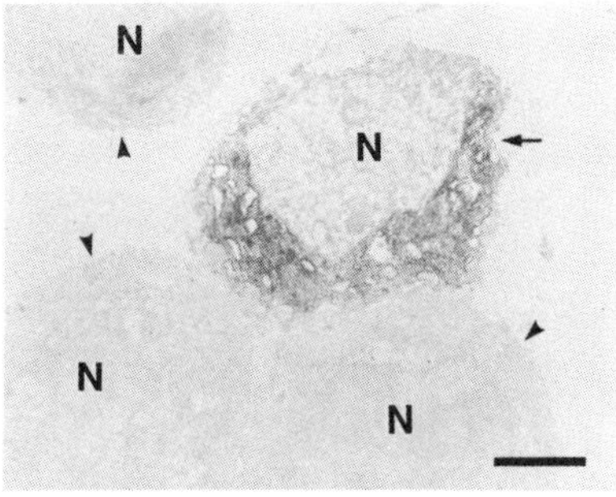

Figure 3. Electron Microscopic Immunocytochemistry of 12-Lipoxygenase-Overexpressing Keratinocytes. Immunoreaction products for 12-lipoxygenase are present in the cytoplasm of 12-lipoxygenase-overexpressing keratinocytes (arrow). Non-transfected keratinocytes are negative (arrowheads). N, nuclei. Original magnifications: x 7800. Bars: 2 μm.

REFERENCES

1. Yamamoto S: Mammalian lipoxygenases: molecular structures and functions. Biochim Biophys Acta 1128: 117–131 (1992)
2. Brash AR: A review of possible roles of the platelet 12-lipoxygenase. Circulation 72: 702-707 (1985)
3. Romano M, Chen XS, Takahashi Y, Yamamoto S, Funk CD, Serhan CN: Lipoxinsynthase activity of human platelet 12-lipoxygenase. Biochem J 296 (Part 1):127–133(1993)
4. Pace-Asciak CR, Reynaud D, Demin P: Enzymatic formation of hepoxilins A_3 and B_3.Biochem Biophys Res Commun 197:869–873 (1993)
5. Takahashi Y, Ueda N, Yamamoto S: Two immunologically and catalytically distinct arachidonate 12-lipoxygenases of bovine platelets and leukocytes. Arch Biochem Biophys 266: 613–621 (1988)
6. Takahashi Y, Reddy GR, Ueda N, Yamamoto S, Arase S: Arachidonate 12-lipoxygenase of platelet-type in human epidermal cells. J Biol Chem 268: 16443–16448 (1993)
7. Chen XS, Kurre U, Jenkins NA, Copeland NG, Funk CD: cDNA cloning, expression, mutagenesis of c-terminal isoleucine, genomic structure, and chromosomal localization of murine 12-lipoxygenase. J Biol Chem 269: 13979–13987 (1994)
8. Hussain H, Shornick LP, Shannon VR, Wilson JD, Funk CD, Pentland AP, Holtzman MJ: Epidermis contains platelet-type 12-lipoxygenase that is overexpressed in germinal layer keratinocytes in psoriasis. Am J Physiol 266: C243-C253 (1994)
9. Yoshimoto T, Yamamoto Y, Arakawa T, Suzuki H, Yamamoto S, Yokoyama C, Tanabe T, Toh H: Molecular cloning and expression of human arachidonate 12-lipoxygenase. Biochem Biophys Res Commun 172: 1230–1235 (1990)
10. Lagarde M, Croset M, Authi KS, Crawford N: Subcellular localization and some properties of lipoxygenase activity in human blood platelets. Biochem J 222: 495–500 (1984)
11. Nugteren DH: Arachidonate lipoxygenase in blood platelets. Biochim Biophys Acta 380: 299–307 (1975)
12. Chang WC, Nakao J, Orimo H, Murota S: Effects of reduced glutathione on the 12-lipoxygenase pathways in rat platelets. Biochem J 202: 771–776 (1982)
13. Hada T, Ueda N, Takahashi Y, Yamamoto S: Catalytic properties of human platelet 12-lipoxygenase as compared with the enzymes of other origins. Biochim Biophys Acta 1083:89–93 (1991)
14. Ho PPK, Walters P, Sullivan HR: A particulate arachidonate lipoxygenase in human blood platelet. Biochem Biophys Res Commun 76: 398–405 (1977)

LEXIPAFANT(BB-882), A POTENT PAF ANTAGONIST IN ACUTE PANCREATITIS

Lloyd D. Curtis

British Biotech Pharmaceuticals Limited
Watlington Road, Oxford OX4 5LY, United Kingdom

1. ACUTE PANCREATITIS

Acute pancreatitis remains a poorly understood and hence unsatisfactorily treated condition[1]. In general this is a mild self-limiting disease, which resolves within days and can be managed with non-specific supportive measures and analgesia. In about 25% of cases, however, the disease becomes severe, (defined as pancreatitis associated with organ failure or a local complication such as abscess, pseudocyst or necrosis[2]). It is this severe group that accounts for much of the morbidity and mortality of the disease. The overall mortality is reported at about 8–10%[3], and this occurs almost exclusively in those with severe disease. Published series have shown that organ failure and particularly respiratory failure are the most common complications.

Although it has been claimed in the past that sepsis was the primary cause of death, in a recent UK series organ system failure complicated 92% of deaths[4]. In the US, it has been reported that 60% of deaths occur in the first week and most of these deaths are associated with pulmonary failure[5].

While it is recognized that inappropriate activation of protease enzymes may underlie the features of the disease, clinical trials of antiprotease therapy have been disappointing[6,7]. This may be because by the time the patient is seen, the enzymes have already triggered a cascade of events with release of soluble mediators such as Platelet Activating Factor (PAF), IL-1, IL-6 and IL-8, which continue further granulocyte, macrophage and complement activation, both locally and systemically. In cases of severe disease this inflammatory cascade leads to a classical systemic inflammatory response syndrome, SIRS, similar to that seen in cases of burns or sepsis syndrome[8].

Therapies aimed at resting the pancreas, including nasogastric suction, cimetidine, and somatostatin have failed to demonstrate clinical benefit.

Much interest has surrounded the possible role of PAF in acute pancreatitis. Experimental acute pancreatitis is associated with a rise in pancreatic PAF levels[9] and injection of PAF can induce experimental pancreatitis[9,10]. In addition, pulmonary PAF levels are increased in lung injury associated with experimental pancreatitis[11]. PAF antagonists protect against the haemodynamic effects of experimental pancreatitis[12] and against the pulmonary effects of both experimental pancreatitis[11] and endotoxaemia[13]. There is evidence therefore that PAF is involved in acute pancreatitis both at the early stages of pancreatic inflammation and in the pathophysiology of the systemic features of the condition and that PAF may be an important mediator of the systemic leucocyte activation common to sepsis and acute pancreatitis.

2. LEXIPAFANT IN ANIMAL MODELS

Lexipafant has also proven effective in ameliorating experimentally induced pancreatitis in rats. In this rodent model, pancreatitis was induced by a technique of microvascular ischaemia. A single intraperitoneal injection of lexipafant was found to significantly reduce the rise in serum amylase and to improve the histology score[14]. The drug also reduced the amount of pulmonary damage seen in this model[15].

3. HUMAN TRIALS

So far two double-blind placebo controlled trials have been completed with lexipafant of which one is in press[16] and the other is being submitted for publication. In the first trial 83 patients, (42 lexipafant and 41 placebo) were recruited from 5 centres in the United Kingdom. Lexipafant was administered as a 1 h infusion of 15 mg four times per day for up to three days.

Lexipafant treatment significantly reduced serum Il-8 on day 1 ($p<0.05$), and both IL-6 and serum E-selectin declined on day 1. There was a significant reduction in the incidence of organ failure ($p<0.05$) and in total organ failure score, OFS, ($p<0.05$) at the end of medication (72 h). Seven of 12 patients with severe acute pancreatitis recovered from organ failure with lexipafant; only 2 of 11 patients recovered with placebo and 2 others developed new organ failure, during the 72 h of treatment.

A phase II study of 51 patients with prognostically severe acute pancreatitis was completed by 11 hospitals in the west of Scotland earlier this year. Two phase II/III trials are currently in progress. In the UK a 300 patient, double blind placebo controlled study of lexipafant 100 mg/d for up to seven days in patients with Apache II scores of 7 or greater was started in November, 1994. In the U.S a dose ranging study with three groups of patients being treated with lexipafant 10 mg/d, 30 mg/d and 100 mg/d versus placebo is due to commence recruitment this quarter.

4. CONCLUSIONS

The potent PAF antagonist lexipafant has shown efficacy in animal models of acute pancreatitis and in a phase II pilot study in humans. Two pivotal trials are underway to try and confirm this early promise.

REFERENCES

1. Fernandez-del Castillo C, Rattner DW, Warshaw AL. Acute Pancreatitis. *Lancet* 1993;342:475–479.

2. Bradley EL. A Clinically based classification system for acute pancreatitis. summary of the International Symposium on Acute Pancreatitis, Atlanta, Ga, September 11 through 13, 1992. *Arch Surg* 1993;128:586–590.
3. Larvin M. McMahon MJ. Apache-II score for assessment and monitoring of acute pancreatitis. *Lancet* 1989; :201–205.
4. Heath D, Alexander D, Wilson C, Larvin M, Imrie CW, McMahon. Which complications of acute pancreatitis are most lethal? A prospective multi-centre study of 719 attacks. Gut 1995; 36:3, A478.
5. McFadden D.W. Organ failure and multiple organ system failure in pancreatitis. *Pancreas* 1991;6, Suppl.1:S37-S43.
6. Buchler M, Malfertheiner P, Uhl W, et al. Gabexate mesilate in human acute pancreatitis. *Gastroenterology* 1992;104:1165–1170.
7. Imrie CW, Benjamin IS, Ferguson JC, McKay AJ, Mackenzie I, O'Neill J, Blumgart LH. A single-centre double blind trial of Trasylol therapy in primary acute pancreatitis. *Br J Surg* 1978;65:337–341.
8. Bone RC, Balk RA, Cerra FB et al. ACCP/SCCM consensus conference: definitions for Sepsis and Organ failure and Guidelines for the Use of Innovative Therapies in Sepsis. *Chest* 1992;101:1644–1655.
9. Konturek SJ, Dembinski A, Konturek PJ, Warzecha Z, Jaworek J, Grustaw P, R Tomasxewska R, Stachura J. Role of platelet activating factor in pathogenesis of acute pancreatitis in rats. *Gut* 1992;33:1268–1274.
10. Emanuelli G, Montrucchio G, Gaia E, Gughera L, Corvetti G, Gurbetta L. Experimental acute pancreatitis induced by platelet activating factor in rabbits. *Am J Path* 1989;134:315–326.
11. Zhou W, McOllum MO, Levine BA, Olson MS. Role of platelet activating factor in pancreatitis-associated lung injury in the rat. *Am J Path* 1992;140:971–979.
12. Guillermo A, Lopez Farre A, Gomez Garre DN, Novo C, Romeo JM, Branquet P, Lopez-Novoa JM. Role of platelet activating factor in haemodynamic derangements in an acute rodent pancreatic model. *Gastroenterology* 1992;102:181–187.
13. Chank SW, Fedderson CO, Henson PM, Voelkel NF. Platelet activating factor mediates haemodynamic changes and lung injury in endotoxin-treated rats. *J Clin Invest* 1987;79:1498–1509.
14. Formela LJ, Wood LM, Whittaker M, Kingsnorth AN. Amelioration of experimental acute pancreatitis with a potent platelet-activating factor antagonist. *Br J Surg* 1994;81:1783–1785.
15. Galloway SW, Kingsnorth AN. Lung injury in the microembolic model of acute pancreatitis and amelioration by a platelet-activating factor antagonist. *Pancreas* 1995. In press.
16. Kingsorth AN, Galloway SW, Formela LJ. Randomized, double-blind phase II trial of lexipafant, a platelet-activating factor antagonist in human pancreatitis. *British J of Surg* 1995. In Press.

THE USE OF LEXIPAFANT IN THE TREATMENT OF ACUTE PANCREATITIS

C. McKay, F. J. M. Curran, C. E. Sharples, C. A. Young, J. N. Baxter, and C. W. Imrie

Departments of Surgery
Glasgow Royal Infirmary and Western Infirmary
Glasgow, Scotland

INTRODUCTION

Acute pancreatitis is a common condition which may varies in severity from a mild, self-limiting illness to a fulminating, rapidly fatal condition. Two phases of illness are recognised. Initially, patients have a systemic illness characterised by varying degrees of organ failure. In the second and third weeks of the illness, the main problems relate to local, pancreatic complications.

Despite many years of study, the pathophysiology of acute pancreatitis remains unknown and it is perhaps not surprising that none of the many therapeutic agents which have been proposed over the years has proven effective in randomised clinical trials.

It has long been thought that acute pancreatitis is mediated by activation of proteases within the pancreas and the resulting release of proteases into the circulation, lymphatics and peritoneal cavity[1,2,3]. This process was thought to cause the release of vasoactive peptides, both locally and systemically, and both activated proteases and vasoactive peptides were thought to mediate the haemodynamic and systemic organ effects commonly associated with the disease. However, unlike experimental pancreatitis, clinical acute pancreatitis is not associated with overwhelming of natural antiproteases[4,5]. This finding, coupled with the failure of antiprotease therapy in clinical trials[6–10], has led to new areas of research into the pathophysiology of the disease.

Many of the systemic manifestations of acute pancreatitis are similar to those seen in patients with sepsis. A hyperdynamic circulation is seen in both conditions[11,12] as are various degrees of respiratory and renal impairment. Respiratory insufficiency is particularly common in acute pancreatitis with adult respiratory distress syndrome (ARDS) accounting for the majority of early deaths. In sepsis, there is now considerable evidence that the systemic illness is the result of uncontrolled activation of mononuclear phagocytes[13] and the consequent release of increased quantities of pro-inflammatory cytokines including tumour necrosis factor alpha (TNFα), interleukin-1 (IL-1), Interleukin-6 (IL-6) and interleukin-8 (IL-8).

Many of the effects of these cytokines are potentiated by platelet-activating factor (PAF) which also plays a role in the autoregulation of cytokine production[14]. There has been much recent interest in the possible role of these mediators in the pathophysiology of acute pancreatitis and particularly of the early, systemic illness seen in these patients.

MONONUCLEAR PHAGOCYTE ACTIVATION AND CYTOKINES IN ACUTE PANCREATITIS

In 1988, Rinderknecht first proposed the hypothesis that cytokines may play an important role in the pathophysiology of acute pancreatitis[15]. He cast doubt on the conventional view that activated pancreatic proteases were responsible for the systemic manifestations of acute pancreatitis and suggested that excessive stimulation of neutrophils by phagocytosed cell debris may lead to production of harmful quantities of free oxygen radicals, leukotrienes and the newly described cachectin (tumour necrosis factor) in a situation analogous to septic shock.

Until recently, little was known of the role of these inflammatory mediators in acute pancreatitis. Early studies were restricted to the measurement of plasma levels of cytokines and met with varied success. It was convincingly demonstrated that IL-6 levels are increased in patients with complicated or fatal acute pancreatitis[16–18] and IL-6 levels have been reported to have prognostic significance early in the course of the disease[16]. IL-6 may therefore be responsible for the acute phase response in acute pancreatitis[17] and the commonly observed rise in C-reactive protein[18].

The first attempt to investigate this hypothesis was by Banks and colleagues. who measured serial plasma levels of TNFα in 27 patients with acute pancreatitis[19]. They found raised levels of TNFα in some patients, including two patients who died of their illness, but found no statistically significant difference in TNFα levels between patients with mild and those with severe disease. In those with raised levels, TNFα levels peaked within the first three days after admission but many patients had TNFα levels which were similar to those observed in healthy controls. The authors concluded that it was unlikely that TNFα played an important pathophysiological role in acute pancreatitis, although they did find evidence of neutrophil activation. Subsequently, Exley and co-workers[20], in 38 patients with prognostically severe acute pancreatitis taking part in a therapeutic trial of fresh frozen plasma, measured serum levels of TNFα during the first week of admission. They found detectable levels of TNFα on admission in 45% of non-survivors compared with 23% of survivors and overall, the levels of TNFα were higher than those reported by Banks et al. In addition, endotoxin was detected in 91% of non-survivors on admission. These studies demonstrate the problems associated with plasma TNFα measurement. TNFα is thought to act primarily at a paracrine level and a cell-bound form has even been described[21]. Circulating levels therefore represent "spillover" into the circulation of excess TNFα and the absence of TNFα in the circulation does not imply that TNFα is not involved in a disease process. Circulating TNFα inhibitors rapidly bind TNFα[22,23] and can interfere with its detection in assays which further complicates the interpretation of these studies[24]. In addition, TNFα has a relatively short circulation half-life[25] so that transiently raised levels may be missed.

In order to overcome the problems associated with plasma cytokine measurement, we examined directly the secretion of these pro-inflammatory cytokines by peripheral blood monocytes, isolated from patients with acute pancreatitis[26]. Cytokine secretion was compared between patients who developed systemic organ complications and those who had an uncomplicated course. We found that peak levels of secretion of IL-6, IL-8 and TNFα were significantly higher in those patients who developed systemic complications providing strong evidence of a role for these mediators in the development of the systemic complications of acute pancreatitis.

PLATELET-ACTIVATING FACTOR AND ACUTE PANCREATITIS

Platelet activating factor (PAF) is a lipid mediator formed from membrane phospholipid by the action of phospholipase A_2. It is produced by several cell types, including activated macrophages, granulocytes and platelets and has potent effects on neutrophil activation, platelet aggregation and endothelial cell function[14]. PAF has been shown to induce the release of TNF from monocytes[27] and the effect of PAF on neutrophil superoxide radicle production is greatly enhanced by pre-incubation with TNF[14]. PAF antagonists have also been reported to ameliorate the toxic effects of TNF in an experimental model[28].

Much interest has surrounded the possible role of PAF in acute pancreatitis. Experimental acute pancreatitis is associated with a rise in pancreatic PAF levels[29] and injection of PAF can induce experimental pancreatitis [29,30]. In addition, pulmonary PAF levels are increased in lung injury associated with experimental pancreatitis[31]. PAF antagonists have been reported to be beneficial in ceruelin-induced pancreatitis and to reduce superoxide radical production[32]. In addition, PAF antagonists protect against the haemodynamic effects of experimental pancreatitis[33] and against the pulmonary effects of both experimental pancreatitis[31] and endotoxaemia[34]. There is evidence therefore that PAF is involved acute pancreatitis both in the early stages of pancreatic inflammation and in the pathophysiology of the systemic features of the condition and that PAF may be an important mediator of the systemic leukocyte activation common to sepsis and acute pancreatitis.

PAF ANTAGONISTS IN ACUTE PANCREATITIS

West of Scotland Study

In order to investigate the therapeutic potential of a platelet activating factor antagonist in patients with acute pancreatitis, a prospective, randomised, placebo-controlled trial was conducted between 11 hospitals in the West of Scotland. The aims of this study were twofold;

Firstly to assess the therapeutic potential of the PAF antagonist, Lexipafant (British Biotech Ltd, Oxford, UK) in acute pancreatitis with particular reference to the development of systemic complications.

Secondly, to assess the effect of Lexipafant on the activation of mononuclear phagocytes and neutrophil polymorphs, thought to be responsible for the development of systemic complications.

Patients were selected for admission to the study by the presence of one or more of the following:

- Admission APACHE II score > 5
- C-reactive protein >120mg/l
- Three or more positive Glasgow criteria

Table 1. Lexipafant in acute pancreatitis: results

	Placebo (n=24)	Lexipafant (n=27)
Change in organ failure score from baseline (Mean (SD))	0.17 (1.97)	−1.33 (1.44)*
No of patients with organ failure at admission	9	16
No of patients with organ failure at end of treatment	8	7
Ninety day all cause mortality	6	4

*P=0.005, Wilcoxon signed rank sum test.

Patients were randomised to receive either placebo or Lexipafant (100mg per day) for up to 7 days or until full clinical and biochemical resolution.

Organ failure scores were calculated each day for eight days after admission and patients followed up to discharge.

The results of this study are summarised in table 1 above.

There was a significant reduction in the number of patients with organ failure after 7 days in the Lexipafant treated group which was mainly accounted for by a marked reduction in the number of patients developing new organ failure following entry to the study in the treatment group.

Liverpool Study

A similar trial has been reported by Kingsnorth and colleagues in Liverpool, UK. In this study, patients with acute pancreatitis of every grade of severity were included and a similar effect of Lexipafant on organ failure scores was observed.

SUMMARY AND CONCLUSIONS

The pathophysiology of systemic organ failure in acute pancreatitis has been the subject of debate for many years but there is growing evidence that increased production of pro-inflammatory cytokines plays an important role. From this work and from the results of studies in experimental pancreatitis there exists a rationale for the use of PAF antagonists in the treatment of acute pancreatitis. Two pilot studies have now demonstrated a beneficial effect of the PAF antagonist Lexipafant on acute pancreatitis which may lead to an important advance in the treatment of these patients. A multicentre trial aiming to recruit 300 patients with severe acute pancreatitis is now underway in the UK, the results of which will be awaited with interest.

ACKNOWLEDGMENTS

The West of Scotland study was funded by British Biotech Ltd

The authors are grateful to the consultant surgeons and their staff at the following hospitals for their co-operation in this study:

Glasgow Royal Infirmary, Western Infirmary, Gartnaval General Hospital, Stobhill Hospital, Victoria Infirmary, Southern General Hospital, Law Hospital, Monklands Hospital, Vale of Leven Hospital, Hairmyres Hospital, Inverclyde Royal Hospital.

REFERENCES

1. Frey CF, Wong HN, Hickman D, Pullos T. Toxicity of haemorrhagic ascitic fluid associated with haemorrhagic pancreatitis. Arch Surg 1982;117:401–404.
2. Ohlssen K, Tegner H. Experimental pancreatitis in the dog. Demonstration of trypsin in ascitic fluid, lymph and plasma. Scand J Gastroenterol 1973;8:129–133.
3. Amundsen E, Ofstad E, Hagen P-O. Experimental acute pancreatitis in dogs. Hypotensive effect induced by peritoneal exudate. Scand J Gastroenterol 1968;3:659–664.
4. Balldin G, Ohlsson K. Demonstration of pancreatic protease-antiprotease complexes in the peritoneal fluid of patients with acute pancreatitis. Surgery 1979;85:451–6.
5. Wilson C, Shenkin A, Imrie C. Role of the protease-antiprotease balance in peritoneal exudate during acute pancreatitis. Br J Surg 1991;78:78–81.
6. Medical Research Council of the United Kingdom. Death from acute pancreatitis: Multicentre trial of glucagon and aprotinin. Lancet 1977;ii:632–635.
7. Imrie CW, Benjamin S, Ferguson JC, McKay AJ, MacKenzie I, O'Neill J, Blumgart LH. A single-centre double-blind trial of Trasylol in primary acute pancreatitis. Br J Surg 1978;65:337–341.
8. Leese T, Holliday M, Heath D, Hall AW, Bell PRF. Multicentre clinical trial of low volume fresh frozen plasma therapy in acute pancreatitis. Br J Surg 1987;74:907–911.
9. Leese T, Thomas WM, Holloday M, Attard A, Watkins M, Neoptolemos JP, Hall C. A multicentre controlled clinical trial of high-volume fresh frozen plasma in prognostically severe acute pancreatitis. Ann R Coll Surg Eng 1991;73:207–214.
10. Buchler M, Malfertheiner P, Uhl W, Scholmerich J, Stockman F, Adler G, Gaus W, Rolle K, Beger HG and the German Pancreatic Study Group. Gabexate mesilate in human acute pancreatitis. Gastroenterology 1993;104:1165–1170
11. Bradley EL, Hall JR, Lutz J, Hamner L, Lattouf O. Haemodynamic consequences of severe pancreatitis. Ann Surg 1983;198:130–133.
12. Beger HG, Bittner R, Buchler M, Hess W, Schmitz JE. Haemodynamic data pattern in patients with acute pancreatitis. Gastroenterology 1986;90:74–79
13. Tracey KJ, Lowrey SF. The role of cytokine mediators in septic shock. Adv Surg 1990;23:21–56.
14. Braquet P, Paubert-Braquet M, Bourgain RH, Bussolino F, Hosford D. PAF/cytokine auto-generated feedback networks in microvascular immune injury: consequences in shock, ischaemia and graft rejection. J Lipid mediators 1989;1:75–112.
15. Rinderknecht H. Fatal pancreatitis, a consequence of excessive leukocyte stimulation? Int J Pancreatology 1988;3:105–112.
16. Heath DI, Cruickshank A, Gudgeon M, Jehanli A, Shenkin A, Imrie CW. Role of interleukin-6 in mediating the acute phase protein response and potential as an early means of severity assessment in acute pancreatitis. Gut 1993;34:41–45.
17. Leser H-G, Gross V, Scheibenbogen C, Heinisch A, Salm R, Lausen M, Ruckauer K, Andreesen R, Farthmann EH, Scholmerich J. Evaluation of serum interleukin-6 concentration precedes acute-phase response and reflects severity in acute pancreatitis. Gastroenterology 1991;101:782–785.
18. Viedma JA, Perez-Mateo M, Dominguez JE, Carballo F. Role of interleukin-6 in acute pancreatitis. Comparison with C-reactive protein and phospholipase A. Gut 1992;33:1264–1267.
19. Banks RE, Evans SW, Alexander D, McMahon MJ, Whicher JT. Is fatal acute pancreatitis a consequence of excessive leucocyte stimulation? The role of tumour necrosis factor. Cytokine 1991;3:12–16.
20. Exley AR, Leese T, Holloday MP, Swann RA, Cohen J. Endotoxaemia and serum tumour necrosis factor as prognostic markers in severe acute pancreatitis. Gut 1992;33:1126–1128
21. Kriegler M, Perez C, Defay K. A novel form of TNF/cachectin is a cell surface cytotoxic transmembrane protein. Ramifications for the complex physiology of TNFα. Cell 1988;53:45–53
22. Fernandez-Botran R. Soluble cytokine receptors: their role in immunoregulation. FASEB J. 1991;5:2567–2574.
23. James K, van de Haan J, Lens S, Farmer F. Preliminary studies on the interaction of TNFα and IFN with alpha 2 macroglobulin. Immunology Letters 1992;32:49–58.
24. McIntyre CA, Chapman K, Reeder S, Dorreen MS, Bruce L, Rodgers S, Hayat K, Schreenivasan T, Sheridan E, Hancock BW, Rees RC. Treatment of malignant melanoma and renal cell carcinoma with recombinant human interleukin-2: analysis of cytokine levels in sera and culture supernatants. Eur J Cancer 1992;28:58–63.
25. Blick M, Sherwin S, Rosenbum M et al Phase 1 study of recombinant tumour necrosis factor in cancer patients. Cancer Res 1987;47:2986–2989

26. McKay C, Baxter JN, Imrie CW. Systemic complications in acute pancreatitis are associated with increased cytokine release by monocytes. Digestion 1994;55:316–7.
27. Bonavida B, Mencia-Huerta JM, Braquet P. Effect of platelet activating factor (PAF) on monocyte activation and production of tumour necrosis factor (TNF). J Allergy Appl Immunol 1989;88:157–160.
28. Sun X, Hsueh W. Bowel necrosis induced by tumour necrosis factor in rats is mediated by platelet activating factor. J Clin Invest 1988;81:1328–1331.
29. Konturek SJ, Dembinski A, Konturek PJ, Warzecha Z, Jaworek J, Gustaw P, R Tomaszewska R, Stachura J. Role of platelet activating factor in pathogenesis of acute pancreatitis in rats. Gut 1992;33:1268–1274.
30. Emanuelli G, Montrucchio G, Gaia E, Dughera L, Corvetti G, Gubetta L. Experimental acute pancreatitis induced by platelet activating factor in rabbits. Am J Path 1989;134:315–326.
31. Zhou W, McOllum MO, Levine BA, Olson MS. Role of platelet activating factor in pancreatitis-associated lung injury in the rat. Am J Path 1992;140:971–979.
32. Dabrowski A, Gabryelewicz A, Chyczewski L. The effect of platelet activating factor antagonist (BN 52021) on ceuelin-induced acute pancreatitis with reference to oxygen radicals. Int J Pancreatol 1991;8:1–11.
33. Guillermo A, Lopez Farre A, Gomez Garre DN, Novo C, Romeo JM, Braquet P, Lopez-Novoa JM. Role of platelet activating factor in haemodynamic derangements in an acute rodent pancreatic model. Gastroenterology 1992;102:181–187.
34. Chang SW, Fedderson CO, Henson PM, Voelkel NF. Platelet activating factor mediates haemodynamic changes and lung injury in endotoxin-treated rats. J Clin Invest 1987;79:1498–1509.

60

EFFECT OF A POTENT PLATELET-ACTIVATING FACTOR ANTAGONIST, WEB-2086, ON ASTHMA

A Multicenter, Double-Blind Placebo-Controlled Study in Japan

G. Tamura,[1] T. Takishima,[2] S. Mue,[3] S. Makino,[4] K. Itoh,[5] T. Miyamoto,[6] T. Shida,[6] and S. Nakajima[7]

[1]Tohoku University
[2]Chest Institute of Technology
[3]Sendai University
[4]Dokkyo University
[5]University of Tokyo
[6]Japan Asthma and Allergy Clinic
[7]Kinki University, Japan

1. INTRODUCTION

Platelet-activating factor (PAF) has been suggested to be an important chemical mediator in bronchial asthma[1]. When given by inhalation, PAF produces short-term bronchoconstriction in both normal[2] and asthmatic subjects[2,3], increases airway reactivity in normal volunteers[4,5], stimulates leukotriene and thromboxane A2 production in asthmatics[6], and also worsens gas exchange in mild asthma[7]. In aminal models, it has been reported that exogenous PAF causes bronchoconstriction, microvascular leakage, mucous hypersecretion, an increase in bronchial responsiveness, and eosinophil recruitment into the airways of guinea pigs[1]. Thus, PAF induces various pathophysiologic changes relevant to bronchial asthma.

Consequently, potent and specific PAF receptor antagonists, such as WEB-2086[8,9] (apafant) and UK-74,505[10,11] (modipafant), have been developed and examined for their clinical efficacy in asthma. However, to our knowledge, there are no clinical reports that potent PAF receptor antagonists have a beneficial effect on asthma.

In this study, we examined the efficacy of a potent PAF receptor antagonist, apafant, in the treatment of mild and moderate asthmatics.

2. METHODS

2.1 Protocol

This was a multicenter, double-blind, placebo-controlled parallel group study performed in Japan. It consisted of two phases: a 2 wk-baseline period during which patients were required to demonstrate symptomatic asthma, and an 8 wk-treatment period.

2.2 Subjects

Males and females aged 17 to 77 yr participated in the present study after giving informed consent. The study was approved by the local ethics committees of all 59 institutes taking part in the study.

The total number of patients participating in the study was 221 patients with mild and moderate asthma, who received concomitantly bronchodilators and/or corticosteroids of less than 10mg/day of oral predonisolone or 800μg/day of BDP. As much as possible, we tried not to change these drugs, except for rescue use of inhaled β2 agonists.

2.3 Drug Administration

Using 40mg tablets of apafant and/or identical oral placebo tablets, patients who belonged to the low-dose (L) group were administered one active and one placebo tablet 3 times a day (after breakfast, after supper, and before going to bed). Patients who belonged to the high-dose (H) group were administered two active tablets in the same fashion. Patients who belonged to the placebo (P) group were administered two placebo tablets in the same fashion. These tablets were administered in a double-blind fashion.

2.4 Items for Assessment

During the period of the study, patients were required to fill out a diary card recording symptoms and treatments. Symptoms recorded on the diary card were rated by the patient for severe, moderate, and mild attacks, wheeze and cough. For assessment of clinical improvement, these scores and the treatments were evaluated.

For analysis of PEF, mean values of each morning and evening PEF per every 2 weeks were calculated in 104 cases who were measured for morning and evening PEF during the period of the study. Each PEF value recorded was the best of three efforts.

2.5 Safety Measurement

During the baseline period and at the end of the treatment period, venous blood and urin were taken for a safety check. An electrocardiogram was performed in the same manner.

2.6 Composition of Patients

A total of 221 patients (P group:75, L group:74, and H group:72) with asthma were enrolled, of whom 199 (P group:67, L group:65, and H group:67) completed the treatment period. Before unblinding the data, data from 26 patients were excluded due to poor compliance and violations of the protocol. Consequently, 173 patients (P group:55, L group:59, and H group:59) were accepted for assessment of clinical improvement. For analysis of PEF, data obtained from 104 patients (P group:31, L group:35, and H group:38) out of 173 were used.

2.7 Statistical Methodology

Intergroup data was analysed by Dunn's multiple comparison and intragroup data was analysed by Wilcoxon's one-sample test. All statistical tests were two tailed, and a p value of less than 0.05 was considered statistically significant. To examine the relationship between the changes in daily variability and the variability during the baseline period, linear regression analysis was used.

3. RESULTS

3.1 Analysis of Clinical Improvement

In the background of the 173 patients accepted for the assessment of clinical improvement, we found no differences in age, types of asthma, severity of asthma, or use of steroids among the 3 groups.

The final clinical improvement of all 173 cases is shown in Fig. 1. In the high-dose group, 11.9% of 59 patients had excellent improvement, 45.8% had more than good improvement, and 69.5% had more than slight improvement. In the low-dose group, these figures were 8.5%, 40.7%, and 62.7%, respectively. The placebo group was 3.6%, 30.9%, and 45.5%. In terms of the final clinical improvement, we found statistically significant differences between the high-dose group and the placebo group.

Fig. 2 shows the final clinical improvement based on the type of asthma, severity of asthma, and use of steroids. In the chronic and perennial types of asthma, moderate asthma, and asthmatics treated with steroids, we found statistically significant differences in the clinical improvement between the high-dose group and the placebo group, although in paroximal and seasonal types of asthma, mild asthma, and asthmatics treated without steroids, we found no differences among the 3 groups.

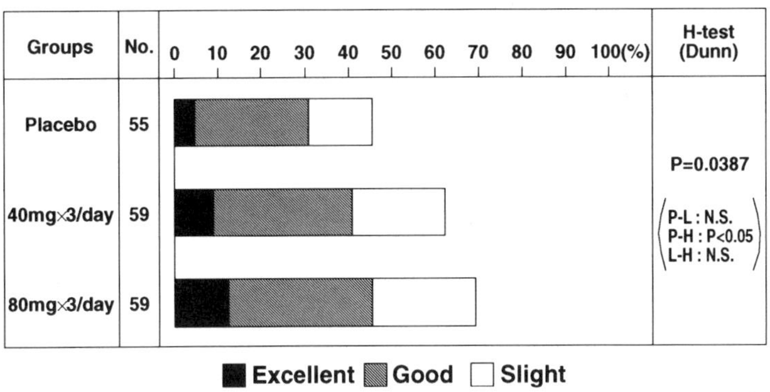

Figure 1. The final clinical improvement of all 173 cases. The clinical improvement was evaluated by symptoms and treatments obtained from patients' diaries.

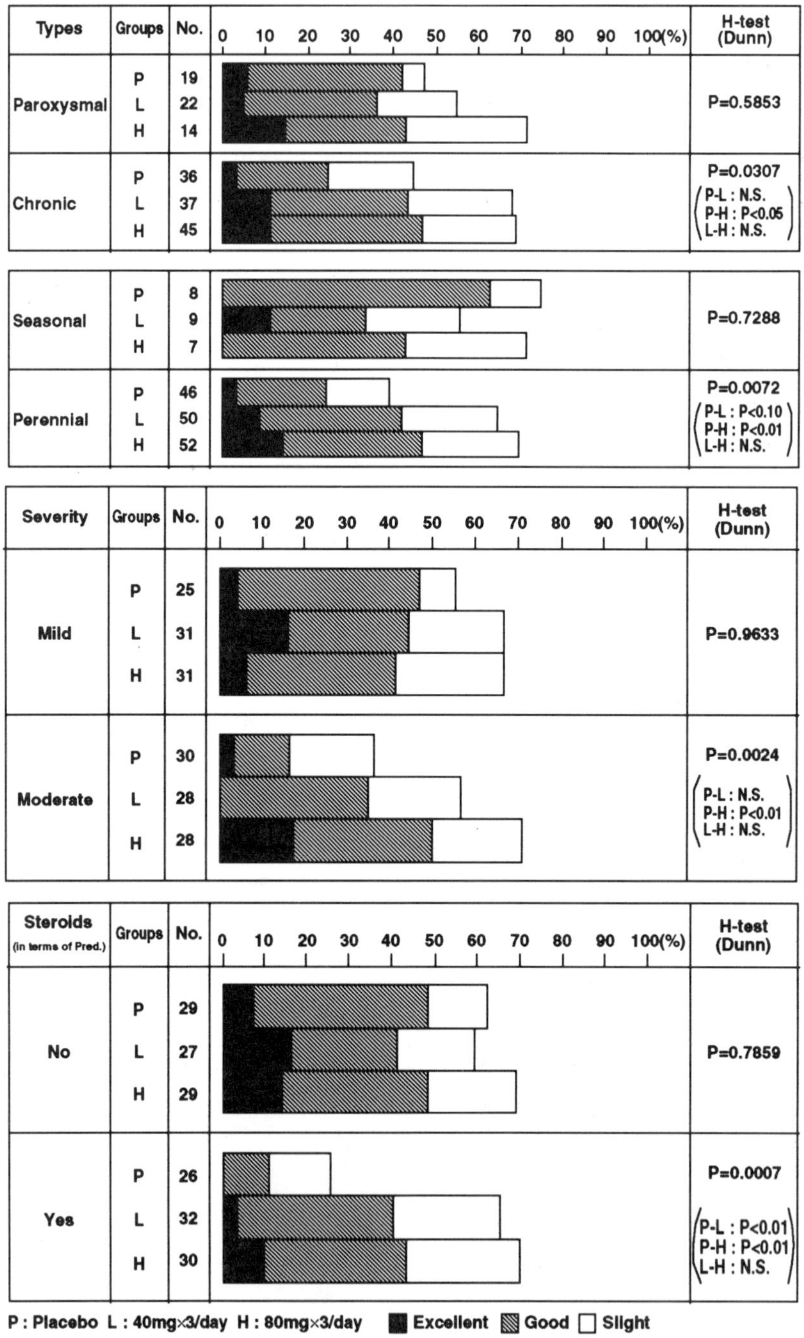

Figure 2. The final clinical improvement based on the type of asthma, severity of asthma, and use of steroids.

3.2 Safety Data

The adverse effects were evaluated in 215 patients. The placebo group had nausea(2 cases), vomiting, stomach-ache and pain in the tongue, the low-dose group had edema, discomfort in the epigastrium, arrhythmia, chest tightness and pollakiuria, and the high-dose group had hemorrhage under conjunctiva bulbi, nausea, a sense of fullness, diarrhea, eruption, and itching. These adverse effects were not serious. In addition, mild increases in GOT and GPT were found in 2 patients of each treatment group. We found no other serious biochemical or hematological changes in the 3 groups. Therefore, we concluded that apafant was well tolerated and safe.

3.3 Analysis of PEF

In the background of the 104 patients measured for morning and evening PEF, there were no differences in age, types of asthma, severity of asthma, or use of steroids among the 3 groups.

Fig. 3 shows the changes in PEF of all 104 patients. Absolute values of changes in PEF were calculated as follows: Mean values of PEF per every 2 weeks during the period of drug administration - those during the baseline period. As shown in Fig. 3, only the high-dose group showed gradual improvement in the morning and evening PEF, in contrast to the low-dose group and the placebo group. Especially the PEF values at 7 to 8 weeks-treatment in the high-dose group were significantly higher than those in the baseline period and the placebo groups at 7 to 8 weeks-treatment. The increases in mean morning and evening PEF values in the high-doses group at 7 to 8 weeks after the treatment were 31.1 l/min and 29.2 l/min, respectively.

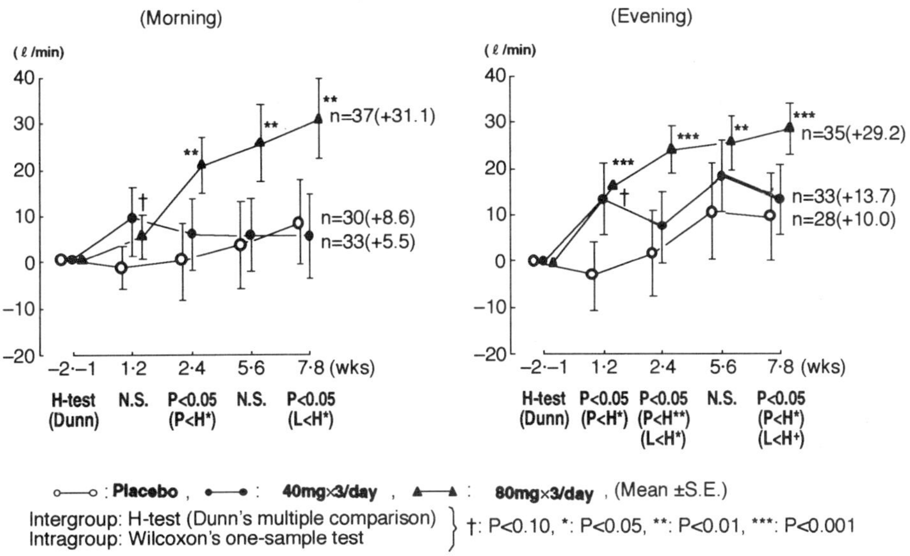

Figure 3. The changes in PEF of all 104 cases. Absolute values of changes in PEF were calculated as follows: Mean values of PEF per every 2 weeks during the period of drug administration - those during the baseline period.

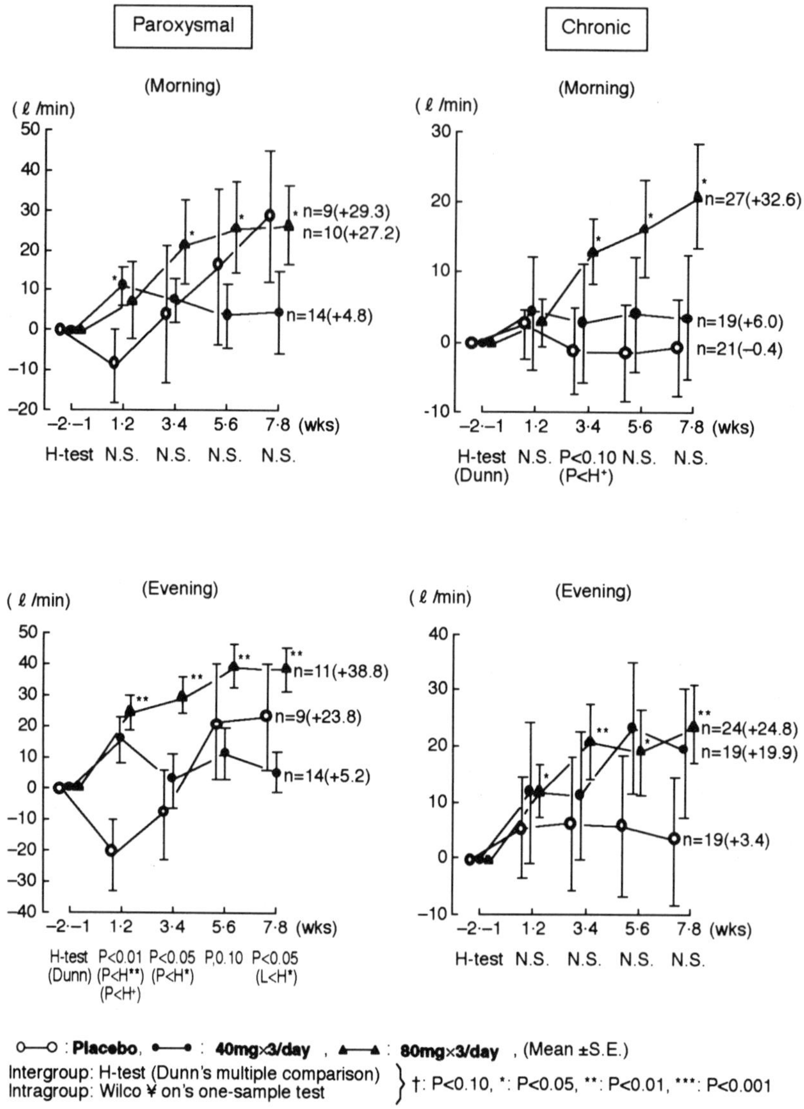

Figure 4. The changes in morning and evening PEF based on the type of asthma.

Fig. 4 shows the changes in morning and evening PEF based on the type of asthma. As shown in the figure, in patients with the chronic type of asthma, only the high-dose group showed gradual improvement of the morning PEF, but not the low-dose and placebo groups. In contrast, in asthmatics of the paroxysmal type, we found no differences in the changes in PEF among the 3 groups.

Fig. 5 shows the changes in morning and evening PEF based on the severity of asthma. As shown in the figure, in patients with moderate asthma, only the high-dose group showed gradual improvement of the morning and evening PEF, but not the low-dose and placebo groups. In contrast, in mild asthmatics, we found no differences in the changes in PEF among the 3 groups.

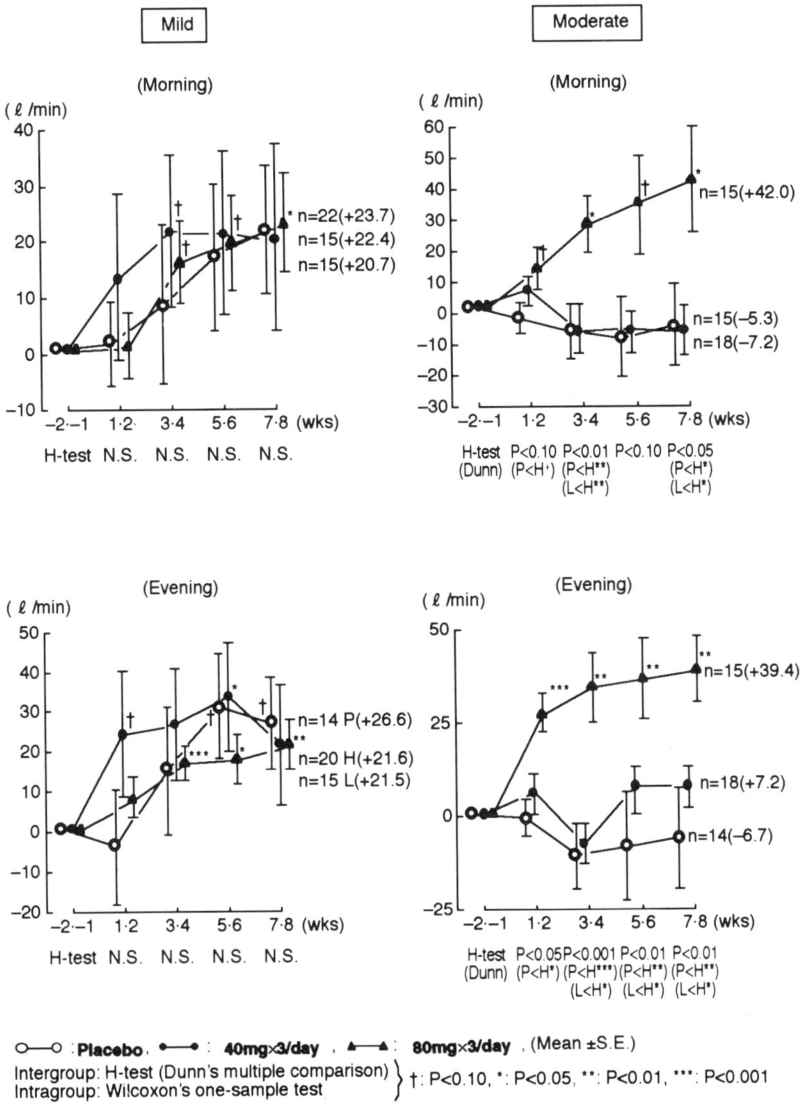

Figure 5. The changes in morning and evening PEF based on the severity of asthma.

Fig. 6 shows the changes in morning and evening PEF of patients with steroid-dependent and -independent asthma. As shown in the figure, in steroid-dependent asthmatics, only the high-dose group showed gradual improvement of the morning and evening PEF, but not the low-dose and placebo groups. In contrast, in steroid-independent asthmatics, we did not find any differences in PEF among the 3 groups.

Fig. 7 shows the relationship between the changes in daily variability and the variability during the baseline period. The changes in daily variability were calculated as follows: The daily variability at the 8 wks-treatment - the variability during the baseline period. The variability during the baseline period was expressed as the percentage of evening PEF - morning PEF to the evening PEF. As shown in the figure, especially the regression line in the high-dose group leaned downward toward the right and there was a

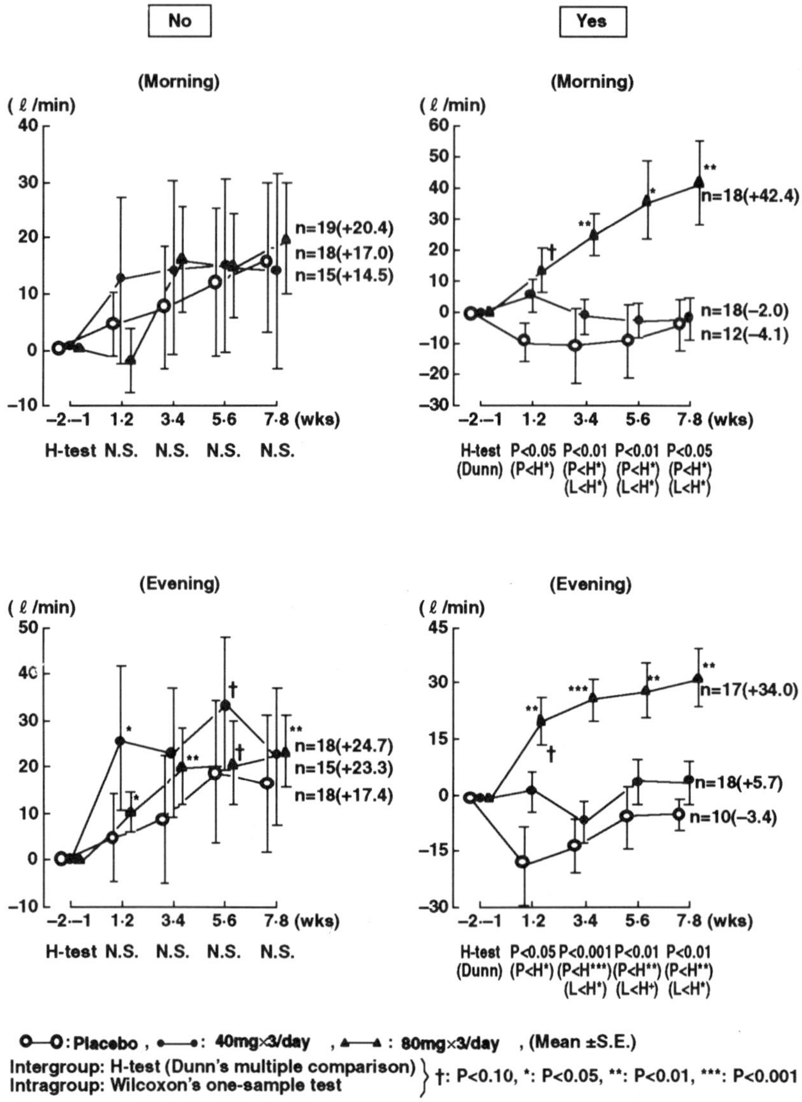

Figure 6. The changes in morning and evening PEF of patients with steroid-dependent and -independent asthma.

dose-dependency in the degree of the slopes among the 3 groups. Thus, this finding supports that apafant may improve airway narrowing detected by PEF in mild and moderate asthmatics.

4. DISCUSSION

In this multicenter, double-blind placebo-controlled parallel group study, we showed that the final clinical improvement after 8 wks-treatment with 80mgx3/day of apafant was significantly higher than that after the treatment with an inactive placebo, and that the

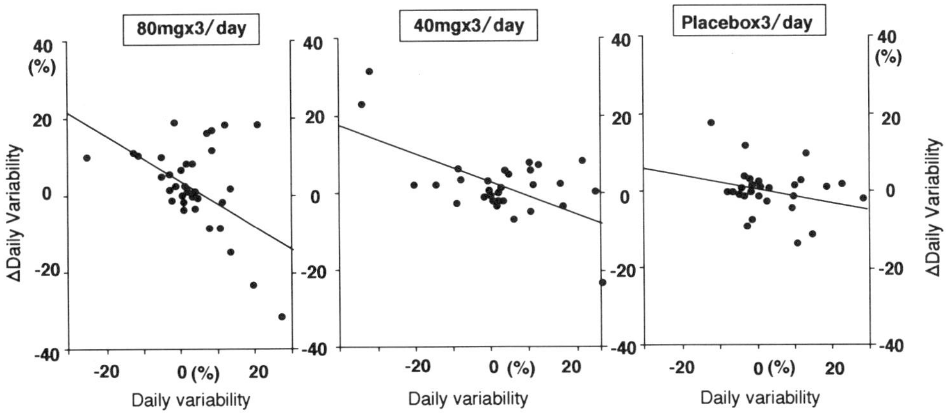

Figure 7. The relationship between the changes in daily variability and the variability during the baseline period. The changes in daily variability were calculated as follows: The daily variability at the 8 wks-treatment - the variability during the baseline period. The variability during the baseline period was expressed as the percentage of evening PEF - morning PEF to the evening PEF. We found that there was a dose-dependency in the degree of the slopes among the 3 groups.

mean values of morning and evening PEF at 7 to 8 wks-treatment with 80mgx3/day of apafant were significantly higher than those in the baseline period and those of the placebo group at 7 to 8 wks-treatment. Thus, 8 wks-treatment with 80mgx3/day of apafant significantly improved not only subjective symptoms but also airway narrowing objectively determined by PEF in mild and moderate asthmatics. In addition, there were no differences in the patients' background among the 3 groups, and the adverse events profile revealed that apafant was well tolerated and safe. Therefore, we suggest that 80mgx3/day of apafant may prove to be a useful agent in the treatment of asthma and that PAF may be one of the chemical mediators in deteriorating asthma.

In a clinical study[12], it has been reported that oral treatment with apafant of 40mgx3/day does not reduce the requirement for inhaled steroids in atopic asthmatics. One of the major reasons for the difference from our finding may be that their study examined the sparing effect of corticosteroids and selected asthmatics in a more severe state. As shown in the results, however, apafant improved the changes in daily variability of PEF in a dose-dependent fashion and apafant of 40mgx3/day did not significantly improve asthmatic symptoms and changes in PEF compared with those of the placebo group. In addition, as previously reported, 1 week-administration of 40mgx3/day of apafant does not completely inhibit PAF-induced platelet aggregation (Data on an investigational brochure of apafant, pp61, Nippon Boehringer Ingelheim, 1993), although the treatment can effectively prevent PAF-induced bronchoconstriction in humans[13]. Therefore, we suggest that 40mgx3/day of apafant may prove not to be useful in the treatment of asthma.

In a clinical study[14], it has been reported that one week of treatment with apafant of 100mgx3/day does not attenuate allergen-induced early or late responses or airway hyperresponsiveness. In addition, other PAF receptor antagonists have been reported not to inhibit the responses[15,16]. This may be because the effects of PAF antagonism would take longer to improve than could be demonstrated in an acute study. Because of the one week of treatment, however, the amount of PAF released by inhaled allergen may be too little to produce bronchoconstriction[16], although PAF could be released by inhaled allergen[17,18].

In conclusion, our findings suggest that only 8 weeks-treatment with apafant of 80mgx3/day may prove to be a useful agent in the treatment of asthma and that PAF may be one of the chemical mediators in deteriorating asthma, especially in asthmatics of the chronic and perennial types, in patients with moderate asthma, and in asthmatics concomitantly treated with corticosteroids.

REFERENCES

1. P. J. Barnes, K. F. Chung, C. P. Page. Platelet-activating factor as a mediator of allergic disease. J Allergy Clin Immunol 1988 ; 81 : 919–934.
2. F. M. Cuss, C. M. S. Dixon, P. J. Barnes. Effects of inhaled platelet activating factor on pulmonary function and bronchial responsiveness in man. Lancet 1986 ; 2 : 189–92.
3. R. Louis, T. Bury, J. L. Corhay, M. F. Radermecker. Acute bronchial and hematologic effects following inhalation of a single dose of PAF. Comparison between asthmatics and normal subjects. Chest 1994 ; 106 : 1094–9.
4. B. B. Vargaftig, J. Lefort, M. Chignard, J. Benveniste. Platelet-activating factor induces a platelet dependent bronchoconstriction unrelated to the formation of prostaglandin derivatives. Eur J Pharmacol 1080 ; 65 : 185–92.
5. M. G. Kaye, L. J. Smith. Effects of inhaled leukotriene D4 and platelet-activating factor on airway reactivity in normal subjects. Am Rev Respir Dis 1990 ; 141 : 993–7.
6. I. K. Taylor, P. S. Ward, G. W. Taylor, C. T. Dollery, R. W. Fuller. Inhaled PAF stimulates leukotriene and thromboxane A2 production in humans. J Appl Physiol 1991 ; 71 : 1396–402.
7. M. A. Felez, J. Roca, J. A. Barbera, C. Santos, M. Rotger, K. F. Chung, R. Rodriguez-Roisin. Inhaled platelet-activating factor worsens gas exchange in mild asthma. Am J Respir Crit Care Med 1994 ; 150 : 369–73.
8. W. S. Adamus, H. Heuer, C. J. Meade, G. Frey, H. M. Brecht. Inhibitory effect of oral WEB 2086, a novel selective PAF-acether antagonist, on ex vivo platelet aggregation. Eur J Clin Pharmacol 1988 ; 35 : 237–40.
9. W. S. Adamus, H. Heuer, C. J. Meade, H. M. Brecht. Safety tolerability and pharmacologic activity of mulitiple doses of the new platelet activating factor antagonist WEB 2086 in human subjects. Clin Pharmacol Ther 1989 ; 45 : 270–6.
10. V. A. Alabaster, R. F. Keir, M. J. Parry, R. N. de Souza. UK-74,505, a novel and selective PAF antagonist, exhibits potent and long lasting activity in vivo. Agents Actions (Suppl.) 1991 ; 34 : 221–7.
11. B. J. O'Connor, S. Uden, T. J. Carty, J. D. Eskra, P. J. Barnes, K. F. Chung. Inhibitory effect of UK-74505, a potent and specific oral platelet activating factor (PAF) receptor antagonist, on airway and systemic responses to inhaled PAF in humans. Am J Respir Crit Care Med 1994 ; 150 : 35–40.
12. D. P. S. Spence, S. L. Johnston, P. M. A. Calverley, P. Dhillon, C. Higgins, E. Ramhamadany, S. Turner, A. Winning, J. Winter, S. T. Holgate. The effect of the orally active platelet-activating factor antagonist WEB 2086 in the treatment of asthma. Am J Respir Crit Care Med 1994 ; 149 : 1142–8.
13. W. S. Adamus, H. O. Heuer, C. J. Meade, J. C. Schilling. Inhibitory effects of the new PAF acether antagonist WEB 2086 on pharmacologic changes induced by PAF inhalation in human beings. Clin Pharmacol Ther 1990 ; 47 : 456–62.
14. A. Freitag, R. M. Watson, G. Matsos, C. Eastwood, P. M. O'Byrne. Effect of a platelet-activating factor antagonist, WEB 2086, on allergen induced asthmatic responses. Thorax 1993 ; 48 : 594–8.
15. L. M. Kuitert, K. P. Hui, S. Uthayarkumar, W. Burke, A. C. Newland, S. Uden, N. C. Barnes. Effect of platelet activating factor antagonist UK-74,505 on the early and late response to allergen. Am Rev Respir Dis 1993 ; 147 : 82–6.
16. R. G. Townley, R. Eda, R. J. Hopp, A. K. Bewtra, M. S. Gillen. The effect of RP 59227, a platelet activating factor antagonist, against antigen challenge and eosinophil and neutrophil chemotaxis in asthmatics. J Lipid Mediat Cell Signal 1994 ; 10 : 345–53.
17. M. Chan-Yeung, S. Lam, H. Chan, K. S. Tse, H. Salari. The release of platelet-activating factor into plasma during allergen-induced bronchoconstriction. J Allergy Clin Immunol 1991 ; 87 : 667–73.
18. J. A. Burgers, P. L. Bruynzeel, H. J. Mengelers, J. Kreukniet, J. W. Akkerman. Occupancy of platelet receptors for platelet-activating factor in asthmatic patients during an allergen-induced bronchoconstrictive reaction. J Lipid Mediat 1993 ; 7 : 135–49.

61

EX VIVO INHIBITION OF PAF-INDUCED β-THROMBOGLOBULIN RELEASE IN MAN BY ABT-299, A POTENT PAF ANTAGONIST

Daniel H. Albert, Terrance J. Magoc, Hollis D. Kleinert, Eugene Sun, Ana E. Reyes, George W. Carter, and James B. Summers

Abbott Laboratories
100 Abbott Park Road, Abbott Park, Illinois 60064-3500

1. INTRODUCTION

Although still unproved, the therapeutic potential of PAF antagonists for the treatment of diseases with an inflammatory component has been recognized for at least fifteen years. Over this period numerous antagonists encompassing an array of structural classes have been discovered that exhibit moderate to excellent intrinsic potency and are generally effective in animal models of PAF-mediated diseases[1, 2]. However, PAF antagonists for the most part have not proven effective in human disease[3–7]. To date, the most encouraging results come from the recent report that WEB-2086, when given at higher doses than studied previously to patients with mild to moderate asthma, resulted in improvement in lung function parameters[8]. In sepsis studies, improvement in survival of a subset of gram-negative patients receiving BN-52021 has also been reported[9]. A trend toward reduced mortality was also observed in follow-up studies, although the effect did not reach statistical significance[10]. These modest improvements observed in asthma and the suggestion of a positive effect in sepsis suggest that the currently available antagonists may lack sufficient potency to be optimally effective and support the need for more potent antagonists for human studies.

We have recently described ABT-299, an aqueous soluble prodrug that is rapidly converted *in vivo* to a novel, highly potent, specific PAF antagonist[11]. These properties make ABT-299 a promising candidate for further studies in human diseases that are potentially PAF-mediated. To that end, phase I clinical trails have been conducted to evaluate the safety and pharmacodynamic profile of ABT-299 in humans. Bioactivity of the drug was monitored in these studies by evaluating PAF-induced platelet degranulation *ex vivo* in whole blood via measurement of secretion of β-thromboglobulin, a platelet-specific alpha granule constituent. This communication reports the extent and duration of the inhibitory effect of ABT-299 on this PAF-induced platelet response in man.

2. METHODS

2.1 Study Design

Double blind, placebo controlled trials were conducted at U-Gene Clinical Research, Utrect Netherlands. Healthy male volunteers, age 18 to 40, were divided into three rotating groups and administered intravenously single escalating doses, beginning at 0.8 mg, of ABT-299 (6–7 subjects) or placebo (3–4 subjects). Blood samples were taken prior to and after dosing for drug level and *ex vivo* platelet degranulation analysis. Blood from subjects receiving placebo, 0.8 mg and 70 mg ABT-299 taken 30 minutes prior to dosing and 0.5 and 12 hours after dosing was used for *ex vivo* β-thromboglobulin analysis. Blood from the 70 mg group taken 0–24 hours post dose was used for determination of drug levels.

2.2 Secretion of β-Thromboglobulin in Whole Blood

Peripheral blood was collected and mixed with ACD anticoagulant (6:1, vol/vol) . Within 30 minutes after collection, aliquots of the blood were incubated with 10, 100 or 1000 nM PAF or vehicle (0.2% BSA in normal saline) for 15 minutes at 37°C and then placed in ice-water for 10 minutes. The blood samples were centrifuged (1000 g for 15 minutes) and the plasma was removed and stored frozen until analyzed. In preliminary experiments to establish the potency of the active component of ABT-299 (A-85783), blood obtained from healthy volunteers was preincubated with vehicle (0.2% BSA in normal saline) or A-85783 at the indicated concentration for 15 minutes at room temperature and then challenged with PAF and processed as above. Concentration of β-thromboglobulin in plasma was measured using commercially available immunoassay kits according to the manufacturer's recommendations (Diagnostica Stago, American Bioproducts Co., Parsippanny, NJ).

2.3 Drug Plasma Levels

Plasma samples, stored frozen, were thawed, adjusted to pH 8.5 with 0.5M Na_2CO_3 and extracted with 10 volumes hexane:ethyl acetate (1:1). Extracts were dried, resuspended in acetonitrile (0.1% acetic acid) and analyzed by high performance liquid chromatography (HPLC) using a C18 reversed phase column (5 x 0.46 cm, 3 μ Spherisorb S30DS2, Regis, Morton Grove, IL). The column was eluted with a linear gradient of 10 mM tetramethyl-ammonium perchlorate and 10% methanol in acetonitrile (40:60–50:50) over 20 minutes and monitored at 270 nm. The concentrations of A-85783 and its pyridine-*N*-oxide metabolite were determined by comparison of their peak areas on the chromatogram to peak areas from an external calibration curve.

3. RESULTS

3.1 Competitive Antagonism of PAF-Induced β-Thromboglobulin Release in Blood

When incubated with human peripheral blood, PAF caused an increase in plasma β-thromboglobulin (figure 1). The release, which is complete within 15 minutes (not

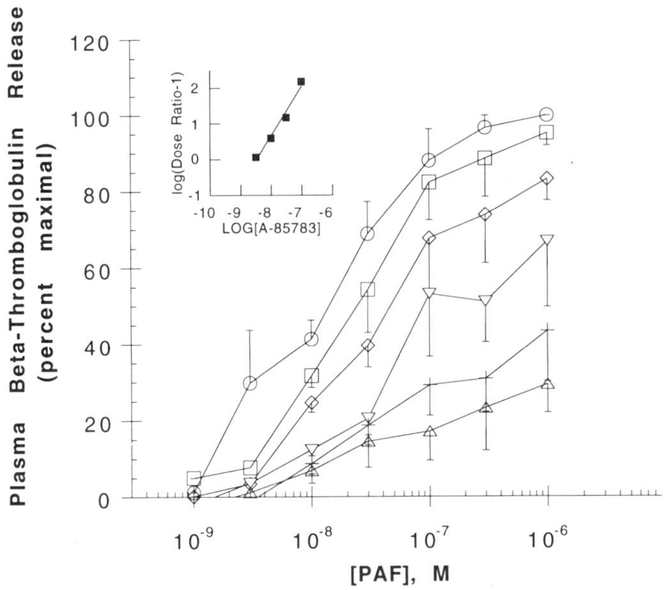

Figure 1. Inhibition of PAF-induced β-thromboglobulin release in human blood by A-85783. Blood was pre-incubated in the absence (circle) or presence of 3 nM (square), 10 nM (diamond), 30 nM (inverted triangle), 100 nM (cross), or 300 nM A-85783 (triangle) and then incubated with PAF. The release of β-thromboglobulin was monitored and expressed as percent of maximal response to PAF in the absence of drug (318 ± 160 U/ml). Values are the means ± standard deviation of 4 independent experiments. Schild analysis (insert) yielded an A_2 value of 3.3 nM (slope = 1.4).

shown), is dependent upon the PAF concentration. In the studies represented in figure 1 the potency (EC_{50}) of PAF for inducing release in blood samples from four donors ranged from 10 to 15 nM (mean 13.9 ±2.0 nM). Pretreatment of the blood with A-85783 inhibited the release of β-thromboglobulin in a concentration-dependent fashion (figures 1). The rightward shift in the potency of PAF in the presence of increasing concentrations of the antagonist is indicative of competitive inhibition. From Schild analysis of the rightward shift, the A_2 value for A-85783 for functional inhibition of the PAF response was calculated as 3.3 nM. In other experiments, A-85783 (10 μM) did not cause release of β-thromboglobulin and did not inhibit release in response to thrombin (0.3 U/ml). Taken together these results indicate that in human blood A-85783 selectively and competitively inhibits functional activity coupled to the platelet PAF-receptor.

3.2 *Ex Vivo* Inhibition of β-Thromboglobulin Release following ABT-299 Administration

Intravenous administration of ABT-299 lead to pronounced inhibition of PAF-induced platelet activation, as measured by *ex vivo* β-thromboglobulin release in blood (figure 2). With the highest PAF concentration used in the study (1 μM), release of β-thromboglobulin in blood from subjects in both treated groups (0.8 and 70 mg) was markedly inhibited, compared to the pre-dose values, by 87 and 100%, respectively, 12 hours after drug administration. There were no significant differences in release in the placebo group at the two sampling periods.

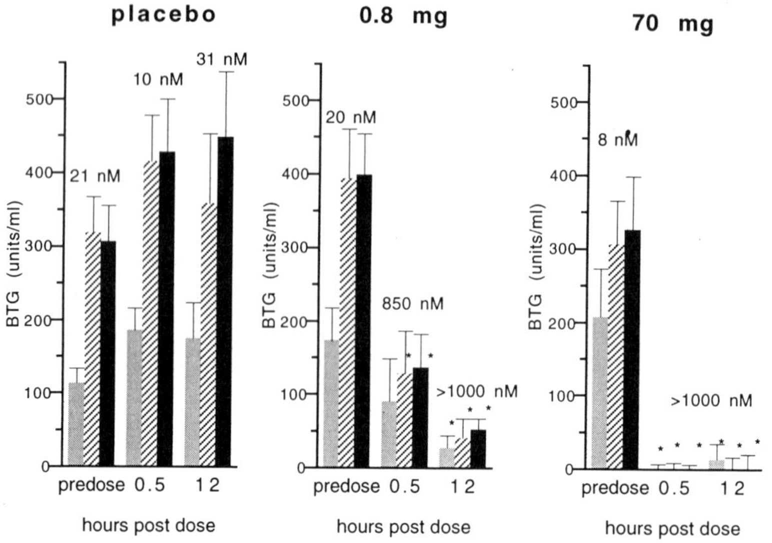

Figure 2. *Ex vivo* inhibition of β-thromboglobulin release in blood from subjects receiving placebo or ABT-299. Blood was taken from human subjects receiving placebo (n = 9), 0.8 mg (n = 7), or 70 mg (n = 7) ABT-299 at the indicated times after dosing and incubated with either buffer or with 10 (thatched), 100 (stripped) or 1000 nM PAF (solid). Plasma β-thromboglobulin was determined and expressed as net release. Asterisks indicate a significant difference ($p < 0.5$) between predose value. Estimates of the PAF EC_{50} are shown above the bars.

Inhibition of release was reflected by a shift in the PAF concentration-response relationship in blood from subjects receiving ABT-299. Thus the estimated EC_{50} for PAF-induced release was shifted from 20–30 nM in the placebo group and predose treated groups to values >1000 nM 12 hours after administration of ABT-299. The extent of the shift was dose-dependent: 50 fold for the 0.8 mg dose *vs* >125 fold for the 70 mg dose.

3.3 Pharmacokinetic Analysis

Results of HPLC analysis of plasma obtained from subjects receiving 70 mg ABT-299 are illustrated in figure 3. Plasma levels of the active moiety of ABT-299 were >100 nM for the first hour following intravenous administration then decreased to an average of 7 nM 12 hours after dosing. Based on these data, the elimination half-life of A-85783 was calculated to be 7.1 hours. Elimination of A-85783 was coupled with the appearance of significant levels of a metabolite that was identified, based upon comparison with authentic standards, as the pyridine-*N*-oxide of the parent compound.

4. DISCUSSION

PAF is well recognized as a potent stimulus of platelet activation, a feature that has proven useful for *ex vivo* evaluation of bioactivity of PAF antagonists in several clinical studies[4,6,12]. Platelet shape change, aggregation, degranulation and release of granular constituents are all consequences of PAF mediated platelet activation[13]. Previous investigators have relied upon platelet aggregation as an index of platelet activation. The aggregation response to PAF is well characterized and straightforward to monitor, but generally re-

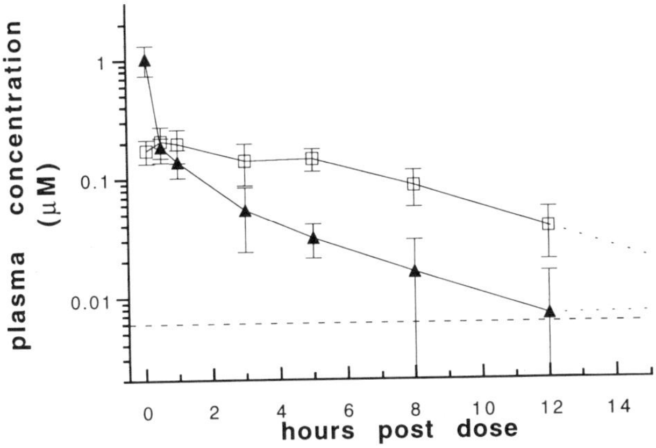

Figure 3. Plasma levels of active drug and metabolite. Plasma levels of the active drug (diamond) and metabolite (square) were measured by HPLC over 24 hours following administration of ABT-299 (70 mg). Values are means ± standard deviation (n=7). Concentration of the drug and metabolite were below the minimal detection level (7 nM) at 24 hours post dosing.

quires timely on-site analysis of the response. Platelet degranulation, an alternative index of platelet activation, was employed for assessing the pharmacodynamics of ABT-299. The degranulation process can be monitored by measuring the appearance in plasma of platelet-specific granular constituents in response to PAF. In the present study, release of β-thromboglobulin, an alpha granule constituent, served as an index of this process in whole blood and of the ability of ABT-299 to block the release as an *ex vivo* indication of the drug's bioactivity. The procedure employed required minimal on-site processing, provided high throughput, and afforded the opportunity to utilize a range of PAF challenge concentrations for each sample.

The potency and competitive nature of the active component of ABT-299 for inhibiting PAF-induced platelet responses was evident from the results of experiments in which the drug was added to blood from control subjects and then challenged with PAF. The derived A_2 value (3 nM) is consistent (within five fold) of the potency of A-85783 for inhibiting PAF binding to isolated human platelets (K_i = 1 nM), and PAF-induced calcium flux (4 nM) and elastase release (0.6 nM) in isolated human neutrophils[11]. Taken together, these results provide additional evidence of the ability of the active component of ABT-299 to functionally antagonize PAF mediated responses in highly relevant cell populations and, in the case of platelet degranulation, in the complex biological milieu represented by whole blood.

ABT-299, when administered to human subjects, greatly attenuated the *ex vivo* platelet response to PAF throughout the course of the study. Bioactivity, particularly at early time points following drug administration, is not surprising, given the potency of the drug against PAF-induced cellular responses. However, the duration of activity observed in this study is somewhat longer than one would expect from the plasma levels of the parent drug. The mean plasma level of A-85783 12 hours after dosing 70 mg ABT-299 was 7 nM. This concentration, based upon the A_2 value of the drug for inhibiting platelet degranulation in blood, should shift the PAF EC_{50} approximately 4 fold compared to the >100 fold shift observed at 12 hours. This apparent enhancement of potency may be at least in part due to the presence of bioactive metabolites. The pyridine-*N*-oxide of the parent is present at a relatively high concentration (38 nM) 12 hours after dosing ABT-299. This metabolite, which has been previously detected in studies with a variety of non-human species, is essentially equipotent to the active drug in terms of intrinsic binding potency and potency for inhibiting PAF-induced responses *in vivo*[14].Thus it is likely that the *N*-oxide metabolite contributes to the bioactivity of ABT-299 observed in the current studies.

Interestingly, when given at an almost 100 fold lower dose (0.8 mg), ABT-299 still provided significant antagonism of the platelet response for at least 12 hours. Although pharmacokinetic analysis is not available for this dose group, it is reasonable to assume that plasma levels achieved following the 0.8 mg dose were significantly lower that those from the 70 mg dose. This unexpectedly long duration of activity at this dose may be a result of the high affinity of the antagonists for the PAF receptor. Similar long duration's have been reported for other PAF antagonists with high intrinsic potency[15,16].

In summary, the release of β-thromboglobulin in blood samples challenged with PAF has proven to be a useful tool for evaluating the bioactivity of ABT-299 in clinical studies. Administration of this PAF antagonist resulted in pronounced inhibition of PAF-induced degranulation for at least 12 hours. These results, along with pharmacokinetic parameters generated in the study, may be useful in future PAF antagonist studies designed to examine the relationship between *ex vivo* activity and activity at target organs/tissues.

REFERENCES

1. Koltai, M., D. Hosford, P. Guinot, A. Esanu, and P. Braquet. "Platelet activating factor. A review of its effects, antagonists and possible future clinical implications." *Drugs* 42 (1991): 9–29, 174–204.
2. Summers, J.B., and D.H. Albert. "Platelet Activating Factor Antagonists." In *Advances in Pharmacology*, edited by J.T. August, M.W. Anders, F. Murad and J. Coyle, 67–168. New York: Academic Press, 1995.
3. Freitag, A., R.M. Watson, G. Matsos, C. Eastwood, and P.M. O'Byrne. "Effect of a PAF antagonist WEB-2086, on allergen induced asthmatic response." *Thorax* 48 (1993): 594–598.
4. Wilkens, J.H., H. Wilkens, J. Uffmann, J. Bovers, H. Fabel, and J.C. Frolich. "Effects of a PAF antagonist (BN-52063) on bronchoconstriciton and platelet activation during exercise induced asthma." *Br. J. Clin. Pharmacol.* 29, no. 1 (1990): 85–91.
5. Kuitert, L.M., K.P. Hui, S. Uthayarkumar, W. Burke, A.C. Newland, S. Uden, and N.C. Barnes. "Effect of the platelet-activating factor antagonist UK-74,505 on the early and late response to allergen." *Am. Rev. Respir. Dis.* 147 (1993): 82–86.
6. Bel, E.H., M. De Smet, T.H. Rossing, M.C. Timmers, J.H. Dijkman, and P.J. Sterk. "The effect of a specific oral PAF-antagonist, MK-287, on antigen-induced early and late asthmatic reactions in man." *Amer. Rev. Resp. Dis.* 143, no. 4 part 2 (1991): A811.
7. Dermarkarian, R.M., E. Israel, M.A. Rosenberg, A. Jansen, M.R. Danzig, J. Fourre, and J.M. Drazen. "The effect of Sch-37370, a dual PAF and histamine antagonist, on the bronchoconstriction induced in asthmatics by cold, dry air isocapnic hyperventialation." *Am. Rev. Respir. Dis.* 143, no. 4, part 2 (1991): A812.
8. Tamura, G., T. Takishima, S. Mue, S. Makino, K. Itoh, T. Miyamoto, T. Shida, and S Nakajima. "Effectiveness of a Potent Platelet Activating Factor Antagonist, WEB-2086, on asthma: a multicentere, double-blind, placebo-controlled study." *Eur. Respir. J.* 7 (1994): 752.
9. Dhainaut, J.F., A. Tenaillon, U.Y. Letulzo, B. Schlemmer, J.P. Solet, M. Wolff, L. Hillzapfel, F. Zeni, D. Dreyfuss, J.P. Mira, J.M. Crétien, V. Lavergne, and P. Guinot. "Platelet-activating factor receptor antagonist BN 52021 in the treatment of severe sepsis: a randomized, double-blind, placebo-controlled, multicenter clinical trial. BN 52021 Sepsis Study Group." *Crit Care Med* 22, no. 11 (1994): 1720–1728.
10. Dhainaut, J.F., A. Tenaillon, M. Hemmer, P. Damas, Y. Letulzo, P Radermacher, M.D. Schaller, J.P. Solet, M. Wolff, L. Holzapfel, F. Zeni, J. Motin, J.P. Mira, F. de Vathaire, J.M. Crétien, J. Marsais, M.L. Gourlay, and P. Guinot. "Confirming phase III clinical Trial to study the efficacy of a PAF antagonist, BN-52021, in reducing mortality of PTS with severe gram-negative sepsis." *American Journal of Respiratory and Critical Care Medicine* 151, no. 4 (1995): A447.

11. Summers, J.B., D.H. Albert, S.K. Davidsen, R.G. Conway, J.H. Holms, T.J. Magoc, G. Luo, P. Tapang, D.A. Rhein, and G.W. Carter. "ABT-299, A Potent Antagonist of Platelet Activating Factor." In *Adv. Prost. Leuko. Res.*: Raven, 1994.
12. Adamus, W.S., H.O. Heuer, C.J. Meade, and J.C. Schilling. "Inhibitory effects of the new PAF acether antagonist WEB-2086 on pharmacologic changes induced by PAF inhalation in human beings." *Clin. Pharmacol. Ther.* 47 (1990): 456–462.
13. Mustard, JF, RL Kinlough-Rathbone, and MA Packham. "Platelet activation—an overview." *Agents Actions* Suppl; 21 (1987): 23–36.
14. Davidsen, Steven K., James B. Summers, David J. Sweeny, James H. Holms, Daniel H. Albert, Richard G. Conway, and George W. Carter. "Active metabolites of indole pyrrolothiazole PAF antagonists." *J Med. Chem.* (in press).
15. Herbert, J.M., G. Valett, A. Bernat, P. Savi, J.P. Maffrand, and G. Le Fur. "SR 27417, a highly potent, selective and long-acting antagonist of the PAF receptor." *Drugs Fut.* 17, no. 11 (1992): 1011–1018.
16. Thompson, W.A., S. Coyle, K. Van Zee, H. Oldenburg, R. Trousdale, M. Rogy, D. Felsen, L. Moldawer, and S.F. Lowry. "The Metabolic Effects of PAF Antagonism in Endotoxemic Man." *Arch. Surg.* 129 (1994): 72–79.

A PHASE II TRIAL OF AUTOLOGOUS BONE MARROW TRANSPLANTATION (ABMT) IN ACUTE LEUKEMIA WITH EDELFOSINE PURGED BONE MARROW*

William R. Vogler,[1] Wolfgang E. Berdel,[2] Robert B. Geller,[1] Joel A. Brochstein,[3] Roy A. Beveridge,[4] William S. Dalton,[5] Kenneth B. Miller,[6] and Hillard M. Lazarus[7]

[1]Emory University, Atlanta, Georgia
[2]Freie Universtät, Berlin, Germany
[3]Hackensack Medical Center, Hackensack, New Jersey
[4]Fairfax Hematology-Oncology Associates, Annadale, Virginia
[5]University of Arizona, Tuscon, Arizona
[6]New England Medical Center
Tufts University School of Medicine, Boston, Massachusetts
[7]Case Western Reserve University, Cleveland, Ohio

1. INTRODUCTION

Alkyl-lysophospholipids are analogs of naturally occurring lyso-phospholipids initially synthesized as immunologic modifiers to render macrophages more cytocidal to tumor cells [1]. However, early experiments indicated that these compounds had direct anti-tumor activity [2]. In addition, it was found that these compounds were selectively toxic to tumor cells and had less effect on normal cells [3–5]. The most active of these compounds was 1-O-octadecyl-2-O-methyl-rac-glycero-3-phosphocholine, formerly termed ET-18-OCH_3 and more recently named edelfosine. If indeed the compounds were selectively toxic to leukemic cells and spared normal cells, particularly normal bone marrow progenitor cells, they might prove effective in purging leukemic cells from remission marrows in patients with acute leukemia who were candidates for ABMT. Incubation of human leukemic cell lines and freshly obtained leukemic cells from patients demonstrated that the compounds were active whether measured by viability, tritiated thymidine incorpora-

* Supported in part by National Institutes of Health grants CA-29850 and P30CA43703 and Atlanta Leukemia Research, Inc.

tion or clonogenic assays [6–8]. When a sensitive cell line (HL60) was mixed with normal marrow cells and incubated with varying doses of edelfosine for 1 to 4 hours it was found that a 4 hour incubation with 50 μg/ml eliminated clonogenic leukemic cells with no effect on normal marrow progenitor cells [9]. In a murine leukemic model in which a remission marrow was simulated by mixing WEHI 3/b leukemic cells with normal murine marrow cells and exposing the mixture in vitro to edelfosine prior to injection intravenously into lethally irradiated littermates resulted in a dose responsive prolongation of survival with apparent cures [10]. Similar results were obtained after freezing and thawing the mixture prior to injection [11]. Following these results a phase I clinical trial was initiated and the results have been published [12]. It was found that a 4 hour incubation at a dose of 75 μ/ml was the optimal dose for purging. Based on these results a phase II trial was designed with participation by multiple institutions.

2. MATERIALS AND METHODS

2.1 Purging Agent

Edelfosine was a gift from Dr. R. Nordström of Medmark Pharma (Grünwald, Germany). It was supplied lyophilized in 75 mg sterile vials. For reconstitution it was dissolved in RPMI 1640 medium prior to use.

2.2 Patient Population

Eligible patients were those with a documented diagnosis of acute leukemia, myeloblastic (AML) or lymphoblastic (ALL) provided they were in complete remission (CR), did not have an HLA compatible marrow donor, had an ECOG performance status of 0, 1 or 2, a normal bone marrow of at least 40 % cellularity with less than 5 % blasts within 2 weeks prior to harvest at least 1 month after CR. In addition, patients could have no severe organ dysfunction: left ventricular cardiac ejection fraction of 50 % or greater (if less than 50 % clearance by a cardiologist), pulmonary function of at least 75 % of normal, a serum creatinine of 2 mg/dl or less, a serum bilirubin of 2 mg/dl or less and SGOT, SGPT or alkaline phosphatase of less than 2.5 times upper limits of normal and no central nervous system disease. Patients had to be able to undergo general, epidural or spinal anesthesia.

All patients or their legal guardian signed informed consents to participate in the protocols approved by the Investigational Review Boards of the participating institutions.

2.3 Bone Marrow Harvest Procedure

Under anesthesia multiple aspirates of bone marrow were obtained from the posterior and anterior iliac crests and the sternum in order to obtain a minimum of 3×10^8 mononuclear cells per kg of ideal body weight. This allowed 1.3×10^8 cells/kg to be reserved as back-up (unpurged) marrow sample. The harvest and procedure for concentrating the cells have been previously described [12]. The cells were suspended in RPMI 1640 medium (Life Technologies, Inc., Grand Island, N.Y.) containing 10 % of a 5 % albumin solution (Hyland, Glendale, CA) at a concentration of 2×10^7 cells/ml. Edelfosine was added to give a final concentration of 75 μg/ml. To prevent clumping DNAase (Sigma, St.Louis, Mo.), 500 u/ml, was added. The cells were incubated at 37° C for 4 hours with frequent agitation. The cells were then centrifuged at 258g for 10 minutes, resuspended in RPMI 1640 with 10 % albumin to a final concentration of 8 x 10^7 cells/ml and cryopreserved with 10 % dimethylsulfoxide (Research Industries, Miduale, Ut.) as previously described [12]. Thawing was done at the bedside and marrow infused through a central line.

2.4 Treatment Regimens

In AML two preparative treatment programs were used. Originally the ablative procedure consisted of busulfan (1 mg/kg orally every 6 hours for 4 days) followed by cyclophosphamide (50 mg/kg daily for 4 days) followed by a day of rest prior to marrow infusion (protocol ED1a). At one institution the protocol was modified. The busulfan schedule was the same but the cyclophosphamide dose was changed to 60 mg/kg daily for 2 days, followed by GM-CSF (5 μg/kg by continuous intravenous infusion daily for 5 days), and cytosine arabinoside (1500 mg/m^2 daily for the last 2 days of the GM-CSF infusion. The marrow was infused after a day of rest (protocol ED1b).

In ALL intrathecal methotrexate was given 13 days prior to transplant. Total body irradiation (TBI) (1320 cGy) in fractionated doses 3 times per day was given over 3 days followed by etoposide (60 mg/kg) intravenously on day -3 prior to transplant (protocol ED2). Supportive care included prophylactic antibacterial and antifungal antibiotics, transfusion support and nutritional support. The use of recombinant hematopoietic cytokines was optional.

2.5 Statistical Methods

Kaplan-Meier plots were used for survival and disease-free survival (DFS) . The student t-test was used for comparative data.

3. RESULTS

From March of 1992 through January of 1995 57 patients with acute leukemia underwent ABMT with edelfosine purged marrow. The ages ranged from 2 to 65 with a median of 33. There were 28 females and 29 males. Table 1 shows the remission status at the time of transplant for each of the 3 protocols. As can be seen about half of the patients on protocol ED1a were transplanted in first remission compared to 81 % on protocol ED1b.

The infusion of the edelfosine purged marrow was generally well tolerated. One patient, who was febrile on the day of infusion, received hydration and mannitol prior to the

Table 1. Patient population. Remission status at time of ABMT

	Protocol			
Remission	ED1a	ED1b	ED2	Total
1st CR	12	22	2	36
2nd CR	7	5	2	14
3rd CR	4	0	1	5
4th CR	0	0	2	2
Total	23	27	7	57

administration of 450 ml of thawed marrow developed ventricular bigeminy at the end of the infusion. She became hypoxic and hypotensive, required intubation and vasopressors, but recovered and remains in remission. It was believed that fluid overload was the major contributing factor contributing to this adverse event.

Back-up bone marrow was infused in 6 patients between days 29 to 69 because of no evidence of engraftment on examination of blood and marrow. Four had been treated on protocol ED1b and 3 were transplanted in first CR. One patient had granulocyte recovery by day 39 but was infected with aspergillus and because of granulocyte counts of less than 500 and high fevers the unpurged marrow was infused on day 56. Another patient without evidence of marrow recovery developed Candida Krusei sepsis and his back-up marrow on day 29. A repeat marrow 3 weeks later showed relapse of his leukemia. Another first CR transplant patient had high fevers post transplant but began to show some evidence of recovery by day 21 but remained febrile. A day 28 bone marrow biopsy was hypocellular and he was started on GM-CSF without improvement and his back-up marrow was infused on day 55. He died of infection on day 90 without marrow recovery. The other patient treated on protocol ED1b in second CR developed a probable fungal infection one week after transplant. GM-CSF was started on day 16. A bone marrow biopsy on day 28 was hypocellular. The back-up marrow was infused on day 35 without recovery and the patient died on day 53. Two patients on protocol ED1a received back-up marrow. One transplanted in second remission had no evidence of engraftment by day 40, received her back-up marrow on day 43 but died of infection and renal failure on day 69 without evidence of recovery. The other patient transplanted in third CR developed progressive pulmonary infiltrates post transplant. A day 28 marrow biopsy was interpreted as aplastic and back-up marrow and peripheral blood stem cells were infused along with G-CSF followed by a partial marrow recovery. However, the patient relapsed at day 106 and died one month later.

Table 2 shows the time to marrow recovery. The median time for granulocytes to reach 500/μl was 33 days and for platelet recovery to 25,000/μl was 46 days. There were no significant differences in recovery time between protocols ED1a or ED1b or between those transplanted in first remission or later. There were 9 patients who failed to recover granulocytes, 8 have died and the median survival was 53 days. One was at day 28 at the time of last follow-up and had not recovered as yet. There were 19 patients who failed to recover platelet counts, 14 have died, 3 died before day 42, 6 relapsed before day 120 and 5 remained aplastic. Five are alive and 4 have no evidence of recurrent leukemia. The median survival for this group was 116 days.

Of the 27 deaths in AML patients 19 were secondary to complications arising after relapse. Three patients succumbed to fungal infections, 3 of bacterial sepsis, 1 of veno-occlusive disease and 1 of gastrointestinal bleeding and renal failure.

Table 2. Days to marrow recover

	Median	Range
Granulocytes/μl		
> 500	33	14–72
> 1000	43	17–84
Platelets/μl		
>25,000	46	20–675
>50,000	53	22–203

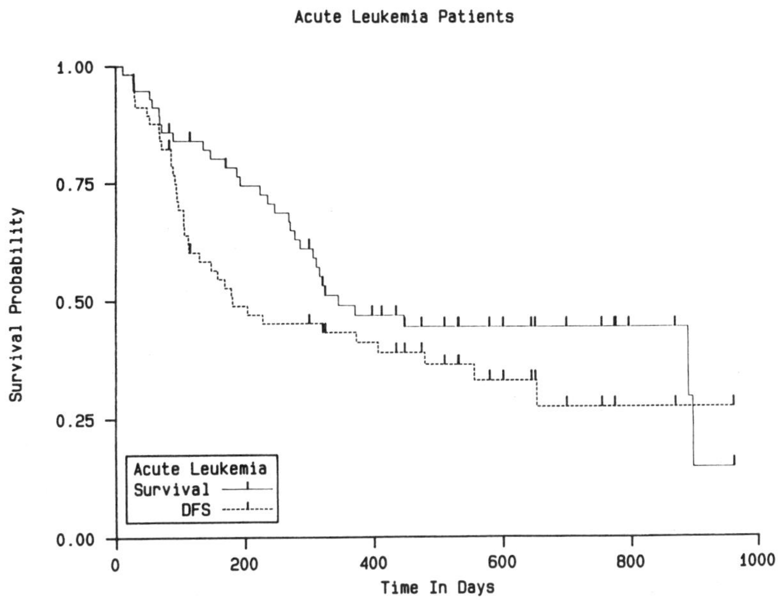

Figure 1. Kaplan-Meier plots for overall survival and disease-free survival for all patients.

Figure 1 gives the Kaplan-Meier survival and DFS plots for all of the leukemic patients. Forty-seven % are alive for a median survival of 371 days and 37 % remain in CR with a median DFS of 181 days. Of those transplanted in 1st CR 61 % are alive and 47 % are disease free. In contrast, only 19 % of those transplanted in a later remission are alive and disease free. The difference in survival is significant ($p < 0.01$) but the difference in DFS is marginal ($p < 0.10$).

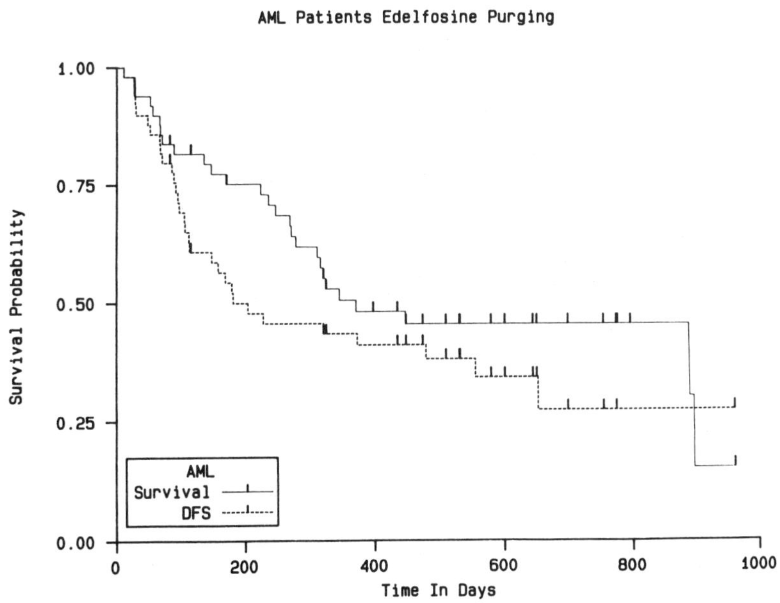

Figure 2. Kaplan-Meier plots for overall survival and disease-free survival for AML patients.

Of the 7 with ALL 2 who were transplanted in 4th remission and 1 in third remission have died. One of 2 transplanted in second remission is alive and free of disease at 300 days and 2 transplanted in first remission are alive. One relapsed at 407 days and one remains in remission at 869 days. All of the deaths occured after relapse.

Figure 2 displays the survival and DFS for the AML patients. Forty-six are alive for a median of 371 days and 38 % are disease free for a median of 204 days. Comparing the two treatment schedules, ED1a and ED1b, and correcting for the number of patients transplanted only in first remission, there is no significant difference in survival (55 % versus 59 %). Figure 3 illustrates the similarity in DFS of the two programs.

4. DISCUSSION

The value of purging in ABMT in acute leukemia has not been proven as no randomized clinical trials have been reported. Efforts to establish the effectiveness of purging have been derived from data collected by autologous transplant registries. Gorin et al [13] summarized the European experience in AML and concluded that those patients transplanted within the first six months of the first remission had a significantly greater chance of relapse if they received unpurged marrow. It is generally agreed that ABMT in subsequent remission results in a lower DFS which is consistent with our observations. Studies indicate that 40 to 60 % of patients transplanted in first remission with purged marrow may be long term survivors but most of these trials have not matured enough to permit certainty of the ultimate outcome. One exception has been the report by Laporte et al [14] who reported 10 years of experience of ABMT using mafosfamide purged marrow. At 8 years they reported a DFS of 58 % for 64 AML patients and 56 % for 29 ALL patients transplanted in first remission. AML patients transplanted in second remission had a DFS

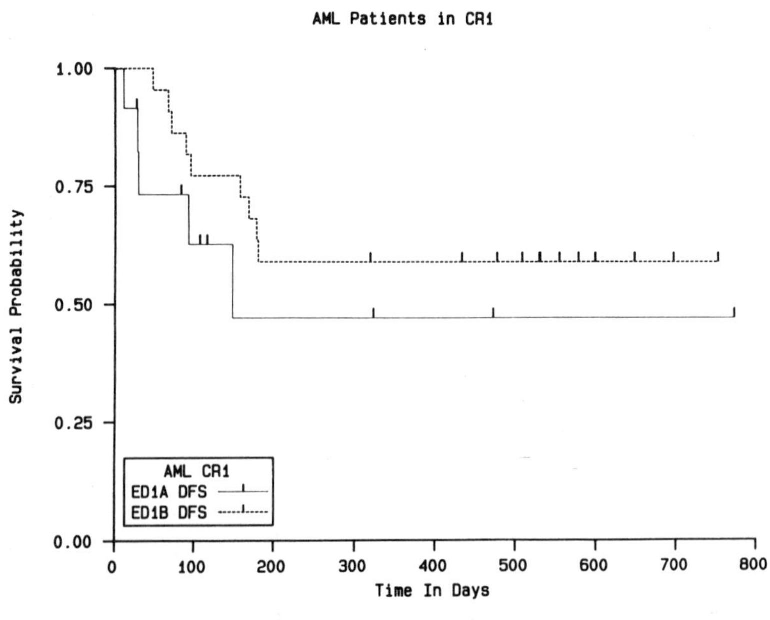

Figure 3. Kaplan-Meier plots for disease-free survival for patients treated on protocols ED1a and ED1b.

of 34 % at 5 years. Although seemingly better than our results we had 3 patients transplanted beyond second remission. The numbers are too small to show a significant difference. Since between 20 and 30 % of patients with AML are apparently cured by conventional chemotherapy it would seem that ABMT gives a better result. However, there is considerable bias in considering the selection of patients for transplant. They must have achieved a remission with conventional chemotherapy, be in good clinical condition, have adequate marrow cellularity to harvest, have no underlying infections or organ dysfunction and are usually of a younger age. Thus comparisons are hazardous.

This phase II trial indicates that edelfosine purging is easy to perform, safe, and gives a result not dissimilar from other purging trials. The time to granulocyte recovery to 1000/μl appears to be somewhat longer (43 days versus 30 days) and platelet recovery to 50,000/μl shorter (53 versus 90 days) compared to mafosfamide purging [14]. Other approaches such as purging peripheral blood stem cells are being pursued making it unlikely that a phase III trial of marrow purging will ever be done.

5. SUMMARY

Alkyl-lysophospholipid compounds which are selectively cytotoxic to neoplastic cells and relatively sparing of normal marrow progenitor cells are appealing as purging agents to rid remission marrows of residual leukemic cells. A multi-institutional phase II study was conducted in 57 patients with acute leukemia (50 AML and 7 ALL) in which remission marrows were purged *in vitro* and reinfused after ablative chemotherapy. The median time for granulocyte recovery to 500/μl was 33 days and for platelet recovery to 25000/μl was 46 days. The overall DFS and survival was 37 % and 46 % respectively. Transplantation in first remission gave a better survival than transplant in a subsequent remission.

REFERENCES

1. Munder PG, Weltzien HU, Modolell M: Lysolecithin analogs: a new class of immunopotentiators. Immunodeficiency and immunostimulation, in Miescher PA (ed): Seventh international symposium on immunopathology, Basel/Stuttgard, Schwabe & Co, 1976, p 411
2. Munder PG, Modolell M, Bausert W, Oettgen HF, Westphal O: alkyllysophospholipids in cancer therapy, in Hersh EM (ed): Augmenting agents in cancer therapy, New York, Raven Press, 1981, p 441
3. Andreesen R, Modolell M, Weltzien HU, Eibl H, Common HH, Lohr GW, Munder PG: Selective destruction of human leukemic cells by alkyl-lysophospholipids. Cancer Res 38:3894, 1978
4. Modelell M, Andreesen R, Pahlke W, Brugger U, Munder PG: Disturbance of phospholipid metabolism by the selective destruction of tumor cells by alkyl-lysophospholipids. Cancer Res 39:4681, 1979
5. Andreesen R, Modollel M, Munder PG: Selective sensitivity of chronic myelogenous leukemic cell populations to alkyl-lysophospholipids. Blood 54:519, 1979
6. Vogler WR, Whigham EA, Somberg LB, Long RC, Winton EF: The effect of alkyl-lysophospholipids on tritiated thymidine incorporation and clonogenicity in vitro of normal and leukemic human cells. Exp Hematol 12:569, 1984
7. Verdonck LF, Witteveen EO, van Heugten HG, Rozemuller E, Rijksen G: Selective killing of malignant cells from leukemic patients by alkyl-lysophospholipid. Cancer Res 50:4020, 1990
8. Dulisch I, Neumann HA, Lohr GW, Andreesen R: Clonogenicity of normal and malignant hematopoietic progenitor cells after exposure to alkyl-lysophospholipids. Blut 51:393, 1985
9. Okamoto S, Olson AC, Vogler WR, Winton EF: Purging leukemic cells from simulated remission marrow with alkyl-lysophospholipids. Blood 69:1381, 1987

10. Glasser L, Somberg LB, Vogler WR: Purging murine leukemic marrow with alkyl-lysophospholipids. Blood 64:1288, 1984
11. Vogler WR, Somberg LB, Glasser L: Effect of cryopreservation on purging of leukemic marrow with alkyl-lysophospholipids. Exp Hematol 15:360, 1987
12. Vogler WR, Berdel WE, Olson AC, Winton EF, Heffner LT, Gordon DS: Autologous bone marrow transplantation in acute leukemia with marrow purged with alkyl-lysophospholipid. Blood 80:1423, 1992
13. Gorin NC, Aegerter P, Auvert B, Meloni B, Goldstone AH, Burnett A, Carella A, Korbling M, Herve P, Maraninchi D, Lowenberg R, Verdonck LF, de Planque M, Hermans J, Helbig W, Porcellini A, Rizzoli V, Alesandrino EP, Franklin IM, Reiffers J, Colleselli P, Goldman JM: Autologous bone marrow transplantation for acute myelocytic leukemia in first remission: a European survey of the role of marrow purging. Blood 75:1606, 1990
14. Laporte JP, Douay L, Lopez M, Labopin M, Jouet JP, Lesage S, Stachowiak L, Fouillard L, Isnard F, Noel-Walter MP, Pene F, Deloux J, Van Dan Akker J, Bauters F, Najman A, Gorin NC: One hundred twenty-five adult patients with primary acute leukemia autografted with marrow purged by mafosfamide: a 10-year single institution experience. Blood 84:3810, 1994

INDEX